The Anglo-Maratha Campaigns for India

This is a cross-cultural study of the political economy of warfare in South Asia. Randolf G. S. Cooper combines an overview of Maratha military culture with a battle-by-battle analysis of the 1803 Anglo-Maratha Campaigns. Building on that foundation he challenges ethnocentric assumptions about British superiority in discipline, drill and technology. He asserts that these campaigns, in which Arthur Wellesley served with distinction, represent the military high-water mark of the Marathas who posed the last serious opposition to the formation of the British Raj. He argues that the real contest for India was never a single decisive battle for the subcontinent. Rather it turned on a complex social and political struggle for control of the South Asian military economy. The author shows that victory in 1803 hinged as much on finance, politics and intelligence as it did on battlefield manoeuvre and war itself.

RANDOLF G. S. COOPER is a Visiting Fellow at Wolfson College, University of Cambridge.

The Anglo-Maratha Campaigns and the Contest for India

The Struggle for Control of the South Asian Military Economy

Randolf G. S. Cooper

CAMBRIDGE
UNIVERSITY PRESS

CAMBRIDGE UNIVERSITY PRESS
Cambridge, New York, Melbourne, Madrid, Cape Town, Singapore, São Paulo

Cambridge University Press
The Edinburgh Building, Cambridge CB2 8RU, UK

Published in the United States of America by Cambridge University Press, New York

www.cambridge.org
Information on this title: www.cambridge.org/9780521824446

First published 2003
This digitally printed version 2007

A catalogue record for this publication is available from the British Library

Library of Congress Cataloguing in Publication data
Cooper, Randolf G. S.
The Anglo-Maratha Campaigns and the contest for India: the struggle for control of the
South Asian military economy / Randolf G. S. Cooper.
    p.   cm.
Includes bibliographical references and index.
ISBN 0 521 82444 3 (hardback)
1. Maratha War, 1803 – Campaigns.   2. Maratha War, 1803 – Finance.   I. Title.
DS475.3.C66   2003
954.03'12 – dc21   2003051518

ISBN 978-0-521-82444-6 hardback
ISBN 978-0-521-03646-7 paperback

This book is dedicated to the memory of my
maternal grandparents:
        L. David Eliezer LeVine
        and
        Anastasia Ivanoff

Contents

Maps

Acknowledgements

Approximately one-third of this book's content can be traced directly to material written during my years as a graduate student. The remaining two-thirds developed slowly after I left school as archival references were cobbled together to answer a research puzzle that I abandoned in 1992. This work reflects the help and support of people encountered during the evolution of an intriguing investigative odyssey.

While I was studying for my Master's degree in South Asian Studies at the University of Toronto, Professor N. K. Wagle gave his time generously. His painstaking translations of Marathi documents written in Modi script gave me first hand insight into the Marathas' approach to warfare. Wagle also urged me to venture out on the conference circuit and it was there during the mid-1980s that I met Dirk H. A. Kolff of the Netherlands when we were placed together on a panel to discuss South Asian warfare at a symposium held in Heidelberg. Kolff's pioneering study of the North Indian military labour market was brilliant and my development of material on the South Asian military economy owes much to him. It was intriguing to know others recognized that war had traditionally been an economic driver in South Asia. While in Germany, the late Gunther D. Sontheimer of Heidelberg University was instrumental in convincing me that I should develop my South Asian conflict analysis theories at a major British university. Professor Sontheimer's dedication to classical Indology, the Marathi language and Maratha culture, were nothing short of inspiring and his untimely loss to the field in 1992 remains evident.

At Cambridge I received continuous intellectual support for my efforts. My doctoral studies there were aided by funding derived from a PhD fellowship in South Asian Military History granted by the Social Sciences and Humanities Research Council of Canada. Gordon Johnson, my thesis supervisor, encouraged me to expand the scope of the work beyond the technicalities of late eighteenth-century South Asian military affairs – transforming it into a consideration of war and society in more general terms. He showed great wisdom and insight with regard to the analytical potential of the historic model. Dr Johnson's influence profoundly affected the way in which this 'Maratha war story' became a vehicle for exploring cross-cultural conflict analysis and for that I will always be indebted to him. A world-class library and the chance to exchange ideas with distinguished imperial historian C. A. Bayly, as well as numerous visiting scholars such as John Richards, further augmented the benefit of the Cambridge experience. In those days Hew Strachan was at Corpus Christi College and his comments were very helpful in balancing my views on the British Army's approach to colonial warfare. Lionel Carter, who at that time headed the specialized research collection at the Cambridge Centre of South Asian Studies on Laundress Lane, showed me great kindness by searching-out holdings that I requested.

While in Mumbai, I benefited from the advice of Dr Sanjiv P. Desai who expedited my access to primary source material in the Government of Maharashtra's Department of Archives. Dr Aroon Tikekar, long an ardent defender of the Marathi vernacular press, took great personal interest in my well being when I was in Mumbai. In addition to making sure I was fed a hearty Maratha home-cooked meal, he gave me access to his personal library containing several historic works on the 1803 Anglo-Maratha Campaigns. Aroon allowed me to photocopy some rare sources that I could not have duplicated elsewhere. C. R. Mankame of Mumbai came to my aid by locating out-of-print Maratha histories published in the dying days of the British Raj.

His Royal Highness the Maharaja of Jaipur was gracious in letting me spend several days interviewing Yaduendra Sahai, Director of the Maharaja Sawai Mansingh II Museum. We met for long hours in the abandoned cannon foundry of the Jaigarh Fort discussing casting techniques and investigating the virtually intact production facilities. The Jaigarh's collection included two extremely rare artillery pieces. One proved to have been a battalion gun used by Mahadji Sindia's sepoys at Lalsot while the other was a laminated 'bar gun' with a hexagonal iron bore sleeved in bronze; the latter matching the description of the type used by the Marathas at Delhi, Agra and Laswari. In Udaipur, Abdul Hakim provided complete access to his arms factory where muskets are still produced by hand in the traditional fashion and the techniques demonstrated there explained much about smallarms proliferation in eighteenth-century India. While in Udaipur I also had the benefit of examining the historic arms collection of the Parihar family, owners of the Century Arms Corporation.

The Indian National Archives in Delhi yielded some of the more obscure political references in this work, but by far the greatest collection of East India Company (EIC) documents rests in the British Library's

Oriental and India Office Collection (BL: OIOC) formerly known as the India Office Library and Records (IOL). The staff of the OIOC always took time to answer my naïve questions about the evolution of their catalogue system. In particular I wish to thank Timothy Thomas and Anthony Farrington who were never less than friendly when my queries sparked a distraction from their busy activities.

Distinguished Napoleonic historian David Chandler invited me to Sandhurst where we discussed military history as a discipline. Dr Chandler was correct in saying that academic pressure should not compel one to remove 'the messy business of killing' from military history because that is what war is all about. Elsewhere in Britain, I gleaned specialized information from antiquarians and Wellingtonians. Over the years Anthony S. Bennell, a former Ministry of Defence historian and author of several books and articles on Anglo-Maratha affairs, offered advice on Arthur Wellesley's campaigns in India. Bennell's detailed knowledge of Wellesley's diplomatic record in South Asia remains unrivalled. Dr C. M. Woolgar, archivist of the Wellington Collection at the University of Southampton's Hartley Library, showed me a number of unique Maratha items from the Collection including captured Maratha maps. I am grateful to the Controller of Her Majesty's Stationery Office for permission to quote from Crown Copyright material among the Wellington Papers. Anthony J. Mitchell, of Britain's Arms and Armour Society, gave of his time and knowledge to help me learn about regional distinctions in South Asian edged weapons. The recently deceased Robert Wiggington, a tremendous authority on the arms of Tipu Sultan, exchanged ideas on the South Asian vs. European origin of weapons found in Mysore.

I owe a collective thanks to the staff of the National Army Museum (NAM) in Chelsea who helped in such a great variety of ways. The NAM is a major repository for records of the British Indian Army, a marvellous collection. Dr Alan J. Guy, Assistant Director (Administration), was the man who first urged me to visit the museum and he suffered through the reading of several early drafts of material. Dr Peter Boyden, Assistant Director (Collections), went out of his way to ensure that I had access to all the pertinent archival records and he was particularly good about notifying me when newly acquired material held promise. Jenny Spencer-Smith, Head of the Department of Fine and Decorative Art, helped to locate etchings and sketches of the Anglo-Maratha Campaigns – a fascinating source of iconographic evidence. Spencer-Smith was also responsible for suggesting the picture used on the dust jacket of this book. She recommended a depiction of former Maratha mercenary James Skinner as a man who symbolized the continuity to be found within the

South Asian military economy of the nineteenth century. The staff of the NAM's Department of Weapons arranged special viewings of relevant military technology including both South Asian and British smallarms as well as artillery. The latter ranged from EIC battalion guns to 'visiting pieces' such as one of the Sikh Sutlej guns. The late Peter Hayes, as department head, helped to create a departmental environment in which scholarly investigation ranked equally with collecting. Michael Baldwin, Peter's immediate successor, also made items available for my research, while Martin J. Hinchcliffe donated hours of his personal time with regard to explaining weapons fabrication techniques that he learned for the purposes of historic conservation and restoration.

One individual that deserves special mention is David F. Harding, a former British officer who served in the 10th Princess Mary's Own Gurkha Rifles. His unit, known as '10 GR' prior to amalgamation into the Royal Gurkha Rifles, was the regimental heir to the 10th Madras Native Infantry who fought so courageously at the Battle of Assaye in 1803. David F. Harding is the world's leading authority on smallarms of the EIC and his knowledge of the Company's military policies is encyclopaedic. He repeatedly demonstrated his willingness to share his knowledge while engaging in a healthy intellectual exchange on the origins and history of the British Indian Army. Harding's four-volume study of EIC smallarms should be considered as essential by future historians attempting to tackle the meaning and significance of Company policies related to weaponry and its use as described in the phrase 'discipline and drill'.

I was very fortunate in receiving words of encouragement that sustained me over the years that it took to complete this work. Field Marshal John Chapple had to listen to one of my more dreadful lectures at the NAM. Yet he urged me to struggle on with what came to be known generically as 'the Maratha book'. There were a number of old soldiers who were particularly instrumental in developing the combat leadership aspects of this study. I label them 'old soldiers' because it is an epithet they wear with quiet, distinguished pride. Lieutenant-Colonel Patric J. Emerson, Secretary of the Indian Army Association, shared the benefit of his service with the Bombay Grenadiers but he also provided the names and addresses of other veterans with relevant Indian Army service. Of those contacts the most helpful was the late Lieutenant-Colonel P. M. W. Doyle MC, Chairman of the Mahratta Light Infantry Regimental Association. 'Paddy' Doyle proved a vital link in establishing the longevity of Maratha military tradition and identity. His accounts of leading Maratha troops also helped me to define the difference between leadership issues and Maratha officership issues. Another one of the veterans was C. F. E.

Wilmot of the 9th Lancers (later known as the 9th/12th Lancers) who saw events as only a senior non-commissioned officer (NCO) could. 'Charlie', who passed-away as a pensioner at the Chelsea Hospital soon after I finished my PhD, joined the British Army at age fifteen, serving as a 'boy soldier' with the 9th in various postings across South Asia. His experiences as a lancer as well as his observations on having drilled with the *Nizam* of Hyderabad's mounted units were very useful.

Janos Marffy produced all but one of the maps for this book and his clear and concise artwork helped to clarify the confusing detail of fire and movement that plagued many formerly complex battle maps of the 1803 Anglo-Maratha Campaigns. Tamzen Le Blanc provided technical support for a number of problems that occurred as material was transferred between different types of computer operating systems. Mildred Gallot in Grambling, Louisiana, kept faith over the years – reminding me in friendly but firm terms that 'it is time to finish the book'. Ron Matthews of Cranfield University was always positive and willing to discuss the finer points of defence economics. Dr Ramayyar Subramanian helped me by hand-carrying a number of sources from India. Fain Brock of Austin, Texas, showed me how to juggle various professional commitments while Harry Lucas Jr, founder of the Educational Advancement Foundation, granted me time away from an important research project so I could complete last-minute revisions for Cambridge Press.

I also wish to acknowledge the long hours of research help provided by my mother Esther S. Cooper. With more than thirty years of career experience in libraries Esther was called upon numerous times in England, India and North America, when the bibliographic trail went cold. While my wife Sylvia suffered through numerous holidays virtually alone while I spent hours rewriting the notes I gathered on research trips that masqueraded as 'vacations'.

It is impossible to list all those who gave of their time and knowledge but by the same token it must be said that not everyone noted here may have cooperated so fully if they had been able to visualize the finished work. In the final analysis this creation remains truly interdisciplinary as an exercise in cross-cultural conflict analysis; Maratha history, British history, military history, economic history as well as the history and philosophy of science have all managed to find accommodation within this study. Researching and writing this book proved to be a stimulating challenge and the conclusion of the process brings not a sense of relief but rather an odd emptiness – as if one had lost a friend. And as the result of my commitment to that friendship I can truly say that I accept full responsibility for all mistakes, errors and omissions in this book.

A note on transliteration and references

During the past two centuries there have been many changes in the English spelling and pronunciation of South Asian names. The variety stems in part from poorly enforced historic guidelines for transliteration. Some of the earliest efforts were based on English-language perceptions of phonetic spellings, a technique fraught with hazard since variations in British regional accents were enough to distort the transference as often found with the seemingly random insertion of the letter 'r' in eighteenth-century British accounts of battles in India. The written statement 'Sindia returned to Pune' can be dated to some degree by the transliteration used. If we look on a modern map of Maharashtra we see the city of Pune, which at various times served as the Marathas' capital. However the British knew Pune in the eighteenth century as Poonah and during the mid-nineteenth century the spelling Poona became standard. As for Sindia in the above statement, it is a clan name and without specific individual reference. In the absence of either direct or contextual evidence we do not know if it is likely to have been a reference to Mahadji Sindia or his successor Daulat Rao Sindia or perhaps another kinsman. The Sindia clan name has also appeared in English-language texts as Shinde, Scindia, Shindia, Sindhea, Scindhia, Sindhia and Scindeah.

An effort was made to stress consistency in the main body of the text; therefore a statement that 'Sindia's troops departed from the vicinity of Pune' remains easily traceable by a constant repetition of the spellings Sindia and Pune throughout the chapters of the book. However, the original British spellings remain intact in all quoted source material and references to aid future researchers in the location and cross-referencing of documents. Those looking for sources dealing with the history of Pune would be advised to consider British documents filed under Poona as well as Poonah.

In this book bibliographical references and other information traditionally associated with academic footnotes will appear in the endnotes. The footnotes that do appear at the bottom of pages are limited to geographical finding aids (longitude and latitude), alternative transliterations for place names, or in some instances a quick elaboration on the identification of individuals.

Abbreviations used in the references

BL   British Library
BL: OIOC   British Library Oriental and India Office Collections
NAI   National Archives of India
NAM   National Army Museum, Chelsea

Introduction

Focusing on the 1803 Anglo-Maratha Campaigns

While studying history as an undergraduate I came across a rather grandiose volume of world history – the title of which I have long since forgotten. It was one of those ungainly texts approaching the size of a coffee-table book. Having just 'discovered' South Asian history, I quickly denounced this pretentious tome as too elementary in its abbreviated coverage of the subcontinent. The offending volume skipped easily from the Mughal Empire to the British Raj with heavily illustrated pages that showed greater concern for continuity than content. The text contained brief summaries of political events implying that the Mughals controlled India from 1526 until 1857 and then the British apparently stepped in as imperial rulers from 1858 to 1947. Leaving aside the long history of the East India Company (EIC) prior to 1858, there were still fundamental problems with this story. If modern South Asia had grown out of a seamless transition of imperial power, why had the British fought such a long series of wars there? There were three Anglo-Bengal Wars, four Anglo-Mysore Wars, at least three Anglo-Maratha Wars, as well as Anglo-Sikh and Anglo-Afghan Wars, which suggested the transference of power in South Asia was not analogous to passing the baton of governance in a relay race.[1]

As my studies in South Asian history progressed at the University of Toronto, I realized that I knew little about the Maratha people that featured prominently in the lectures of N. K. Wagle. The professor's observations on Brahmin dominance of the Maratha administrative system helped me to realize that Westerners had largely overlooked the more secular aspects of Maratha military history. Prior to lengthy tutorials with Wagle, my elementary understanding of the Marathas had not developed much beyond the feeble textbook descriptions that summarized the Marathas' identity as the 'Marathi-speaking Hindu people native to the state of Maharashtra'. In the long run, that less than adequate cultural definition proved to be another gross generalization of humanity on a par

1

with the sweeping world history that I found offensive. And so it was that I came to read more about Maratha history.

Maratha military history drew my attention because it reflected the struggle of a people who at various times posed a military challenge to the Mughals and the British. However, I did not feel I could do justice to a serious study of Mughal–Maratha conflict beyond the tactical level. But in contemplating that option I became fascinated with the way one's own culture influences the perception of armed conflict among others. Within my studies, 'cross-cultural conflict analysis' came to mean an analysis of war's dynamics as influenced by the presence of competing cultures. The conflicting cultures might be ethnic, racial, religious, national, political or any combination thereof; although for me the most challenging case studies were those that featured conflict between competing military cultures. Having acknowledged my limitations in trying to address Hindu–Muslim strife as a cross-cultural factor in Mughal–Maratha warfare, I turned specifically to look at the Maratha military challenge to the British in South Asia. It appeared that Britain's military efforts in 1803 had a dual nature. There were counterinsurgency operations in disputed areas of control, but there was a much more serious conventional war for the cloak of the Mughal Emperor; a story often downplayed in modern British histories for one reason or another.

By the time I went up to Cambridge to begin a PhD, I recognized that my interest in cross-cultural conflict analysis had moved on to a very specific question. What role does cultural conditioning and cultural perception play in the formulation of war plans and the prosecution of war? Numerically based conflict analysis, or 'game theory' as used in the Pentagon's war planning, caught my attention. To me it represented the contemporary cultural arrogance of strategic assumption and a portion of my thesis argued that 'game theory' and other mathematically based analysis systems remain less than ideal for military scenarios that feature cross-cultural conflict. In other words, linking your military response to assumptions about your enemy's actions (i.e. strategy and tactics) is dangerous if you come from a different cultural background than that of your opponent. What seems to be your enemy's next logical move or a 'sure thing' in his projected strategic behaviour may actually be a reflection of your own cultural conditioning. People of differing cultures do things in different ways and that includes waging war. While there may be similarities that make certain assumptions safe, there may also be differences that negate the logic of planned military actions. If you wage war against an opponent from a different culture, it is never safe to assume that the 'givens' that govern your behaviour also govern those of your enemy.

The Anglo-Maratha Campaigns of 1803 provide an interesting case study where military victory obscured the degree to which a Western power misread an Asian opponent. That the British were the military winners in 1803 is not disputed. Rather it is the continuing misrepresentation of their Maratha opponents and the explanation of how victory was attained that are contested. The events of 1803 are now two centuries old, the dust stirred by Britain's retreat from empire has settled and it is time to question what passes for the inherited wisdom. For me the Anglo-Maratha Campaigns of 1803 hold a greater historical significance for two specific reasons.

First, I believe these campaigns represent the misunderstood 'high-water mark' of Maratha military power.[2] One should not become attached to the notion that the Maratha military forces of 1803 were 'Hindu' armies. A proto-national model would be more appropriate. A model based on the realization that collectively the Maratha armies of 1803 were quite secular and not dissimilar to the armed forces of modern India in being composed of military professionals from across the subcontinent. The Maratha powerbrokers of that era were interested in victory and their military effort drew men from the broadest military spectrum – one that included Hindus from every caste, Muslims, Sikhs and Christians. In that respect the Maratha armies of 1803 competed directly with the British for the loyalty of soldiers needed for the projection of power within the contest for India.

Following 1803 there was no indigenous South Asian military hope of driving the British back into the three EIC enclaves established as the Bengal, Bombay and Madras Presidencies. By 1804 Bengal and Madras were joined and their territories expanded towards eventual inland link-up with Bombay.[3] The 1803 campaigns also saw the extension of British power to Delhi and with that, the British became the 'guardians' of the Mughal Emperor. Possession of the Emperor was all-important because it expedited Britain's imperial ascendance and served as additional political cover for the first half of the nineteenth century. The EIC assumed control of a crumbling Mughal infrastructure but with their vast financial network the British were able to selectively employ Mughal political servants and officials, to solidify their own hold on South Asia. There would be many more years of fighting but it was this imperial transition that would later appear as a seamless handoff of power in those sweeping world-history textbooks that allocate a chapter per civilization.

The Marathas were the last indigenous South Asian power that was militarily capable of not only halting but also rolling back the consolidation process that ultimately produced the British Raj. The Anglo-Sikh Wars (1845–6, 1848–9) and the Anglo-Afghan Wars (1839–42, 1878–80,

1919) occurred after the British had achieved a military perimeter around
the majority of Hindus in India. In subcontinental terms, these later wars
were comparatively localized conflicts which would have had limited in-
terethnic political appeal for Hindus beyond the regional strongholds of
the Sikhs and Afghans. As for the events of 1857, whether you call them
rebellion, mutiny, or the first freedom struggle, they were of seismic proportion in reshaping the already existing political and military order of
British rule. Despite the potential appeal of a Pan-South Asian resistance
in 1857–8, the British were still able to draw on vast numbers of soldiers
who continued to serve them loyally for one reason or another. The departure of the British from South Asia would ultimately have to wait for
a more profound shift in world order.

Second, the historic misreading of the Maratha military challenge and
the portrayal of British victory in 1803 – as something inevitable or part
of a conflict process that was determined by so-called 'Western military
superiority' – serves as an example to demonstrate that cross-cultural
conflict analysis remains a particular weakness for Westerners. And I
submit that our analytical failure has contributed to the construction of
dangerously ethnocentric strategic theories to support a Western version
of the world's military history.

This story of the 1803 campaigns shows in a unique way that even
though the British won, it was not for those military reasons we might
have assumed from reading William McNeill, Paul Kennedy or Geoffrey
Parker.[4] Western authors have consistently ignored the depth of the South
Asian historic record and arranged explanations with a cultural bias that
upholds Western military culture and its own special brand of imperialism. We manipulated our interpretation of events to make them appear
as logical in the imaginary court of human history, or better yet, scientifi-
cally inevitable. Despite being culturally pressed for more serious revision
in the 1980s, we clung to the idea that 'Western military superiority' was
self-evident in military victory and subsequent political ascendance. A
belief in a superior Western innovation, technology, discipline and drill
formed the backbone of a revised theory about the European Military
Revolution: an expanded argument that saw the 'rise of the West' in imperial terms as having been derived from a European 'Military Revolution'.[5]
However, the theory's Western fondness for a Social Darwinist approach
to the clash of military cultures is an embarrassing racial carry-over from
the nineteenth century.[6]

Chanting the culturally chauvinistic mantra 'the military rise of the
West' has dulled our senses and left us ill equipped to analyse military
cultures that we find foreign. Oddly enough, when foreign military cultures seemed technologically similar to our own, we had a tendency to

derogatorily dismiss them as if they were shabby imitations of our cherished 'Western way of war'.[7] Those who believe that the 'rise of the West' was somehow inevitable as the result of a Military Revolution have taken far more comfort from their uncontested theories of technological ascendance than from theories of defence economics concerning the clash of international systems and market dominance. However, the time has come to question whether the historic record really leads to a rational explanation of dominance predicated on technological determination – meaning it is time for a 'reality-check' on the argument that Western ascendance was determined by supposed Western military superiority in the form of weapons, drill and doctrine.

Unfortunately the danger of culturally distorted Western military romanticism continues to linger. We in the West still want to believe in explanations of superior technology, discipline and drill, because they continue to suit our cultural and political purposes in the twenty-first century. Attributing the rise and fall of empires to a European 'Military Revolution' has become something of a prerequisite to accepting the 'Revolution in Military Affairs' (RMA) that emerged in the final decade of the twentieth century.[8] It was as if we needed a Western military version of the past to underpin a new 'high-tech' vision of Western cultural superiority in the future.[9]

Some scholars have been willing to question the basic European context of the revised Military Revolution thesis.[10] That has left the door open for others to challenge whether there ever was such a thing as a true Military Revolution; if indeed the phrase refers to anything other than a round of accelerated military evolution. Any military organization of historic magnitude is constantly at war with itself in balancing stagnation with military evolution. Soldiers and armies are bound by their military traditions as well as their inherited hierarchies in the form of political purpose and organizational behaviour – which may be as simple and all pervasive as rank. Yet at the same time they are constantly seeking evolution and innovation in strategy, tactics and especially technology, to provide the military answers for current conflicts as well as future wars. But being in a dual state of stagnation and evolution precludes them from being truly revolutionary. Not even the twentieth century's greatest purpose-built revolutionary armies – the Red Army of the USSR and the People's Liberation Army of China – could shake the stagnation/evolution dichotomy. They were held fast in spite of periodic dedicated attempts to be 'more revolutionary' or to renew their revolutionary credentials in a technological and military context.[11] As for the unrepentant cultural chauvinism of those who continue to advocate technological determination as an explanation for Western ascendance,

I can only say that the 1803 Anglo-Maratha Campaigns demonstrate – with a surprising reverse example – that technological innovation and superior firepower were never the absolute guarantors of military success in war.[12]

Although the book that follows is in large part an attempt to redress the historic and cultural imbalance, it is also an examination of how information is lost or misconstrued as it passes from one cultural setting to another. Westerners have used a very rigid and predictable model for South Asian conflict analysis, a biased model that has downplayed the legacy and meaning of South Asian warfare. But part of our problem is that we have dealt with large blocks of time in order to make analysis easier. Going back and studying the wars on an individual battle-by-battle basis is mandatory if we are going to revise the gross generalizations that were made about South Asian military culture and experience. I feel there is good reason to retell the story of these campaigns in the light of new discoveries about the manner in which the British achieved military victory in 1803.

The Maratha military challenge

To a great extent this book deals with Maratha military culture and the challenge it presented to the British. However, this portion of the text is intended to provide the reader with some idea of how the events of 1803 fit into the larger picture of South Asian military history.

The extension of European conflict

The British 'conquest of India' came about as the result of a rather lengthy series of economic, political and military events stretching over more than two centuries from the founding of the EIC in 1600. This has caused some historians to rethink the once popular theories of decisive battles for control of the subcontinent, which are now much more open to debate. While economic rivalry was always a potential trigger for violence between competing European powers in South Asia, it could at times be tempered by shared defence concerns or a desire for the maintenance of peace between home governments. But during the eighteenth century the South Asian extension of European wars was facilitated by overall improvements in France's as well as Britain's ability to organize and equip indigenous defence forces that were interchangeable with European troops in the order of battle. This point is crucial in trying to understand all of the Western hyperbole surrounding the use of so-called 'European discipline and drill' by South Asian troops. By instructing South Asian soldiers in the latest version of standard procedures for their armies, competing European powers were merely ensuring a compatible level of *interchangeability* between their home armies and those of their colonial military forces. A number of mid-eighteenth-century South Asian armies already utilized both indigenous and European military organizational systems and South Asians already had all the personal warrior attributes they needed prior to this latest round of standardization aimed at European political objectives. The widespread imposition of a nation-specific European theory of military organization (i.e. Britain or France) enabled Europeans to use South Asian soldiers more effectively in terms of European colonial force structure and its tactical deployment towards attaining European political objectives in a colonial setting. These mid-eighteenth-century European institutions did not set a historic precedent for the introduction of either discipline or drill in South Asian armies.

During the eighteenth century South Asian colonial armies held enormous potential for extending the ability of England and France to wage war in Asia.* A colonial army consumed military resources but it was still far more cost-effective than shipping an all-European force to the far side of the globe. South Asia's extreme climate had provided the Europeans with an initial reason to look for indigenous allies who could prosecute their wars more effectively, while South Asian leaders sought European allies who might help tip the balance of power in local struggles.[†] It was often a mutually symbiotic relationship nurtured by the joint quest for military advantage. And although large portions of South Asian society had been militarized since Vedic times, the Anglo-French rivalry of this period would feed directly into larger South Asian regional power struggles so that it became hard to distinguish influence from impetus and cause from coincidence.

As the EIC grew, in terms of financial and military power, it became a more sophisticated civil–military mechanism capable of managing British wars in Asia. Within India there were three separate Presidencies, Bengal, Bombay and Madras, each with its own Governor, Council and army. But Bengal was the largest and wealthiest of these Presidencies and after implementation of Pitt's India Act (1784), the Governor of Bengal became the Governor-General with powers placing him above the other Governors and providing scope for coordinating the segmented governmental apparatus on the ground. Nonetheless, the separate Presidencies retained their distinctive military identities via their individual armies

---

* The First Karnatak War (1744–8) was a part of the War of the Austrian Succession (1740–8).
† i.e. The Second Karnatak War (1749–54).

and that enabled the 'Honourable Company' to amass a wealth of tactically based regional knowledge as well as develop military specializations suited to specific combat environments. Over the years the British had also dispatched a number of the King's troops to India and although they were often used to spearhead assaults they were not generally as knowledgeable about 'in-country' operations. From his seat of government in Bengal the Governor-General had senior military authority and he coordinated various EIC military deployments as well as operations involving His Majesty's troops. The tremendous distance between London and India meant that the Governor-General operated for months at a time without direct guidance and he often had to act on his last available orders or a set of principles interpreted from policy guidelines.

Anglo-Maratha conflict

By the second quarter of the eighteenth century the Marathas could be said to have controlled 75 per cent of the subcontinent.[13] And with the EIC gradually pushing the economic and political hinterlands of its three Presidencies ever inward, a clash was inevitable. Once the British began to engage the Marathas militarily, the conflict process flared sporadically for two generations, stretching from the Battle of Aras on 18 May 1775 until the Siege of Asirghar on 7 April 1819. In fighting this series of Anglo-Maratha wars the EIC in effect accelerated the disintegration of the Maratha Confederacy that was the indigenous heir to India's military fortunes.[14] Those unfamiliar with South Asian history often find it difficult to trace the Anglo-Maratha Wars as they overlap with the Third and Fourth Anglo-Mysore Wars in which the Marathas were British allies and neutrals respectively. The sequencing of the major wars looks something like this:
- First Anglo-Maratha War[15] (1775–6, 1779–82)
- Third Anglo-Mysore War (1789–92)
- Fourth Anglo-Mysore War (1799)
- Anglo-Maratha Campaigns of 1803
- Holkar (Maratha) Campaign of 1804–5
- Maratha and Pindari War of 1817–19

The Anglo-Maratha Campaigns of 1803 were directed against those Maratha leaders who opposed the Treaty of Bassein,* a Subsidiary Alliance agreement signed in desperation by the Maratha *Peshwa*† (Prime Minister) Baji Rao II. For his part, the *Peshwa* probably saw the document as a politically expedient and cost-effective means of temporarily obtaining British troops to protect himself – a marriage of convenience

---

* Aka Treaty of Bassain, Treaty of Bessain.   † Aka Peshwah, Peishwa, Peishwah.

if you will. Ironically, the *Peshwa* needed protection from the leaders of the Sindia and Holkar clans that formed part of his own broadly based political network. The most militarily powerful Maratha clans menaced Baji Rao II with their armies because the *Peshwa* was, in their opinion, an ineffectual and loathsome political figure. Baji Rao II apparently saw his own signature on the Treaty of Bassein as a ploy in a larger game of control, which he believed he could win. The naively optimistic *Peshwa* apparently hoped to use the British as temporary allies in 1803 to help claw back political power – after which time these foreigners could be dispensed with and crushed.

British imperial historians have tended to see the battles of 1803 as springing from the unwillingness of independently minded Maratha leaders to peacefully acknowledge their subordinate status under the Treaty of Bassein as signed by the *Peshwa*. The formula for a British intervention in 'native affairs' was not new and the Treaty of Bassein was in keeping with the basic tenets of the Subsidiary Treaty Alliance System. A similar treaty with the *Nizam* of Hyderabad had apparently been successful but the *Nizam* did not have to contend with multiple princely retainers each with greater military power than his own. Governor-General Richard Wellesley's[*] supporters have asked us to believe that, if the Treaty of Bassein had worked as intended for the EIC, it would have seen the various competing Maratha clan factions brought peacefully together under a rather weak but manageable *Peshwa*. By signing the Treaty of Bassein, Baji Rao II surrendered his autonomy and unknowingly gave Richard Wellesley a political cover to meddle further in Maratha affairs. Ever since the impeachment trial of former Governor-General Warren Hastings[†] the scrutiny of Governors-General had been aimed at detecting abuses of power that might suggest 'high crimes and misdemeanours'. The Treaty of Bassein was a godsend to Richard Wellesley in that it allowed him to explain his expansionist Maratha policy in terms of political involvement on behalf of his reluctant ally the *Peshwa*. This could be used to help portray the events of 1803 as a just and legal war against opponents of the treaty as long as one accepted the premise that *Peshwa* Baji Rao II was the legitimate Maratha 'ruler'. But real Maratha power lay elsewhere and the British soon began to despise the *Peshwa* as well.

Governor-General Wellesley's main Maratha enemies in 1803 were Daulat Rao Sindia,[‡] the Maharaja of Gwalior, and Raghuji Bhonsle II[§] of Nagpur.[16] Raghuji possessed a magnificent mounted force in 1803 and

[*] Aka the Earl of Mornington, Lord Mornington, the Marquess Wellesley.
[†] Found not guilty by the House of Lords in 1795.
[‡] Aka Dowlut Row Sindia, Daulat Rao Shinde, Dulat Row Scindiah, Dowlut Rao Scindhia, Daulat Rao Schinde, Dowlet Row Scindeah, Daulat Rao Sindhia (b.1780–d.1827).
[§] Aka Raguji Bhonsle, Maharajah of Berar, Rajah of Berar, Berar Rajah, 'the Berar man'.

some tenacious infantry units in the form of battalions from Hindustan and the Persian Gulf. But the most formidable enemy army, the one that could march on short notice, take land, occupy it, hold it, forcing the British to pay the maximum price, was the army of Daulat Rao Sindia. The 'regular corps' of Sindia's army featured sepoy battalions that were often indistinguishable from those of the EIC in uniform, drill and ethnic origin. The third principal Maratha powerbroker of this period was Jeswunt Rao Holkar,* the Maharaja of Indore. However, Holkar abstained from participation in the battles of 1803 in an apparent hope that Sindia and the British would fight each other to a weakened point that he (Holkar) might be able to exploit.[17]

The British strategy for dealing with the Marathas in 1803 depended heavily on a two-prong projection of power into the interior of the subcontinent. The main British infantry forces were divided between Commander-in-Chief (C-in-C) General Gerard Lake's[†] 'Grand Army' in the northern theatre (Hindustan) and Major-General Arthur Wellesley's 'Army of the Deccan' in the southern theatre. Arthur was the younger brother of Richard Wellesley – Britain's Governor-General of India. However, Arthur went on to gain greater European fame in the Peninsular War and he is most often remembered for his famous victory over Napoleon at Waterloo in 1815.[‡]

Although this overview makes it sound as if Hindustan[§] and the Deccan[‖] were the only campaign areas in 1803, there were three other operational areas that were essential to the British war effort. Unfortunately there is neither time nor space in this book to do justice to these other military events. Two actions centred on securing coastal regions. Colonel Murray captured Sindia's port of Broach in Gujarat[18] on the west coast while Raghuji Bhonsle's maritime province of Cuttack in Orissa on the east coast of India was taken in a pincer movement launched by troops from Bengal and Madras.[19] The seizure of the Maratha ports completed the British effort to seal the coastline of the subcontinent in

---

* Aka Yeswant Rao Holkar, Jaswunt Rao Holkar, Yesvant Rao Holkar, Jaswunt Row Holkar.
† The rank distinctions and chain of command in India were never simple during the period of EIC rule. The Bombay Presidency and Madras Presidency in 1803 each had their own C-in-C but Lake was supreme and held the King's commission. He was Britain's paramount soldier in India during 1803.
‡ Arthur Wellesley became Viscount Wellington in 1809, Earl of Wellington in February 1812, Marquis of Wellington in August 1812, was promoted to Field Marshal in 1813 and received the title Duke of Wellington in 1814. He was made Master-General of the Ordnance in 1818 and capped his career by becoming Prime Minister in 1828.
§ In its most simplistic form – Land of the Hindus – North India.
‖ Aka Dakhan, Dekhun. Defined historically as the subcontinent's tablelands. In relation to North India – the south-central plateau between the western and eastern *ghats* (hills).

1803; a military move that paid lasting benefits in their subsequent South
Asian wars. The great Maratha naval tradition of coastal raiding, once up-
held by clans like the Angrias, was history.[20] And with increased freedom
of movement in the shipping lanes, the British enjoyed further logistical
advantage as men and war materiel were transferred between EIC bases
without fear of Maratha interception. The third major action of 1803 was
the landlocked operation in Bundlekund aimed at creating a buffer zone
and further disconnection between Maratha forces in the interior of the
subcontinent.

In Wellington's shadow

The Anglo-Maratha Campaigns of 1803 have not been well covered in
either Western or South Asian military history. Over the years a variety
of Anglo-Maratha battle narratives have emerged but they have tended
to be polarized along the lines of cultural difference that leave us wanting
a more balanced picture. In South Asian military histories, 1803 is often
treated as an anomaly that pales next to the Maratha military exploits of
the seventeenth century. As far as British military history is concerned,
the Napoleonic Wars take precedent. Most students of military history,
who have heard of the Anglo-Maratha conflict of 1803, are familiar with
the events in the southern theatre as typified by the daring exploits of
Arthur Wellesley at the Battle of Assaye. Arthur's later rise to become
principal commander in Europe and ultimately Prime Minister speaks
volumes about selective memory and the fashion in which history is culled
for political purpose. The northern theatre under General Lake remains
underreported.

Assaye was destined to be remembered as the key battle in establishing Arthur Wellesley's reputation for offensive combat leadership in India.
Technically speaking, the pursuit of Dhundia Waugh was Arthur's first independent combat command, but it was not to be referenced in the same
way as Assaye. A career built on counterinsurgency carried no panache in
the first quarter of the nineteenth century. Within the dominant British
military culture of that period reputations were built on conventional operations and the warrior ideals associated with traditional Western warfare. Critics could not downgrade Assaye to the lowly status of a bandit-
control programme. As a neo-classical battle, on the banks of the Kailna
River, Assaye was a natural candidate for epic glorification. Indeed its
similarity to Alexander the Great's defeat of Persian King Darius's forces,
along the Pinarus River at the Battle of Issus in 333 BCE, meant that the
analogy was simply too tempting to resist. The historic interpretation of
the British victory at Assaye took on a life of its own. Over time, it was

rewritten and sanitized so that it could be conveniently retrofitted to the later image of Arthur. The accounts were polished to befit the Duke of Wellington as Britain's pre-eminent soldier – the 'Iron Duke'.

Following Waterloo and the defeat of what was commonly acknowledged in Britain as the international Napoleonic threat, there were those who re-read Assaye as representing the historic continuum of Britain's international battle against France. Certain aspects of Assaye contributed to a very confusing picture and the presence of some key British soldiers in both Europe and India was taken as supporting evidence in the theory that the Anglo-Maratha Campaigns of 1803 were the definitive South Asian extension of the Napoleonic Wars. However, any such argument is simply too frail to withstand investigation. Granted, Napoleon had diplomatic correspondence with Mysore in the 1790s.[21] And three French agents – Courson, Durhone and Dauble – were seized in the *Peshwa*'s capital wearing native garb on the eve of war in 1803.[22] But by no stretch of the imagination could that be portrayed as seriously active Napoleonic participation in the Maratha Campaigns of 1803.[23] In reviewing the military and political accounts of Anglo-Maratha conflict, it would seem that the greatest impetus to the Napoleonic 'makeover' of 1803 occurred in the three years immediately following Waterloo. The greatest written source guiding that process and influencing later histories was Major William Thorn's memoir.[24] Thorn's work became the *de facto* official history of 1803 and that helped elevate his opening remarks on the Napoleonic significance of the war.

Thorn believed that the stability of Britain in the Napoleonic Wars was based upon the 'plains of Hindoostan' and the consequent emancipation of Europe began, though mortals could not see it, in a train of operations extending across India.[25] Thorn was convinced that Napoleon considered himself in competition with Britain for the resources of India. Thorn subscribed to the theory that revenue-bearing lands, assigned to French mercenaries in Maratha service, constituted a French state in North India.[26] He saw this as a French footing 'in the central and most fertile part of India, where they must be said to have ruled absolute over the counsels of the principal states, and in particular over the descendants of the imperial house of Timur'. Or, in more contemporary words, a French satellite state that in turn controlled the region by controlling the remnants of the Mughal Empire.

What was Thorn's motive for writing such a Eurocentric interpretation of events? Was he merely a Francophobic veteran? Was he seeking to elevate the importance of his Indian theatre experience by letting his readers know that the 1803 Maratha operations represented a serious war – a battle against a larger European threat and not some damn sepoy adventure? Or, were his introductory remarks a more market-oriented approach to selling books linked to Wellington, an effort in which his memoir was laced with anti-French references to enhance its sales potential in the post-Waterloo world of British military history?

Regardless of his motivation or intellectual objective, Thorn's words did reinforce the popular belief that the Marathas were South Asian pawns in the hands of a European power. It was a line of reasoning which did nothing to enlighten readers with regard to the Marathas' military tradition or their ability to challenge one of the greatest military superpowers of the day. Thorn related the victory over the Marathas back to the central front and he made it sound as if India's importance was that of underpinning Europe. Arthur Wellesley, as the Duke of Wellington, was the link that brought European power full circle, from England to India to Europe and back to England after victory over Napoleon 'who, in his aim at universal dominion, experienced that check in the interior of India, which, by a circuitous train of events, led to his total overthrow on the fields of Waterloo'.[27]

In the final quarter of the nineteenth century – as wide-scale European fratricidal warfare slipped from Britain's collective memory – national interest in India and the empire experienced a renaissance.[28] With the golden age of the British Raj there came another imperial makeover of Arthur Wellesley's actions in 1803. In a review article published on 6 October 1887, *The Times of India* spoke of Assaye as 'Wellington's greatest victory, which made the English masters of India'.[29] During that later Victorian era, specifically the pre-Boer War period, it was comforting for Britons to see themselves as masters of India by virtue of their military heroism. Indeed virtue seemed a more desirable handmaiden for military endeavour during those years of ubiquitous brushfire wars in Asia and Africa. And a particularly military form of cultural imperative fuelled the nation's desire to uphold heroes whose deeds could be seen as exemplifying those military virtues that underpinned a greater British imperial identity.

An overview of the text

To understand the British victory of 1803 in terms of this cross-cultural study it is necessary to start with a much broader analysis of Maratha military culture. The first chapter is devoted to establishing a more accurate understanding of the Maratha military profile from 1600 to 1800. Using that as the foundation for continuing reassessment, the second chapter moves forward on a comparative basis to look at what shaped British military attitudes towards the Marathas prior to the 1803 campaigns.

The third and fourth chapters deal with the southern and northern campaigns of 1803 respectively. While the fifth chapter is an attempt to pull together new research that shows there was far more to the story of 1803 than most historians had previously realized. The sixth chapter reviews the traditional explanations for British victory and ranks them according to the evidence presented in the preceding chapters.

1 Maratha military culture

Introduction

This chapter employs a thematic approach that attempts to conform to the chronology of Maratha history as closely as possible. It examines the confusion surrounding Maratha military culture by discussing issues such as discipline and drill, the role of infantry, light horse tactics, the misnomer of 'guerrilla warfare', artillery doctrine, smallarms, and the use of European mercenaries in Maratha armies. This section is intended to provide the reader with information that can be compared with Western military perceptions of the Marathas as found in chapter 2. The discourse reveals that there were several long-term patterns in Maratha warfare that went largely undetected with the result that we often see erroneous historical generalizations about the nature of Maratha armies.

This book differs from others in that it depicts Maratha military culture as directly linked to the political economy of war. Central to this chapter is the concept of the South Asian military economy. Research revealed a dynamic and vibrant economic environment with a distinct military orientation; a military economy that revolved around the ever-present possibility of armed conflict. While some would argue that the label 'military economy' is an unnecessary sub-division of the word economy, I would point to our contemporary institutionalization of defence economics as a formal discipline.[1] Modern economic markets are characterized by trade in resources, commodities, goods and services, as well as financial products. The historic South Asian military economy was also identifiable by its component parts. Human resources, weapons, logistical services, credit instruments for military financing – they all had a relative value that was in large part determined by the dynamics of the market place. It would be inappropriate to label this as an early military industrial complex. However, it did include manufacturing facilities and consumed resources on a pervasive commercial scale that bridged many socio-economic divisions as well as geographic regions in South Asia.

The Marathas, as active participants in this military economy, evolved a military culture that was responsive to market forces. It probably began with small Maratha warrior bands that had a societal predisposition to organizational adaptability and military survival. They soon took service in relatively cosmopolitan local armies. Upon consolidation, those local armies proved to be the building blocks of bigger regional armies. And in time the regional armies began to serve more than the purpose of security forces, they became institutions of political control in one form or another. The South Asian military economy eventually evolved into a network of subcontinental power. By 1803 dominance of that economy was a prerequisite to any serious bid for political supremacy or the balance of empire.

The foundations of Maratha military culture

Maratha warrior traditions reflect an age-old Vedic inspiration. And since the days of simple cattle raids, Maratha hero stones have silently extolled the virtues of bravery, skill at arms and the judicious employment of extreme violence. Some hero stones depict ancient battles in Maharashtra and there are two interesting examples at Bavde* and Akluj[†] that portray local forts under siege.[2] The carvings clearly show the disciplined ranks of well-drilled troops moving in tactical linear formations. The Akluj stone is particularly noteworthy because it demonstrates the use of combined operations with disciplined infantry and two types of mounted troops. The fort's defenders are shown as having launched an attempt to break the siege ring tightening around them. Their breakout force featured a vanguard of lancers flanked to the left by tightly packed infantry. This leading line of infantry, with shield and sword, stood shoulder-to-shoulder in formation presenting a shield-wall to the attacking siege force. Behind the line of shields, archers were waiting to unleash their arrows. To thwart this effort the besieging force countered the defenders' cavalry charge with horsemen of its own. As this was being done, the besieging army flanked right, using elephants in echelon to trample and sweep the breakout force's left in an attempt to roll-up the defenders' infantry line. The craftsman who carved the stone chose to freeze in time the screaming agony of a defending infantryman of the line. His open mouth hauntingly depicted as a circular depression and his right leg shown in the process of being crushed beneath an advancing elephant in full stride.

The ancient hero stones provide recorded evidence of South Asian warfare's sophistication in the pre-European period.[3] This is the type of evidence so often overlooked by Western writers who asserted that the

---

* 16°+, 73°+.   † 17°+, 75°+.

Europeans introduced military drill and discipline to India; some insisting it did not arrive until the middle of the eighteenth century. They were quite willing to accept the frescoes of Rome and Greece, the friezes of Assyria or even the stele of Mesopotamia, as proof of superior military drill and discipline in the so-called Western world.[4] But the iconographic evidence of South Asia, and the hero stones of Maharashtra in particular, are every bit as much proof of drill and disciplined formations.

The Marathas emerged militarily as a loosely knit group of clan-oriented powerbrokers. Their alliances were based on familial as well as economic military considerations and the transitory nature of Maratha clan relations contributed to the search for military advantage. Maratha clan armies grew in an atmosphere that featured a results-oriented approach to warfare established years before Europe's 'Age of Empire'. The cultural and physical environment of Maharashtra helped to spawn the growth of military alliances as smaller bands of warriors sought employment together. In this military setting leaders could act as broker/dealers in arranging contractual service for their bands.[5] Many went beyond that in becoming players in the greater South Asian military economy and the fortified strongholds of Maratha clan leaders in the Deccan were used for bouts of territorial exploitation. However, their forts also provided opportunities for the Marathas to participate as specialist troops in the large dynastic armies that came to be a regular feature of Deccani* history during the pre-European period. The fourteenth-century rise of the Bahmani kingdom yielded far-ranging military employment and the widespread use of firearms meant an increase in the infantry soldier's value in warfare as well as the building of improved fortresses. And with the sixteenth-century demise of the Bahmanis there came greater military competition and opportunity in the emergent form of five growing sultanates. The two lesser sultanates were those of Imad Shahi of Berar and the Barid Shahi dynasty of Bidar. They paled in military power compared to the three major sultanates, the Nizam Shahis of Ahmadnagar,† the Adil Shahis of Bijapur and the Qutb Shahis of Golconda.[6] With the increased nucleation of power there was another corresponding increase in militarization.

By 1485 local Maratha clans had already established a favourable reputation as infantry garrison troops in the hill forts near Pune.‡ Word of their ability spread as the bitterly contested forts eventually fell to Malik Ahmad, founder of the Nizam Shahi dynasty.[7] Enemies who fought well were not to be wasted; good soldiers were highly valued within the South

---

* Aka Dakhani; meaning of, or from, the Deccan.
† 19°+, 74°+. Aka Ahmednagar, Ahmednuggur, Ahmednaghur.
‡ 18°+, 73°+. Aka Poonah, Poona, Punah.

Asian military economy and the Marathas found service with their former opponents. The Nizam Shahi dynasty was the first of the three Deccani sultanates to emerge from the Bahmani kingdom's decline and Malik Ahmad, under the more popularly known title of Ahmad Nizam Shah,* based his military power directly upon the fortress of Ahmadnagar. This massive fortification had a deceptively low profile set into the Deccani countryside and it proved to be a most dynamic military labour spot market for the Maratha chiefs who travelled there to offer various military services. These Marathas were independent mercenary agents who sold their skills and those of their kinsmen without the impediment of nationalism or sectarian prohibition.[8] For his part, Ahmad set a personal example in weapons drill, discipline and military learning by way of the martial arts that he promoted.[9]

Maratha infantry were used to defend the chain of fortifications that added backbone to the Nizam Shah's territorial claims, while more transitory Maratha light horsemen found increased employment during target-specific campaign seasons. The latter group of Marathas experienced cyclical opportunities to renew their contractual relationship as the 'Ahmadnagar army took to the field twice a year at the time of the early and the late harvests, to plunder the country near Daulatabad in order if possible to reduce the fort by famine'.[10] Some historians would have you believe that campaign seasons were rigidly applied across South Asian culture. They envisioned a predictable world in which South Asians did not campaign in the monsoon because of the rain nor at harvest time owing to agricultural labour requirements. The Marathas, keen soldiers seeking to earn their pay when the opportunity presented itself, proved willing and able to campaign at any time of the year. If there was work to be had on campaign with the Nizam Shah, why lose the chance for military or financial gain? A harvest presented a military opportunity but one had to time it carefully. In an ideal world you let the enemy finish the fieldwork, to minimize your effort and maximize your return. You had to strike before he got the bulk of the grain stored in his fortress. A field of stubble burned easily and left him with nothing – not even a blade of grass for cattle forage. And if you had the enemy's grain, or denied him access to his fields while your camp followers harvested it, then there was a better chance of starving him into submission.

The harvest season was tied to the monsoon cycle, which also provided regular access to the international military arena as reflected in the navigation patterns that brought a variety of vessels to Maharashtra's coastline.

---

* Aka Ahmad Bahri Nizam Shah (reign title). He captured Daulatabad in 1499, thereby consolidating power before his death in 1508. The Nizam Shahi dynastry of Ahmadnagar lasted until 1636.

Ships from Africa, the Persian Gulf, the Red Sea and Europe served to expand opportunities to incorporate Islamic, African and European military science.[11] Ahmad Nizam Shah was a regional powerbroker who attracted the attention of the larger world. By serving him the local Maratha *sardars* learned more about the ever-expanding world of military science. The fortresses of the Deccani sultanates, as centres of military employment, offered chances for Maratha soldiers of fortune to learn how the great armies of the day functioned in an administrative as well as military capacity. The Marathas stationed at Ahmadnagar had an opportunity to see the Portuguese military men who visited the fort openly as 'respected friends' of the Nizam Shah. And although the sultanate's eventual adoption of Shi'ism might have offended Catholic Europeans, it did lead to increased ties with Persia and that in turn meant a chance for Maratha soldiers to learn more about Persian military affairs. Western India was growing exponentially with regard to military skills, the volume of firepower and exposure to the international arena.

Ahmadnagar was also noted for its artillery and it was not unique in that regard. Increased military competition in western India had bolstered the quest for additional firepower. The correspondence of Portuguese officials in the first half of the sixteenth century makes it apparent that they considered South Asian artillery superior to their own.[12] South Asian metal smiths, with literally thousands of years of collective experience, could produce quantities of metal that surpassed that of historically contemporary Europe. This was particularly evident in the production of steel and the alloying of brass* or bronze.[13] Centuries of casting experience was utilized to manufacture temple ornaments and religious paraphernalia in India and these production techniques readily leant themselves to casting artillery. The ease with which traditional Indian craftsmen could be integrated into military production was quite natural. Within market towns and cities, socially ranked metal workers were often found in proximity to local wheelwrights making parts for bullock carts and both groups found employment in manufacturing artillery. Production grew steadily under the patronage of local powerbrokers that tended not to interfere with the free-market forces of the South Asian military economy.[14] This free-market attitude towards military technology flourished in an India unrestricted by state monopoly or centralized control over artillery.[15]

The manufacturing of quality artillery was more than just a cottage industry on a regional scale and the scientific aspects of South Asian

---

* Metallurgically speaking, virtually all 'brass cannon', regardless of nationality, are actually bronze.

artillery production should not be ignored. In general, historians have failed to consider the combined elements that would have contributed to making western India an ideal historic home for the evolution of artillery. Advanced mathematics was essential to the effective development of artillery and the Hindu science of numbers was augmented with contributions from Arab Muslim mathematicians and then readily applied to ballistics. In addition, the calculations for alloying and estimating molten metal mould-volume requirements demanded precision. The coming together of Hindu metallurgy and mathematics with Arab as well as Persian contributions represents the type of military science that South Asians were never given credit for in Western texts. As for Hindu astronomy, it was used in the weather forecasting needed to set the dates for the arrival of the monsoon.[16] Bronze cannon casting was best done before the monsoon since barrels cast in that season were more likely to be flawed, which meant they would have to be melted down and recast at a later date. During the monsoon the air's moisture content directly influenced the formation of bubbles, which lodged between the mould's interior surface and the molten metal. Bubbles of that description – which we now know to have been hydrogen – would manifest themselves as pitted surface flaws as the metal cooled to reveal the location of gas trapped between the mould and the barrel.[17]

To a great degree sixteenth-century warfare in the Deccan escaped detailed European observation, but it seems that the early South Asian lead in artillery quality and volume of firepower was sustained in the transition from siege artillery to massive field artillery. The issue of how to attain mobility for 'great guns' in the Deccan was resolved by creating transportation corps with large numbers of elephants and camels that offered tremendous advantages over horses. South Asian leaders could also purchase the services of thousands of menial labourers within local labour markets. In 1546 Ibrahim Adil Shah attacked the field army of Ahmadnagar as it lay encamped on the banks of the Bhima River.[18] The capture of 170 pieces of artillery with tumbrels and 250 elephants indicates that theories of artillery-based fire superiority had already affected manoeuvre warfare in the Deccan. Increased firepower, or the multiplication of force if you will, paid dividends in the reduction of the number of men needed with smallarms. Men with such skills were moderately expensive in most armies owing to their training time and smallarms procurement costs.[19] While the Europeans could also fabricate large cannon, they tended to use them as rather immobile siege guns.[20] Without sufficient brute strength or some form of mobility the European siege guns were too unwieldy as field pieces. Forty years after the battle on the Bhima, Venetian traveller Caesar Frederick saw some of Ahmadnagar's later generation of artillery. He observed that they had been cast in sectional components that added to their pack mobility as they could be dismantled for animal transport and then reassembled for deployment; a great advantage for campaigns in the hill country or crossing monsoon-swollen rivers. Once again elephants were to prove an environmental asset. Although some Western authorities have dismissed South Asian guns fabricated in sectional rings as being later Ottoman pieces, these were cast in India and fully operational in Ahmadnagar's army when Caesar Frederick saw them in 1586. He noted, 'Though they were made in pieces the guns worked marvellously well.'[21]

Establishing a military profile

Arguably the most skilful soldier/statesman to govern Ahmadnagar was Malik Ambar* (d. 1626). A black Habshi[†] slave known in some texts as 'the saviour of Ahmadnagar', Malik Ambar employed large numbers of Marathas using *bargi-giri* predatory light horse tactics (see below, p. 26), deploying them to great effect against Mughal army supply lines. And it was that light horse tactical experience which later became associated with the *modus operandi* of Maratha mounted units in the Deccan. Unfortunately this is also where some of the historic confusion begins, as historians have tended to merge Maratha ethnic and military identity with just a single aspect of their military profile.

A clan-based group of 'horsemen for hire' could all be of one ethnic group and some Marathas served as cavalrymen, but by far the greatest numbers were infantry and garrison troops. The existence of Maratha clans serving Malik Ambar in the first quarter of the seventeenth century, and the coexistence of some predominantly mounted Maratha bands in that period, cannot be projected forward as a safe assumption that 'Maratha armies' were all ethnically Maratha or that all true Maratha armies were mounted practitioners of *bargi-giri* warfare. One would not accept the same cultural romanticism in the British army if someone were to postulate erroneously that English armies were all ethnically English or all truly English armies were cavalry-based since the days when the Roundheads and Cavaliers faced each other on horseback during the English Civil War. It was true that the most powerful *sardars* of clans like the Bhonsle's and Nimbalkar's were ethnic Marathas. But their 'rank-and-file' subordinates – the soldiers from other ethnic communities who served them – became 'Maratha forces' by association. Therefore, when

---

* Aka Malik Amber.
† Aka Hubshi, Hubshee; cited in *Hobson-Jobson* as an Abyssinian or Ethiopian.

we refer to a Maratha army in the evolution of South Asian military history we must consider that these armies were addressed as Maratha for the sake of identifying them in terms of categorization.[22] Unfortunately the mythology of 'pure Maratha armies' is often polarized or made worse by partisan portrayals of the Mughals as 'alien oppressors' or an 'army of occupation'. In truth, the Indianization of the Mughals was well under way by the time of Akbar (1542–1605). And given the historic date of the arrival of Islam in the Deccan and the social integration of the five Deccani sultanates, the word 'oppressor' is misplaced as a cultural generalization in this context. The Marathas sold their services as professional soldiers and their potential employers' religion presented no insurmountable barrier to contractual dealings.

Maratha *sardars* who changed employers repeatedly among the Deccani sultanates, joining one side and then another, were mercenaries in the strictest historic sense of the word in that they served various masters for a price. As broker/dealers in the military labour market, bartering the services of their troops, they exerted leverage via collective bargaining to obtain the best employment terms available.[23] Beyond wages, the negotiable points of service might include percentages of enemy plunder, equipment as well as animal replacement costs and establishing who was responsible for ammunition expenses. That Maratha *sardars* bargained determinedly over such points tells us about their appreciation of defence economics and what they considered to be inclusive vs. exclusive costs in their approach to war as a business opportunity within the South Asian military economy.

An entrepreneurial approach made seventeenth-century Maratha military powerbrokers both competitive and effective. When they saw a military need that they could meet for a price, in effect an economic military opportunity, they responded. *What do you need; soldiers, guards, watchmen – offensive or defensive?* You could have any or all within the military labour market. *What kind of men? What are you willing to pay? Do you want some peon goons to break a few villagers' heads and collect a few debts?* Perhaps you are a serious player who needs heavily armed forces and professional soldiers. *What about infantry with good quality matchlocks, are they more suited to your budget?* There was also a chance to broker animals as well as men. *Do you need Deccani cavalry with local ponies that can survive the drought on a handful of dried jowari* [millet] *a day?* Arabian horses were also available in the port cities but they were better kept for breeding because extended campaigns in the Deccan called for endurance. The prospective military employer might have sufficient funding to hire a small army complete with regular infantry, heavy artillery and cavalry accompanied by grasscutters to find the best fodder for good horses in the worst of times. *Are you paying cash or a percentage of the spoils after the campaign?* It was all on offer, it was all there for hire – at the right price.

As seventeenth-century Maratha military entrepreneurs expanded their forces they were bound to acquire the specialities and services of non-Maratha people. Some Maratha coastal clans were keen to exploit the opportunity of incorporating the peripheral non-Marathas who had gained military expertise in service to one of the many foreign powers who prowled along the western coast. Outcasts often made good soldiers because foreign military employment offered them upward social mobility, status as warriors, cash and weapons. This appears to have been true of coastal tribals, from Bombay to Gujarat, who had been exposed to smallarms as the result of African and Arab recruiting for local wars. The Europeans, who eventually followed, also sought to augment their limited military manpower. Maratha *sardars* soon found it profitable to recruit so-called Topasses* from Goa and the other Portuguese colonies. Many of these soldiers were of mixed South Asian and Portuguese heritage, some having been raised as Christians or converted to Christianity. Their indoctrination in contemporary European military science, particularly that of the artillery and the infantry, served to add another dimension to Maratha military capability.[24] The peripheral incorporation of people living on both the physical and cultural margin of Maratha society was an on-going process that spanned several centuries. The drawing-in of specific social groups, tribes, clans and castes often reflected their collective military occupational specialities as directly related to weapons proficiency, tactical skills, manufacturing capabilities, or their role in facilitating logistics. Making weapons, feeding armies, serving as soldiers and scouts, these groups often represented some form of temporary advantage for expanding Maratha armies. If the Maratha clan armies had been exclusionary, with regard to ethnicity, religion, caste or level of social integration, they would have denied themselves the competitive advantage found in the diverse range of military human resources that western India had to offer.[25]

The Shivaji legacy

Shivaji[†] Bhonsle (1627–80)[‡] remains the most famous Maratha warrior in history. Like other great warrior-statesmen, such as the Duke of

---

* Aka Topaz, Toupas, Toepass. There are conflicting stories associated with the origin of the name.
† Aka Shivajee, Shavajee, Shiwaji, Sheevajee, Sewajee.
‡ Some histories place Shivaji's birth as 1630.

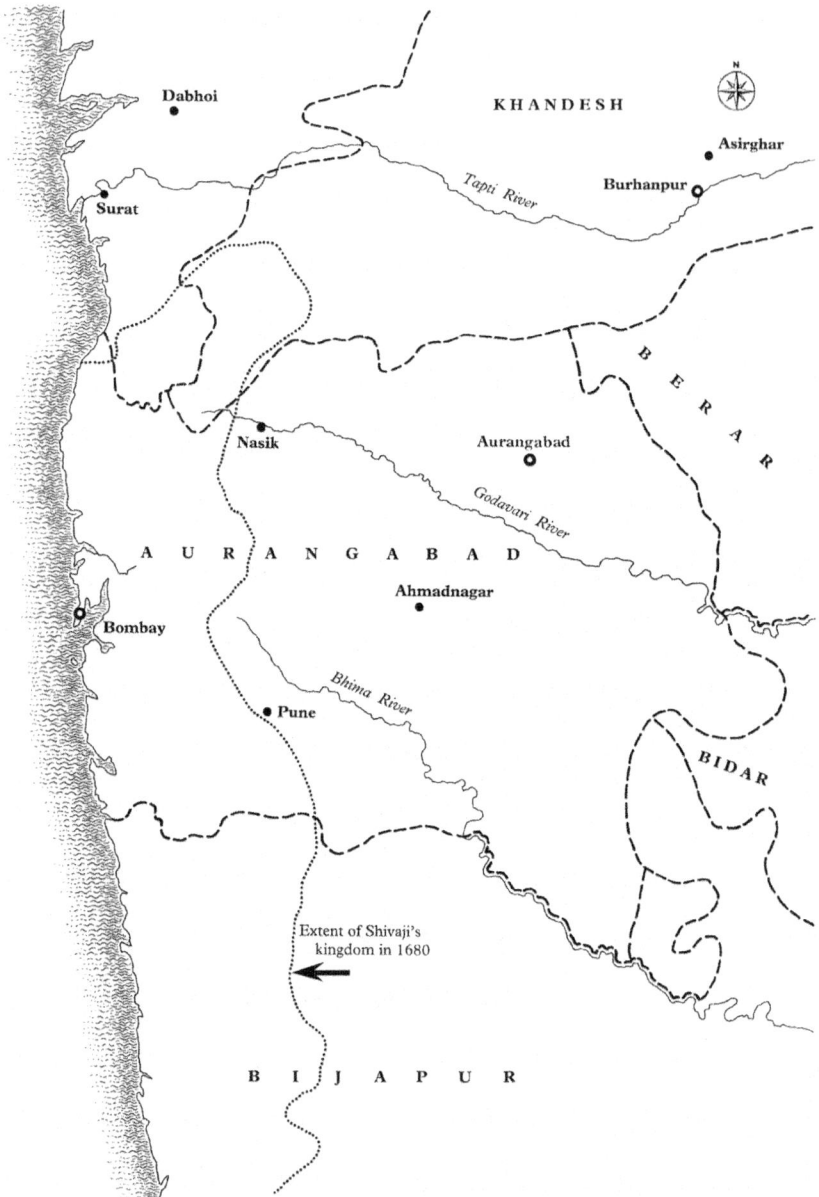

Map 1 The Western Deccan as it would have appeared to Mughal administrators. Note the extent of Shivaji's coastal access. Based on I. Habib, *An Atlas of the Mughal Empire* (Delhi, 1982, 14A)

Wellington, we find that Shivaji's accomplishments have been 'made-over' or expropriated by those seeking to enhance the cultural or political objectives they advocate. During the twentieth century Indian Nationalists, Hindu separatists and Maharashtrian patriots all had reason to present selective portraits of Shivaji's military profile in history. *Chhatrapati*** Shivaji is often referred to as the 'warrior king' and credited with founding the Maratha Empire. Despite once popularly cited lineage claims that stretch to the Rajput stronghold of Udaipur, the evidence indicates an immediate paternal linkage to the territories surrounding the Ahmadnagar fortress. In fact Shivaji's father Shahji was directly connected to Ahmadnagar by way of his service to the Nizam Shahi dynasty and the Mughals.[26] And it is interesting to note that many of the Mughal accounts testify to the elder Bhonsle's ability as a commander of infantry forces rather than just bands of light horsemen.

Light horse raiding was well known in South Asian warfare as a means of exploiting logistical scenarios relating to the replenishment of supplies. Mounted 'cattle raiding' was common in Maharashtra in ancient times and was glorified in the earliest hero stones.[27] But light horse raiding was known throughout recorded history in both Asia and the West. Alexander the Great witnessed it in both the Mediterranean and Indus theatres of war and the Romans encountered it throughout their empire.[28] Men in forts must eat and men on horses have mobility; while some tactics may be more popular with a given commander or specialized unit, there was little new under the sun when it came to fixed positions vs. mobility in siege operations. And the Mughal records do note that it was customary in Deccani warfare to use well-mounted horsemen for a reconnaissance in force.[29] The seventeenth-century imperial records clearly state the Mughals also used light horse tactics for logistical interdiction to ensure that no grain found its way into rebel-held forts.[30] There was no Maratha monopoly on this tactic and no historically contemporary claim on this *modus operandi* when the Mughals fought the Marathas. The Mughals had used light horse raiding extensively to plunder grain delivery convoys by *banjaras*[†] in an effort to starve out Shahji Bhonsle's allies in Daulatabad.[31]

Historians continue to leave the impression that Shivaji's later 'Maratha' army was overwhelmingly composed of light horsemen. To support their variation on the theme of a predominantly Maratha mounted tradition they draw an evolutionary line of descent directly from the previously mentioned Malik Ambar and his command of Ahmadnagar's military forces. They speak of 'Malik Amber's masterful use of

---

* Lord of the Umbrella, aka King of Kings.
† Aka Bunjaras, Binjarries (from Binjarry), Brinjaris, Bunjarras. An itinerant socio-ethnic group who specialized in the trafficking of grain.

guerrilla warfare, which was known as *bargi-giri* in Dakhani. This consisted of not meeting the enemy main force in the field, but cutting off supplies, manoeuvring for the best position, and using superior mobility to strike vulnerable locations far from the battle site.'[32] We are then further assured that these tactics 'were elegantly used by Shivaji in the 1650s'. There was no explanation offered as to how the Persian term *bargir*, defined as meaning 'a soldier who enlisted without a horse', came to be synonymous with avoidance of pitched battle, logistical interdiction and raiding techniques emphasizing surprise and mobility.[33] If it were a simple linkage to military employment one might ask: why does it have no tactical counterpart in *silladar*, one who 'brought his own horse and equipment'? Noting of course that the vast majority of irregular light horsemen who crowded the military labour market had their own horses, granted many acquired by plunder.

During the nineteenth century W. Irvine, then considered an authority on Mughal warfare, speculated on a relationship between the equipment holder and the *bargir* as 'burden taker', taking up the horse as a contractual burden of service that the rider assumed.[34] However, an alternative interpretation is offered here. A more appropriate linkage of terminology to tactical evolution would seem to stem from an earlier interpretation of the Persian term *bargir*, which emphasized 'a load taker', 'a baggage horse'.[35] Since *bargi-giri* doctrine is often associated with the raiding of supply units and the baggage train, it is not unreasonable to see the harassment and interdiction of logistical lines as the taking of the baggage animals ('a baggage horse' or other beasts of burden such as camels or oxen – this was a more basic burden to take than that of contractual service). Alternatively, the raider might capture loads of supplies, thereby becoming 'a load taker'. If we juxtapose this alternative interpretation with the description of the tactic witnessed and identified by the Mughals, who recorded it as *barjijari*, a more broad-based but precise understanding may be derived. The profile of *bargi-giri* tactics that emerges from this period may be generically labelled as that of 'irregular light horse'. The key elements of this doctrine hinged on mobility and speed, with tactical attention being paid to reconnaissance, logistical interdiction and diversionary attack. They avoided pitched battle unless confident of numerical superiority or relief by troops of the army to which they had become attached.

Probing patrols, when backed by large waves of horsemen, provided the option of information gathering or limited engagement in the context of a reconnaissance in force. Dismounting to fight a pitched battle was not usually encouraged. That deprived them of speed and negated the psychological impact of surprise while sacrificing the advantages offered

by mobility. One of the reasons *bargir* horsemen shunned direct engagement unless they possessed numerical superiority had to do with their
lack of heavy artillery for fire support. Their light artillery usually took
the form of *bana* (rockets).[36] Tactical countermeasures to thwart *bargi-
giri* attacks included the deployment of mobile 1 pounder and 2 pounder
artillery mounted directly on the backs of camels or elephants. The so-
called 'camel gun' was of great use in protecting mounted foraging parties because it could keep up with horses on the gallop. Although a
1–2 pounder may not sound particularly powerful, in comparison to
standard field guns, it was greatly superior in ballistic performance to
the ounce of lead fired by many smallarms. The 'camel gun' was easily
capable of dropping a horse and rider at a hundred yards.[37]

Many twentieth-century Maratha histories portrayed Shivaji as a brilliant 'guerrilla leader' who sprang from the bosom of Maharashtra. According to this genre of portraiture he, along with his hardy Maratha
horsemen, ran the Mughal armies ragged. Shivaji, as the great man
in Maratha history, supposedly engraved the benchmark of Maratha
'guerrilla warfare' upon the Deccan for Muslim and *feringhi* (foreigner)
alike to see. It therefore deductively followed that all subsequent successful Maratha military efforts must have been in that 'guerrilla' vein
and that any subsequent Maratha military failures could be dismissed as
deviation from his example. Or, so the story goes.

The linkage of Maratha warfare to the emotive phrase *guerrilla war*
was firmly established during the Indian Nationalist fervour of the 1920s
and 1930s when a number of Maratha histories, both South Asian and
Western in origin, repeatedly referred to Shivaji's traditional Maratha
way of war as being that of the guerrilla.[38] In the years following Britain's
departure from India, it was more than acceptable to use the politically
charged term 'guerrilla'. And the epithet was almost mandatory among
certain Hindu freedom fighters if one intended to show respect for the
historic continuity of their struggle. The problem was that as the popularity of Shivaji edged towards deification it lessened the chances of an
accurate historic revision that would give him the credit he rightly deserved as Maharashtra's most versatile military leader. A simple guerrilla
label deprives Shivaji of the homage owed to him as an infantry strategist and naval commander who challenged both Mughal and European
power.

Several authors of Maratha military studies, such as S. N. Sen, were
not primarily trained as military historians and apparently used the term
'guerrilla warfare' because they assumed that all 'hit-and-run' tactics constituted guerrilla warfare. Others knew full well they were building on
the iconography that catered to a form of saffron-clad Hindu liberation

theology. But it would be equally misleading if this work were to paint a stark picture where all those who used the term 'guerrilla warfare' were either blissfully ignorant of terminology or craftily engineering history. There were, no doubt, those who simply saw merit in using a simple military analogy to explain the success of a distinctly Maratha hero. It was a time when India needed her own heroes and consideration of Nelson or Wellington was nothing less than cultural treason.

Yet the guerrilla epithet is a generalization that does not sit comfortably with the historic evidence. It fails to meet either (a) the historic model derived from the word *guerrilla*'s nineteenth-century evolution, meaning 'small war' as coined in the Peninsular War 1807–8, 1809–14;[39] or (b) the contemporary model extolled by guerrilla warfare's greatest architects during the twentieth century.[40] *Bargi-giri* tactics, often construed as guerrilla warfare, were merely one facet of the Maratha mounted tactical repertoire and not limited to the Marathas. Furthermore, there is no evidence that any major Maratha military leader, including Shivaji, ever used *bargi-giri* tactics to the mutual exclusion of infantry and artillery in positional warfare. Marathi Professor Ian Raeside's use of the term 'guerrilla tactics', in an extensive military translation of later Maratha history, emphasized Marathi-language linkages rather than Persian-based terms such as *bargi-giri*. But Raeside's rendering did not clarify the issue in that it continued to perpetuate the association of raiding tactics with a doctrinal classification of guerrilla warfare. He quoted Dattaji Sindia as saying, 'We will take up guerrilla tactics.'[41] The corresponding footnote stated, 'The Marathi phrase *ganimi kava* literally means "enemy (i.e. Maratha) trickery!" It was first made famous under Shivaji who used his mountain troops to make lightning raids on the more static Muslim forces.'[42]

The simultaneous existence of many types of conflict including irregular or unconventional warfare in a larger struggle can be confusing. Sam Sarkesian, who studied guerrilla warfare in both a military and an academic context, made an extraordinary effort to correlate and distil the models of war that we could specifically identify as 'guerrilla'. 'Despite the ambiguities, some characteristics are common to all these efforts: the use of force; the objective of changing the composition of government; revolutionary goals; organisation; and the fact that the participants are apt to appear to be civilians and avoid conventional battle tactics.'[43] Sarkesian's notation of guerrillas appearing in civilian garb, as opposed to military uniform, is a rather significant and lasting form of physical distinction. Shivaji's troops are identified in several historically contemporary accounts as being in Maratha 'uniforms'.[44]

The assembled troops of Shivaji's armies were not formally indoctrinated with regard to a unifying political or religious ideology. Shivaji began by augmenting a series of existing hill forts, to provide a backbone for his authority.[45] This network of forts permitted communications and intelligence to be disseminated in a fashion that extended control over an increasing amount of territory. Mounted Maratha units could provide reconnaissance or double-up as a mobile assault force to execute 'hammer-and-anvil' strategies in conjunction with conventional infantry. However, the forts were blatant obstacles to Mughal authority and when Maratha raids provoked Mughal counterattacks, these bases provided a fallback position with all the benefits of defensive fixed positions. The strategic dilemma for the Mughals was how best to counter a multi-front Maratha threat; to go on the offensive or to increase their defensive profile? If they spent their time and resources defending their own possessions then there was no positive movement towards destroying the root problem – the Maratha enemy who lay safely embedded in a well-established fortification system. By the same token, if the war was carried to the Maratha hill forts it meant the potential over-extension of logistical lines in terrain that did not lend itself to rapid re-supply or rescue. *Bargi-giri* tactics were effective in that type of scenario because they compelled the invading force to expend precious resources in order to ensure logistical security. For the Mughals a strictly defensive war left the Maratha command and control infrastructure intact while an offensive war was an operational nightmare since individual sieges could last a year or more and re-supply lines were vulnerable to monsoon degradation as well as hit-and-run tactics.

In time, Shivaji's control was projected to the sea where coastal fortifications served as bases for his navy and allowed the completion of new 'exterior' logistical lines.[46] Western authors have often built on the erroneous belief that 'Hindu powers' in India did not possess navies because they were supposedly afraid to cross the proverbial 'great dark waters'.* This was a standard story in the construction of martial race mythology and it was used as a justification for large-scale Muslim recruiting during the Victorian era. But in Shivaji's attack on the Sidi Yakub's† fort of Danda Rajapur‡ we have an example of a military operation that defies Western stereotypes on two counts. Although the sea protected this fortress, there was an adjacent rock island. Shivaji's forces launched an amphibious assault and secured the island.[47] Siege guns were then landed and a bombardment commenced. The siege

---

* Later, under the leadership of Khanoji Angria, the Marathas waged a successful naval campaign against the British.
† Aka 'Seedy Yacoot', Siddi Yakub. The Sidis were of African origin and their naval bases on the western coast of India posed a serious problem for Shivaji.
‡ Aka Dunda Rajapur.

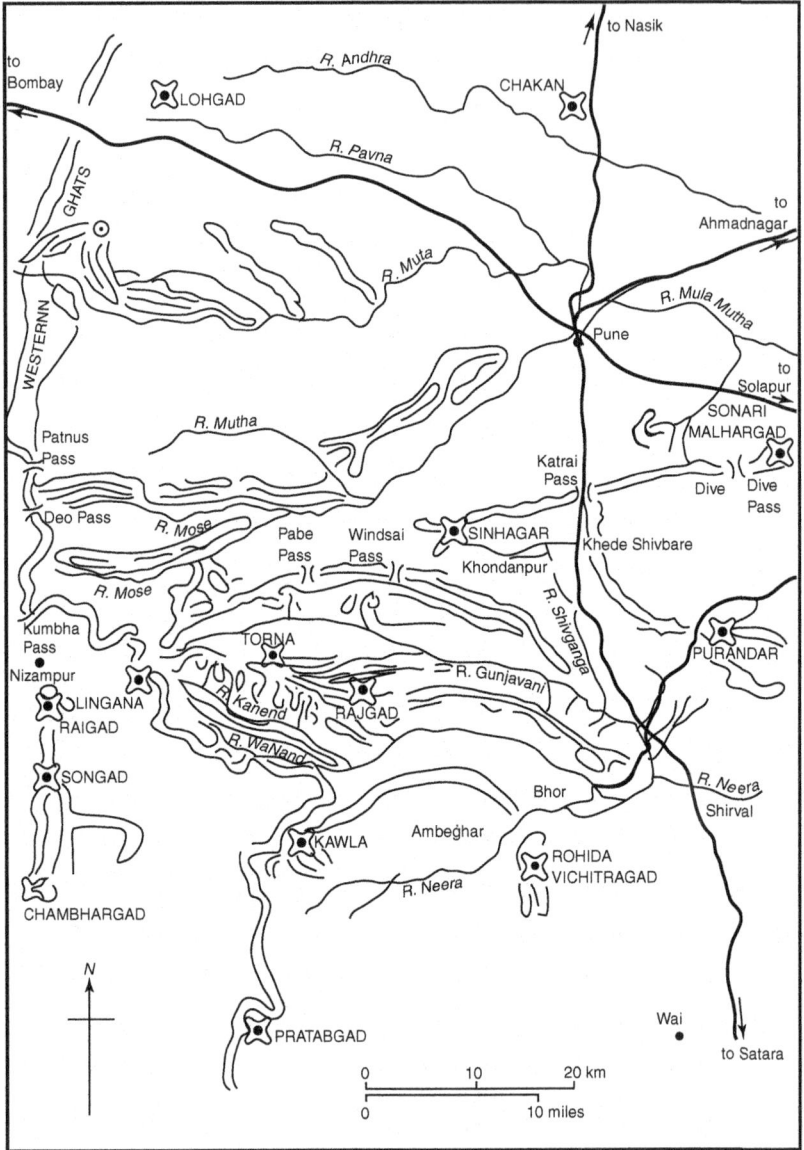

Map 2 Main roads and forts of the Pune region *c.* 1660, based on a map from Stewart Gordon, *The New Cambridge History of India II. 4, The Marathas, 1600–1818* (Cambridge University Press, 1993), p. 72.

progressed for two years, which underscored the Marathas' skill in weatherproofing their guns and fighting through the monsoons.[48] As discussed in subsequent chapters of this book, the monsoon myth has been slow to die.

In accordance with established Maratha military culture, Shivaji extended the practice of hiring foreign mercenaries to his navy. By 1659 he had a fleet of twenty warships, which also refutes the guerrilla label. Shivaji was concerned about the potential of a Portuguese sea-borne threat and the easiest way for him to gain insight into his rival's military knowledge was to embrace it by hiring Portuguese personnel out of their naval service. Shivaji greeted the Portuguese seamen with open arms and their numbers soon grew. Rui Leitao Viegas was trusted to such a degree that he was appointed to command Shivaji's fleet. This alarmed the Portuguese authorities because they were worried Shivaji would unleash his Portuguese officered ships against Lisbon's ally 'the Sidi of Danda-Rajapur'. This triggered a series of covert negotiations and the Portuguese urged Viegas to defect from Maratha service. When he 'came in' Viegas returned with 300 Portuguese mercenary sailors who had been serving in Shivaji's navy.[49] The Portuguese were to prove just as mercenary as the Marathas in their attitudes towards survival in the South Asian military environment and they too engaged in shifting alliances out of political and military necessity. These shared employment ethics posed no long-term obstacle to economic opportunity and within six years the Portuguese were selling ammunition and grain to Shivaji as he fought to resist an attack by Mughal forces.[50]

Shivaji's army operated along a clearly defined model of organizational behaviour in which one can see a modern defence triad consisting of infantry, cavalry and artillery. That Shivaji had attained infantry and artillery capability equal to European standards is suggested by his systematic elimination of European forts that lay along the line of march to Surat.[51] European mercenaries serving in the Mughal forces received first-hand intelligence about the Marathas from English and Dutch survivors of Shivaji's sack of Surat in 1664.[52] As we saw in the rise of Ahmadnagar, the Marathas had already begun a prolonged phase of positional warfare by the time they came to the attention of the Deccani sultanates. And Shivaji's utter dependence on conventional infantry warfare is well documented in his treaty settlement patterns with the Mughals. The 1665 Treaty of Purandhar emphasizes the surrender of specified fortresses coinciding with territorial concessions. And if we apply the criteria of Maoist struggle, we can see that Shivaji was negotiating not as a guerrilla or rebel with an irregular army but rather as a conventional rival to the Mughal central authority. Despite the loss of twenty-three forts by Shivaji in the

Treaty of Purandhar, twelve forts remained, including his sanctuary at Rajgarh.[53] This formed part of his non-negotiable fortified core position and Shivaji continued to exploit the resources available to him in a bid to remain militarily competitive with the Mughals.[54]

Jai Singh, the Rajput general who served as Mughal Emperor Aurangzeb's C-in-C, sent a letter of protest to the Portuguese envoy to Bijapur, Padre Gonsalo Martez. Jai Singh objected to the Portuguese mercenaries who served in Shivaji's army.[55] Shivaji used experienced cannon casting technicians from Goa to assist in his manufacture of weapons. We learn from Jai Singh's letter that 'Shivaji had granted important posts in the army to the officers in charge of the manufacture of guns'.[56] This protest made specific reference to the Portuguese who served Shivaji's army as opposed to his navy. Records in Lisbon also included a proclamation of 19 May 1668 by the Viceroy of Goa, Joao Nunes da Cunha, which ordered all Portuguese nationals in the service of 'Delhi, Bijapur and Shivaji armies to return to Portugal'.[57] The use of a proclamation to remove European mercenaries from Maratha service is worthy of historic note. It suggested there were certain liabilities in depending heavily on such transitory human resources and the issue returned to haunt the Marathas, as we shall see later.

The price of *pindari** employment

A great deal of the Marathas' military reputation stems from their employment of *pindaries* – predatory hordes that were often retained after they had paid a tax called *palpati* for the right to plunder. From time to time Shivaji used *pindaries* in a role analogous to 'sack men', an advanced wave spreading confusion before the attack of his infantry. Shivaji's directive regarding *pindari* entitlements, that they be allowed to retain plundered coins of copper but not silver, suggests there was a need to specifically regulate *pindari* activity. Later he issued extensive regulations to hold their predatory actions in check because they were a liability when accompanying the army through the territories of allies.

As a military option they were readily available on the military labour market and did not require a major initial investment. Many of those *pindaries* unable or unwilling to pay *palpati* were retained on the promise of receiving a percentage of the goods they plundered. It was more cost-effective for many lesser Maratha *sardars* – acting in the capacity of military broker/dealers – to hire *pindari* armies on commission rather than to utilize precious financial resources to maintain a standing infantry

* Aka pendara (plural pendaras), pindary.

with expensive standardized uniforms and equipment. *Pindari* horse-men reported with their mounts and weaponry, usually consisting of a *tulwar* (sword), lance and a small shield. The *pindaries* could easily number up to 10,000 men per band and when massed for large op-erations figures of 50,000 were not uncommon, their wives and camp followers often swelling the reported numbers of the 'Maratha armies'.[58]
The persistent use of *pindaries* by Maratha military commanders also explains the descriptions of so-called 'robber bands' reported to ac-company Maratha troops. From a doctrinal standpoint, *pindaries* of-ten assumed duties associated with the *bargi-giri* tactics of light horse cavalry:

(1) destabilization by the nature of their attack, creating the general effect of chaos, leaving their enemies frozen in a defensive posture long after they had melted away;

(2) isolation of enemy units via harassment and interdiction of logistical and communication lines;

(3) intelligence observation and surveillance, or as a reconnaissance in force intended to provoke an armed reaction and then report back on the size and capability of the opposition;

(4) logistical procurement when attached to Maratha infantry moving across enemy territory. Ranging far and wide they could carry away enemy food and fodder, thereby helping Maratha infantry live off enemy lands.

The *pindaries* were well suited to picking off stragglers and tying down outlying pickets rather than charging bodies of infantry. These tactics heavily emphasized the physical isolation of enemy forces. The *pindaries* were never intended to charge on-line in the same fashion as European heavy cavalry. That was not their role in the Maratha order of battle. It would be a waste of resources to have the lance-carrying *pindaries* attack infantry formations where they might be slaughtered by massed small-arms' or artillery fire. Their reluctance to squander themselves against conventional forces would be all too evident in later battles with the British. However, *pindari* military behaviour was greatly influenced by an individual desire to survive. These were not heavily indoctrinated mem-bers of a specific unit with a high level of élan or esprit de corps. They took no oath to die for some lofty political or national cause. To a great degree it was everyman for himself and if you were foolish enough to die for someone else's concept of honour you could not enjoy the spoils of plunder. It was an ethic not condoned or recognized as legitimate by later European armies imbued with racial and nationalistic rhetoric that reinforced the creed of *duty, honour, country.* The *pindaries* were not trained to die *en masse* for the 'good of the cause' as their cause was often nothing more than seeking to survive as best they could by plundering as a profession.

One major drawback in making use of the *pindaries* was that the Marathas, as their employers, became too closely associated with them in the eyes of others. This contributed to the misconception that the historic norm for a Maratha army was that of a *pindari* army. Some foreign observers came to believe that the predatory horsemen were Maratha 'regulars'. The confusion was complicated by the presence of other Maratha light horsemen, which might include *bargis* or *silladars*. In some cases *silladars* who lost their sponsors had to resort to *pindari* employment to earn a living. Few foreigners understood the subtle differences. This misunderstanding over the tactical use of Maratha light horsemen and the identity of *pindaries* is continuously evident in later European reports concerning the criminal and 'despicable behaviour' of Maratha light horse and their perceived 'cowardice' in not charging British infantry in formation.

Peshwa Baji Rao I

The ability of Shivaji Bhonsle to adequately divide his attention between the civil and military sectors of his administration is probably lost to history in the sense that the two became so intertwined that we cannot readily extricate one from the other. Following the death of Shivaji in 1680, the Marathas lacked a unified vision of the political process. This later yielded unfortunate consequences with regard to the clear delineation of civil and military power between the *Chhatrapati*, the *Peshwa* (Prime Minister) and the *Senapati* (Commander-in-Chief).[59] The three offices were never intended to function as an administratively equal triumvirate but the three segments did lend themselves to the clear division of civil and military power. The persistent problem for subsequent Maratha leaders, however, was the tendency of powerful men to amalgamate both civil and military powers under their own control.

Baji Rao I (1700–40) became *Peshwa* in 1720 and he eventually eclipsed Shivaji's grandson *Chhatrapati* Shahu in a manner that signalled the political demise of Maratha royalty. Baji Rao was a capable administrator but one could argue that his real forte lay in being an architect of Maratha military expansion. Baji Rao's pre-emptive military move, against a brave but inexperienced young *Senapati* named Trimbuk Rao Dabhade, was made at the Battle of Dabhoi* in 1731. This showdown constituted the distillation of a Maratha civil war into a single battle sparked by a dispute between the *Senapati* and the *Peshwa* over prospective campaign

---

* 22°+, 73°+. Aka Debhoi.

KASHMIR

PUNJAB

DELHI

ROHILLAS

RAJPUTS

JATS

AWADH

Sindia

BENGAL

Gaikwad

MARATHAS

Holkar

Bhonsle

NORTHERN CIRCARS

Peshwa

HYDERABAD

Bay of Bengal

MYSORE

KARNATAK

COCHIN

TRAVANCORE

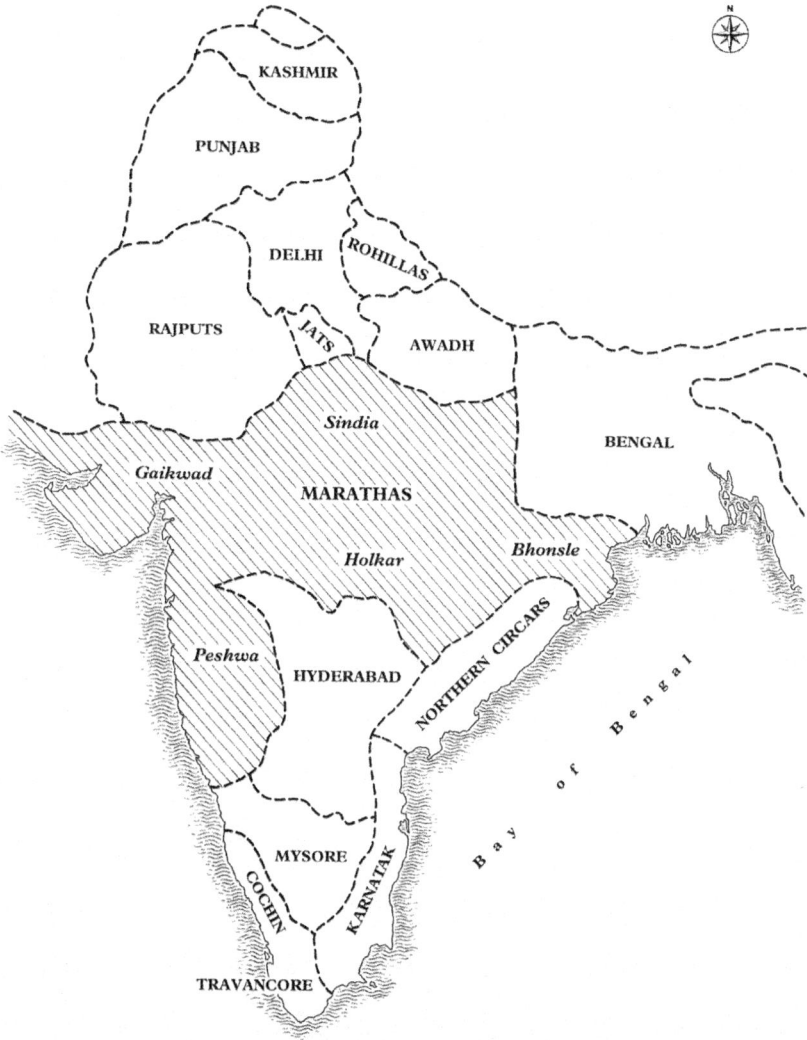

Map 3 Mughal successor states *c.* 1730, based on a map from John Keay, *A History of India* (London, 2000)

territories. The major clan armies of the day were in attendance and there were over 50,000 combatants on the battlefield. Alliances of convenience were still being struck on the first day of combat and the bargaining stakes were high with several hundred pieces of artillery, massed ranks of both flintlock-armed regulars and matchlock-carrying provincial troops in abundance. It was, by anyone's standards, a great eighteenth-century infantry battle. However, it escaped European notice because of its location, timing and circumstance.[60]

Within the Battle of Dabhoi, two key elements can be cited as moving Baji Rao I's doctrine out of the realm of neo-classical warfare and placing it directly into the category of proto-modern warfare:

• an integrated weapons systems approach to the volume of fire theory;
• the tactical field manoeuvre of advancing under covering fire that was called upon to utilize the above theory against enemy forces.

The volume of fire theory, as employed by Baji Rao I's forces, relied on smothering enemy opposition with a blanket of projectiles; the degree to which this aspect of Maratha doctrine differed from that of the Europeans in 1731 has been virtually unexplored by military historians seeking to understand linkages to later Anglo-Maratha wars such as the 1803 campaigns. This was a distinctly different doctrine from the standpoint of the application and volume of integrated firepower as found in infantry warfare at that time. Europeans in this period tended to use artillery for its long-range reach or to blast holes in enemy infantry lines through which they wanted to advance. That was the early European choreographing of fire and movement that relied on artillery to prepare the way for the advance of infantry. In contrast, the Marathas were proponents of a more all-encompassing approach to the application of firepower. Baji Rao I utilized smallarms and artillery firing anti-personnel rounds in a concert of firepower designed to create a lethal envelope of fire to engulf his opponents. This *Peshwa* ensured that his light and heavy artillery, as well as his smallarms fire, were integrated to produce interlocking fields of fire. Maratha artillery doctrine sought to convert entire enemy formations into targets rather than let battles degenerate into an exaggerated series of individual combats. The epitome of this thinking was to take an enemy's most valued battlefield asset – his infantry – and turn it into his greatest target liability.

A few large guns integrated into a Maratha unit and firing anti-personnel rounds, such as canister or grape shot, could give a disproportionately greater capability against those enemies with relatively modest organic firepower.[61] A siege gun might be used to hurl an 18 pound cannon ball against a fortress. But if you used that 18 pounder as a field gun, to unleash large grape shot, it could significantly lengthen the killing zone achieved by a 6 pounder firing smaller grape shot. By interlocking the fields of fire, using the larger to pick up where the smaller had left off as it were, you could blanket increasingly larger targets as you extended the range to deal with enemies moving into or away from your position. Long-range large-calibre field guns, in effect great guns (*topha thorli*) which the Europeans would have reserved for siege work, extended the curtain of fire beyond the usual European limits of effective field engagement, while intermediate guns (*jangi topha* and *jambur*) were the mainstay of defence as used with grape shot or canister rounds.[62] While close-quarters engagements also saw the use of special-purpose mortars (*garbhandi*) to broadcast multiple projectiles in an arch-like pattern.

References in the Maratha royal edict known as the *Ajnapatra* indicate that by the first quarter of the eighteenth century, the Maratha artillery had already settled on using pre-bagged powder charges as standard with their cannon.[63] This pre-packaged approach to loading powder was known in many lands as semi-fixed ammunition and it was a great step forward since it increased loading speed and reduced the number of loading accidents in comparison to the archaic method of ladling powder into the barrel.[64] The higher rate of fire contributed to the Maratha gunners' ability to advance with the infantry when required to do so. The detailing of large numbers of men to each gun crew became commonplace in Baji Rao I's army. Once the field guns were unlimbered they were often far from stationary as drag ropes were used to manhandle field pieces forward under fire, with guns in the 3–6 pounder class, including small howitzers, being pushed ahead with the advancing infantry. This allowed the Marathas to maintain fire continuity with the advancing line. Letters preserved in the *Peshwa*'s official records had noted battles in which *lascars* had to contend with obstacles that often impeded the movement of the guns.[65]

The luxury of having more artillery pieces in the field – particularly mobile large-calibre pieces – increased the Marathas' tactical options and contributed to the development of their doctrine. However, this should in no way be construed as a doctrinal form of technological determination. Possession of the guns did not guarantee victory and the guns themselves did not determine their method of deployment. And it should be pointed out that the European armies had access to virtually the same types of smoothbore muzzle-loading artillery at that time if they had opted to use them in this fashion. Maratha leaders like Baji Rao I determined how they wanted to employ their artillery assets in the hope of increasing their chances for victory.[66]

At Dabhoi in 1731 the Marathas and their allies used flintlocks as well as matchlocks. Several Maratha contingents were composed of what one might term peripheral peoples. Maratha *sardars* from the Gaikwad, Bande and Dabhade clans recruited Kolis and Bhils armed with matchlocks.[67] The coastal Kolis had gained firearms and infantry warfare experience while serving as auxiliary matchlock levies to the Portuguese and to a lesser extent the British. But references to matchlocks seem to cause great confusion for those analysing South Asian infantry warfare in the eighteenth century. Many Western writers have cast a jaundiced eye on matchlocks, suggesting they were intrinsically inferior to the more mechanically complex flintlock. However, the assumption of military superiority, based on possession of a smallarm with a particular type of ignition system, is faulty on two levels of consideration:

- First, the outcome of armed conflict is never pre-determined by the mere possession of what can only be described chronologically as the 'latest' technology.
- Second, saying that a weapon is of one lock system, or the other, does not address ballistic-relevant issues such as velocity, range or projectile energy on impact.

Both flintlocks and matchlocks in this era relied on a black powder* charge to send a projectile down the barrel and through the air. The two locks themselves, flintlock and matchlock, were component ignition lock systems for complete firearms, as in the expression 'lock, stock and barrel'. As locks they could be applied to barrels that were smooth bored (musket) or grooved (rifled) and technically speaking could be applied to either muzzle-loading or breech-loading weapons – although the former were exceedingly more common.

A flintlock's firing mechanism was composed of a trigger assembly, cock (flint holder), lock-plate, frizzen pan assembly, several screws as well as assorted springs of varying size, and the flint. In contrast, the essential matchlock had a trigger bar assembly, side plate, serpentine match holder and touch pan. With fewer moving parts the matchlock could be more easily repaired under field conditions with a minimum of tools and replacement parts; it was less likely to be rendered ineffective from broken springs and missing screws. The matchlock's reduced number of moving parts meant fewer areas with crucial metal-to-metal contact, thereby reducing the amount of lubrication needed under field conditions. This latter point was of particular relevance in prolonged monsoons if weapons were stored improperly or not maintained regularly.

It was cheaper for Maratha powerbrokers to arm men with robust matchlock mechanisms that offered logistical savings.[68] Flintlocks relied on flints imported from England or Europe, although indigenous South Asian agate could have been substituted.[69] However, agate was difficult to shape and cut to size and its hardness had a tendency to damage the flintlock and reduce the effective service life of its frizzen assembly. Each flint had to be cut or knapped to size and was likely to lose peak efficiency if it were not repositioned or reknapped. After as few as ten

---

* Aka gunpowder; a mixture composed of potassium nitrate (saltpetre), charcoal, and sulphur.

rounds it might need adjustment or replacing. In some cases the adjust-ment might take the form of advancing the flint in the screw-tension jaws that held it, or if properly prepared ahead of time, reversing the flint in the holder to present a new striking surface for contact with the steel. There were different sizes of flints for muskets, pistols and car-bines, adding somewhat to the logistical concern. A musket flint might be knapped down to pistol size but a small pistol flint would be short lived if pressed into service in a musket's lock. In contrast, the matchlock had a simple match-cord, which could be fashioned by hand from indige-nous organic material such as cotton, hemp, or the roots of the banyan tree.[70]

The terms matchlock and flintlock, as identification names for types of weapons, only have comparative bearing with respect to the method of propellant-charge ignition. The matchlock introduces the ignition pro-cess by means of a burning match-cord that is touched to the powder in the priming pan as the trigger is pulled, or, as was more often the case, the trigger bar depressed. The flintlock's spring-dependent lock mecha-nism was designed to bring about the sudden contact of flint and steel to deliver a spark to the priming charge. Mechanical complexity was no guarantor of superiority in bringing about the end result of the ignition. The ignition process, once it was complete, resulted in the burning of the main black powder charge, which in turn generated sufficient propellant gasses to force the projectile down the barrel and through the air. The ballistic path or trajectory of the projectile had virtually nothing to do with which means was used to originally ignite the priming charge. The mod-ern diatribe, built on assumptions of 'Western' mechanical superiority, is ballistically irrelevant.

In addition to accuracy and hit probability, other considerations in any serious matchlock vs. flintlock ballistic performance comparison might include:

- The size and nature of the projectile: projectile composition, shape, weight and construction, all affect flight patterns, range and wound channel ballistics.
- The size and type of powder charge: the propellant-to-projectile weight ratio of the charge, oxidization and atmospheric conditions, as well as ramming pressures, all influence results.
- The 'fit' of the projectile to the bore: how closely do the corresponding diameters match in an effort to derive maximum use of the burning propellant's gas pressure? Was paper wadding or a cloth patch used to obtain a gas seal?

To say that certain South Asian troops were armed with matchlocks re-vealed little about the way men fought or which drill standards they followed.[71] While the preceding list suggests technical factors relevant to a meaningful ballistic comparison, the larger question remains: 'What constitutes superior military performance criteria for comparing matchlocks and flintlocks?'

- Is it the rate of fire expressed as the number of rounds per minute that can be fired by the unit on average?
- Is it range? If so, is it maximum range or effective engagement range?
- Is it projectile or 'hit' lethality?
- Does the weapon prove awkward for drill but superior in combat?

As we shall see in later chapters, a number of Maratha infantry units featured a specific integration of forces with flintlock infantry supported by matchlock-armed sharpshooters serving in the capacity of snipers.[72] This overlapping of technologies is not an uncommon phenomenon to those who study how the history and philosophy of science interacts with military technology. Although addressing a European model, Bert S. Hall observed:

Most historians of technology today do not place much credence in simple displacement models of technological change. New machines or tools almost always coexist with their older counterparts for a period of time, a period that may be very long indeed. Therefore, it is the blend of technologies that shapes the context in which the technologies themselves manifest themselves to the historian's eye. One must always be wary of the temptation to attribute everything to a single new entity. Technology can help to explain much, but it cannot explain everything.[73]

British military men in South Asia historically had a healthy respect for South Asian matchlocks, which were often judged to have superior range and velocity over European muskets. These observations are fairly consistent during the eighteenth and nineteenth centuries.[74] Following British setbacks in the First Anglo-Afghan War, the extended range as well as killing power of South Asian matchlocks sparked investigation and changes in the types of weapons issued by the British. Colonel Anderson of Bengal's Ishapore gunpowder mills conducted extensive experiments on EIC gunpowder and 'native' gunpowder. He noted that the slower burning rate of South Asian powders was better suited to the longer-barrelled matchlocks than to the shorter-barrelled European muskets. This led Anderson to observe that South Asian matchlocks and European muskets reached the desired level of ballistic performance with their own respective types of powders. One might also surmise that this accounts for some of the disparaging remarks about results when Indian-made powders were used in European arms. Military historian David Harding's exhaustive research on smallarms of the period led him to elaborate on Anderson's findings in noting:

the small bore and thick barrel of traditional Indian matchlocks allowed them to be used with proportionately larger powder charges, closer to the optimum than was possible with a European musket, without the perceived effects of recoil becoming unbearable. This would help account for the greater range generally attributed to the Indian *jezail* even when smoothbored: the muzzle velocity could be higher, and thus the range longer. Accuracy might well also benefit from the higher velocity, thus extending the effective range.[75]

Harding conducted physical examinations of South Asian matchlocks with particular attention being paid to their barrel dimensions and method of construction. His inspections revealed that physical differences in South Asian matchlock construction lend a scientific explanation to their enhanced performance over the British musket. The majority of such matchlocks have a powder chamber in the barrel, which accommodates air in a manner that would facilitate the powder's burning rate. These chambers also feature a restricted mouth that could serve in a manner similar to a velocity-forcing cone for the propellant gasses. However, Harding's research supports the observation that the restricted chamber mouth formed an 'o-ring', the purpose of which seems to have been to keep the ball at a distance from the powder. Harding's comments, on one such barrel cut into cross-section, explain the significance of this design principle.

The bore diameter was .65in; the chamber mouth was of .375in diameter, but behind this constriction the chamber widened to an irregular shape well in excess of the bore diameter, so that the chamber could hold 250 grains (just over 9 drams) of coarse powder. It would be well nigh impossible for the firer to prevent some air space remaining in such a chamber, and the air space would greatly increase the bore pressure – and thus the muzzle velocity and range.[76]

Our concept of the actual firepower of the matchlock from that period remains underdeveloped because we lack definitive test results using South Asian weapons and their corresponding powders. But modern chronographed tests of historically contemporary European matchlocks revealed them to have at least 75 per cent of the impact or 'knock-down' energy of the most modern twentieth-century assault rifle at 30 metres, a range which approximates many of the engagement distances found in the eighteenth century.[77]

Before leaving the era of Baji Rao I, it is appropriate to comment on the overall impact of having such a capable military leader in the form of an administratively competent prime minister. The amazing territorial growth under *Peshwa* Baji Rao I confirmed the Marathas' ability to project military power on something approaching a subcontinental basis. And historically this could be said to have set the stage for continuing

Maratha claims on the traditional economic heartland of the subcontinent – North India. However, the achievement was largely made possible by Baji Rao's personal bridging of political and military power. Unfortunately, the office of the *Peshwa* was never the same after Baji Rao I and the Marathas' North Indian political option faltered on the borders of Bengal. The cycle that began with Shivaji's birth and ended with Baji Rao I's death reveals the unresolved Maratha issue of civil–military power sharing in a political context. In Shivaji we have a warrior who was drawn increasingly into the role of political leader through the process of kingship and with Baji Rao I we have a prime minister who engineered the expansion of the Maratha state by military means. Despite strengthening the Marathas' military position, Baji Rao I may have inadvertently weakened the political model for the subsequent growth of the Maratha polity. By example of his strong military hand this *Peshwa* may have sent the signal that real political power was there for those Maratha military leaders who were strong enough to take it, as was made evident in his struggle with the *Senapati*. Europeans did not readily grasp that Maratha military culture was destined to outweigh Maratha political unity and when the British eventually played their diplomatic hand it was through the office of the *Peshwa* – a path that ultimately led them to a series of wars with the Marathas.[78]

Mahadji Sindia and the meaning of Panipat*

In 1760 a modernized and sophisticated Afghan army under Ahmad Shah Durrani[†] invaded North India. The Maratha military response was a combined operation that culminated in disaster and the slaughter of thousands assembled at Panipat in 1761.[‡] Afghan *pindaries* held foraging parties in check as they surrounded and isolated the Maratha camp. Without an early and well-organized breakout in force, the Marathas had no offensive option as they slowly succumbed to weakness from lack of food. Maratha casualties escalated under intensive artillery bombardment and eventually malnutrition, disease and desperation compelled the Marathas to give battle under less than favourable conditions. The result was bloodletting on a scale of legendary proportion and the high casualty figures reflect the organized killing of non-combatants as well as families who

---

* 29° +, 76° +. Aka Panniput.
† Aka Ahmad Shah Abdali. Ahmad Shah was a chief of the Abdali clan who rose to be an independent ruler of Afghanistan which led to him being termed *Durr-i-Durran*, 'the pearl of the age', and from that point on the clan name became that of the Durrani.
‡ Estimates vary but it is reckoned the 'Afghan' army totalled 60,000, although half of those were Indian allies. Maratha forces were believed to be 45,000 strong.

had travelled with Maratha forces.[79] Estimates of the 'Maratha dead' at Panipat ranged up to 28,000 and virtually all major clans suffered casualties.[80]

Ahmad Shah's infantry were regarded as disciplined from a tactical standpoint. They had a very unified command structure, although a case for over-centralization could be made. The Durrani forces also had an abundance of camel guns supplemented with specialized medium-bore pieces cast in South Asia.[81] However, Maratha historians usually forgo any scholarly analysis of Afghan 'success' in favour of self-flagellation over the issue of Maratha strategy. Since the Maratha forces were largely composed of clan-based contributions there were a great variety of troops to be found at Panipat. They ran the gamut from ragged *pindaries* to sepoys who fought in the same manner as European regular infantry. Some authors revile the Marathas as having earned their defeat by returning to a so-called medieval style of warfare in which the army became encumbered by its shear volume of baggage and camp followers. This type of analysis often degenerates along the lines of saying, *if only the Marathas had remained true to their fleet-footed bargi-giri tactics . . .* , with the most convincing evidence being Marathi quotations from *sardars* who wanted a greater emphasis put on predatory tactics.[82] Several twentieth-century populists simply said the Maratha loss represented the price of not following 'Shivaji's system of guerrilla warfare' or that the Marathas 'foolishly' experimented with 'Western-style infantry', an explanation lacking credibility given the earlier analysis in this chapter. But a dispassionate non-Maratha analyst might simply say the Maratha army was pinned-down in its camp and starved to the point of malnourishment and weakness. When they tried to break out, an enemy force in more fit condition slaughtered them handily since it had not suffered similar privations. Granted, more aggressive light horse tactics might have aided the Maratha forces logistically at Panipat. But such patrolling and raiding techniques would not have immediately eliminated the threat posed by the enemy's domination of the surrounding countryside or his well-positioned infantry and artillery.

The enormous loss of senior warriors at Panipat posed a dual setback to the Marathas. It was a blow to the Maratha polity and to the informal clan-based *sardar* cadre system. Many of the clans lost more than one family member in the battle. The accumulated casualties among the leading clans weakened the civil administration of the Marathas by depleting the families that served the state. Since there was no clear civil–military division or parallel structure at the clan level, *sardar* casualties had an impact on the subsequent political effectiveness of the clans. The second reason why Maratha *sardar* casualties were of major importance lies in the legacy of Panipat and the failure of the Marathas to institutionalize officer train-ing.[83] During the seventeenth and early eighteenth centuries the appren-ticeship of young men in battle remained, as it had for years, in nucleated warlord environments. The young bloods tended to gravitate towards those *sardars* who were either charismatic, politically powerful or – perhaps more appropriately – natural fighters.

In the wake of Panipat, most of the Maratha leaders in North India were reluctant to spend accumulated revenue on expensive standing armies and huge numbers of *pindaries* were retained for a percentage of plun-der. But Mahadji Sindia, a wounded survivor of the carnage at Panipat, saw the *pindaries* as a threat to the order and economic stability needed to rebuild the military might of his clan. The death of a generation of teacher/father figures, when combined with the Marathas' predisposition for hiring mercenaries, may have eased the way for increasing European officer recruitment. In keeping with Maratha military culture, Mahadji turned to the military labour market to fill the human resource require-ments of his infantry.[84] Mahadji's bid to construct a military force capable of withstanding further Afghan incursions led him to open the door to a new wave of Western mercenaries – veterans of the numerous mid-century wars in South Asia and Europe. Initially they did not merit particular at-tention in official correspondence. They were just foreign soldiers – like so many other mercenaries in Maratha service.

Mahadji knew that North India and in particular the Gunga–Yamuna Doab* held great revenue potential and upon that knowledge he built a respectable army composed of various types of troops from different locations. One of its key features was a 'regular corps' of sepoys. As far as cultural identification, this was an exceedingly heterogeneous army that was in the finest Maratha tradition of equal opportunity employment for skilled warriors – both infantry and mounted. There were Ghosseins from Bundlekund, Brahmins from Bihar, Muslims from Bengal, Deccani as well as Pathan horsemen, some Europeans and also a few soldiers of 'mixed race' from Goa, Calcutta and Agra. Sindia's military forces in North India grew to the point that he was the dominant employer within the Doab's military labour market.

However, as a Maratha, Mahadji was still an outsider to Hindustan. If he was to make a lasting bid for control of the region he needed a means of staking his claim in a manner that yielded greater political legitimacy. As his army grew more proficient his reputation also grew. Then in 1784

* The Gunga and Yamuna Rivers are also respectively known as: Gangu, Ganga, or Ganges River and the Jawan, Jamna, or Jumna River. The land between two rivers in India is often called a *doab*. Writers referring to the 1803 campaigns noted the Gunga–Yamuna doab as simply 'the Doab'.

Mahadji received an 'invitation' to Delhi from Mughal Emperor Shah Alam II with a proposal to 'administer' a greatly reduced Mughal empire as its regent.[85] And as long as Sindia served as the Mughal regent, whether by official appointment or via his balance of military power, his clan could lay claim to the control of Delhi, the Doab and the Emperor. That meant the Marathas held greater political legitimacy than the British in North India.

The rise of de Boigne

Mahadji Sindia's ultimate power was his military strength. Many who have written about that strength put it all down to 'disciplined Western-style battalions' supposedly introduced by the Savoyard mercenary de Boigne. However, there are problems with that rather simplistic analysis. There was no national standard for the formation of a Maratha army, but this chapter has traced the Marathas' historic continuity in models of military organization. Disciplined and drilled infantry had existed in Maharashtra since the pre-European Hindu classical era, as testified to by the earlier noted hero stones. The Marathas had known Portuguese infantry models incorporating concepts such as the 'Spanish square' since the sixteenth century. The firearms drill of Europeans filtered into the areas adjacent to their coastal enclaves and Topasses willingly carried their special knowledge to the greater military labour market – where it was co-opted. During the period from the 1670s to 1740 the EIC's regular troops in Bombay were a mixed group of Europeans and Topasses, a factor which ensured that the 'drill' of that Presidency's regulars was known to both Europeans and South Asians.[86] Manoeuvre warfare and the tactical application of firepower had evolved as a common feature in Maratha conflict and Baji Rao I used massed ranks of flintlock-armed regulars under South Asian officers. Battalions based on French organizational models had served under officers like Ibrahim Khan Garde at Panipat in 1761.

If historians insist that the South Asian adoption of European-style smallarms drill and parade square 'discipline' constitute *military modernization* and/or *Westernization* – a very dubious set of indicators to support a tenuous conclusion – then the proof exists to make the case that those characteristics were found in Maratha armies a half-century before Mahadji's army came into being. And it is entirely incorrect to portray the post-Panipat period as one in which vast numbers of Mahadji Sindia's troops were magically transformed into 'disciplined' and/or 'Westernized' infantry battalions by de Boigne. Furthermore, the British diplomatic intelligence records are quite specific in noting the proliferation of regular infantry in North India and the leading officers commanding it. W. Kirkpatrick, the British Resident to Sindia's court, did not list de Boigne among the influential European figures in 1787.

> It is sometime since Sindhia formed, partly on the model of our detachments at the frontier stations, what are called two camps. They consist each of six battalions of sepoys and, in point of number, a respectable train of artillery. The command of these Brigades was conferred on Meissieurs Lestineau and Vasseut,* the only two Europeans in Sindhia's service that either deserve or possess his confidence.[87]

Disenfranchised Mughal officeholders, particularly those who aspired to retake power by force, resented Mahadji's control of the Doab. A number of them, including Rajput nobles, felt they were being victimized by Mahadji's manipulation of both his office and the Mughal Emperor.[88] This growing resistance to Sindia reached a crescendo with the 'Rajput Rebellion and the Uprising of Ismail Beg Khan' in which Mahadji faced a particularly determined Rajput army fielded by the Maharajah of Jaipur. Although Sindia's infantry were physically in top fighting form, their morale had suffered from their pay being months in arrears. The problem was that Mahadji's revenue returns had been in a shambles since 1785.[89] Mahadji had confiscated a number of *jaidads*† in the Doab and given them to his Maratha *sardars*. But several of those followers were greedy in milking their assigned estates and pocketing the revenue, which deprived the sepoys of their pay on a regular basis.[90] The result of this financial mismanagement was a humiliating military loss as Mahadji's sepoy battalions defected with eighty pieces of artillery at the Battle of Lalsot‡ in 1787. The incident indicated that the overriding weakness of Sindia's army was its financial system.[91] However, this pay scandal could be viewed as having set the stage for the rise of de Boigne as a sound administrator who had a record of religiously paying his sepoys, even if it meant the wages had to come out of his own pocket.[92]

The issue of punctilious payment helped focus Sindia's attention on de Boigne's performance as an officer commanding a battalion at Lalsot. Yet, de Boigne continued to take a secondary role to Lestineau in British intelligence reports.[93] Sindia's army, following the Lalsot defections, was reported to have dropped to just 13,000 men by September 1787 and said to be composed 'chiefly of Maratha cavalry, with a few battalions

---

* Aka in English despatches as Vasseult.
† Aka jeydad. These were territories assigned to individuals or families on the understanding that the revenue would be used for the up-keep or maintenance of troops whose service would be made available to the grantor of the *jaidad*. A number of EIC documents clearly show that in 1803 *jaidad* lands were for the upkeep of infantry troops.
‡ 26°+, 76°+. Aka Lalsont.

of infantry and a few guns'.[94] Ironically, the British lost an opportunity later that year to alter the course of history when Mahadji sought the assistance of British soldiers stationed in Awadh.* He had proposed hiring British troops but Governor-General Cornwallis declined the arrangement despite Mahadji's willingness to accept a binding EIC subsidiary position.[95]

In October 1787 Major William Palmer replaced Captain W. Kirkpatrick as the British Resident with Sindia. This seems to have been crucial to the later British misconception that de Boigne was the creator of Sindia's sepoy battalions because Palmer wrote about de Boigne as if he were responsible for raising the 'regular corps'. In 1790 Palmer's report to Governor-General Cornwallis noted 5,000 sepoys under de Boigne and the disturbing fact that a number of British subjects were to be found as officers in that particular force.[96] Despite the assertion by later Victorian-era historians that de Boigne took over in 1785, it was not until the *Peshwa's* Pune Conference of 1793 that Mahadji set things in motion to give de Boigne a greater hand in expanding the already existing 'regular corps'.[97] It was after the conference that Mahadji returned to Hindustan and established a *jaidad* in the eastern Doab for de Boigne to use for the purposes of supporting additional regular infantry. The 'regular corps' was to be Mahadji's army of last resort and he guaranteed its loyalty with revenue returns estimated to be from Rs. 27–35 *lakhs*.[†98]

Mahadji had already made several organizational changes to his army. These included an increase in recruiting Rajput and Muslim infantrymen, a change that was criticized by some Maratha *sardars* as Sindia's North Indian army began to exhibit greater physical distinctions from his Deccani forces. Changes in uniform were also made as his horsemen gave up the short breeches of the Deccani style and began to wear 'long trousers covering the heel'.[99] As for infantry uniforms, it had for decades been a widespread South Asian military fashion to clothe sepoys in battledress inspired by European uniforms. High-quality South-Asian-made flintlocks were also commissioned for Mahadji's troops and built in local production centres.[100] Regarding additional muskets, Sindia's agents had procured inexpensive surplus European flintlocks in significant numbers following the Third Anglo-Mysore War and rebuilt numbers of EIC muskets decommissioned in Bombay.[101] The definitive British intelligence report on de Boigne's forces, filed with the EIC's Court of Directors in 1794, revealed de Boigne had five musket 'stock makers' per battalion and regularly paid them each Rs. 7 per month.[102]

---

* Aka Oudh.   † A *lakh* is 100,000 in numerical terms.

It would be tempting to say de Boigne reformed the army as of 1793 but that would not be technically true either. British intelligence reports sent to the EIC's Court of Directors showed that when de Boigne was given his *jaidad* in 1793, he controlled only a fraction of Mahadji's forces. A report on the status of Mahadji's troops, filed in 1794, showed the following:

- Jivba Dada Bakshi's* detachment employed in the Marwar succession dispute
- the army of Ambaji Inglia† performing counterinsurgency operations in Mewar
- the army of Gopalrao Bhau‡ battling Holkar's army (later ordered to Bundlekund)
- the standby forces included: Behru Pant Tattya at Panipat, Bappu Malhar at Saharanpore and Appa Khande in Hariana
- Monsieur de Boigne's two brigades augmented with 'Rohila irregulars to act as light infantry' were divided between Mewat and the Doab.[103]

If one still wished to say de Boigne *modernized* Mahadji's army – if indeed such language remains appropriate – the modernization was chiefly that of integrated defence management and military budgeting. It reflected administrative evolution towards greater fiscal responsibility featuring well-regulated defence economics as the key to Sindia's financial recovery. In promoting de Boigne, Mahadji Sindia backed a professional military management specialist. This was in keeping with Sindia's appreciation of administrative solutions to the pressing problems of the day. The delegation of managerial authority over the battalions of the 'regular corps' showed that Mahadji trusted de Boigne's judgement and administrative skill. The Savoyard triumphed in demonstrating the financial gains that could be achieved by applying a stringent 'balance sheet' approach to Sindia's army. Here was an individual who introduced a superior system of defence resource management to make Sindia's army more competitive with other regional military powers while maximizing revenue returns and economic opportunity. Administrative reforms by de Boigne included making the 'regular corps' pay for itself and that complemented attempts to reform the Mughal military economy through effective *jaghir*§ management.[104] But he also applied the mercenary officer corps as an administrative tool to increase overall revenue intake by using his brigade officers as a trustworthy executive management body. Their personal accountability served to modernize economic policy through the reform

---

* Aka Jivvaji Baxy.   † Aka Ambuji Ingle, Ambajee.   ‡ Aka Gopal Bhow.
§ The Mughal-sanctioned right to collect revenue (land tax) from a specified area. Or, revenue-producing lands assigned in lieu of salary with corresponding obligations to supply cavalry.

of an older inherited military revenue system that had stagnated in the hands of Maratha and Mughal powerbrokers.

Monsieur de Boigne's administrative reforms represented a fundamental change in the behaviour of Sindia's army. It became a more modern military force in terms of fiscal responsibility, becoming a value-added facet of defence spending and not a monetary drain on society. With effective military budgeting and finance mechanisms Sindia was freed from reliance on the manpower allocations of *sardars* who often provided unsuitable troops under terms that favoured themselves. This meant de Boigne's reforms constituted a break from what were essentially medieval military management principles. Prior to that, many *sardars* had promised quality troops under existing subordinate *jaidad* agreements but they had kept only a fraction of their infantry quotas – pocketing the pay of the phantom troops. Other *sardars*, some of who received Mughal *jaghirs* as part of Sindia's role as the Emperor's protector, had cavalry obligations. But when call-up notices went out for campaigns they subsequently supplied *pindaries*, which they had negotiated for a pittance. The *pindaries*, many being second rate as 'soldiers', looted the civilian population and there was a corresponding upsurge in reported atrocities. Some civilians were killed and many injured in fighting back against the *pindaries'* extortionate *chauth* (tribute) demands.

The standard budgeting and resource allotment schedules, implemented by de Boigne, brought Maratha military financing out of the dark ages but at the cost of disenfranchising some of the traditional Maratha stakeholders who viewed war not as a social evil but rather as an economic opportunity. That became a further reason for de Boigne to use mercenary officers who were not members of Sindia's *durbar* (court), as they had little social, political or economic incentive to advocate offensive military policies. European and mixed-race mercenaries could be loyal to the system and to Sindia without lobbying for war as an opportunity to increase their purses by plunder. Ironically, the efficiency of Sindia's army increased with the exclusion of certain *sardars*; particularly those who withheld support from Sindia in favour of their own economic interests or those of their allies in the shifting alliance system that characterized Maratha military culture and the political system of the *durbar*. The account books and ledgers of de Boigne's brigades were open for inspection by the Maratha officials of Sindia's *durbar*. Under this new regime of military accountancy South Asian financial products and credit facilities were extensively integrated into the system. There was a proliferation of revenue *credit transfer* notes in the form of *hundi* (in this case, promissory financial notes), which reduced the necessity for cash on hand, thereby lessening the temptation of *sardars* to plunder internally.[105]

As for the economics of weapons procurement, Mahadji saw the sense in allowing de Boigne an increased executive role. Sindia placed five existing arsenals under de Boigne's authority to take advantage of centralized management, thereby bringing greater efficiency to procurement and delivery. In turn, de Boigne utilized Major Sangster of Scotland as a rotating superintendent to oversee quality control and ensure logistical specifications. Uniformity and standardization were critical to the issue of ammunition supply. The ammunition for a 6 pounder from one foundry had to be interchangeable with that of another. Sangster had previously been the superintendent of a gun foundry for the Maharani of Dholpur (circa 1783) and was accustomed to working with South Asian technicians and craftsmen. Artillery and munitions centres in Agra, Mathura,* Delhi, Gwalior, Kalpi and Gohand began co-ordinating their production. In accordance with stringent resource accounting procedures, Sindia's commanders had to pay for artillery ammunition out of budget allocations when it was not provided by stipulation (i.e. specific 'state' campaign purposes).

Artillery ammunition remained expensive at the point of consumption owing to the high costs related to its delivery weight. Subcontracting the more labour intensive aspects of handling and production became common. Hammering short chisel-cut pieces of iron bar produced small grape shot for field pieces. The irregular shape, of beaten as opposed to cast grape shot, did not adversely affect ballistic performance of these projectiles, which were intended to provide wide dispersion for more effective *blanketing* of the target. Labour was in great abundance and this technique of grape shot production made economic sense. It circumvented the immediate need for large numbers of metal-casting furnaces that consumed vast amounts of high-energy fuel such as wood. Large round-shot (cannon balls), however, were cast to ensure a suitably tight fit in the bore so as to make maximum use of the propellant gasses generated by the black powder charges.[†] Many of these round shot were cast at Gwalior owing to available iron deposits. At Agra, a powder mill was built which utilized potassium nitrate and sulphur from Bikanir. The growing South Asian military economy drew craftsmen as well as unskilled labour alike and Sindia's arsenals often relied on pre-existing industries to supply trained workers. Some of the manufacturing personnel had been employed in weapons production for generations with metalworking and leatherworking castes in great demand. References to women

---

* 27°+, 77°+. Aka Muttra.
[†] Depending on the type of ammunition and how it was to be used, an additional propellant gas seal was added in the form of a wooden base plate or *sabot* which was held to the round shot by means of two 'criss-crossed' iron bands nailed to the *sabot*.

and children also indicate these production centres were not unlike their European counterparts in taking advantage of labour sources across a broad spectrum of society. Sindia had come to dominate the regional military economy through his relationship with the military labour market and the industrial sector. While his army paid cash wages to soldiers, civilian workers found greater income-earning potential generated by his burgeoning defence procurement contracts.

The new Maharajah Sindia

Mahadji Sindia, the 'old *Patel*', died in 1794. However, in the brief period of time following the 1793 Pune Conference, de Boigne's financial reforms showed a clear advantage. Revenue-producing lands set aside for the 'regular corps' did not always meet the highest tax return predictions but they were far and away stronger sources of consistent funding than those areas subjected to unregulated *chauth* demands. The 'regular corps' was meticulous about collecting taxes in an orderly fashion and then providing security so that other transient 'tax collectors' could not ride in and claim additional funds. Areas without such security might be subjected to a never-ending series of extortionate demands from whatever band of horsemen happened to ride by. The 'regular corps' encouraged stability and became an economic facilitator in its own right. The military labour market was thriving and that meant the further growth of the cash economy. Battalions as well as individual soldiers became the consumers of goods and services, paying for them with cash. Greater job security and increased financial liquidity had come to an extensively militarized North India. The battalions became human links in the money pump needed to sustain a functional economic recovery. After years when military invasion and upheaval had stifled prosperity this represented a chance for the regularization of the regional economy. This was the first time the Maratha military presence in the Doab could be seen as an adjunct to economic growth rather than a financial drag on the system.

In a move corresponding to EIC policy, some companies of the 'regular corps' were loaned-out to Maratha allies for the purposes of re-establishing order. The 'regular corps' had proven very effective at putting down insurgencies rooted in the numerous forts of Hindustan. But de Boigne wisely kept several companies back from these contractual deployments so that he could cope with any threat to his own power base. The 'regular corps' gained an intimate working knowledge of local conditions in the Doab and their specialized equipment, like battalion howitzers firing explosive or incendiary shells, gave them an advantage in dealing with the stubborn 'mud forts' that dotted the map. This efficiency in tax collection, counterinsurgency and revenue administration attracted a growing jealousy among some Maratha *sardars*. A number of them coveted the battalions of the 'regular corps' as military tools for their own application. Others did not give a damn about the troops; they wanted the revenue and lands that funded the battalions. A cautious de Boigne remained out of the intrigue that grew in the *durbar* following Mahadji's death. The mercenary leader had correctly surmised that he could not risk being marginalized in the internal feuding of the *durbar* where he would be at a disadvantage as an outsider. It was de Boigne's decision to maintain a low political profile and he decided his only chance of retaining his lucrative appointment in service was to stick to what he knew best – soldiering.

As a defence manager with tactical experience de Boigne was a strong believer in the merits of light infantry troops. Sindia's light infantry were results-oriented combat troops. They were not men groomed for the pomp and ceremony of the parade ground but rather hasty tactical deployment where their skirmishing style lent itself to *fire and movement* tactics or doggedly determined holding actions. Disputes over Mahadji's succession continued to motivate de Boigne's efforts to ensure that the light infantry as well as grenadiers were well equipped and maintained at fighting strength. By mid-1795 his correspondence demonstrated not only his continued dedication to the role of light infantry in the order of battle, but also its needs in the form of specially manufactured equipment that had been commissioned. In writing to Second Brigade commander Major Gardner, de Boigne outlined his commitment to strengthening the Light Infantry Battalion.[106] He despatched 126 men from his 'Dolapore Battalion' accompanied by two South Asian officers and five South Asian 'non-commissioned officers' (NCOs with a rank equivalency of sergeant). These men were intended to bring the unit known as 'Sindia's Light Infantry Battalion' up to strength and they were equipped with the battalion's purpose-built 'Light Muskets'.[107] Major Gardner was instructed to make sure the Light Infantry Battalion had more than adequate fire support. He was authorized to transfer additional artillery from his unit to the Light Infantry as de Boigne arranged to co-ordinate delivery of newly manufactured cannon. General de Boigne wrote to Gardner, 'Having got 8 field pieces from Goualier and 9 getting ready at Palwol, you'll soon get the number you may want.' At that time, each of the 'regular corps' battalions usually had two 3 pounders which advanced with the infantry line; that was in addition to two 6 pounders for heavier fire support and a howitzer which served multiple purposes in firing both shot and shell – a total of five guns per battalion. The battalion howitzer, with its high angle of fire, often proved invaluable for dropping rounds on indirect targets behind walls or taking out enemy troops sheltered behind hills. Later, under de Boigne's successor Perron, the light 3 pound field guns would be phased-out in favour of additional 6 pound guns to further increase battalion firepower. Historically contemporary British battalions only had two guns* and the EIC's army followed a parallel track in up-grading firepower by replacing the feeble 3 pounder with the heavier but more versatile 6 pounder.[108]

Mahadji's death had left a leadership vacuum that was temporarily filled by the elder Maratha statesman Nana Farnavis.[†] During this time de Boigne strictly adhered to his policy of loyalty to the leader as an individual. But he was also loyal to the South Asian soldiers who served under his command. In writing to acting brigade commander Captain Robert Sutherland in 1795, de Boigne rationalized that it was better to have a well-trained South Asian physician than a European as the brigade doctor owing to the needs of the troops and perhaps more importantly the morale of South Asian officers. In the passage that follows we see de Boigne viewed serving the needs of his men as an extension of his service to the 'old man', a specific reference to Nana. 'Many of the Commandants and other Native officers having often expressed a wish to have a Country Physician, I have thought best to oblige them in serving the old man, it may often retain officers to their duty rather than go home when they are sick.'[109] General de Boigne made sure brigade doctors in Sindia's service were paid more than 'native physicians' in EIC service. The Second Brigade's doctor was listed as 'Okim KievinoolaKaun' [*Hakim Khalifa-Naula Khan?*] and he was appointed with the salary of Rs. 125 per month. Despite de Boigne's contributions to the welfare of the troops, he continued to have enemies in the *durbar*.

Mahadji Sindia's grandnephew Daulat Rao Sindia eventually rose to take control. Daulat Rao was not a seasoned military veteran who could command the respect of old comrades. Politically he was a neophyte, certainly no match for Maratha officials who had negotiated head-to-head with South Asian and European career diplomats. The new Maharaja Sindia had to be careful of those members of his *durbar* that supposedly acted in his best interests. There were deep divisions within his court, reflecting more than just a simple generation gap. Various social, religious and clan differences accentuated competing political and military factions. Some of the more cunning members of the *durbar*, seeking to

---

* Although modern civilian use of the term 'gun' implies smallarms, the historic and military context is that of gun meaning artillery piece or cannon.
† Aka Nana Fanavis, Nana Fadnivis. Nana Farnavis, arguably the most talented of elder Maratha statesmen, intervened in several political affairs between Mahadji Sindia's death and Daulat Rao Sindia's consolidation of power.

retain favour with Daulat Rao Sindia, subtly suggested that it was not in the Maharajah's best interests to keep regular infantry under European officers. Certain individuals began to pressure Daulat Rao to dismantle de Boigne's administration and hand out its component parts as if they were rewards. There were those who openly resented the European mercenary officers controlling a powerful military, others were jealous of the riches they earned as prize money while campaigning. Daulat Rao needed to keep the support of his kinsmen but he also saw that de Boigne's battalions were his best guarantee of safety from enemies like Holkar and the *Nizam* of Hyderabad, or, for that matter, rival *sardars* in his own camp. It was apparent by now that the more ambitious members of the *durbar* were not only willing but also able to launch a coup. Under de Boigne's strong example, the 'regular corps' became highly supportive of the new Maharajah. The European mercenary officers swore personal oaths of loyalty to Daulat Rao and they became his counterpoise to those rivals who sought to depose him. In addressing his fellow European mercenaries, de Boigne was careful to stress exemplary behaviour and loyalty not to Maratha service, but directly to Daulat Rao. The Savoyard wisely counselled the European and Euro-Asian mercenaries that they served at the Maharajah's pleasure. The issue for de Boigne came down to a behavioural code that he summed-up as integrity and he believed it was wanting in some of Daulat Rao's *sardars* whom he perceived as too anxious to carve out their own petty states.[110] He wrote to Scottish mercenary Robert Sutherland in November 1795: 'I have no doubt of these Mahrattas Chiefs being well satisfied of my administration, Interest to the welfare of the Prince and the good of the Service; they may learn by it what it is to be a man of integrity.'[111]

In lieu of personal respect, which he could not readily command among all the members of the *durbar*, Daulat Rao Sindia could parade his 'regular corps'. This had a tendency to give even the most adventurous *sardars* cause to think twice before risking an outright power play. General de Boigne and his fellow European mercenary officers played a historical role analogous to the Ottoman Janissaries and other mercenary elites in history who served in effect as the 'palace guard'. As alien outsiders to the system they could not easily usurp their prince because they lacked popular support or any shred of ethnically based clan loyalty. But the European mercenaries had no desire to usurp Sindia. Life was good. The frequency with which they were called out on campaign meant there was ample opportunity to fatten the prize fund in which they held shares and they were growing richer than their counterparts in EIC service. Certainly they were much further ahead than officers in King George's army when it came to prize money and additional earnings. But how could

Daulat Rao keep favour with leading Maratha *sardars* and members of the *durbar* while retaining the European-led infantry battalions that served as his personal army? The answer was a compromise, although it merely delayed resolving the question of who should control the army and its revenues.

On Christmas Day 1795 de Boigne wrote again to Robert Sutherland who had by then been confirmed as brigade commander following Major Gardner's death. General de Boigne told Sutherland to await the posting of new Maratha 'pundits' or civil observers serving inside the brigades as regulatory auditors ensuring accurate revenue collection and handling. These new Maratha auditors, loyal to Sadeshau Bhau, were to be 'exchanged' for the existing regulatory personnel that had been previously appointed by Lackwa Dada – one of the senior most members of Sindia's court. This was part of a proposed checks and balances system designed to placate competing factions in the *durbar*. Yet this was not necessarily a complementary system of supervision. Rather, this signalled a deepening rivalry in which the serious jockeying for position was well underway and bloodshed was a real possibility.* From this point forward there would be an increasingly hostile struggle as Lackwa Dada and his clique sought to fend off rival factions in a turf war over who should oversee the brigades' accounting procedures.

Sutherland was given explicit orders from General de Boigne. He was instructed to cordially receive the Bhau's auditors. But that was only to be done after Lackwa Dada had agreed to this Sindia-sanctioned political compromise. To the Europeans it amounted to an invitation for argument as they had been ordered to let some of their most ardent *sardar* rivals review the accounts. The mercenaries were in a precarious situation in trying to second-guess which faction would prevail. Since they had always survived by remaining loyal to the Maharajah, the compromise over administrative over-sight powers meant a potential crisis in confidence. It was difficult to discern if either Lackwa Dada or Sadeshau Bhau would prove ultimately loyal to Daulat Rao Sindia. What if there was a showdown between Lackwa and 'the Bhau'? What if Lackwa were to dig in his heels and refuse to remove his men? In this case de Boigne thought it best to err on the side of caution with regard to his relationship with Lackwa Dada. Sutherland was told, 'You are to act in this case as Lackoo Dadah shall direct you, and no further, but you shall exchange them [the pundits] immediately and without any further instructions from me, if he orders you so to do.'[112]

---

* Sadeshau Bhau was eventually called upon to take to the field for Sindia against Holkar in the Battle of Hadespar fought outside Pune in 1802.

The crisis passed but the damage had been done by the worsening of a poisonous atmosphere that was to prevail. It was time for de Boigne to put greater effort into his retirement plans and they soon took on the appearance of an escape from Maratha service. In truth de Boigne had applied to leave service and return to Europe before Mahadji had died, but his departure for Calcutta was delayed until 1796. As long as Mahadji Sindia was alive he had safely shielded General de Boigne from the political battles of the *durbar* and in effect that action had helped to delineate civil–military affairs. When Mahadji died it became evident that there would be no long-term job security for a foreigner as C-in-C of the Sindia clan's army. General de Boigne had remained aloof from the inner workings of the *durbar* and without Mahadji to shield him it was clear the Savoyard would have to depart or risk being sucked in to the increasingly bitter rivalries.

The shifting alliances, that had characterized Maratha political ascendancy in the seventeenth and eighteenth centuries, continued to be a hallmark of Maratha military culture. And with de Boigne's departure came a new round of infighting, as the Maratha *sardars* remained divided over who should replace de Boigne and have control over revenue collection as well as the 'regular corps'. Several members of the *durbar* coveted the position, but Daulat Rao knew there was wisdom in selecting another foreigner for the same cultural reason de Boigne had been retained for so long. The European mercenaries were not ethnic Marathas and as outsiders, devoid of cultural claims on power, they served a vital role in the checks and balances over the struggle for control of the clan and its lucrative hold on North India. But who best to take de Boigne's position? The decision came down to either Perron the Frenchman or Hugh Sutherland from Scotland. Ultimately General Perron succeeded in winning the coveted position as Daulat Rao's new C-in-C.

Perron faced a difficult position as the *sardars* renewed their demands for greater financial disclosure from the mercenary-led battalions. The potential loss of foreign mercenary allegiance was beginning to grow as the Europeans saw themselves pushed out of the system that they had fought so hard to make work efficiently. When revenue returns from brigade holdings slowed unexpectedly as the result of drought, retaliatory raiding from the Punjab and insurgency in Hariana, Maratha overseers accused the European mercenaries of withholding funds. As money dwindled to a trickle in some outlying areas, the mercenary officer corps argued they should have the first cut of the funds to meet priority defence requirements. The mercenaries believed the army's needs should come before those of greedy account overseers who tended to work for the enrichment of their immediate masters and not Daulat Rao. The officers sought to ensure that their battalion's budgetary allocations were guaranteed ahead of treasury deposits, while the *sardars* elevated the stakes by levelling the charge of attempted embezzlement from the treasury. In early 1801, Dutch mercenary John Hessing admitted to his Scottish brother-in-arms Sutherland that he would have great difficulty in explaining his revenue shortfalls to the Marathas. Extensive plundering, by Sikh warrior bands and freelance Irish mercenary George Thomas, had caused great shortfalls in the income generated by several territories.[113] The increasingly pedantic demands of Maratha overseers adversely impacted the foreign mercenary's morale. The pressure was building and something had to give way. Meanwhile, in Calcutta, the British were actively monitoring the reverberations of discontent. They saw this as more than just a Maratha squabble since the political stability of North India's remaining Mughal infrastructure was at stake. If there was to be another Anglo-Maratha war, it might make a great deal of military sense for the EIC to strike while its opponents were distracted.

Conclusion

As demonstrated earlier in this chapter, the location of Maharashtra and the history of her people provided a unique set of geographical and cultural factors that influenced Maratha military evolution. That is not to suggest this was a historic case of geographic determination. However, it can be said that Maratha military culture does reflect the Marathas' relationship with the greater South Asian military environment. The key patterns of Maratha military culture were set well before European contact in the so-called 'Age of Sail'. But, owing to the nature of South Asian cultural interaction, Maratha military history cannot be physically or ethnically segregated or placed in a test-tube as 'pure Maratha' and studied in isolation. Maratha armies in the seventeenth and eighteenth centuries were, contrary to popular belief, extremely cosmopolitan. Clan-based Maratha armies featured a pragmatic approach to survival as well as battlefield tactics and those attitudes were partially fostered by their reading of events in clan-oriented terms. If one branch of one family could control one area, by occupying a strategic position such as a hill fort in the *ghats*, the clan leaders were not likely to quibble about the ethnicity of the garrison or whether their tactics were essentially Deccani or Mughal.

The pre-European periods of confrontation and collaboration among Maratha clans added to the competitiveness of their military culture as they strove for dominance. Yet Europeans did not always share the same military value system – a system in which survival and profit could be counted as equal in value to combat victory in the context of war as an economic opportunity. In simple language, the Marathas played to win whenever they fought but their definition of winning was not always the same as that found in Western military culture. There remains an undefined historic difference in Western and South Asian military value systems. In some cases South Asian military leaders, serving as broker/dealers for their men in the military labour market, 'sold out' but kept their reputations intact as responsible commanders who maintained the most precious military commodity they had to deal with – the lives of their men. Their strategies and tactics emphasized cost effectiveness in terms of resource expenditure regardless of whether those resources were human, technological or environmental. Opportunity and survival became watchwords in the businesslike world of the Maratha military entrepreneur. Maratha culture embodied a realistic form of dedication to survival within a South Asian military environment that featured a different set of motivational factors as found in the South Asian military economy. It did not necessarily endorse dying for a transitory political cause. War was many things. It was, in essence, the physical manifestation of conflict. But it was also a big business opportunity for those who were a part of the South Asian military economy by birth, choice, or accident.

Artisans and tradesmen in the weapons business, merchants selling everything from woollen uniform cloth to grain, *banjaras* and *beasties*; they were all willing to use war as an economic opportunity. The South Asian military economy was an equal-opportunity market place. From the Maratha *sardar*, or the Rajput prince, to the lowliest of tribal *burkundauze* (matchlock men), the South Asian military economy was an inclusive societal market place of incorporation.[114] It gave status to the 'warriors by birth' and it also served to integrate and legitimize those who lived on the periphery of society. For peripheral people with complementary military and religious cultures it was more than an economic opportunity; it was a chance for social up-grading. A lowly peasant, with a matchlock or blade for hire, could pass himself off as a warrior when soldiers were needed. A criminal with a horse could find ready employment as a *pindari*. War represented a means of getting out of the bind encompassed by the day-to-day drudgery of survival. It was a chance to move up, it was a chance to make money. Of course it was also a risk, but it was a culturally acceptable risk to many. They could choose to believe a death in battle was symbolic of the larger religious struggle in daily life. The ultimate struggle might find its way to being called martyrdom. And the martyrdom of the warrior was hailed in the religious texts of Hindus and Muslims alike – although one did not want to look too closely at the causes of a given conflict – it was simpler to say a warrior's martyrdom could be judged as an honourable death.

The South Asian military economy had a cultural sanction and to a limited degree a religious sanction that appealed to many stakeholders. Who was *not* a warrior? Rajputs, Mughals, members of the *Khalsa*, or do you put it in religious terms of Hindus, Muslims, Sikhs? The Europeans failed to recognize that South Asian military culture was every bit as valid as their own; every bit as dedicated as their own. The Western desire to see Indian armies as backward and locked in time was myopic to say the least. But eighteenth-century British military Orientalists also failed to see that the impetus for cultural change did not necessarily have to come from one specific geographic direction and that military culture could precipitate social change; not all military progress moved from west to east and not all political power rested in the hands of politicians.

The information in this chapter has been presented to underscore the assertion that there were three basic points about Maratha military culture as evident in the period from 1600–1800.

- *Mercenary employment was a regular feature of Maratha armies as it was in most major armies during that time.* But in the case of the Marathas, their acceptance of mercenaries extended from the lowliest rank to that of the commander-in-chief. Cost-effective combat performance was more important than social, religious or physical difference. European mercenaries had served Shivaji and they had also deserted him. Changing sides was something that the Marathas, including the Bhonsles, had also done regularly. There was no shame in being on the winning side because that was where the greatest profits were to be had within the South Asian military economy.

- *Maratha armies sought and employed the competitive advantages offered by technology as well as doctrine.* If a given weapons system killed more enemy troops or could be used as part of a strategy to win more wars, then it was worthy of consideration. The Marathas advocated a doctrine of artillery-based fire superiority long before the Europeans as can be seen in their willingness to carry heavy artillery into the field and use it in an anti-personnel context. The existence of sophisticated armies in other Asian countries (like China) that failed to evolve a similar artillery doctrine suggests there was more to this aspect of Maratha military evolution than just the adoption of 'force multiplier' systems to cope with numerically larger enemy armies.

- *Infantry warfare was not new, or an experiment, for the Marathas in 1803.* The Marathas had an infantry warfare tradition that pre-dated Shivaji. Maratha infantry held the fortresses that served to anchor Shivaji's political and military claim to western India. His combined operations often featured several types of mounted troops operating in conjunction with professional infantrymen. By the end of the second quarter of the eighteenth century the Marathas' offensive doctrine had firmly evolved under *Peshwa* Baji Rao I from neo-classical to proto-modern by emphasizing tactical infantry manoeuvring to take advantage of Maratha leadership in the utilization of artillery.

Maratha armies had used massed infantry firing techniques for generations before de Boigne arrived on the scene. During the administration of Baji Rao I the tactical use of both the infantry column and the line were widespread with the latter predominating at the Battle of Dabhoi (1731).[115] In the past many authors have portrayed de Boigne's impact on Mahadji's forces as nothing less than a military revolution in the 'transformation' of the 'Maratha army' with the introduction of 'European discipline and drill'. But those descriptions were essentially derived from nineteenth-century British military myths offered up as an explanation to answer the nagging question of how the Marathas – a supposed nation of 'freebooters' – could pose such a credible military challenge to what was a historically contemporary superpower.[116]

Monsieur de Boigne's real contribution to Mahadji Sindia's regime was that he provided a means of establishing fiscal responsibility by placing the army on a resource-accounting basis that acknowledged the economic interdependence of the military and civil society. In the period from 1792 to 1794 the European and mixed-race mercenaries in Sindia's North Indian establishment became the favoured executives in a new, more effective, management system. They came to represent a new level of control over a more fully integrated civil–military economy. This civil–military integration of Hindustan, at that time, had three points of interdependence:

- The infantry of the 'regular corps' served at the discretion of Maharajah Sindia and it was used to uphold his rule of law. It was used to defend not only Sindia's territorial claims but also to underpin the Mughal civil administration that yielded the bulk of the bureaucratic machinery used in governing North India. This represented the civil–military interdependence of law and order.

- The greater South Asian military economy, that was one of North India's chief employers, met the military's physical requirements in the form of food, housing, ammunition, etc. This was a complex symbiotic relationship extending beyond those tradesmen who simply worked in weapons production. Even the simple cultivators needed the security provided by the army and the army needed to be paid from the revenue derived from territories made prosperous by simple cultivators. This represented the civil–military interdependence of the regional economy.

• Although the land revenues were assigned and fixed under civil law, it was the military that served as the collection agent of last resort when landlords staged insurrections and rebellions, something they had been prone to do since the decline of Mughal central authority. Land revenue was applied to a military infrastructure, which held the 'state' together when insurgency threatened it. The 'regular corps' under de Boigne ensured cost-effective tax-debt recovery and, perhaps more importantly, scrupulous repatriation of funds to the government upon collection.

This represented the civil–military interdependence of 'state' funding. The departure of de Boigne had no impact on the military efficiency of the 'regular corps' that passed intact to Perron's control as Daulat Rao Sindia's C-in-C. It was clearly a professional force held together by the bonds of loyalty that existed between the mercenary officers and their sepoys as drawn from the military labour market. They soldiered together prosperously in spite of the disrupting influence of Maratha clan politics. And most non-Maratha soldiers in Sindia's service shared the dream of living long enough to retire with enough money to live comfortably.

As we shall see in the next chapter, British observers did not always agree on which Maratha military assets posed the greatest military challenge. But two key points were becoming clear. First, Daulat Rao Sindia's military presence in the Doab and his administrative relationship with the Mughal Emperor gave the Marathas a disproportionate amount of political as well as economic clout in controlling North India. Second, the 'regular corps' under General Perron was counted upon to ensure Daulat Rao Sindia's independence and security.

2 British perceptions and the road to war in 1803

Introduction

Having used the preceding chapter to establish the nature of Maratha military culture, this chapter looks at British military perceptions of the Marathas prior to the war in 1803. The work then proceeds to outline British military war aims as established by the Governor-General. The juxtaposition of historic perception and proposed actions in this chapter can then be combined with the campaign descriptions in chapters 3 and 4 to give readers an idea of the differences in the *theory vs. practice* of fighting the Marathas. One by-product of this chapter is a heightened sense of the degree to which British ruling elites could mirror their counterparts in South Asia. The contrasting backdrop provided by Richard Wellesley's political ambition and his brother Arthur Wellesley's military single-mindedness highlights the disproportionate power of the Wellesley family in 1803.

British perceptions of Maratha military culture 1750–1803

What shaped the opinions of the British officers and administrators who in turn shaped military policy towards the Marathas? One would think a great number of their attitudes would have been influenced by the popular military literature of the day. However, the correlation is ill defined. Commercially published British military books in that period (1750–1800) tended to be very subjective military travelogues. The fact that such influential figures as Robert Clive had participated in joint-infantry operations with Maratha allies in the 1750s was largely glossed over and no one formally addressed the concepts of Maratha military modernity or offensive capability.[1] Stringer Lawrence, once viewed as the spiritual father of the British Indian Army, extolled the virtues of the Maratha infantry soldier but his accolades passed from acceptability to obscurity in less than one generation.[2] John Henry Grose, however, did prove to be particularly observant in writing about the Battle of Sugarloaf Rock that took place on 20 June 1753, during the Second Karnatak War eight years before Panipat. It featured well 'dug-in' Maratha infantry forces under Morai Rao.[3] Their trench-works were in the traditional style of Deccani field fortifications, such as one might have seen a century earlier in the siege lines at Ahmadnagar or Daulatabad. The Maratha emplacements included well-excavated infantry firing positions to protect the marksmen whose aimed fire covered the approaches to the trenches that the main force occupied. This was at a time when published accounts of British infantry doctrine extolled the virtues of unflinching 'head up-right' bravery in the face of enemy fire. And we should consider the degree to which the average eighteenth century reader may have misinterpreted Grose's selective depiction of 'native' tactics as being at odds with their own perception of British military tradition. While one was upright and forward moving, the other was lowly in profile and stationary. But the key point in any such comparison was that the examples represent the antithesis of doctrinal comparison; one presenting the historically contemporary view of offensive doctrine and the other exemplifying the ideal of defensive doctrine. Whether such an unfair comparison ever took place, or how often it took place, can never be determined. The point is that even the small amount of positive military press the Marathas received could have been misconstrued by those wishing to cast negative interpretation on the Marathas' military ability.

There is no ready explanation as to why accounts of the Marathas in the third quarter of the eighteenth century neglect to mention the Maratha infantry's proficiency in the Karnatak Wars. Their positive press had quickly fallen from British notice and that may have been the result of the Europeans passing a premature pronouncement of death on Maratha military capability after the disaster at Panipat in 1761. By the time the First Anglo-Maratha War started in 1775, it was difficult to find objective assessments of Maratha military operations. And the British loss to the Marathas at Telegaon* in 1779 was usually omitted from direct reference in subsequent military assessments of the Marathas from 1780 to 1800. Perhaps that can in part be attributed to the degree to which the First Anglo-Maratha War went unanalysed in popular military books of the day. The rare published reports seem to have been produced by Parliament and the EIC.[4]

Among the few authors who touched on Maratha military culture and society, there was a tendency to place remarks in a European context. Major Innes Munro, in an account published in 1789, likened the

* Aka Talegao.

Marathas in the Deccan to the Germans in Europe – an immense empire of subordinate principalities. Given the similarity of the Marathas to the allegedly 'warlike' people of the German states, he pondered 'was it not then absolute madness in us to have quarrelled with such a power, even if we dreaded no other foe at the time but itself?'[5] Ten years later William Henry Tone, in an assessment that borrowed heavily from the work of others, repeated the analogy in summarizing the Maratha 'Empire'. Tone said, 'It would, perhaps, be best described, by resembling it to the circles of Germany, as a military republic, composed of chiefs independent of each other.'[6] Major Rennel, another military scribe, used a variation on the European warrior-nation comparison. He noted the Marathas had become the 'Swiss' of India, a reference to the Swiss mercenaries, 'alternately courted and employed by different parties'; the cultural distinction he made about the Marathas was that they 'usually paid themselves instead of being paid by their employers'. This might have been intended as a comment about the Maratha propensity to view war as an economic opportunity but it seems, in hindsight, to have been rather snide as Rennel was busily engaged in promoting the sale of his military maps and writings to newly arrived EIC subalterns.[7]

Perceptions from the Third Anglo-Mysore War

A number of damaging military observations on the Marathas were sparked by their activities while serving as British allies against Tipu Sultan in the Third Anglo-Mysore War 1789–92. Major Dirom, who was a Divisional General Staff Officer with the duty of Deputy Adjutant-General to the King's Troops in that war, wrote the most widely known book on the conflict.[8] Dirom claimed to have been impressed with the Marathas as individuals, but he elected to open his book with a negative reference to them during the disastrous monsoon retreat of Cornwallis's army in 1791.[9] The Major left no doubt that he blamed the Marathas for having failed to make their critical rendezvous with the British on an appointed day, which in his opinion necessitated the British withdrawal.[10] Cornwallis's men had no choice but to spike* the heavy artillery and abandon the mud-bogged battering train in their aborted attempt to lay siege to Tipu Sultan's fortress of Srirangapatnam.† The Bombay army of Major-General Abercromby, in order to play their part in the siege, had to carve a fifty-mile road system out of the *ghats* with 'infinite labour'. And

---

* To drive a spike or nail down the touchhole of a cannon to render its immediate use impossible.

† 12°+, 76°+. Aka Seringapatam, Seringaputnum, Shrirangpattan.

bitter was an entirely inadequate word to describe how they felt about the Maratha rendezvous fiasco. Abercromby's men had been forced to manhandle their huge iron siege guns through the mud since monsoon sickness had killed most of their draught cattle. The troops were weakened by a scarcity of food and the camp followers were reduced to eating the putrid flesh of dead gun bullocks. Writing about Abercromby's predicament Dirom noted, 'his army, who thought they had surmounted all their difficulties, had the mortification to find their exertions of no utility, and had to return, worn down by sickness and fatigue, exposed to the incessant rains which then deluged the western coast of the Peninsula'.[11] A short-lived redemption of the Marathas resulted from the arrival of their relief force carrying grain.[12] However, British gratitude was quickly soured by the Marathas' penchant for profit taking in this military spot market. The Marathas did not give the grain to their starving British allies; rather they sold it at a cost so far beyond the market price that the EIC's sepoys could not afford it and 'the pay of a subaltern would scarcely feed his horse'.[13] Collectively, the British did not seem to have noticed that the incident dramatically underscored the Marathas' historic ability to wage war more effectively than Europeans when it came to monsoon campaigning.

Major Dirom's volume, published a decade before the Anglo-Maratha Campaigns of 1803, portrayed the Maratha artillery as obsolete and immobile; disparaging comments were made about the Maratha practice of painting names on the guns.[14] Dirom deemed the Maratha gun carriages and their construction to be 'clumsy beyond belief'.[15] The composite wood wheels were of little value in the Major's estimation, but then he may not have considered their durability in the monsoon mud.[16] The Marathas had compensated for the quagmire condition of the roads and their immediate lack of pioneer units by placing up to fifty bullocks in the harness for each gun.[17] They were also very careful about the breed of animal selected for this type of duty. The gun bullock's speed, strength and regional suitability to diet during monsoon conditions were all part of Maratha military knowledge. British 'public carriage contracts' were notorious for providing unsuitable beasts of dubious pedigree and inexperienced British officers were often duped by nefarious procurement agents who stuck them with inappropriate breeds of bullocks that were destined to perish under inclement field conditions. The British did not make significant progress in this aspect of logistics until they captured Tipu Sultan's breeding stock of bullocks in Mysore during 1799.[18]

Dirom singled-out the Maratha infantry for particular repudiation, although to his credit as a foreign observer, he realized that these units were not composed of ethnic Marathas. Their religious as opposed to ethnic identity underpins the observation that they and their officers were drawn from a local military labour market by allied southern Maratha *sardars* circa 1790.

> The Mahratta infantry, which formed part of the retinue that attended the chiefs at the conference, is composed of black christians, and despicable poor wretches of the lowest cast, uniform in nothing but the bad state of their musquets, none of which are either clean or complete; and few are provided with either ammunition or accoutrements. They are commanded by half-cast people of Portugueze and French extraction, who draw off the attention of spectators from the bad clothing of their men, by the profusion of antiquated lace bestowed on their own; and if there happens to be a few Europeans among the officers and men, which is sometimes the case, they execrate the service, and deplore their fate.
>
> The Mahrattas do not appear to treat their infantry with more respect than they deserve, as they ride through them without any ceremony on the march, and on all occasions evidently consider them as foreigners, and a very inferior class of people and troops.[19]

In the above passage several points are worthy of consideration with respect to the British perception of the Marathas and the later course of the 1803 Anglo-Maratha Campaigns. Dirom was quick to state that these Topasses, whom he termed 'black christians', were 'foreigners' to the Marathas and their officers were 'half-cast people' of Portuguese extraction. This would tend to counter any attempts to generically label all Maratha infantry as being composed of Hindu troops and Pathan mercenaries.[20] It also negates any Hindu nationalist bid to portray this as a Hindu force fighting to rid the homeland of Tipu as a Muslim despot. The identification of half-caste and foreign mercenary officers is directly in keeping with mercenary recruitment in Maratha military culture. Note also that these Maratha troops were armed, as were most Maratha infantry of that era, with flintlock muskets.

What Dirom did not specifically establish for his readers was the organizational context of these Maratha troops. Although in fairness it is not certain that he understood what these troops represented within the overall profile of Maratha infantry. These 'despicable poor wretches' were specific clan contributions to an alliance-generated Maratha war effort. They were men sent by Maratha 'chiefs' such as Hari Punt and Pursheram Bhau, experienced players within the South Asian military economy, who knew better than to send their personal guard or best troops. These were expendable members of the military labour market in the context of contingents sent by participating Maratha powerbrokers helping to meet treaty agreements signed by higher Maratha political authorities. The Marathas were reluctant to send their best troops to aid European allies of convenience. Who would risk their finest military assets without a sufficient reward?

Consistently Dirom spoke of 'the Mahratta infantry' or 'the Mahratta artillery' as if they were each homogeneous entities, or symmetrical military branches of service, analogous to a British Regiment of Foot or the Royal Artillery.[21] Inadvertently Dirom was contributing to the stereotyping of all Maratha military forces. When we review the eighteenth-century descriptions we should remember that Maratha military culture had a historic tendency to foster different types of military forces relative to clan assets and prevailing conditions in the military labour market. Without an opportunity to see other Maratha troops, scattered across India from Tanjore to Rajasthan, Dirom could not be expected to give an accurate assessment of all Maratha troops. Dirom was relaying a physically isolated view of a motley Maratha clan contribution to the Anglo-Mysore war effort. However, his comments passed into popular print as if he were reviewing the entirety of India's Maratha troops on parade. For all Dirom knew this could have been a phantom unit brokered specifically for a participation price – a troops-for-hire scam. In other words, a rag-tag group hired on the military labour market with a fraction of the British funds allocated to an allied *sardar* for troop procurement.

Dirom provided a description of the Maratha infantry's shortcomings in supporting Captain Little's attack on Tipu's forces at 'Simoga' on 29 December 1791. The Mysorian enemy was entrenched and their repulse of the ill-prepared Maratha troops had a negative knock-on effect with regard to the British sepoys.

Part of the Mahratta infantry charged at times, when they saw the enemy appearing to give way, but were always beat back, and returned in such disorder as greatly increased the difficulty in forming and leading our sepoys; while the greater part of their infantry, or corps of 300 topasses, on being directed by Captain Little to advance, declared their unwillingness to take their share in the action, and that they had come out entirely unprovided with ammunition.[22]

Several smaller Maratha clan armies were still recovering from leadership losses at Panipat thirty years earlier. There was nothing approaching a national strategy for military regeneration and as indicated previously there was no institutionalized source of *sardar* training to ensure leadership standards, which were apparently lacking at Simoga. If a clan army was destroyed, then its ability to replace, retrain and reequip fallen infantry might largely depend on disposable assets or the ability to procure new funds through land revenue, borrowing or military annexation. The lack of ammunition mentioned by Dirom may have been a logistical oversight. Yet it may also have been an indicator that the Marathas wanted the burden of cartridge expense placed on the British for participation in one of 'their' wars.[23] Providing troops under a higher treaty authority did not mean that Maratha clan leaders would ignore the financial details of the obligation. If the Marathas could save money or make money at British expense, they would. They were veteran entrepreneurs in the South Asian military economy and one way to increase potential campaign profits was to outsource ammunition requirements at an ally's expense. This latter point was one in a long list of complaints that tended to make printed references to the Marathas less than complimentary.

Apparently Dirom believed light horse units to have been the Marathas' most effective contribution in the campaign against Tipu in 1792. South Asian light horse tactics, at that time, had no direct doctrinal parallel in His Majesty's Cavalry or the EIC's Native Cavalry that was formed on European models. The Maratha mounted units successfully intercepted a number of Tipu's convoys and laid ambushes for the Mysore horse. The Deccani Maratha horsemen were specialists in many tactics that were unfamiliar to European horse soldiers. Their unorthodox but effective field craft, such as setting horse snares, gave the Maratha horsemen reputations that were equally feared and envied. Dirom was one of a small but growing number of British observers who saw Maratha light horse and irregular cavalry as potentially the greatest Maratha military asset for incorporation into the EIC's order of battle. As the singularly most popular work on the Third Anglo-Mysore War, Dirom's book became standard reading for many of the British officers who fought in the Fourth Anglo-Mysore War in 1799; men like Arthur Wellesley.

Dhundia Waugh as a cultural reference point

Despite the graphic suggestions made in a handful of later fanciful artworks, Arthur Wellesley did not lead the storming of Srirangapatnam in 1799. He was deprived of a frontline role in the final assault and was relegated to the reserve trenches. Following the successful conclusion of the battle, Governor-General Richard Wellesley caused great commotion by appointing his relatively inexperienced brother Arthur to be Military Governor of Srirangapatnam. During the chaos accompanying the fall of Tipu's fortress, a number of criminals escaped from his prison. Among the most troublesome was a man known as Dhundia Waugh* who was described extensively in British reports as a 'Marhatta freebooter'.[24] Dhundia evaded recapture and proceeded to terrorize Mysore and the

---

* Aka Dhoondiah Waugh, Dhondiah Vaugh, Dundiah Waugh, Dhondia Wagh, Dhondiah Vagh.

southern Maratha country with a 40,000-man army. His force was composed of many demobilized soldiers from Mysore and the EIC, but by far the greatest bulk of his men came from roving *pindari* bands. With Dhundia on the loose, Arthur Wellesley, in his new position of authority in Mysore, was kept busy trying to secure the Maratha–Mysore border. The Marathas, in contrast to their role as British allies in the Third Anglo-Mysore War, had remained more or less neutral in the Fourth Anglo-Mysore War of 1799. Unfortunately, this meant Arthur found it necessary to violate the existing border neutrality agreement with the Marathas several times. From July to September 1799, Wellesley chased Maratha riders back over the territorial boundary to stop them from plundering.[25] Several leading Maratha horsemen, who had been allies ten years earlier, took advantage of Dhundia's insurgency to do a little raiding in Mysore as well. Their plundering was done under the explanantion that it was a compensatory action for past losses to Tipu. This apparent propensity of both allied and enemy Maratha horsemen to engage in plundering helped establish in Arthur Wellesley's mind a direct correlation between the label 'Maratha' and the words 'predatory horse'.

The military governorship of the Mysore territories gave Arthur Wellesley a voice in EIC matters that was disproportionate to his military rank and experience. Arthur, a prolific letter writer, began to draft scores of letters and memorandums on military affairs as he saw them. He often drew together bits of collected wisdom and laid them out in a comprehensive manner approaching that of a position paper. There is absolutely no indication that Arthur ever tried to selectively portray his lengthy memorandums as the work of genius. That was a task for his later biographers. And he should be remembered positively, in those early days, for having acknowledged the experience of many veterans who gave him the benefit of their years of service. In a letter to Lieutenant-Colonel Dalrymple, written by Arthur as he took to the field against Dhundia in June 1800, there was an indication of respect for those who had previously served in this alien environment. Recognition was paid to the fact that this was a different type of operation or 'species' of war.[26] This difference probably reinforced Arthur's belief that the Maratha way of war, as he understood it to have been expressed by Dhundia, contrasted with that of Britain. Despite the apparent difference in military culture, Arthur recognized that Maratha light horse units held the potential to meet particular tactical shortcomings in British mounted operations.

The Dhundia Waugh campaign was an outstanding success for Arthur Wellesley that culminated in a cavalry charge that killed Dhundia. And despite it being Arthur's first independent combat command, he had actively encouraged alliances with Maratha light horse units who joined in the pursuit of Dhundia. Specifically he cultivated a professional military relationship with the Maratha cavalry leader Appah Rao.[27] The experience was to colour Arthur's political as well as military thinking with regard to the Marathas. It was an experience that confirmed the *Marathas = light horse* impression in his mind and perhaps, in retrospect, the campaign of 1800 fell short of being a successful educational experience. It left him ill prepared for the reality of Maratha infantry warfare that he later faced at Assaye in 1803. But the Dhundia Waugh campaign did convince Wellesley that it took a certain amount of parallel structure, or what we could identify in more contemporary terms as force symmetry, to defeat a *pindari* or light horse threat. Arthur also saw that the *pindari* dynamic might be altered by steady employment that regularly paid cash wages.

Preparing for war

In the spring of 1803 Arthur Wellesley wrote to his superiors with the observation that Appah Rao had 3,000 *pindaries* that would cause severe problems for the British if they were released from service. Wellesley knew full well that loyalty was a transitory thing and that today's allies could be tomorrow's enemies. He observed: 'It therefore appeared to me, and to Major Malcolm, to be absolutely necessary that Appah Sahib should retain them in his service.'[28] The thought of allowing Maratha *pindari* allies to plunder on approval was somewhat disturbing to Wellesley. It was true that British troops were authorized to plunder under specific conditions such as when a fortress was captured after surrender terms had been refused. But to factor organized plundering into British doctrine was something different. Having said that, the difference was not so great as to preclude the British from using that option. The alternative to retaining the *pindaries* via plundering rights or commission payments was to pay them a fixed wage. However, it was not always practical to have the EIC pay them just to keep them out of circulation. That was a practice that could be exploited by *sardars* who knew how to manipulate the military labour market by threatening to join the enemy. Therefore, it became expedient to try and have a South Asian ally absorb the cost of their wages. An ally like the *Peshwa*, already bound by the Treaty of Bassein as part of the Subsidiary Alliance Treaty System, was ideal in that he could be contractually obligated to share defence costs. This plan to defray *pindari* hiring costs would theoretically reduce British expenditure while at the same time depriving Maratha enemies of the *pindaries'* services. 'Major General Wellesley, therefore advises that they should be taken into pay at the rate of twenty rupees per month each man; and he guarantees to

Appah Sahib that the Peshwah . . . shall reimburse to him the sums paid to them. Under this engagement the Pindaries must not be allowed to plunder.'[29]

Arthur Wellesley contributed countless letters to the military message traffic that circulated among the EIC's Bengal, Bombay and Madras armies. Arthur held the King's commission, which meant that officers in His Majesty's Army also had to be mindful of what the prolific Arthur wrote. To one and all, he reiterated his true belief that the Marathas were militarily nothing but an assemblage of 'predatory horse'. Arthur wrote specific doctrinal guidelines for the war of 1803 based on the atypical Dhundia Waugh campaign.[30] Threat *perception* was to influence both his tactics and attitude. And as war with Sindia began to look inevitable in 1803, Arthur Wellesley repeated his recommendations for the acquisition of Maratha light horsemen which he perceived as those soldiers most needed to fill a structural gap in the EIC's armies. An examination of Arthur's correspondence in the months preceding the outbreak of war in 1803 indicates that he had apparently been quite successful in incorporating Maratha light horse and *pindaries* into his own regional forces. By April 1803, Wellesley already had over 6,000 Maratha light horse under his command as well as a wide-ranging assortment of Maratha foot soldiers that he picked up in the bargaining process. Several Maratha clansmen were quick to cash in the human assets they held and off-loading local 'infantry' was all part of doing business in a South Asian military economy made nervous by the threat of an approaching war.[31]

Arthur Wellesley was preparing to fight the Marathas as he understood them and, in doing so, he was committing the soldier's sin of preparing to fight the last war and not the next. The Dhundia Waugh campaign featured elements of mobile pursuit that were not representative of sustained infantry warfare in which attrition takes many forms. Arthur had dwelt upon the element of the chase to the point of obsession. Apparently Dhundia had escaped capture on one occasion by crossing a monsoon-swollen river in the nick-of-time. Arthur believed Dhundia could have been apprehended at that moment if only the British had the foresight to bring a pontoon bridge with their column. Arthur had written to Major-General Brathwaite, then the Commander-in-Chief of the Madras army.

I crossed the Malpoorba yesterday at the deepest and most rapid ford that I have ever seen; but I am now encamped on the right bank of the river. It would be of considerable advantage to warfare in these countries if the army were provided with pontoons. If you approve of the idea, I could easily get some made at Seringapatam, where they might be tried. If I had had pontoons on the Malpoorba, Dhoondiah could not have escaped; and it is inconceivable the advantage they would give us over all the Native armies.[32]

It was Arthur's belief that the Marathas could not use vast mounted armies in the monsoon owing to problems he imagined they would have in crossing rivers and so at that time he reasoned the Marathas were unable to fight effectively in the monsoon. The enemy light horse units would presumably be stranded on islands created as the rivers rose and turned into torrents which would isolate the Marathas in pockets of land between the rushing waters. He theorized that the British could advance to these islands of trapped horsemen with marvellous scientific precision using modern prefabricated pontoon technology.

There was ample historic evidence that the Marathas were accomplished at crossing rivers and, for that matter, bridge building. But Arthur Wellesley took it into his head that modern British military technology, in the form of a pontoon train, would prove a decisive advantage in waging war against Maratha enemies. Ironically, Arthur Wellesley had crossed the Wardha River on a Maratha military bridge in 1799 but he had apparently taken little notice of it. Lieutenant Lambton, who accompanied Wellesley on that journey, had observed how Maratha field engineering made good use of fascines.* Lambton recorded that they crossed the river, 'by means of a bridge made of Bamboos belonging to the Camp, a very excellent contrivance. They are bound together fixed to stakes driven into the bottom of the river, and covered with fascines.'[33] Lambton noted the bridge was only three miles away from a battlefield where the Marathas had used similar field-craft in a confrontation with Haidar Ali in 1772. Haidar had selected an excellent position protected by deep ditches designed to thwart a Maratha assault. But Maratha infantry advanced under fire to fill the excavation with load after load of fascines. Having established a bridgehead, they sent their cavalry across to ride through Haidar's infantry who were exposed inside the perimeter formed by the ditches. Lambton also remarked on the local rafts that were made by lashing bamboo canes to large-capacity earthenware pots (*chatties*).[34] This method of construction took far less time than felling trees of sufficient size and lashing them together. Lieutenant Lambton and the young Colonel Wellesley had seen the same sights but the two had come away with different perceptions of the Maratha military environment. Arthur had not taken on board the fact that the Marathas had a long history of military field engineering. Building bridges and crossing rivers was never a problem for them historically and they had excelled at it during their raids aimed at Bengal during the administration of Baji Rao I.[35]

---

* In Europe this usually meant bundles of branches but in South Asia it could be bundles of bamboo.

Arthur's British military culture was very much a pre-Victorian construction that emphasized the need to find 'scientific' answers derived through 'scientific reasoning'. In this case we can detect elements of both technological and geographic determination. If monsoons increased the volume and velocity of South Asian rivers and British units had problems crossing rivers, then any modern tool that allowed them to cross more rapidly could prove decisive in defeating an army that by its very 'native' nature must be unscientific and therefore ill prepared. The technological answers for this pre-Victorian case of 'scientific military reasoning' drew ethnocentrically on the design and engineering principles of the great British industries that reinforced that nation's bid for supremacy in the Industrial Revolution. He wrote, in what became a series of letters on the subject to the C-in-C of the Madras army: 'My idea was to make copper pontoons. These are absolutely necessary for warfare in this country and on the whole of the west side of India. If you send to England for them, orders ought to be given to have them made very light.'[36] Arthur specifically wanted copper vessels fabricated in England. That was a modern Western technological response to the apparently backward foreign military environment he faced. Boilermakers and copper sheet-metal workers were on the leading edge of British technological advancement. They were the mechanics of industrial modernization as the steam era depended upon the boilers they made. The conquest of the environment through scientific reasoning was indicative of a Western mentality that perceived industrial output as synonymous with victory – the triumph of superior industry in an imperial setting. Later, when others failed to see his reasoning, Arthur conceded that Indian-made pontoons would suffice.[37]

As we saw in chapter 1, the Marathas' military culture was one that looked to the immediate human and physical environment to yield tactical answers rather than to pronounce strategic judgements. The Marathas remained open to the adoption of new technology when they saw it as beneficial to their way of war. But they had already developed less cumbersome means of crossing rivers. They may not have had copper pontoons carried on cantilevered wagons, but then they did not need them if history was any indication of how to fight effectively while avoiding the problem of making oneself an unnecessary target for logistical interdiction. As war approached in 1803, Arthur exchanged extensive correspondence about transporting a great pontoon train into the Deccan. In a pedantic obsession with his own theory, Arthur flooded official channels with reports dealing with every aspect of pontoon construction, logistics and deployment.[38] 'I wrote the memorandum upon the subject of the bridge of boats early in April, as soon as I saw a probability of a campaign in this country . . . The rivers will fill between the 14th and 20th of June, and at that time we ought to have the bridge in order to be able to carry on the war in any style.'[39] However, there were several major flaws in his monsoon strategy. Had anyone thought this through to the point of asking: what if there should be a monsoon failure? There was also an unforeseen problem with independent civilian entrepreneurs in the South Asian military economy. Small boats and ferries were found along many waterways and for a price the craft and their operators could be procured. How were they to be taken out of the picture? And then there were those often-overlooked questions regarding the Maratha military tradition. This 'nation of predatory horsemen' was, in reality, a diverse collection of professional soldiers; many of them being accomplished monsoon fighters and tacticians with infantry-relevant experience in military field engineering.

The view from Bengal

In reviewing Arthur Wellesley's military observations on the Marathas, one becomes aware of a regional bias in his reporting. That is not to denigrate his experiences or cast aspersions on his well-intended words concerning the nature of Maratha warfare. It is just that his personal combat experiences prior to 1803 were drawn from the southern half of the subcontinent. He had no first-hand experience in fighting Sindia's 'regular corps' or dealing with the type of fortresses that dotted the Gunga-Yamuna Doab in the north. His limited theatre experience is important in trying to reconcile the different emphasis that existed between his writings and those of his brother Richard Wellesley, whose governmental machinery was based in Bengal. As Governor-General, Richard's focus was tied to a broader view of India's as well as Britain's strategic security. To a great extent that meant Richard's concerns revolved around Calcutta in Bengal as the East India Company's 'in-country' nerve centre. Without presupposing that all the EIC's rivalries were economic rivalries, by virtue of the fact that this was a merchant company with imperial aspirations, let us consider why Bengal had an economic as well as political interest in what the Marathas were doing in the Doab.

In 1803 the Gunga-Yamuna Doab was still seen as one of South Asia's most highly coveted economic regions. It had, since Vedic times, been the economic breadbasket of India; a cornucopia in the sense that it was more than just a source of grain and/or cash to be extracted in the form of taxes. The Marathas could threaten the area with the resumption of large-scale *chauth* raids as had been done decades earlier during the administration of the *Peshwa* Baji Rao I. But by the 1790s Mahadji Sindia had seen that such a retrograde policy was counterproductive in the Doab. A marginal opium-poppy-growing area in Bihar, a desert entrepot in

Rajasthan, a fishing village in Konkan, could all be forced to pay *chauth*. The amount of tribute money they begrudgingly handed over and the regularity with which they could be milked varied considerably owing to prevailing local conditions. However, the Gunga-Yamuna Doab held greater potential and a faster turnaround time as a renewable revenue resource because of the regularity and predictability with which food crops could be grown, taxes assessed and then collected. It was known for the cyclical dependability that had been used to build classical empires like those seated at Kannauj.* The food crops, as commodities, held a dual value to military powers like the Marathas. In times of war entire fields of grain could be written off against tax ledgers, harvested, stored and finally shipped by *banjaras* to keep vast armies in the field. Or the crops, accepted as payment in kind, could be sold on the spot market to provide cash that could be directly spent on military endeavour. But there was a certain economic sense in permitting the inhabitants to grow their crops, sell them and pay their taxes, so that those tax revenues could be collected in cash from the *zamindars* (landlords). Why not let those middlemen expend their energy on individual cultivator collection? The economic and tax-generating predictability of the region had been historically consistent.[40]

No Maratha leader in 1803 could hope to reinvent himself as a Shivaji figure, meaning no Maratha *sardar* in 1803 could rise to become a warrior king. But then Shivaji, even at the peak of his political and military power, could never lay claim to the Gunga-Yamuna Doab. The great Shivaji had to seek political accommodation with the Mughal Emperor Aurangzeb. Was there not greater wealth, and arguably greater political power as well as prestige, for a clan leader like Daulat Rao Sindia to be the puppet master of the Mughal Emperor? As long as Sindia served as the Mughal regent, whether by official appointment or *de facto* military status, his clan could lay claim to the control of Delhi, the Doab and the Emperor himself. That meant the Marathas held greater political legitimacy than the British in North India. Shah Alam[†] II was portrayed in numerous British reports as evoking the utmost sympathy from those officers and dignitaries who visited him. Having been blinded by the Afghans during an earlier captivity, living a pitiful existence on a niggardly Maratha pension, the old man allegedly had only the joys of poetry to sustain him between the rare visits by genuine well-wishers. If Richard Wellesley wanted to win support from London for a campaign against the Marathas in 1803,

---

* 27°+, 79°+; Aka Kanauj, Kanouge, Canouge. The seat of a great empire under Harsha in the seventh century.
[†] Aka Shah Alum.

then he had to portray the Marathas as holding India hostage in the form of the Mughal Emperor.[41]

Within the parameters of British political strategy in 1803, Shah Alam II could be 'liberated from the Maratha yoke' but he could never be restored as an independent ruler. There were greater gains to be had in freeing him from Sindia and then serving as his benevolent guardian. That was not only useful in terms of credibility among subject peoples but it helped keep an entire series of contractual agreements in place when it came to revenue collection and military service. If the Gunga-Yamuna Doab was to be acquired by the EIC, then it was important to maintain the continuity and facade of a Mughal presence. To push aside, overthrow, or reject the Mughal political infrastructure was to appear decadent, foreign and as an opponent of South Asian cultural sensibility as well as tradition. Why increase resistance and chances of non-cooperation among existing stakeholders? The old Mughal political and economic infrastructure had its deficiencies but a thinly stretched EIC, drawn further out along the Gunga-Yamuna corridor, needed to assume an existing revenue and control system to meet its needs until gradual transition could be implemented. And to ensure control of the Doab it would be necessary for the British to keep the Delhi–Agra axis intact once the Marathas were removed. Since the time of the great Mughal Emperor Babur, that axis and the lands that bordered it were known as the *key to Hindustan* in terms of the two things that counted – money and power.[42] There was ultimately more to be had in exercising patience when it came to the process of watching over the demise of the Mughal power. The EIC would ultimately derive greater benefits in sitting by the bedside of the Mughal Empire and waiting for this imperial entity to die a lingering, though nonetheless decisive and seemingly natural death.[43]

Major Thorn later wrote succinctly about the importance of the Doab campaign as seen from the vantage point of Richard Wellesley's Bengal. But in doing so he injected an overriding note of pathos as if to offer a human distraction to counter those who still questioned the legality of Richard Wellesley's actions.

The importance of this branch of the war will be evident, from the objects to which the particular attention and energies of his Excellency were applied. These were, the destruction of the power of the French party established on the banks of the Jumna, under Monsieur Perron; the extension of the British frontier, in the possession of Agra, Delhi, and a chain of posts on the right bank of the Jumna . . . But the attention of the British government at this time was drawn in a forcible manner to the abject condition of the aged emperor of Hindoostan, over whose person and authority the Mahrattas and the French exercised an absolute control.[44]

British war aims and military objectives in 1803

The decision to go to war was not taken lightly in 1803 and there was a prolonged period of negotiation during which it remained uncertain whether Sindia and Bhonsle would stand down and acknowledge the political authority of the *Peshwa*.[45] Since it remained unclear whether there would be peace or war, Governor-General Richard Wellesley had to issue orders imparting dual authority. His brother Arthur was to be given full diplomatic as well as war-making powers in order to deal more directly with Sindia and Bhonsle over the issue of whether they would accept the *Peshwa*'s leadership under the Treaty of Bassein, or go to war.[46] Richard was still quite willing to negotiate but he made the alternative quite clear to his brother.

> It is probable that the state of the rivers will afford great advantages to your army, and will embarrass the enemy in a considerable degree, if hostilities should commence during the rainy monsoon. In this event, I direct you to use your utmost efforts to destroy the military power of either, or of both chiefs, and especially of Scindiah; and to avail yourself of every advantage, which circumstances may offer to the utmost extent of the strength of your army. It is particularly desirable that you should destroy Scindiah's artillery, and all arms of European construction and all military stores which he may possess.[47]

The Governor-General went on to tell his brother that in the event of war he should 'take proper measures for withdrawing the European officers' from Sindia's service and that Arthur was 'at liberty to incur any expense' for that purpose.[48] Richard Wellesley made it understood that, if fighting broke out, Arthur's actions should be aimed at neutralizing Maratha military power and bringing about the physical isolation of Maratha military threats. The vast amount of discretionary power was made necessary by the fact that the Deccan or southern theatre, as Arthur's sphere of operations, provided the primary trigger but not the primary objective of the war.

Although British generals had extraordinary civil as well as military authority in colonial theatres of war, the enormity of Arthur Wellesley's mandate seemed to exceed the boundaries of a judicious EIC administration; all this of course with the blessing of elder brother Richard.[49] Arthur was given sweeping powers because much of the territory that constituted the 'area of operations' was not yet officially under British legal control and nothing was to be left to chance with regard to the waging or conclusion of this war. It was one of those aspects of the conflict which suggest in hindsight that the successful prosecution of colonial warfare required the delegation of sufficient civil and military authority to those fighting the war. Taking the proper steps to document precedent as well as authorization was imperative. Richard Wellesley needed to deflect criticism, like that levelled at former Governor-General Warren Hastings, for waging an 'illegal' war of aggression.[50] Delays in communication, ranging from weeks to months, also meant that the granting of sufficient power was a sensible move. The degree of civil–military authority wielded by British theatre commanders in 1803 drastically reduced the delay, which an over-centralized command in Calcutta would have necessitated. Although it proved successful in this case, Governor-General Richard Wellesley was later subjected to criticism for having bestowed supreme war and peace-making powers – in effect governmental authority – on his brother.[51]

Richard wrote to Arthur, 'I will seize Agra, Delhi, take the person of the Mogul under British protection, and occupy the Doab, together with Cuttack, at the earliest practicable moment after I shall have learned that you deem hostilities inevitable.'[52] To a large extent Arthur was left on his own to work out the issue of sensible military objectives in the Deccan. The young Major-General had decided that his first blow would be an attack on the fortress of Ahmadnagar.[53] That would help secure the communication and logistics links between Bombay and Pune while providing a forward depot from which to launch operations against Maratha mounted units making incursions into either the *Peshwa*'s or *Nizam*'s allied territories.[54] As for destroying Sindia's regulars, that was much more problematic since the largest number of his battalions were in the North protecting his greatest assets.

When it became apparent that Daulat Rao Sindia and Raghuji Bhonsle would not willingly fall into a subordinate line behind the *Peshwa* under the Treaty of Bassein, war became inevitable. Governor-General Richard Wellesley outlined a very specific set of North Indian military objectives in a despatch to C-in-C General Lake:[55]

- Seize all of Sindia's land in North India between the Gunga and Yamuna Rivers.
- Take the Mughal Emperor out of Maratha protective custody and place him in British protective custody.
- Form an alliance system with the Rajputs, and other 'lesser states', in the area west of Delhi.
- Occupy the territory of Bundlekund.*

The proposed alliance system with the Rajputs and 'lesser states', if completed, would eventually provide the British with a defensive alliance system west of Delhi that we would recognize today as a bulwark of satellite states. These would not be initiated along the immediate lines of the

---

* Aka Bundlecund, Bundlekhand.

Subsidiary Treaty Alliance System that embraced the *Nizam* and *Peshwa* –
that would come later. This first step, in courting the 'country powers' of
the region, was intended to look like a mutual defence pact aimed at protecting Rajasthan from Maratha and Sikh incursions. However, history
would prove the British to be reluctant when it actually came to spending
their own blood and treasure to protect their new allies.[56]

The occupation of Bundlekund formed part of a grand strategy of containment as the territory lay astride one of the traditional Maratha invasion corridors. And although Daulat Rao Sindia's clan had focused their
attention on Delhi and its environs for two generations, the Marathas had
been known to use the Bundlekund corridor as a passageway to expedite
the flow of their troops towards Calcutta, which in 1803 still carried the
scar of a defensive fortification known as the 'Maratha ditch'. Any lasting
peace for the British would depend on hemming in the Marathas and
that meant denying them an ability to shift troops regionally from the
southern to the northern theatre. The Deccan–Hindustan–Bengal corridors through Malwa and Bundlekund had to be controlled if complete
containment was to be achieved. If left untouched, the Marathas could
exploit secondary road systems. The permanent safety of Bengal from
Maratha-sponsored invasion depended on being able to guarantee the
disconnection of the traditional invasion corridors that criss-crossed the
land.

Richard Wellesley also advised that 'the following circumstances would
require immediate attention':[57]
- the removal of conventional infantry forces under Daulat Rao Sindia's
  C-in-C General Perron,
- possession of the Maratha-controlled forts south of the Yamuna River
  so as to facilitate a strategy of containment and control of the Doab.

The troops under Perron would be offered a chance to defect and 'come
over' to the British. But if they did not exercise that option, then the
British theatre commanders were authorized to begin search and destroy
operations aimed at 'reducing' Maratha military capability. As for Richard
Wellesley's reference to forts on the south side of the Yamuna, India was
a hostile environment for the British in more ways than one. Climatically,
militarily and in terms of potential political adversaries, permanent control of the Doab would be problematic for the British unless they held the
area by way of a string of forts. The forts on the south side of the Yamuna
River, particularly the great forts of Delhi and Agra that formed anchor
points for the Delhi–Agra axis, held more than a defensive significance.
As Mughal imperial fortresses, both locations had developed as administrative centres and they might fall more readily to the British if 'liberation' of the Mughal Emperor was successful. Agra was also a well-known

weapons production centre. Another major consideration was the role of these two forts in guarding communications and logistical traffic on the Yamuna River that flowed east towards Bengal. The river would easily lend itself to a British transportation and communication system linking Delhi to Calcutta. It had become necessary to systematically acquire forts in the 1803 Doab campaign for three reasons:

- to neutralize main centres of military resistance which could not be by-passed without creating a potential threat to the security of British rear areas,
- to build a contiguous logistical chain that allowed the army to move forward from strength to strength,
- to serve as defensive strongholds in the event of strategic withdrawal or to exert power when adjacent urban areas rebelled or came under attack.

North India's Gunga-Yamuna Doab – the cradle of Hindustan – was the lynchpin for the greatest Maratha military challenge to British expansion in 1803. And, as can be seen from the evidence presented in chapter 4, this military high-water mark was entirely in keeping with infantry-dominant patterns of Maratha military culture. Sindia may have operated outside the Maratha political norm as seen from Pune but his army held the balance of political power in North India. The removal of Sindia's 'regular corps' was mandatory if Richard Wellesley was to prevail in plans to seize the Doab and eliminate opposition to the Treaty of Bassein. The bulk of infantry battalions as well as the most crucial fortresses, those that were the backbone of Sindia's power in 1803, were in the north. The military importance of the southern front should not be denigrated to that of a 'sideshow of the war'. But it can be stated categorically that the strategic importance of the Deccan was secondary to the elimination of Maratha military power in North India owing to the disproportionate political as well as economic value embodied in Hindustan and the remaining Mughal infrastructure.

Many young British officers in Arthur Wellesley's Army of the Deccan agreed with their opposite numbers in Sindia's service – Pune was not a place to die for in 1803. However, Arthur Wellesley was optimistic. He had fought Dhundia Waugh and won. His correspondence, right up to the eve of battle in 1803, reflected a great confidence that this 'nation of freebooters' could be brought to heel. And when the word arrived that Daulat Rao Sindia and Raghuji Bhonsle were not willing to stand down and acknowledge the *Peshwa*'s signature on the Treaty of Bassein, Arthur thought it was time to teach these 'predatory horsemen' a lesson. But there were others in the British defence establishment, men like Lieutenant-Colonel John Ulrich Collins, who had seen Sindia's infantry in action and remembered that the earlier Anglo-Maratha campaigns were not a walk-over.

Arthur Wellesley, in the company of several officers including the very observant John Blakiston, set off to meet with Collins. The observations offered by the aged Colonel should have been welcomed as valuable intelligence. After all, the old soldier had just departed Daulat Rao Sindia's court in his capacity as official British Resident. But the young British officers accompanying Arthur Wellesley were green in terms of experience as well as manners. They laughed at Collins' eccentricities although the wizened EIC soldier had seen more than his share of killing, having been cited for personal bravery under fire. An apparent victim of fashion, Collins' seemingly archaic mid-eighteenth-century uniform was accentuated with lace and elaborate piping, which appeared hysterically comical to the young know it alls. The sight of that 'costume' combined with his South Asian entourage – one of whom was designated to carrying a parasol – underscored the Colonel's nickname of 'King Collins'. For his part, Collins was respectful and courteous but he did not suffer fools lightly. Arthur and, for that matter, his brother the Governor-General in Bengal were too young and too ambitious in Collins' estimation. Nepotism was not a pretty sight and it was all the more visible in the smallness of British India. Collins had waded through Arthur's numerous letters on how to fight the Marathas' *pindari* armies. The Colonel tried to persuade Arthur that Sindia's army was different and the conventional threat that it posed was very real. A single one of Sindia's five brigades, when accompanied by its artillery park, packed enough firepower to maul Wellesley's Army of the Deccan. Upon the conclusion of their meeting, Blakiston heard Lieutenant-Colonel Collins address Major-General Arthur Wellesley with measured politeness – his promotion still rankled the old soldier. 'I tell you, General, as to their cavalry [meaning the enemy's], you may ride over them wherever you meet them; but their infantry and guns will astonish you.' The statement haunted Blakiston who confided in his memoirs: 'As, in riding home afterwards we amused ourselves, the General among the rest, in cutting jokes at the expense of "little King Collins", we little thought how true his words would prove.'[58]

3   The Deccan Campaign of 1803

Introduction

Many histories of this campaign religiously follow the story laid out in the Duke of Wellington's correspondence. This narrative refrains from over-reliance on those dispatches, instead seeking to use 'Wellington material' found in a variety of other locations. The following account differs from many other versions of the 1803 Deccan Campaign in seeking to build on chapter 1's observations about the greater South Asian military environment. Therefore, it should come as no surprise to see a secondary theme concerning Arthur Wellesley's continuing military education in the context of age-old South Asian military cycles – a cross-cultural view if you will. With regard to chronological sequence, chapter 2 led the reader up to the eve of war in August 1803. Chapter 3 backtracks slightly and opens in the spring of that year with Major-General Arthur Wellesley adapting to the reality of having to make a rather deep thrust into uncertain territory.

Our noble ally the *Peshwa*

Arthur Wellesley assembled the nucleus of his Army of the Deccan at Srirangapatnam in February 1803. According to the *Military Reminiscences* of James Welsh, then a captain in the Madras Presidency's army, there was precious little time for Wellesley to come to terms with the essential drill that his men would need before marching: 'he manoeuvred his future army, and taught us that uniformity of movement, which was afterwards to enable him to conquer foes twenty times as numerous'.[1] The troops then proceeded to Harihar.* Having rendezvoused with additional forces there, the immediate objective was to escort *Peshwa* Baji Rao II to Pune and oversee his reinstallation in the capitol. The return of the *Peshwa* to Pune was calculated by the British to send a political as well as military

---

* 14°+, 75°+. Aka Hurryhur.

Baroda

Narmada River

Baroach

Asirgarh
Gawilgarh

Burhanpur

Tapti River

Surat

Argaum

Nagpur

Wardha River

Ajanta

Assaye

Godavari River

Bassein

Ahmadnagar

Bombay Panvel

Pune

Bhima River

Hyderabad

Krishna River

Goa

Tungabhadra River

Harihar

Arabian Sea

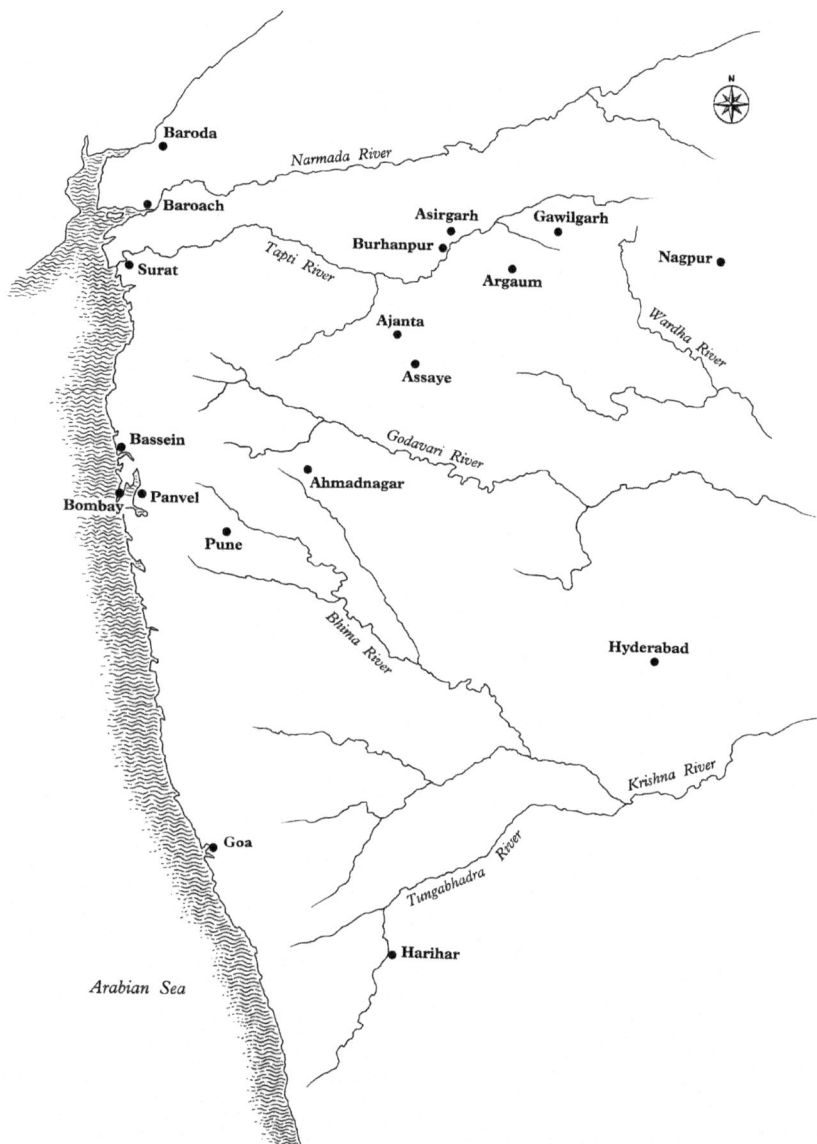

Map 4 An overview of Arthur Wellesley's area of operations during his
1803 Deccan Campaign

message to those Maratha leaders who despised the Treaty of Bassein. If the Marathas would not fall into line behind their Prime Minister, the British were poised to attack them.

For their part, some of the British officers on the march were still trying to reconcile the political process with the military reality of their situation. Was it wise to support such a weakling as Baji Rao II and would it not be better to abstain from involvement in these 'native disputes'? At the very least one might consider staying out of the situation until the on-going Sindia–Holkar rivalry had weakened one or both to the point of making this task easier. James Welsh proved remarkably frank in his assessment and admission:

> Very much in the dark with regard to Indian politics, we had naturally concluded, that as we came to succour the Peishwa, his friends would be our friends, and his foes our likeliest opponents; but here we reckoned without our host, for the man we were now to attack was not Holkar who had deposed him, but Scindiah, who had upheld him, and actually suffered a defeat, near Poonah,* in his cause! Having never troubled my head with the intricacy of state affairs, I have never learned the real cause of this war.[2]

Few British officers had yet to form a first-hand opinion of the *Peshwa* as a military ally. However, as the days passed on the march to Pune, more of the *Peshwa*'s men joined the column. To the chagrin of Wellesley and his staff, not all of the Maratha horsemen under the *Peshwa*'s General Gokale looked like gentlemen.[3] Some had served in the EIC's native cavalry and been discharged, but a number of others were obviously independent Maratha light horsemen accompanied by *pindaries* who had ridden and plundered for various masters in the past. Gokale himself had annoyed the British in 1799 by plundering on the Mysore–Maratha frontier.[4] Money, as is often the case, was the great fixer. And while the arrangement with Gokale was no more extortionate than many contracts within the South Asian military economy, it was on a grand scale that challenged British commitment to 'stay the course'. If one did not agree to set a price for the services of these Maratha light horsemen they would leave the alliance and plunder the lands of anyone they chose to – including their own *Peshwa*.[5] Many increases in pay allowances were justified with the rationalization that it was cheaper to retain these dubious Maratha horsemen at an inflated price than deal with them as *pindaries* if they were released from service. The threat of leaving British service became the new cash cow for Gokale and his cronies, effectively surpassing the rewards of *chauth* raids. It made some of the young British soldiers uneasy to think that there was no visible means to distinguish

---

* Sindia vs. Holkar in the Battle of Hadespar fought outside Pune in 1802.

the light horse auxiliaries of Holkar or Sindia from those of the *Peshwa*. What did the enemy look like? What did an ally look like? How could one tell the difference?[6] John Blakiston described how the column's military as well as civilian stragglers became prey for the Maratha *pindaries* that he called the 'Looty Horse', noting them as 'semi-thieves and semi-soldiers'.[7]

As the grand procession inched closer to Pune, the handiwork of Holkar's forces was evident from the large tracts of land that had been burned. Famine was to prove a particular problem in the dryness and desolation of Maharashtra in 1803.[8] The countryside was a tinderbox and the danger of fire was so great that Wellesley banned smoking cheroots for fear that their careless disposal could spark a grass fire that would eliminate the meagre amounts of fodder that remained standing.[9] The British 'Army of the Deccan' was logistically dependent on bullocks to carry its food supply as the lengthy column snaked its way along. But there was so little fodder growing that Wellesley's men were forced to feed those animals thatching from the roofs of deserted houses.[10] The soldiers' hellish impressions of the countryside were confirmed at the end of a day's march when they found temperatures inside their tents as high as 120 degrees Fahrenheit.[11]

At that time Holkar's forces, under Amrut Rao, still controlled Pune and they threatened to torch it rather than have it occupied by the puppet *Peshwa* and the British.[12] That Holkar thought so little of Pune as the *Peshwa*'s 'capital' is reflected in the fact that he had no qualms about burning it to the ground. As the British column plodded on during the third week of April 1803, intelligence reports arrived which stressed that the destruction of Pune was not an idle threat and that the cost in human, economic and political terms was too high. Cavalry were dispatched without infantry in a dash to Pune to save it from destruction. Holkar's forces under Amrut Rao withdrew to avoid being drawn into a holding action, which would have given Wellesley's main infantry force time to come up. By staging a strategic withdrawal from the region, Holkar's forces avoided dragging their leader into the war that would soon be declared. In a carefully orchestrated event, the *Peshwa* entered the city on 13 May 1803 with Arthur Wellesley's escort.

The cavalry dash to Pune was the stuff of romantic novels but militarily it was of no consequence to the outcome of the war. 'Saving' Pune, as part of the British effort to restore the *Peshwa*'s authority in 1803, had propaganda value. The city's loss would have been a political embarrassment to an interventionist guardian whose role included upholding the props of their ally. The puppet regime's symbolic retention of a traditional political centre meant something to the identity of the 'old order', but where were the retainers in the drama that was unfolding? To see the Pune of 1803 as the cultural, spiritual and political heart of Maharashtra is to retrofit a vision that would warm Tilak's heart.* But the Maratha 'polity' of 1803 was very much a warlord state with fortified strongholds and revenue sources well beyond the depleted heartland. Sindia's standing armies garrisoned a northern archway from Rajasthan to Awadh. That meant the real war, for Britain in 1803 India, was far from Pune. The Marathas knew that political and military value were two different things in the priority ranking inherently found in pragmatic military strategy. Holkar and Sindia had claims on Pune that conflicted with those of the *Peshwa*. But Sindia and Holkar retained identities and power bases outside the city and its districts. The *Peshwa* was not so lucky – he was tied to Pune and, regrettably for him, tied to the British.

Arthur Wellesley was compelled to make the cavalry dash to save the city so it could serve as the stage for the *Peshwa*'s performance in accordance with the British script. The threatened burning of Pune and, more symbolically, Wellesley's bid to 'save' it were easily spun into a thread that would be used to weave a rather mystical cloth. As in the European fairy tale known as 'The Emperor's New Clothes', this covering was illusory and the sceptics saw the *Peshwa* as politically naked. It was a fabric that provided political cover only to those who would believe; such as those who believed that the *Peshwa*'s British escort showed commitment to 'restore a native situation' that had spiralled out of control. When Captain Welsh reached Pune, with Wellesley's force, he remarked on the prosperity to be found. Welsh noted that Pune appeared 'a place of great wealth, and to concentrate all the trade of empire'. But beyond the travelogue observations, he had begun to question the veracity of the cover story that he and his fellow officers had also been asked to accept as an official explanation. He said the wealth of merchandise on display in Pune 'seemed to give a flat contradiction to the reports, which had induced the General to make a forced march'.[13]

Far to the north, some of Maharaja Daulat Rao Sindia's forces were reported to be heading south across the Narmada River.† Although very distant, some British intelligence sources judged the movement to be a posture that also threatened Pune. The British made ready in case Sindia and Raghuji Bhonsle combined forces in an attack on the city. If we look at the correspondence of political agents and senior soldiers during May–June 1803, we see that there was still a hope that a negotiated

---

* A reference to Bal Gangadhar Tilak, the Maratha Brahmin who advocated Maratha cultural revival at the dawn of the twentieth century.
† Aka Narbada River, Narmadda River.

settlement might prevail. However, negotiations did not halt the very real preparations for combat. Governor-General Richard Wellesley asked Lieutenant-Colonel Collins, Resident at Sindia's court, to see what the Maharaja's intentions were. When asked if he would offer opposition to the Treaty of Bassein, Sindia replied that he would have to consult with Bhonsle.[14] This was deemed an inappropriate response. The fear was that the request for consultation was a stalling tactic to buy time while Maratha troop mobilization could be completed. The supposed proximity of the Maratha forces to Pune was quickly being elevated as a reason for a pre-emptive British military action in the Deccan.

Arthur Wellesley and his army departed Pune on 4 June 1803 and headed towards the medieval fortress of Ahmadnagar, which had figured so prominently in Maratha military history. Sindia's regular infantry in the Deccan were thought to have been reduced to 10,000 or perhaps as many as 12,000. He accurately judged his holdings in the north to be the real target of the war and had reportedly sent 4,000 to 5,000 of his best troops from the Deccan, under Chevalier Dudrenec, to reinforce his 'Hindustan army' that was already believed to number 16,000 to 17,000 regular infantry.[15] Sindia and his remaining Deccani regulars were believed to be at Burhanpur, a fort with a position that would allow them the option of deploying against either of the British allies: the *Nizam* of Hyderabad or the *Peshwa* in Pune.[16] But if the regulars were not to be found in Ahmadnagar, why did it merit target priority?

In many British accounts the advance is portrayed as a clever forward deployment intended by Arthur to bring the greatest advantage if war were to break out. To give the Major-General his due, he had become something of a forward-planning specialist as evidenced by his drafts for a similar stance against Goa during the 1801 French invasion scare.[17] But in truth there was more to this plan than the input of Arthur Wellesley. The British had been compiling an intelligence file on Ahmadnagar for months and its target priority status was a foregone conclusion. Items such as scaling ladders were ordered in advance of an anticipated escalade.[18] In Wellington's *Supplementary Despatches*, there is a 'Memorandum On Ahmednuggur Fort' dated June 1803 and attributed to Arthur Wellesley. It specifically mentions intelligence gained from 'A man of the pettah'.[19] The *pettah* was the walled village that lay outside the fort beyond the moat.

Ahmadnagar was Sindia's possession but it was not a primary offensive threat to the British. Ahmadnagar held potential in British plans that sought to guard Pune while projecting power further into the Deccan. It was a logistical target that, if seized, could serve multiple purposes in the British strategy of containment. First, it was a well-fortified stronghold, which could be used for the storage of vast quantities of grain and war material if one had the capital and supply networks to deliver such commodities in a famine year like 1803. Second, a well-provisioned Ahmadnagar would lend itself readily to British efforts to defend and pacify the *Peshwa*'s as well as the *Nizam*'s territories. As a base for counterinsurgency operations against *pindaries* and light horse units, it was ideal. *Pindaries* might prey on *banjara* convoys headed for the fort. But they were powerless to roust out an infantry garrison with defensive artillery support. Third, it would make an excellent staging area for operations aimed at cutting the Deccani link in the historic Maratha invasion routes leading to North India. The routes led from Pune to Ahmadnagar to Aurungabad, from Aurungabad to Burhanpur through Malwa to Agra. At Agra they entered into the Doab where they picked-up the link to Bengal via the old Grand Trunk Road.[20] Fourth, if the Maratha situation became an all-out war, Ahmadnagar was a defensible position where British field units could rest in relative comfort while they recovered from the toll disease, climate and Marathas inflicted upon them.

Daulat Rao Sindia had wisely refrained from extensively garrisoning Ahmadnagar with his 'regular corps' because they held the potential to defeat British infantry in combat on an open battlefield. He realized Ahmadnagar's strategic strengths were not without tactical shortcomings.[21] Sindia knew that it would be a waste to station his best 'southern' brigade, under Pohlman, in a fortress where they could be cut off and starved into submission. And Perron, as Sindia's C-in-C, was not about to send any more troops to the Deccan while the riches of Hindustan were at stake. Using spies in Sindia's camp, Lieutenant-Colonel Collins was later able to confirm to Governor-General Richard Wellesley that Sindia 'had received intelligence of the design of the Hon'ble General [Arthur] Wellesley to commence an immediate attack'.[22] In the absence of a full-time garrison from the 'regular corps', Ahmadnagar's main defence force was composed of 1,200 Arab mercenaries.[23] The ethnicity of those mercenary troops was not in the least bit unusual in the context of the South Asian military labour market. Arab mercenary troops had been a regular feature of Deccani armies, at least since the sixteenth century, and they were still to be found in service with the *Nizam* of Hyderabad as late as 1908.[24] Arab mercenaries had a good reputation as marksmen skilled at aimed fire using their own particular long-range matchlocks.[25] They were also known to excel at defending fortifications. While viciously capable on defence, those in Ahmadnagar – without supporting cavalry acting in a combined role featuring aggressive patrolling and logistical interdiction – could easily be surrounded and their supply lines cut. Too much of Ahmadnagar's defence had been left to the thickness of its walls rather than a proactive defensive strategy using combined operations in the style of Shivaji or his father.

Word of the failed diplomatic negotiations with Sindia reached Arthur Wellesley at Walki* on 6 August 1803. The British force had camped there, en route to Ahmadnagar, to wait for a break in the monsoon rain, which happened to be unexpectedly heavy at that time.[26] Sindia and Bhonsle failed to either support the Treaty of Bassein or withdraw their troops from what were deemed to be sensitive areas. Some EIC officials thought Sindia would sue for peace at the final moment. But there were great cultural differences in threat perception. If Sindia 'backed down' it would send the wrong message to his old rival Holkar. And as we shall see later in this study, appeasement of the British was detrimental to the balance of power within Daulat Rao Sindia's own *durbar*.

Confident that all his supporting troops were in place, Arthur Wellesley ordered his army's offensive deployment to commence against Sindia on 6 August 1803.[27] Siege planning had been completed and Wellesley and his force arrived at Ahmadnagar on 8 August just as a letter requesting surrender was delivered to the fort's *killadar*† (commandant). The proposal was refused, as was a subsequent offer of protection to the *pettah* outside the fortress, if it were to accept occupation.[28] The 1,200 to 1,500 Arabs in the *pettah* were by then supported by a body of some 3,000 'Marathas' according to Captain Welsh. Presumably the latter were Maratha light horsemen who were noted by others as having been camped between the town and the fort, but they soon dispersed.[29] Captain Thorn of the 25th Dragoons, who was not in attendance owing to his presence in the northern theatre, was alone among the veterans of the war in saying a battalion of 'Scindiah's regular infantry' were encamped with the horsemen.[30]

Arthur Wellesley set his plan in motion with an initial attack on the *pettah*. It would provide ideal cover and concealment for the second stage of operations, which would be the actual siege of the fortress itself. The story of the brief struggle for the *pettah* is often accompanied by a quotation attributed to the previously mentioned Maratha General Gokale: 'These English are a strange people, and their General a wonderful man: they came here in the morning, looked at the Pettah wall, walked over it, killed all the garrison and returned to breakfast! What can withstand them?'[31] But in fact the assault was not that simple and it got off to a rather bad start. Neither field reconnaissance nor covert intelligence had revealed that this particular *pettah* had been constructed (for some unknown reason) without a rampart behind its exterior curtain wall. In other

---

* 18°+, 74°+. Aka Walkee.   † Aka *quiladar*.

words, there was no 'ledge' or 'catwalk' inside the top of the *pettah*'s walls where you would have expected to find one. Wellesley advanced his men in three columns of considerable size to make a three-pronged escalade.[32] The centre column used their scaling ladders to gain the top of the 18-foot high walls but this left them exposed to a blistering aimed fire from the Arab troops in the adjacent towers. Since there was no rampart, that meant no footing or place to step onto once a British soldier had crossed over the top of the *pettah* wall – just a straight drop down the other side to the earth below. British troops scampered up the ladders but found nowhere to alight. They began to *bunch up* as they reached the top of the precipitous wall – there was simply no clear forward objective. Welsh recorded that Wellesley lost fifteen men killed and fifty wounded as the British were driven back down the ladders and forced to withdraw under fire. Five of those dead were officers and they had been killed in the first ten minutes before their column's retreat to safety. Wellesley's enlisted men had been well served by officer leadership under fire but his officers had been let down by poor intelligence.

Artillery fire support from the main fortress of Ahmadnagar, directed towards relieving those besieged in the *pettah*, caused a British artillery elephant to run amok in one of the other columns, which added to the delay and aided the enemy's cause.[33] The men of the third column, which included Welsh, then reasoned that the scaling ladders had to be planted by one of the *pettah*'s bastions as only a tower of that sort would provide sufficient opportunity for the scaling parties to climb inside the walls. The incentive to speed up the ladder was great owing to the accuracy of Arab sharpshooters firing from the bastions. The almost comic aspects of the incident did not escape Welsh's observation. 'Captain Vessey was then a very stout heavy man; but what impediment, short of death, can arrest a soldier at such crisis?'[34] After some time two European companies had gained access to the town via the ladder up the bastion's corner, but they found themselves engaged in urban warfare as house-to-house fighting ensued with the defenders. A counterattack was made by a group of Arab troops against the grenadiers of His Majesty's 78th Regiment but they were repulsed and momentum shifted decisively in favour of the British. The confused fighting in the close-quarters of the old town did contribute to the *friendly fire* death of one British soldier as a party of troops were trying to batter their way through a gate as their comrades were attempting to open it from the other side. Wellesley's casualties, for the seizure of the *pettah*, were listed by Thorn as 30 killed in action and 111 wounded.[35] Grant Duff differed in saying 28 killed and only 22 wounded but he did go to some pains to point out that 6 of the dead were European officers.[36] In other words, roughly a fifth or 20 per cent of Wellesley's killed-in-action were officers, which did not bode well for his ability to maintain command and control of his enlisted men in future battles.

Arthur Wellesley reconnoitred the main fortress on the evening of 9 August while some of his troops seized another minor outer works within 400 yards of its walls.[37] During the night a four-gun battery was quickly assembled with the aid of Captain Johnson of the Engineers and Captain Heitland of the Pioneers. The Maratha artillerymen on the fort's walls fired illumination rounds or carcasses through the night.[38] These projectiles continued to burn brightly after they fell to earth. They were not intended as incendiaries in this case but rather to light up the British gun pits so that the Arab marksmen could snipe at the pioneers as they laboured to bring their batteries into service. As a precaution the British posted additional pickets that night to thwart Maratha foot patrols probing the lines.[39]

John Blakiston noticed that British engineers operating in India evolved their own field craft that did not strictly follow the accepted standard text for artillery battery construction as evolved by Müller.[40] They expediently modified the European practices by incorporating the human and geographic elements of the South Asian military environment. Blakiston, an engineer himself, observed that the availability of human resources in Indian war zones meant that many more labourers than usual were employed in construction, with human chains being formed to convey baskets of earth. This permitted some leeway in selecting the locations from which construction material was obtained, based on the relative ease with which preferred types of soil could be extracted and conveyed. As a result – or so Blakiston believed – maximum efficiency was derived from the engineer, as it was his expertise and not his muscles that were taxed. This efficient, though by European standards unorthodox, utilization of Indian 'resources' impressed Blakiston since it meant that an artillery battery could be erected in as little as one night. This was also facilitated with the use of sandbags, previously issued as empty to the men, which they filled in the field for building 'hasty redoubts'.[41]

On the morning of 10 August 1803 Wellesley's battery opened fire. The British plan was to initially use artillery to neutralize the Maratha guns mounted on the wall that had been selected for breaching. Captain Welsh indicated that there were approximately sixty pieces of ordnance in total mounted along the walls and on the bastions. These ranged in size from 12 to 52 pounders.[42] The most noteworthy piece was the tower-mounted *Maha Laxmi*,* a bronze gun that measured 22 feet in length

---

* Aka Mahalachmi, Mahaletchmee.

and delivered a 17-pound ball. Once the guns had been silenced, above the location selected for attack, the process of weakening the walls with sustained fire could begin. But the firing soon caused the fort's *killadar* to request a cease-fire for negotiations. However, Arthur Wellesley refused a cessation of the bombardment until hostages had been delivered to the British camp.[43]

The surrender was timely as the British were short on ammunition for their breaching guns and no one had worked out how to attack the fort with scaling ladders that were too short for use inside the fort's ditch.[44] There was also the crucially underreported problem of how to plant them successfully in what was estimated to be ten feet of water. Wellingtonians continuously downplay the depth and historic water capacity of the ditch, which was obviously underestimated in the June 1803 'Memorandum'.[45] Suffice to say Major-General Wellesley saved a great deal of time and effort by falling back on a much more traditional means of attaining his goal – a negotiated financial settlement. Arthur was learning that virtually everything of military value had a price in the context of the South Asian military economy.

On 12 August 1803, the Maratha forces in Ahmadnagar capitulated to the British and the garrison marched out. Many of the 1,400 troops took up plundering about 70 miles from Ahmadnagar where they were involved in an action about a month later at 'Kurjet Koriagaum'.[46] Maratha historian G. S. Sardesai wrote that the original commandant was a European who had been bribed and defected at the commencement of the British attack.[47] He further stated that the fort's Brahmin *killadar* had no choice but to seek terms of capitulation. Jadunath Sarkar, relying on the extensive Marathi-language records in *Aitihasik Lekh Sangraha* edited by Khare and Khare, added some detail. 'In Ahmadnagar fort there was an Englishman in Sindhia's *paltan*; he turned traitor and delivered the fort to Wellesley. The *quiladar*, Malhar Kulkarni of Chambhargonda, made terms to save his own property.' 'The two battalions in the garrison turned disloyal. The English seduced the artillerymen, hence there was treachery.'[48] In an interesting post-script Sardesai wrote that the *Peshwa*'s flag was then flown over the fort to conceal its 'foreign conquest'. Regardless of which version of events one chooses to follow – British or Maratha – it should be remembered that it was the *pettah* at Ahmadnagar and not the fortress proper that fell to Wellesley by escalade.

The loss of war materiel in Ahmadnagar was a blow to Sindia's Deccani infantry units. But the fortress also held one of Sindia's 'palaces' stocked with luxury goods such as silks, satins, furs, cash, telescopes and 'two electrifying machines'.[49] The abundance of goods proved too much of a temptation for Wellesley's troops and looting broke out. Captain Welsh had posted European sentries to prevent just such an occurrence but they too joined in the orgy of plunder.[50] Arthur Wellesley had seen this type of behaviour before when Tipu's capital of Srirangapatnam fell. His less than sympathetic but obviously biased attitude towards re-establishing control was evident in the general order (GO) he issued. 'A troop of native cavalry to be sent immediately to drive all the camp followers who are down near the fort of Ahmednuggur back into camp, and to prevent any more of them from going towards the fort. The troop above ordered is to cut down any follower who does not instantly retire to camp.'[51] This makes it quite clear that native troops under his command would be used to summarily execute any South Asian camp followers who disobeyed the order to disperse. The same GO continued in a moralistic tone: 'Major-General Wellesley is convinced that there is no good soldier in this detachment who would infringe this capitulation . . . and he will punish with the utmost severity any person who may be found plundering in the fort of Ahmednuggur.'[52]

Wellesley then personally ordered two sepoys to be hanged in a bid to regain control of his men and leave no doubt, regarding discipline, in their minds.[53] Welsh, who as an EIC officer sympathetic to his Indian comrades, noted in his work, 'two of our Native soldiers were instantly seized and hanged, in the only gateway, *in terrorem*; though the Europeans escaped'.[54] Major R. B. Burton's 1908 official Intelligence Branch account of Wellington's campaigns quotes an eye-witness to the hanging as saying: 'a measure which it must be confessed created some disgust at the moment, but which, at the outset of the campaign, was perhaps a necessary example for the sake of discipline, and a proper indication of the British character for justice and good faith'.[55]

The fortress was now to serve as an EIC administrative centre as well as a military post. Captain J. G. Graham was appointed as the Collector of Ahmadnagar District that held some of the Sindia clan's oldest traditional revenue-producing lands, which predated the clan's hold on the Doab.[56] Armed revenue-collection units (*sebundi*) were quickly dispatched to the dependent constituencies, which were to be considered as immediately within the Company's jurisdiction.[57] The fort's grain and foodstuffs were highly valued owing to the soaring market prices that prevailed in the famine districts. While some of the fort's military stores were directly incorporated into EIC inventories others were doled out to the *Peshwa* and the *Nizam* to relieve equipment deficiencies that they suffered. Arthur Wellesley also made sure that artillery officers fixed a 'valuation' on the captured Maratha cannons so that his Deccan Army's prize fund could be fattened.[58]

After having garrisoned Ahmadnagar with loyal troops, Arthur Wellesley led his army across the Godavari River.* Wellesley headed north towards Aurangabad[†] while Colonel Stevenson and the *Nizam*'s Subsidiary Force was stationed at Jafarabad[‡] northeast of the city. The object of their actions was to intercept the Maratha light horse units that had entered the *Nizam*'s territory.[59] Daulat Rao Sindia and Raghuji Bhonsle had sent their *pindaries* south after penetrating the Ajanta[§] hill range. The Maratha horsemen slipped easily between Stevenson and Wellesley who, although they had worked in tandem since the days of the Dhundia Waugh campaign, had trouble coordinating their movements on anything but the most rudimentary of terms. Arthur Wellesley heard criticism about this 'division' or splitting of his forces for several months. Several experienced veterans considered it a foolish risk of command and control capability in a land notorious for poor communication as well as deficient intelligence. Thomas Munro, among others, thought it imprudent to divide an army that could conceivably become engaged and defeated on a piecemeal basis. With almost prophetic timing, Wellesley's tag-team partner found himself scurrying to ascertain enemy light horse positions. Having been outflanked, Stevenson became tied down in fending off the predatory attacks by Sindia's *pindari* foragers who sought out EIC convoys to sustain their rapid and light-footed thrust into the region.[60]

Wellesley reached Aurangabad and received reports about the Maratha movements to the southeast. He feared they would cross the Godavari River, which he now realized was easily forded by any number of means during the diminished monsoon of 1803. Once across the river the light horsemen could make a rush towards the *Nizam*'s capital of Hyderabad. In an attempt to reposition himself between the Marathas and Hyderabad, Wellesley dashed south to the Godavari again and hugged the north bank as he followed the river eastward in his advance to contact. Mountstuart Elphinstone, a key member of Wellesley's staff, expressed optimism in his private letters that Hyderabad would be saved from *pindaries*. But he also knew there were reports that Sindia had fresh infantry somewhere in the Deccan.[61] Was this really just as Arthur had predicted, a nation of predatory horsemen who melted away under the pressure of a sustained drive? Arthur Wellesley's aggressive patrolling tactics limited the Marathas' *pindari* option. The Marathas returned to a position north of Jalnapur.[||] Grant Duff, who had unprecedented access to the Maratha papers of the Bombay Presidency as well as interviews with many of

---

* Aka Godavary River, Godavery River.   [†] 19°+, 75°+.
[‡] 21°+, 77°+. Aka Jaffeirabad.   [§] Aka Adjanta, Adjuntee, Adjante.
[||] 19°+, 75°+. Aka Jalna, Jaulnah, Jaulna, Jalnapore.

the campaign's veterans, indicated that Sindia and Bhonsle were not in agreement on strategy.[62] Lacking relevant military experience against the British at the theatre level, the two Maratha leaders seemed torn over the type of resistance to offer. There were allegations that Sindia preferred regular infantry engagement while Raghuji Bhonsle supposedly favoured more 'predatory' mounted operations. This was a perfectly logical division of opinion reflecting the respective strengths of each Maratha leader.

One additional benefit to Wellesley's southern movement was the protection of two convoys of grain and treasure sent to Ahmadnagar by Lieutenant-General Stuart from Mudgal.*[63] *Pindari* survival depended upon mobility, which often meant freedom from cumbersome convoys of *banjara* grain merchants. When opportunities were lost to plunder enemy convoys it meant that food and fodder had to be plundered from the villages in the area. However, like many parts of Maharashtra, the small villages here were often constructed with stout walls.[64] In several instances *pindari* looting had alienated the local population and civil defence had emerged as a by-product of cyclical plundering. Local resistance could waylay the rapid sweep of horsemen if they succumbed to the temptation of trying to take fortified villages. But if *pindaries* wanted to take a *sebundi*-defended walled village they needed artillery. The smaller bands of Deccani *pindaries* travelled light most of the time, with nothing larger than smallarms or the odd camel gun. This meant they had to skirt fortified villages, as it was too risky to waste time and energy on defended hard targets. A well-directed fire from *sebundi* served to deter half-hearted efforts. Captain William Thorn noted the degree to which popular resistance had become galvanized when he wrote: 'the spirit of the people was generously roused against them, and their parties were in many places defeated by the peons, or irregular infantry, stationed in the different villages for the collection of the revenue'.[65] Major John Malcolm saw the evolution of fortified towns in a philosophical light. He reflected on their prevalence in a Deccani environment plagued by predatory horsemen and commented: 'Every evil produces its own remedy.'[66]

Local villagers were loath to surrender the meagre crops they had harvested. Years of campaigning in this area by Sindia and Holkar meant there were no surpluses. To resist meant the possibility of mutilation or death, but to surrender food supplies during the famine of 1803 meant sure starvation for the entire village. The famine was so severe that it depopulated portions of the region. In the *taluk* (subdivision) of Nevasa, in Ahmadnagar District, only 21 villages out of 180 were listed as

* 16°+, 76°+.

inhabited in the 1803–4 census.[67] The district records for Ahmadnagar reveal that between 5,000 and 6,000 people died of starvation in 1803. An emergency-feeding programme, rushed into operation by Arthur Wellesley at Ahmadnagar, was reckoned to have saved 5,000 souls. Many of those refugees had flocked in from out-lying districts. They were keen to accept manual labour assignments in exchange for the grain that Wellesley generously offered as payment. It was a mutually beneficial relationship in the South Asian military economy. The Major-General needed to strengthen the fortress and the refugees needed to eat. If they had been paid in cash they would have had little purchasing power in local bazaars where hyperinflation had caused grain prices to skyrocket. With their help Wellesley restored the damage to the fort and then had a glacis* built around Ahmadnagar to augment the existing fortifications.

Seeing that there was little to be gained by leaving their light horse-men in the area, Sindia and Bhonsle ordered their men to pull back and regroup north of the Ajanta Pass.[68] From there the pindaries could be put to better use in screening the movement of Maratha infantry moving south as well as raiding Wellesley's communications and logistical lines. The arriving pindaries told about their encounters with the British. Sindia then decided to commit his infantry regulars to battle since light horse tactics were proving increasingly ineffectual. However, with the move-ment of Sindia's regular battalions south of the Ajanta Pass there was no turning back. As early as July 1803 Barry Close, the British Resident with the *Peshwa* at Pune, had written to Collins, the Resident with Sindia, lay-ing out the scenario for main force engagement. 'I cannot but adhere to the opinions which I have communicated to the Hon'ble Major-General Wellesley, namely, that if Sindia venture to pass the Ajunta Ghat, he ought to be attacked by the combined forces under Colonel Stevenson.'[69]

The march to Assaye†

The spirits of Wellesley's sepoys sagged as they marched further into the interior. Their particular mission was now fully transformed into a 'search and destroy' operation aimed at the 'Maratha enemy'. But the only signs of opposition were fleeting glimpses of pindaries who made life miserable by harrying the army on the march. While Wellesley's despatches paint a picture of Arthur as logistical genius, Elphinstone's letters make it clear that by mid-September 1803 the Marathas' pindaries were able to prey effectively on the Major-General's foraging parties. They even had the

---

* A gentle bank or slope devoid of cover for the attackers.
† 20°+, 75°+. Aka Assye, Asai.

audacity to rob 250 of Wellesley's Mysore horse while foraging – disgracing them by relieving them of their mounts. The incident underscored the intelligence vacuum in which the British operated. Elphinstone admitted he did not know where Sindia's forces were but that the *pindaries'* presence suggested the main enemy force could not be far away.[70]

The sepoys of the Madras Presidency, under Wellesley, were far from the comforts of their South Indian homes. The supply of 'Mangalore red' and brown rice, dispatched through the Bombay Presidency, was nourishing but not appetizing to those accustomed to white rice.[71] Stevenson and Wellesley's correspondence noted a particular quest for *basmati* white rice. An 'alien' diet, a hostile environment – several discouraging elements combined and morale suffered. The regimental order books revealed an increase in cases of unauthorized absence before the Battle of Assaye. The growing proximity of a major confrontation apparently did little to encourage any devotion to duty. Arthur Wellesley presided over the court martial of several Madrassi troops for desertion. Shaik Daud of the 1st Company, 1st Battalion of the 10th Regiment Madras Native Infantry (1/10th Madras NI), was one of those sentenced by Wellesley to be shot.[72] This most drastic of sentences was not to be ruled out despite a shortage of trained men. Wellesley apparently considered it as a valuable lesson in discipline to his remaining troops, one that outweighed the value of a handful of sepoy lives. Mohamed Ishock, who served with Shaik Daud, deserted between the nights of 9 and 10 September 1803. Wellesley could have sentenced Mohamed to death under the Articles of War, but the General 'let him off' with a mere 1,000 lashes. It is no wonder that the British soldiers, both Indian and European, feared their commander as much as the enemy. It was this type of motivational factor that contributed to their amazing performance at Assaye.[73]

After Sindia and Bhonsle had united their infantry and cavalry forces at Bhokardan* on 21 September 1803, their campsites dotted the landscape as far away as Jafarabad to the east. In the meantime Arthur Wellesley's forces had linked up with Colonel Stevenson on that same day at Badnapur.† Wellesley and Stevenson had decided to maintain the separation of their forces and act in concert as they moved north in parallel fashion to effect a pincer movement designed to bring the Maratha main force to battle on 24 September. In accordance with this plan, the two British officers set off on the morning of 22 September with Stevenson's detachment taking the 'left' while Major-General Wellesley and his men took the 'right'. Wellesley's path, and at times it was literally that,

---

* 20°+, 75°+. Aka Bokerdun.　　† Aka Budnapur, Budnapoor, Budnapore.

was intersected by a low belt of east–west running hills tapering off to an undulating plain in Bhokardan District, northward – above his position.[74]

On 23 September Wellesley proceeded to Nalni* having put in a lengthy march of 21 miles that day. Some authors say that two *banjaras* were then intercepted on their way to sell grain in the Maratha camp. And that upon questioning they revealed that the combined armies of Sindia and Bhonsle were much closer than Wellesley had thought. Other sources have it that the Major-General was surprised to receive information from his *harkarrahs* that the Maratha main force was camped at a location believed to be only 6 miles away. This intelligence failure may have been the result of an inaccurate cultural reading of the greater South Asian military environment.[75] Arthur Wellesley erred in counting on the *harkarrahs* who had travelled with him up from Srirangapatnam. It was not that they were unqualified but the fact that they were from Mysore that rendered them worse than useless for the purposes of obtaining local covert intelligence. They spoke an alien language and could not travel without attracting unnecessary attention. Being regionally distinct southerners they were immediately identifiable to the Marathas, as glaringly distinct from the local population as any European. It was easy for Maratha sympathizers to feed the Mysore *harkarrahs* disinformation since they were quickly spotted as 'Wellesley's men'. This forced him into relying on local *harkarrahs* whom he did not know.[76]

Twenty-four-year-old Mountstuart Elphinstone, who served Arthur in the triple capacity as Marathi interpreter, secretary and EIC executive intelligence officer, provided some of the most revealing comments.[77] His observations are crucial to our understanding of the tactical intelligence failure in the Deccan.[78] Elphinstone's correspondence reveals that he agreed with Wellesley in that both men felt they had been saddled with an ineffective and dangerous policy in the form of the *harkarrah* system.[79] The 'native intelligencers' brought with the army from Mysore were a liability in the Maratha country.[80] To make matters worse, Elphinstone used some spies supplied on the recommendation of Colonel Close's Brahman advisors in Pune. These men failed to infiltrate the inner workings of Sindia's camp, being relegated to the menial work of grasscutters, and it was feared at least one was hanged as a message to others. Elphinstone commented that he thought these spies were especially selected 'for their stupidity'.[81] But Elphinstone was also part of the problem when it came to the failure of Arthur Wellesley's field intelligence. Elphinstone had found it much easier to understand the 'Hindustani'† language he had

---

* 20°+, 75°+. Aka Naulnair, Naulnye, Naulniah, Nulni.
† Identified by Colebrooke as the language we have come to know as Urdu as opposed to the modern Hindi.

originally encountered in North India during his first years in India. His fallback or contingency language was 'Moors' (Persian) while Marathi was his weakest language, which he *spoke* with great difficulty in 1803. By his own admission his study schedule called for only spending one hour a day on intelligence, two hours a week on Persian and four hours a week on Marathi; his 'most troublesome appointment' was the one hour a week allocated to being 'Interpreter of all tongues'. He later confessed to Strachey:

knowing that I must interpret, whether well or ill, and not having much anxiety about my reputation as a Hindustanee, I interpret quite coolly, and have the use of all my senses and all my language. But my stock of Hindi is really too small. I cannot readily understand all that is said to me, much less say all that I ought to express. I mean in talking to Mahrattas, which is my common employment. I even find a difficulty with Deckanee Mussulmans. Their words, their song, and their phrases are so different from the Hindustani of Gilchrist,* that he is of no use to me.[82]

An exasperated Wellesley, dealing with *harkarrah* disinformation, would later write: 'Allow me to assure you, my dear Sir, that these hircarrahs are not to be believed: they never bring any intelligence that is worth hearing, and when they circulate their false reports they do infinite mischief to our cause. I shall be obliged to you if you will be so kind as to desire your hircarrahs to confine their reports to yourself.'[83] Blakiston thought it unlikely that one could get the 'truth' from a paid professional source intelligence system that rewarded the delivery of information *per se*.[84] The system, once adopted, was fraught with hazard. General Stuart, C-in-C of the Madras army, warned that it might be unwise to discharge any of the *harkarrahs*, after the British had retained them, as they had seen the internal workings of the Deccan army at the staff level.[85] The *harkarrahs*, a supposed intelligence asset, held the potential to become an intelligence liability.

Although Captain Welsh was not present at the Battle of Assaye, having been sent to Ahmadnagar, he recorded a somewhat different version of events leading to the 'discovery' of the Maratha army. He noted that Wellesley's forces had captured two men on horseback. They revealed that the Maratha 'regular corps' was camped nearby but the light horse units accompanying them were preparing to move off. More startling was the news that rather than being 12 miles off, as Wellesley apparently thought, they were only 5 miles away from him.[86] Wellesley saw an opportunity for a pre-emptive attack that would be lost if he waited a day to link up with

---

* Gilchrist was used, at that time, as the EIC's standard translation dictionary and transliteration guide.

Stevenson. Better to risk an immediate action against the unsuspecting Maratha infantry than to hesitate in such a manner that would allow their light horsemen a chance to interdict Stevenson's approach and thwart any British union of forces. If Arthur opted to withdraw, the light horsemen might block his line of retreat and force him to go to ground in an unfavourable location not of his choosing. The real threat in that scenario was not the Maratha horsemen but the danger that it would allow time for the Marathas to reposition their infantry and guns. Wellesley placed his baggage and stores in the town of Nalni and after posting a battalion of sepoys and 400 additional men from the native ranks to guard the baggage, he set off to engage the Marathas.

Wellesley apparently believed that if he did not go on the offensive, at that moment, then he risked losing the opportunity for victory and inactivity would certainly jeopardize the safety of his force. If he went into a defensive stance in the town of Nalni, while waiting for Stevenson, the Marathas might simply withdraw northward into the Ajanta range. If, on the other hand, Wellesley were to lead his entire force out to engage the enemy, and try and hold them until Stevenson arrived, he risked exposing an elongated baggage train to the depredations of the *pindaries*. Similarly, if he were to try and withdraw from the area, his rear guard and slow-moving baggage column would fall prey to the Maratha horse that would seek a holding action until their artillery and infantry could be brought up. In that regard, Wellesley's decision to leave the baggage guarded was a logical move intended to increase his mobility and reduce his target profile.

The Marathas knew Wellesley was in the vicinity but Marathi sources indicate Arthur's decision to fight came as a complete 'surprise'.[87] Sindia's men did not think Wellesley would attack until the following day (24 September). Previous Maratha intelligence reports, delivered earlier on the morning of 23 September, indicated Wellesley's advancing movements as well as Colonel Stevenson's location. The Marathas had made a decision not to take immediate action. The Maratha high command assumed that their English opponent would not attack with such small numbers, which were the result of Wellesley having split his forces with Stevenson. Surely the logical British move was to wait until Stevenson and Wellesley had rejoined their forces. Only a madman would risk an engagement of this magnitude with just one-half of his total available forces. The Marathas had further reasoned there was additional merit in waiting until Stevenson and Wellesley were reunited because then they could be finished off in a single battle that would make the job easier. The Marathas were correct in reading the dictates of preservationist logic, but incorrect in assessing Arthur Wellesley's fear of failure. His name and reputation were probably worth more to him at that moment than his life. Arthur had suffered a great loss of face among EIC personnel for a failed night attack in the Fourth Anglo-Mysore War (1799) while he was still a colonel.[88] He had come away from that experience a true believer in one mode of action when facing a South Asian opponent – attack, attack, attack. His belief in the need for immediate offensive action was further confirmed by his memories of the Dhundia Waugh campaign where his opponent had narrowly escaped at the river and Dhundia's death had only been achieved with a dramatic cavalry charge. At Assaye in 1803, we see Arthur Wellesley as the champion of an offensive battle strategy that demanded a direct infantry assault on Sindia's regulars and their guns; it stands in stark contrast to his later performance as the Duke of Wellington at Waterloo in 1815.[89]

The standard story has it that Wellesley rode out and stopped some-where about a mile south of the Kailna River* to view the Maratha encampment that stretched before him.[90] If one considered all the Maratha light horsemen, camp followers, logistical personnel and auxil-iary *pindaries*, it could be said to have stretched for upwards of 6 miles. But the time had come for Wellesley to deal with Sindia's regular infantry and that portion of the camp was fairly compact and readily recognizable. Bhonsle's forces were positioned nearer to the centre of the Marathas' ex-tended camp and they were largely superfluous to the action that was about to unfold.[91] Sindia's 'regular' or sepoy battalions had selected a camp area at the eastern extremity of the main position. They were readily identifiable by appearance, demeanour and separation within the Maratha camp line. This battle is often said to have pitted Wellesley against 'the entire Maratha army', but this action focused on what was the main force threat – Sindia's sepoy battalions. They were nestled in the fork formed by the Kailna River and the Juah River† which both ran roughly west-to-east.

Wellesley and his men reached the site of the Maratha camp at Assaye about one o'clock in the afternoon. The components of Wellesley's force were listed as 1,200 European and native cavalry combined, 1,300 European infantry and artillery, as well as 2,000 sepoys. The total of the force under Wellesley's command, as listed by William Thorn, was approximately 4,500 men.[92] Thorn's account, which often passes as an official history, states that Sindia's regular battalions were 'under the com-mand of two French officers, named Pohlman and Dupont‡'.[93] Jadunath Sarkar, echoing the research of many before him, listed Pohlman as a

---

* Aka Khelna River, Kaitna River, Khaitna River.   † Aka Jooah River, Juha River.
‡ Apparently this is a reference to Dutch Captain John James Dupon.

Hanoverian.[94] The Marathas were reported to have had sixteen battalions of regular infantry composed of Pohlman's 6,000-man brigade, Dupont's brigade of 2,500 men and four battalions – belonging to that most capable of female military leaders the Begum Sumru – which added 2,000 more men. Welsh stated that a number of Maratha 'orderly books', which were later captured on the field of battle, indicated the Maratha regulars numbered 10,800.[95] There were also Sindia's irregulars and the troops of the Maratha artillery park. Last, but not least, there were by most estimates 30,000 to 40,000 irregular light horse and *pindaries* in the area stretching from Assaye to Bhokardan. Despite the temptation to dismiss the latter figures as exaggeration, the numbers were indeed attainable in that region and the Maratha force is usually rounded off to a figure of 50,000 Maratha troops of all descriptions.[96] We should, however, remember that 'troops' is a generous description as some *pindaries* were mounted horse minders who held 'loot ponies' to carry away the spoils and they were never destined for combat as such.[97]

Arthur Wellesley briefly summarized his opening moves:

> Although I came first in front of their right, I determined to attack their left, as the defeat of their corps of infantry was most likely to be effectual; accordingly I marched round to their left flank, covering the march of the column of infantry by the British cavalry in the rear, and by the Marhatta and Mysore cavalry on the right flank.[98]

Sindia's battalions were occupying what some have termed a 'tongue of land' or we might say a small *doab* tapering into an eastern terminus (>). It should have been a tactical advantage to have the Juah as the northern camp boundary and the Kailna to the south. Looking southwards from the Marathas' vantage point, the left or eastern-most end of the Maratha line seemed safe enough since it was protected by the village of Assaye. Arthur Wellesley spotted the hamlets of Pipalgaon and Warur* adjacent to each other near the point where the rivers joined, which was off to his right as he looked in a northerly direction. This was to the enemy's far left as they were facing in a southerly direction watching Wellesley's approach. Arthur supposedly reckoned there was a safe means of crossing the river at that point owing to the closeness of Pipalgaon and Warur to each other, so he directed his men there.

In surveying the battlefield during the third week of September 1990, I took note of the riverbeds, depth of water, historic position of the village foundations and physical evidence of high-water marks. Indeed some buildings may have been closer to the water in that era as indicated by

---

* Aka Pipulgaon and Warud.

abandoned foundations. But as late as 1990 they still seemed perilously close to the water for such 'rolling to flat' terrain and that testifies to these streams not often being raging torrents in monsoon season. And we know that 1803 was a drought year. Many secondary sources indicate the riverbanks were too steep and this was the only place Wellesley could have crossed. That is incorrect. I made it a point to walk the banks of both rivers – above as well as below Assaye – entering both the Kailna and the Juah rivers at more than one place, crossing and recrossing both rivers by wading through water that varied from knee to waist deep in some holes. The banks of the Kailna certainly were generally accessible for men and horses although if crossing with cannon the fords would be fewer.[99]

It would seem Wellesley's crossing point at the hamlets was quite logical for another reason, which is blatantly obvious once you actually see the site. Wellesley's angle of approach was critical to concealing his exact intentions, purchasing time to bring his men up and perhaps, most importantly, denying the Maratha artillery a clear sight picture for aiming purposes. He could never have survived a direct northerly drive across the river head-on into Sindia's infantry. It would expose his entire column to Maratha artillery fire, in effect concentrating their target for them. While a parallel linear approach from the south would necessitate crossing the river on an extended front, a guaranteed disaster for coordinating the command and control of an assault line. Moving trained men across the parade square on-line is relatively simple but on the broken, irregular ground of a watercourse it would mean regrouping after crossing, halting and standing in formation on the enemy's side of the water, and waiting while the line reformed. The on-line crossing of a watercourse was also to be discouraged in the presence of enemy cavalry who could move through waist-deep water with ease on horseback. Even if they did not kill a man on the first pass, being knocked down into the water would jeopardize the immediate serviceability of a soldier's musket and paper-wrapped cartridges.

If Arthur Wellesley's force stood a chance of beating Sindia's regulars it was imperative to get across the water and onto the small *doab*, winding up beside Sindia's men on the same turf. Then, and only then, could he begin tactical manoeuvring in earnest. If you walk the approach and wade the river you can see he was doing the only thing he could – he was placing as much concealment as possible in the Maratha line-of-sight before advancing his force. The hamlet structures of Pipalgaon and Warur, barely mud walls with thatched roofs, would offer scarcely more than partial concealment for a force of 4,500 men. If the Marathas had opened up with 24 pounder iron round shot or 5.5-inch howitzers firing explosive shell, there would have been no effective protection from the hamlet's mud walls. But protective cover and visual concealment, as most infantrymen know, are two different things. And in trying to make the best of a bad situation, Wellesley was cooled-headed and correct in opting for every advantage of visual concealment he could find; anything to reduce his men's target profile was a good idea. Wellesley could have crossed in several locations but it would have made much more tactical sense to use the screen provided by the hamlets of Warur and Pipalgaon. The late Wellingtonian author Jac Weller took a reasoned approach to providing us with some idea of how big a target Wellesley's force presented before he reached the critical river crossing. 'A column of six and a half battalions moving on a front of half companies at quarter distance with fourteen guns would have been more than 1,000 yards long, probably about 1,400 yards in an area like this.'[100] It is known, from other examples in this battle as well as later in this campaign, that Wellesley's men made use of irregularities in the ground to provide cover and concealment while waiting.

Wellesley by this time was a seasoned veteran of river crossings in India – there was certainly no need of a pontoon bridge here. He knew that the natural tendency of troops about to wade into a watercourse is to slow down as they enter the water cautiously in order to 'feel' their footing underwater. Those who follow tend to press forward faster than their brethren can traverse the obstacle, with the result that there is considerable bunching-up. Keeping one's interval has been a concern since the age of classical warfare and the problem was accentuated with the introduction of artillery as clusters of men numbering in the hundreds make particularly attractive 'area targets'.

Contemporaneous Marathi sources indicated that the Maratha army was not expecting combat that day, let alone that moment, and some elements were not battle ready.[101] The long-range field and siege artillery was still limbered when 'Wasli Sahib' attacked. Although Wellesley's *harkarrahs* were a liability as noted, Marathi sources have been cited which concede an equal intelligence failure on the part of Sindia's and Bhonsle's men. They were caught unprepared to move the big guns while their artillery bullocks were out grazing.[102] The surprise approach by Arthur Wellesley left them able to bring only 80 to 90 of their guns into service, with the larger calibres remaining limbered and therefore unavailable for immediate response as the attack began. This would mean that 25–33 per cent of their firepower, 30 to 40 of the later estimated 120 pieces, were not used in the action. Jadunath Sarkar's analysis also implied that morale in the Maratha ranks was extremely low owing to pay being far in arrears and much hardship being inflicted by the famine.

He interpreted the Marathi records as saying the Maratha troops were 'starving' and in such a state of mutiny that they threatened Sindia's *diwan* with personal violence. Some units were apparently quite prepared to sit out the battle in a form of military labour protest. Sarkar went on to say that despite claims of Wellesley facing 'myriads' of Marathas, he only directly confronted five battalions.[103] Certainly there is a discrepancy between the number of South Asians found in the initially detected extended Maratha camp and the number of 'enemy troops' who actively engaged Wellesley.

The Maratha gunners were not going to let Wellesley cross the Kailna River unopposed and they opened a distant harassing fire, which was probably intended to slow Wellesley while the Maratha infantry formed up. Owing to the distance and nature of their British target, situated in the undulating clefts of the stream bank, they initially loaded iron round shot or, as they are popularly known, cannon balls, which held the potential of skipping if they grazed the ground. The object was to drop a grazing shot right among Wellesley's men. The Maratha artillery crews lobbed a few opening rounds towards the ford between the hamlets. As luck would have it, some of the first cannon shot found their mark. As one cannon ball grazed in, the rebound distracted the men who ducked. They looked up to see Wellesley's Brigade-Major Campbell knocked from his horse by a hit in the leg. Moments later another Maratha cannon shot scored a devastatingly hideous psychological victory. Wellesley and his immediate staff were supervising the river crossing. The next incoming round missed the General but it decapitated Wellesley's dragoon orderly as he tarried in the streambed of the Kailna between the hamlets of Warur and Pipalgaon.[104] Some of the English authors insist the round was from an 18 pounder although that is not substantiated by Maratha accounts that say their heavy guns were still limbered. The estimation of calibre may have been based on the magnitude of horror that the round inflicted – but the skull offers little resistance to cast iron moving at a velocity of 500 feet per second and the damage could have easily been done by a smaller field gun. The sight of the headless dragoon haunted many: 'the body being left kept its seat by the valise, holsters, and other appendages of a cavalry saddle, it was sometime before the terrified horse could rid himself of the ghastly burden'.[105]

By entering at the eastern or narrowest end of the battlefield, Wellesley was positioning himself in such a way as to funnel the Marathas towards his men if they should attack. It was the most confined section of the battlefield and at a right angle to the original Maratha deployment, which was an east-to-west line parallel to the Kailna River. This meant that: (a) he stood a slight chance of 'rolling up' the Maratha line from the eastern end if he could move out quickly enough, or more logically (b) the Marathas would have to expose themselves in a 'change of front' by swinging their line at a ninety degree angle and redeploying to face him in a parallel north-to-south line. This was not simple given the length of their pre-existing line running east-to-west and the extent to which it would have to be condensed and reconfigured to form a new line running north-to-south to face Wellesley. The Maratha artillery also had to reposition. It was a test in precision marching for Sindia's sepoys and it left them dangerously exposed to Wellesley on their eastern flank as they marched under shouldered arms *wheeling left* to 'change front'.

There is incomplete agreement on the 'time-line' of the battle. Some sources say the Marathas subjected the British to a brutal two-hour cannonade and others have the British making use of all available terrain to avoid presenting themselves as a mass target until the last possible moment. One of the earliest modern attempts to develop a 'time-line' approach to analysing this battle dates back to 1912 and an article by Brevet Lieutenant-Colonel Bird.[106] In his version of events, Bird placed a limitation of two hours from the time Wellesley's troops first saw their river-crossing objective to the point at which they were all across and assembled on-line.

Wellesley formed up his troops into two lines of infantry and a third line of cavalry in support; they were arrayed in parallel fashion, the lines stretching from the south-to-north. Mounstuart Elphinstone's letter, written the day after the battle, clearly noted that the Maratha advance began as Wellesley was still positioning his men. 'While we were forming, the enemy's infantry and guns advanced on us, and their cannonade was very destructive.'[107] Some British observers hinted that Wellesley's judgement might have contributed to their rapidly mounting casualties. 'Wellesley had not given the enemy credit for being able to change their front with so large a force, without falling into disarray. On perceiving the alteration in the enemy's position, he saw that it was necessary to extend his front.'[108] Blakiston noted that Wellesley was forced to change his deployment at least twice to avoid the enemy's tactical manoeuvring and the rapid redeployment of their artillery that seemed to acquire and dispose of new targets with machine-like precision.[109] Following their change of front the Maratha infantry, who at Wellesley's approach had formed an extended line east-to-west parallel to the north bank of the Kailna, now ranged north-to-south from Assaye towards the Kailna. The Marathas had executed their manoeuvres with parade-ground precision, in effect hemming Wellesley into the eastern end of the battlefield so that he was now trapped in the narrowest portion of the crotch formed by the fork in the rivers.

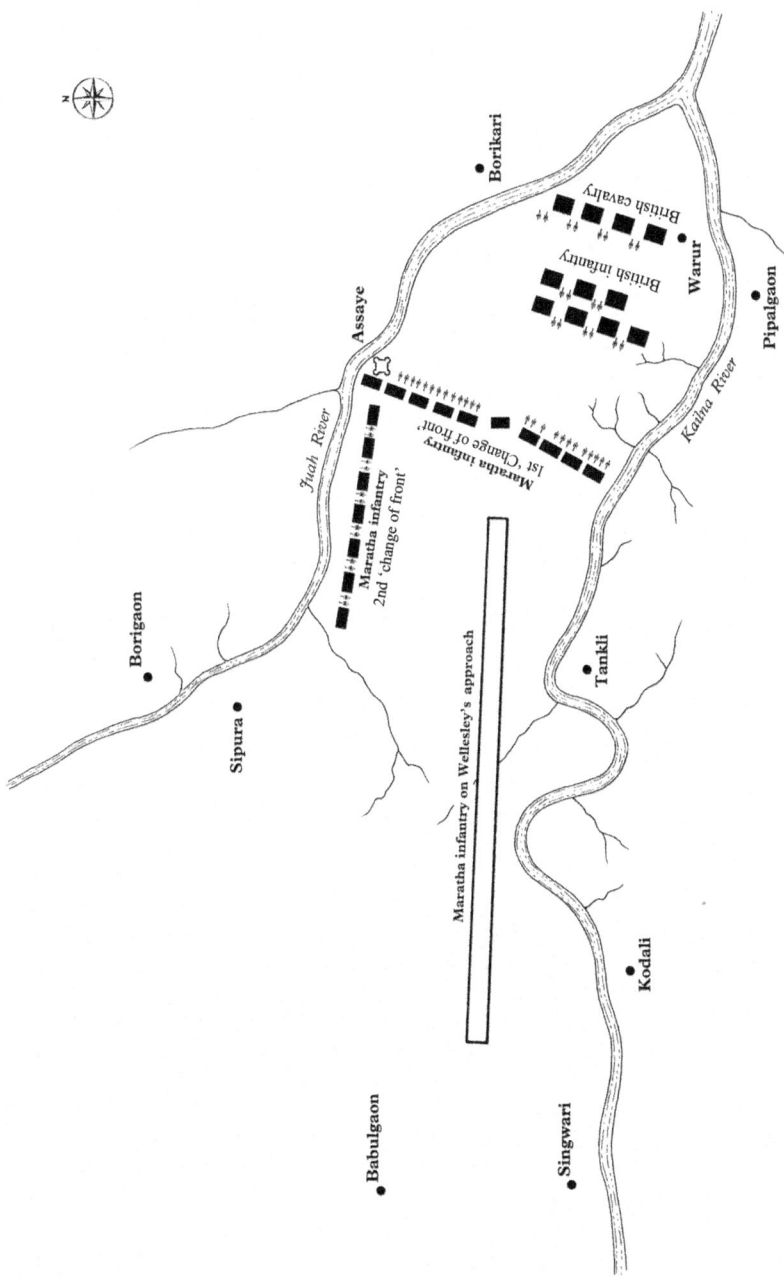

Map 5 The Battle of Assaye 23 September 1803, based on a map that appeared in Major W. Thorn, *Memoir of the War in India*, 1818

At the same time that Sindia's sepoys formed their front line facing Wellesley on a north–south axis, additional Maratha infantry and guns formed a second line, or fall-back position, running east-to-west. This second Maratha line had its eastern-most end anchored at Assaye and faced south so the men had their backs to the Juah river. The Marathas' two newly formed perpendicular lines met roughly at Assaye as a ninety-degree pivot-point, not unlike the way doors pivot or 'swing' on a hinge. Wellesley, seeing the way the Marathas' primary north–south line traversed the small *doab*, was forced to merge his two infantry lines into one long line to face the Marathas in such a way that would deny them the opportunity of out-flanking him on the northern end of the battlefield beside the Juah River. The Madras Native Infantry (1/10th, 1/8th, 1/4th, 2/12th), so often omitted from the Eurocentric renditions of the battle, were posted in the centre section of the British line. The concept was a British line running south-to-north composed of native infantry in the centre with European or, more specifically, Scottish troops of the 74th and 78th Regiment on either end.

John Blakiston's recollections of the battle remain among the most objective of published British accounts. He was critical of Wellesley at times but more importantly, from the standpoint of a cross-cultural analysis, he was unswayed by racist definitions of what constituted 'manly' behaviour by the sepoys under fire. From his position in the saddle, Blakiston saw the Madrasi sepoys in the centre of the line singled out for destruction. The Maratha artillery was apparently hoping to literally blow out the centre of the British line so that Wellesley's men could be easily divided into two isolated pockets of resistance. Blakiston witnessed the Maratha artillery doing their best to render good men as casualties. British troop strength had to be maintained somehow if there was to be any hope of victory:

At this time the fire of the enemy's artillery became, indeed, most dreadful. In the space of less than a mile, 100 guns, worked with skill and rapidity, vomited forth death into our feeble ranks. It cannot, then, be a matter of surprise if, in many cases, the sepoys should have taken advantage of any irregularities in the ground to shelter themselves from the deadly shower, or that even, in some few instances, not all the endeavours of the officers could persuade them to move forward.[110]

Blakiston also recorded that the sepoys were using their knapsacks as a form of protective cover during the most violent moments of the cannonade. The Europeans usually had their packs carried for them and were deprived of such cover, although he noted that even if they were available, European troops may not have had the 'ingenuity' to do this.[111]

That the sepoys of the Madras army paid an inhuman price was evident in Arthur Wellesley's personal appeal in the pension claim of the Indian officer known as 'Burry Khan'. The *soubahdar* was wounded so badly at Assaye that, seventeen months after the battle, bone splinters continued to be pulled from his shoulder. He was no youngster and Wellesley argued that 'he has served the Company forty years, and that he was a soubahdar of the 1st class. He was also a man of good character, and well connected in his corps; and five of his relations, commissioned and non-commissioned officers in the battalion, were killed in the battle of Assye.'[112]

Arthur Wellesley's infantry forces kept their position and waited helplessly until the British artillery could be brought across the river and up on to the battlefield where it was hoped they could challenge the enemy's guns. Of the sources consulted, Colonel John Grahame's twentieth-century account stands alone in saying that Wellesley's guns stuck in the mud approaching the river's ford.[113] Apparently most British gun crews had managed to cross the watercourse and come into action, returning fire at 400 yards, but with little effect other than to incur the wrath of the Maratha artillery. Each second that it took the British to load and fire their pieces was a potential eternity as they faced not only greater fire superiority but guns served by crews whose ranks had not yet been thinned by casualties. Counter-artillery fire was a Maratha specialty and their experienced gunners had the advantage in the number of guns, amount of ammunition and number of men per gun crew. They had also managed to bring some of their 9–12 pounder guns into action by then and this gave them a decided advantage in projectile 'throw weight' over the British battalion 6 pounder guns. The Maratha camp held numerous tumbrels of ammunition and the Maratha gunners had a wide selection of projectiles and charges to choose from.[114] Their British targets were now well within the optimum range for what the British knew as French-style 'large grape shot'. It could not only smash troop formations and kill draught animals but splinter British artillery carriages and immobilize Wellesley's guns so they could not advance with his line. Elphinstone, in writing about the moment, said: 'the enemy who were advancing on us, and beginning to get near us, renewed and redoubled their cannonade, which had slackened. It was no longer ineffectual, for it knocked down men, horses, and bullocks, every shot.'[115]

The Maratha salvoes compounded the confusion of trying to execute manoeuvres amongst the infantry and casualties among the British gunners were soaring by the minute.[116] The British guns could not easily be moved forward or fired, owing to the number of men and draught animals killed, so they were abandoned.[117] Assaye veteran Colin Campbell avoided blaming Wellesley, but 'most of the gun bullocks were killed and some of the corps, I think waited too long wishing to bring forward their guns'.[118] His statement made it apparent that command and control of the British artillery had broken down under sustained enemy fire. Campbell also noted that the Marathas had begun firing chain shot as well. Although often associated with sea battles, the whirring sound of chain shot scything through the air meant the Marathas were determined to hole Wellesley's infantry line. Elphinstone observed: 'it was found that many of our guns could not be dragged for want of hands. The General then told them to limber up, but the bullocks were killed. He then ordered them to be left behind, which was done, but not immediately, and all the time the men were getting knocked down very fast.'[119]

The advance of Arthur Wellesley's infantry was dictated by the immediate need to drive the Marathas from their guns. But delivering sufficient volleys of musketry to silence the guns would mean having to stand fixed in the *beaten zone* of Sindia's artillery. An advance with the bayonet on the Maratha guns was the only logical manoeuvre left to neutralize them and get Wellesley's men out of this killing field. It also meant that Wellesley could try to reclaim, to a certain degree, the momentum of the battle as the Maratha line would have to respond to his infantry's advance in a manner which would limit the Marathas' immediate options and command decisions. In other words, by ordering his infantry forward Wellesley stood some chance of shaping the further course of battle rather than allowing it to overwhelm him in a situation where inaction would mean annihilation by the Maratha artillery.

The 78th Regiment, in the southern portion of Wellesley's line, had a relatively straightforward but far from easy objective lying before them. They had to march across the killing zone in the hope of taking out the southern end of the Maratha line that extended to the Kailna. In the northern sector of the battlefield EIC Lieutenant-Colonel William Orrock was supposed to advance with the infantry pickets supported by the 74th Regiment. With the British cannon neutralized, their gun crews and draught animals now dead or incapacitated by life-threatening wounds, Wellesley is reported to have told the 74th to 'get on' without their supporting artillery.[120] Owing to Wellesley's earlier order to redeploy in an extended fashion, from two shorter lines of infantry to a single elongated infantry line, Orrock was to have angled off towards Assaye and then to have faced half left to correct his deployment by bringing his men into alignment with those in the southern segment of the British line. By most accounts based on Wellesley's records, Wellesley supposedly told Orrock to avoid Assaye itself. But for reasons unknown, Orrock is said to have 'got it wrong' and gone off on a tangent that conformed to his earlier angle of deployment to the northern portion of the battlefield, never turning the men half left to complete the intended straightening of the line needed for a uniform north–south linear axis of advance. Intentionally or not, Orrock led the pickets and the 74th directly towards the stout-walled village of Assaye, which contained large concentrations of artillery as well as Maratha irregular infantry. This meant that rather than engaging the Marathas as a unified line, Wellesley's infantry was split into two main actions: (1) the advance of the 78th and the Madras Native Infantry on the south-central half of the Maratha line and (2) the oblique march of the 74th towards Assaye on the northern portion of the field.

Wellesley's men in the southern and central section of the field advanced at a steady pace, fired a volley and then drove through the Maratha line, apparently having to bayonet the Maratha gunners who fell by the wheels of their cannon.[121] Some proved more agile than others and ducked under their guns only to lunge out with their famous broad-bladed Maratha spears and 'Rajput half-pikes' slashing at what they could.[122] With the first line of the Maratha artillery penetrated, the Madrassi troops and the kilted members of the 78th pressed forward with the bayonet to engage the Maratha infantry who had been posted behind the guns. Surprisingly the assault evoked a temporary rally among the Marathas, but they were eventually driven off towards their fallback position, that second perpendicular line of defence running west along the Juah River from Assaye.[123] Bird's account seems correct in depicting Wellesley as having ordered the 78th on his left, as well as his Madrassi troops in the centre, to wheel right and march on the Marathas reforming on their second line along the south bank of the Juah.[124] In completing that marching order, however, the British troops exposed their backs to hundreds of Maratha 'dead' that littered the fields. Scattered among the corpses lying by the cannon were Maratha artillerymen who had merely pretended to be dead. After the British troops had safely passed, a number of the Maratha gunners, who were still quite alive, jumped to their feet and turned their cannon on the backs of the line that had passed them. Blakiston, with an eye for technical detail, examined the guns after the battle and found they had been 'laid' according to textbook standard with an elevation of less than zero degrees for point-blank discharge.[125] The close-quarters discharge of canister shot cut many British troops down and threatened the cohesion of the assault by breaking the line further into small uncoordinated clumps of soldiers who could not be controlled easily on the chaotic battlefield.[126] The ruse also gave cause for the Maratha sepoys to halt their retreat towards the Juah line – they faced south once again and readied their flintlocks. Wellesley's infantry were now trapped between two walls of Maratha fire – volleys of musketry to their front and artillery fire off to their rear.

Back in the northern sector of the battlefield, the murderous fire coming out of Assaye was slaughtering Orrock's pickets and the Highlanders of the 74th. Some of these men had found their path was obstructed by the thorny bushes known as 'milk hedge'.[127] The surviving pickets, who had been in the vanguard, fell back to seek shelter in the advancing ranks of the Highlanders.[128] Wellesley ordered Lieutenant-Colonel Maxwell and the 19th Dragoons to the extreme right or north end of what had become his second line.[129] It was from that vantage point, while patiently awaiting orders for the cavalry to advance, that Dragoon Sergeant Swarbruck witnessed the butchering of the 74th. Swarbruck left perhaps the most brutally honest account of the carnage among the Highlanders. The majority of first-hand accounts were usually written by officers, which probably reflected literacy rates in the ranks. The value of the sergeant's account, aside from his NCO status, lies in its frankness and employment of accurate terms such as 'retreat'. All of the officers of the 74th were rendered casualties in this action, being either killed outright or wounded. Swarbruck wrote:

Then our infantry was all formed they opened a severe fire of round and grape shot. Our infantry suffered severe, the right brigade charged but was forced to retreat for they were nearly all killed and wounded. The Brave 74th Regiment displayed their bravery to the last moment for the Regt. had only 63 men left when they retreated, non [sic] one single Officer but was killed and wounded lying on the plains except Major Swinton and he was wounded in the back, but he retreated with what few of the regiment that was left. The other two regiments suffered very severe, the enemy then charged our infantry on the retreat and advancing in front of their own park,* gave No quarter to any of our wounded, then only cutting and shooting them as they came up with them.[130]

The 74th were falling back when the Marathas rushed in and cut up the survivors. Much later there would be accusations that Sindia's mercenary officers were guilty of ungentlemanly behaviour on the field of honour. Wellesley wrote that some of his men overheard Sindia's British mercenaries say: 'You understand the language better than I do: desire the jemidar of that body of horse to go and cut up those wounded European soldiers.'[131]

Swarbruck's account of Swinton being wounded in the back would also tend to support an extremely unique eyewitness account that was published in the *Madras Courier* a month later. Although the *Courier* was

---

* Park meaning artillery park.

an official voice for the Madras Presidency, the testimony seems to have been expunged from later official accounts for rather obvious political reasons.

Here our resolution was put to a severe trial, the enemy in a large body of Infantry and Cavalry charged the 74th Regiment on our right and actually cut through them, at the same time that a second body of their Cavalry retook their guns, and our own also which had been left in the rear; dismounted, and after killing some Officers and Men, turned and fired them upon our rear . . .[132]

In other words, some of Wellesley's men were felled by their own cannon, which the Maratha horsemen had seized and fired after penetrating the British lines.

The cavalry detachment under Maxwell, consisting of his available dragoons and elements of the 4th as well as 5th Madras Native Cavalry, were now ordered into action by Wellesley. The survivors of Orrock's failed advance were being hacked unceremoniously to pieces by Marathas wielding razor-sharp *tulwars* (swords). But the approaching British horsemen began absorbing a withering fire of grape shot from Maratha artillery posted around the village. To effectively relieve the infantry of the 74th Regiment, the Marathas had to be driven off the field and their guns along the walls of Assaye neutralized. Maxwell's dragoons headed directly towards the carnage where the crumpled men of the 74th lay dead and dying. Sergeant Swarbruck of the 19th Dragoons was among those who went to their aid, unfortunately having to trample a number of the Scots in order to drive away the last of their tormentors. In rather disjointed but honest prose Swarbruck recalled:

Our brave General then rode up saying now . . . you must make the best of you [sic] cavalry or we shall be done. Our gallant commander of the cavalry then gave the word . . . 3 Cheers, on the 3rd cheer we dashed forward with our brave General with us exclaiming death or history riding over our poor wounded men as they lay bleeding with their wounds.[133]

Having driven off the Marathas, the British horsemen were then able to swing their attention to the line of guns posted around and about Assaye.

Cavalry could be devastating against infantry but it was always a somewhat imprecise tool to use against artillery. Many tacticians in that era believed that the true 'weapon's value' of cavalry lay in the 'shock of the charge'. In an ideal cavalry vs. artillery scenario, enemy gunners would break and run upon sight of horsemen closing at the gallop. Supposedly, the fleeing cannoneers could then be ridden down and sabred. But the Maratha gunners at Assaye knew that their guns were their salvation. The trick was not to lose one's nerve and run off into the open only to be cut down. If the British got in too close to hit with cannon fire then the gun had to serve as a defensible obstacle against the horsemen. Lieutenant Alexander Grant, while serving as Colonel Maxwell's Major of Brigade, misjudged the speed and stopping distance of his mount. Grant went hurtling forward trying to skewer a Maratha gunner only to find his horse tightly wedged between the muzzle and a wheel of a Maratha cannon as it discharged.[134] Captain George Sale engaged another member of a Maratha gun crew wielding either a half-pike or spear 'with which all Sindia's artillery-men were armed'. A second Maratha assailant climbed atop the gun and thrust at the mounted Sale. The result was a diagonal wound across the chest as the blade was deflected by Captain Sale's sternum. Sergeant Strange, near at hand, saw the action and rode to Sale's rescue using his sabre to run through the captain's Maratha opponent. But in leaning out and extending his blade the sergeant exposed himself to danger and another Maratha artilleryman, sheltered under the gun, thrust a spear upward into Strange's abdomen.[135]

An undetermined number of the Marathas were eventually driven from the vicinity of the artillery line 'with great slaughter' down the riverbank and into the Juah.[136] The British cavalry splashed through the watercourse and cut their way through the Maratha fugitives who had fled the line to seek safety on the northern side of the Juah. The cavalry pursued many on the other side but their efforts met with mixed results. Although they would later be credited with saving the day, they had lost considerable numbers themselves. Command and control had broken down badly by the time they reached the opposite side of the Juah. Maxwell's horsemen came dangerously close to going too far and they risked being cut off by the Maratha light horse that waited just further to the north of the riverbank.[137] Maxwell began to call in his horsemen on the north bank of the Juah in an attempt to regain control.

Maxwell pulled his men back across to the south side of the Juah River and regrouped near Assaye. Several authors have made much of the fact that although there were thousands of *pindaries* on hand they did not charge forward to assist the Maratha infantry or Maratha cavalry. But one must remember this was not the manner in which *pindaries* fought or were used and in this case they were deprived of their traditional role. *Pindaries* excelled at harassing and interdicting convoy lines, spreading confusion among armies on the march and performing general reconnaissance. They were not known as the type of mounted troops to attack well-formed infantry – a role more closely associated with regular cavalry. Nor were the *pindaries* ever intended for direct frontal charges against heavily armed European cavalry carrying a sabre and a brace of pistols. So in this case we should not make too much of the *pin-daries* melting-off on the periphery. The thousands of *pindaries* were of minimal use in this type of combat scenario; they could not effectively charge Wellesley's cavalry or his infantry. By leaving his baggage under heavy guard at Nalni, Wellesley had wisely foreseen the *pindaries'* role in the Maratha order of battle and deprived them of suitable tactical employment.

Arthur Wellesley now had to restore order in the central sector of the battlefield where his men had been trapped in a crossfire. The Maratha gun crews at that location had committed what many of the more chivalrous British officers considered to be an act of treachery in feigning death. Wellesley ordered the 78th back to the Maratha artillery position they had fought through, approaching from the west while he, leading members of the 7th Madras Native Cavalry Regiment, approached from the east. The Marathas were eventually driven from the guns or bayoneted on the spot and resistance crumbled as Maratha stragglers made for their fallback position on the second line along the Juah. It was at that point that Wellesley had his horse Diomed piked from under him in the struggle to take the Maratha guns a second time.[138] He remounted and issued new orders. The 7th Native Cavalry were posted with the recaptured enemy guns to prevent them from being carried off by the Marathas.

The British infantry in mid-field were to reform and march northward as a single line to meet the remaining Maratha forces who still held their fallback position along the Juah. Maxwell was to advance at the gallop from the east in an attempt to break the enemy's left. But Maxwell's approach was met with a shower of grape and canister shot which proved extremely accurate. Maxwell was pierced through the body in several places. One projectile was believed to have struck his right hand or arm with such violent impact that the limb was jerked perfectly upwards from his body in a seemingly deliberate hand signal that those immediately behind him recognized as the command to halt. The bulk of the dragoons that followed hot on their heels could not stop fast enough and in this chaos they did not charge through the Maratha line but veered parallel to it as Maxwell fell from his horse dead.[139] What was to have been the final great British cavalry charge at Assaye degenerated into a lethal ride past the muzzles of the Maratha artillery as the dragoons turned, leaderless, southwards after having been raked by yet more canister shot from the Marathas.[140]

The conclusion of the battle from that point was judged to have been fairly anti-climactic by British witnesses. The British infantry advanced with the bayonet on the Maratha line, killing with a renewed purpose. And, as if spent of energy and purpose, Sindia's last line of resistance melted away. Sporadic explosions continued to puncture the air for hours as the Marathas had booby-trapped their abandoned ammunition tumbrels by leaving lengths of slow match* burning among the powder charges.[141] Some fugitives crossed the Juah but many slipped off to the west. Not all the Maratha infantry units present were engaged in the heat of combat and Wellesley's men saw numbers of survivors form up with other non-combatants and move off in good order. The British infantry, both sepoy and European, had prevailed over one of the finest artillery forces of the era. As we shall see in chapters 5 and 6, there was a particular explanation for the collapse of Maratha resistance and morale. Combined British casualties (European and native) were later listed in official records as 428 killed in action, 1,138 wounded in action, and 18 missing in action.[142] This meant total British casualties of 1,584 out of a force of approximately 4,500 – which is how one arrives at the conclusion that Wellesley's victory cost him more than one-third (35.2 per cent) of his force. Maratha losses were reported as an unconfirmed 6,000; however, that would include casualties among their camp followers and those Maratha wounded who were bayoneted by Wellesley's men.

The aftermath

Wellesley's men were too exhausted from their forced 21-mile march and three hours of intense combat to pursue the retreating Marathas. In hindsight it was probably a prudent move since the thousands of Maratha horse still lay beyond the field, capable of isolating anyone who advanced too far without support. A number of the battle's eyewitness authors were wounded, including Blakiston who was hit twice during the action. He received a graze on the wrist and was hit in the abdomen with such force by a piece of 'spent grape', that it knocked him to the ground and left him breathless.[143] The 74th Regiment had been subjected to a high concentration of anti-personnel rounds such as canister shot and grape shot as well as the horror of chain shot. Other members of the unit had been cut up by Maratha horse and then 'ridden over' by the 19th Dragoons. Blakiston recounted the horror of seeing what Maratha *tulwars* could do. Their sharply honed blades possessed a keen cutting edge that dramatically contrasted the shortcomings of British swords and their scabbards.

---

* Slow match was a form of slow burning fuse often used for time-delayed ignition.

On passing over the ground where the 74th and piquets had been engaged, the carnage was dreadful, and the wounds inflicted by the swords of the enemy's cavalry were such as I could have no conception of. This was the only time I ever saw heads fairly cut off. Such a thing could not be done by our cavalry swords in their usual state; for, however good the material may be, the constant drawing in and out of an iron scabbard soon blunts the edge; whereas those of the native horsemen, though seldom of such good stuff as ours, by never being drawn except for use, or for the purpose of being cleaned, are capable of inflicting a wound ten times the depth, particularly when applied in the drawing manner usually practiced by the Indian swordsmen.[144]

The accumulative slaughter inflicted on the 74th Regiment resulted in their being struck from the roster. Elphinstone reckoned they had lost exactly 400 killed and wounded.[145] It ceased to be an effective unit until replacements could be found and the walking-wounded could be nursed back to health.[146] But there had been no shortage of bloodletting on either side. Blakiston had spoken earlier of the necessity of bayoneting the Maratha gun crews to ensure they did not rise again and evidence indicates the Maratha wounded were dispatched in the same way to a large extent.[147] While in terse words Elphinstone admitted: 'No prisoners were taken.'[148] The British wounded lay where they fell and were scattered about the battlefield. Four days later officers were still trying to account for their men and decide whom to send to the field hospital once it was established.[149] Captain Welsh would later remark: 'The battle of Assaye had collected all the birds of prey in the country, a few following the army, and the rest taking possession of the inheritance left them, by their kindest benefactor, man, on the field of battle.'[150] Ironically Sergeant Swarbruck, who gave personal testimony as to riding over the wounded of the 74th Regiment, became one of the wounded left lying on the battlefield. Fortunately his stay on the ground was only a day and he wrote:

parties was sent from each Regiment to pick up their killed and wounded. Some lay 2 and 3 days before they were taken up. I am very sorry to say I had the misfortune to lose my right leg in the charge and my horse killed from under me. And left to the mercy of the field for 24 hours without any assistance of a Surgeon.[151]

But at least Swarbruck was alive. Dead Maratha sepoys littered the field and Elphinstone made reference to most of them being 'stout Bengalees'.[152] Ten days after the battle Elphinstone rode back and noted the Maratha dead still unburied: 'Some of the dead are withered, their features still remaining, but their faces blackened to the colour of coal, others still swollen and blistered.' He then came across the body of a dead Maratha mercenary officer, which he had noted previously because of a disagreement in opinion. Some of the British officers had said the fair-skinned mercenary was French but Mounstuart Elphinstone thought it more likely he was Persian or even a Mughal *mirza*. Now he stumbled across the corpse again. Elphinstone's peculiar sense of romanticism permeated his letter on the horror of the battlefield: 'The Persian I mentioned was perfect everywhere, and had his great quilted coat on; but his face had fallen, or been eaten off, and his naked skull stared out like the hermit's of the wood of Joppa (in the "Castle of Otranto").'[153]

The order to establish the field hospital at Ajanta did not go out until six days after the battle.[154] Assessing the wounded was a gruesome affair with triage being conducted by the superintending surgeon. His evacuation recommendations were based on considerations of who would best survive the move to Ajanta field hospital and not on the criteria of who needed more sophisticated medical care. The removal of the first 297 wounded from the battle zone did not begin until 1 October, more than a week after the battle.[155] The hospital itself was quickly organized with the staff surgeon's servants being pressed into service. The pioneers sent twenty of their men to help unload the wounded. The 1/10th and 2/12th Madras Native Infantry were each ordered to furnish a man to serve dressing wounds. Madeira wine was to be used for medicinal purposes and authorization was granted to purchase 120 bottles at public expense. Having each unit donate one tent, for every ten casualties they sent, provided the shelter necessary to create the field hospital.[156]

In response to emergency requests for more surgeons General Oliver Nicolls, commanding officer of the Bombay military forces, forwarded a proposal to the Bombay military board that they passed immediately. The Bombay Presidency was short of European surgeons so a decision was made to send two Indian medical practitioners of Portuguese extraction. The medical board approved and sent Manuel De Cunha and Francesco Dias. They were to be given a month's advance in pay at an established rate of Rs. 60 per month each.[157] The feeling was expressed that even if their medical skills were not up to a European surgeon's standards, they would be useful in dressing wounds given the volume of casualties and the need to frequently change dressings in that climate.

As early as May 1803 Arthur Wellesley had been formally advised that any major action by European troops in the Deccan would generate problems. Alex Kennedy, staff surgeon to the *Nizam*'s Subsidiary Force that served under Colonel Stevenson, had sent a prophetic 'Memorandum' to Wellesley.[158] 'The desertion of dooly Bearers, an unlooked for increase of sick by keeping the field during the Rains, or, having a number of men wounded, might render the establishment of a field Hospital absolutely necessary – Cloathing, Wine, & European medicines are indispensibly necessary, and if such Emergency is deemed at all probable, they ought to be provided for before hand.' Kennedy's warning about preparing for the flight of dooley bearers was spot-on. The majority of Wellesley's dooley bearers, whose job it was to carry off the sick and wounded, came from the Bombay Presidency. Those dooley bearers, who served with European troops in the Army of the Deccan, found their work constantly exhausting. Europeans were more likely than sepoys to get sick and owing to the tendency of British officers to lead with European troops, their combat casualties were proportionately greater. As the dooley bearers began to desert in droves Arthur Wellesley wrote to the Bombay government seeking a remedy to his plight.[159] For their part, Bombay government officials turned the matter over to the police with instructions to apprehend all fugitive dooley bearers.[160] But the Major-General's problem went from bad to worse with the desertion of 230 dooley bearers in just two nights. This gave Bombay Presidency officials cause to question Wellesley's methods.[161] And it eventually produced an investigation of the matter by the Bombay military board.[162]

Since there was a shortage of dooley bearers, eight bullock carts were pressed into service from the 4th Madras Native Cavalry and the 1/10th Madras Native Infantry. This was not an easy decision to make since the heavily loaded carts, with their lack of suspension, meant that the ride over the primitive road would be bone jarring. This had been among the considerations in selecting the comparatively healthy wounded to be sent to the hospital. Some officers opposed the use of 'wheeled carriages' because it was felt they clogged the roads in such a way that they were a security risk to armies on the march.[163] Nearly 100 other wounded men were individually loaded, tied or draped across bullocks and sent to the hospital.[164] An influential body of eighteenth-century British army surgeons favoured smaller regimental hospitals rather than larger collective hospitals because of the danger of infection and disease posed by higher concentrations of sick and wounded men.[165] Arthur Wellesley's medical staff reflected the greater professional division inherent in the question: what does it mean to send a man to hospital? Some viewed it as standard procedure while others viewed it as a death sentence.

The medical disaster generated by Assaye seemed to drag on forever. More than eight months after the battle wounded sepoys were just arriving in Bombay from the Deccan. Some European survivors, including two men of the 19th Dragoons and forty-one disabled soldiers of the 74th Regiment, were sent to Panvel* in June 1804, almost nine months after the battle.[166] They were then taken by boat for admission to the hospital in

---

* 18°+, 73°+. Aka Panwell.

Bombay.[167] Many of the badly wounded men were amputees and a considerable number still had to be fed by attendants. That was an expense many could not afford on their own, far from home and drawing only a disability wage. As a result, Major-General Wellesley ordered that these poor wounded men, both European and native soldiers of the Madras Presidency, should be granted their field pay while in Bombay but, in a cost-cutting move to save the Bombay Presidency the expenditure, it was to be charged against the Madras Presidency.

Money was always an issue uppermost in a soldier's mind. After all, many had joined the EIC's army for a chance to earn increased *batta* or a share in the legendary riches that supposedly followed successful campaigns such as that against Tipu Sultan in the Fourth Anglo-Mysore War of 1799.[168] But during the 1803 Maratha Campaigns the EIC's legal advisors were often kept busy ruling which spoils of war could be taken as plunder and which had to be disposed of through the process of the prize fund. There was a rivalry between the Company's and the King's troops over how much prize money they would get. The Bombay Presidency's legal advisor Mr J. M. Thriepland had to rule twice on what constituted prize vs. plunder during the invasion of Sindia's port of Broach.[169] 'Plunder' in that case was deemed to be the items which soldiers carried away, while 'prize' money was to be composed of funds derived from the sale of military stores and goods which were not readily transportable, such as old or obsolete artillery. The lengthy decision was difficult and the legal report noted that it was not uncommon in previous wars to draw these distinctions 'before taking to the field'. After the Battle of Assaye some Europeans had sought to withhold cattle that they had taken and not reported. Alex McKay of the King's 78th Regiment was charged with plundering after cattle were found in his possession. Wellesley sentenced McKay to 200 lashes but later he served notice that any man found guilty of plundering would be put to death.[170]

Much of the European plundering at Assaye had started innocently as souvenir hunting, but it got out of hand. The 19th Dragoons had been thrown into such disarray by the ferocity of the action that much of their valued equipment laid scattered on the field of honour. Pistols and swords were highly prized mementos and eventually an inventory had to be conducted among the infantry to see how many trophy seekers were in possession of contraband.[171] Several of Wellesley's infantrymen held kit plundered from their own cavalry and so they were permitted the benefit of an amnesty during which they could turn in illicit booty without punishment. Charges of plunder and theft may partially explain the obsession with wills and the presence of witnesses at the disposal of dead officers' effects as often found in EIC records and diaries.

Breaking the Deccani chain

Accurate and immediate tactical-level intelligence, as opposed to strategic theatre-level intelligence, was a continuing problem for Arthur Wellesley. The Begum Sumru's sepoy battalions, who had accompanied Sindia's regulars to Assaye, escaped the battle unscathed and in the days that followed there was a great deal of concern over their exact location. Both Sindia and Bhonsle still had significant quantities of infantry and horsemen of various descriptions numbering in the tens of thousands. Four months earlier, Lieutenant-Colonel Collins – usually a wealth of information on Maratha troops – had told Wellesley that Bhonsle had ten battalions of infantry commanded by 'Native Officers' with thirty-two pieces of artillery.[172] A Maratha sepoy survivor was brought in and he confirmed that at least ten to twelve pieces of Sindia's artillery had been carried off the field at Assaye and the British feared the Marathas were about to consolidate the forces who had escaped the battle. Colonel Stevenson's *harkarrahs* reported a surviving *campoo* had been joined by another *campoo* from Burhanpur.[173] That would have made sense if there was a Maratha consolidation of forces, but then Stevenson's *harkarrahs* – having dogged Sindia and Bhonsle since the day after the battle – reported that the two Maratha leaders held a final meeting on 29 September 1803 before parting company.[174]

Wellesley had pushed his men in high temperatures and they had proven more than capable. But having suffered severe casualties, it was wise to conserve effort until the next opportunity to challenge the Maratha infantry clearly presented itself. Wellesley reckoned that the Marathas had three military options: (a) stand and fight, (b) move off to menace the *Nizam's* territory, or (c) move to cut British communication and supply from Surat to the west. Arthur could try and second-guess Sindia and Bhonsle but it was impossible to pursue them effectively with his army in a weakened condition. What if the Marathas had split up? Casualties, among horses and men of Wellesley's cavalry, meant he did not have all the mounted resources thought necessary for an immediate active pursuit of Sindia and Bhonsle.[175] The Maratha light horse and *pindari* troops would have an advantage if the Major-General began an extended cross-country pursuit before his troops were up to strength. There were, however, other local targets that cried out for his attention. Burhanpur and Asirghar were part of the historic Deccani chain of forts that included Ahmadnagar.

Asirghar was now worthy of consideration as a secondary logistical centre since Ahmadnagar had been seized earlier. Sindia was said, by Colonel Stevenson's *harkarrahs*, to be headed towards Burhanpur with a large *campoo*, fifteen guns and two howitzers, but exact troop strengths were proving difficult to estimate.[176] Apparently Sindia intended to stow a detachment of his regulars there during the first week of October. But situation reports were conflicting and at times contradictory; some said Sindia had arrived at Burhanpur and, after leaving a token force as well as some wounded, he had pulled out. The British were unable to discern how many men were in the small garrison force or if this was disinformation intended as an ambush lure. Colonel Stevenson was an old soldier plagued by ill health but he had a keen sense of duty and he allowed himself to be used as Arthur Wellesley's eyes and ears. Stevenson wrote to Wellesley that he thought the apparent late September split between Sindia and Bhonsle was another Maratha ruse. Stevenson cautioned Wellesley that the British should not be diverted from their intention to move against Burhanpur.[177]

Colonel Stevenson then received an intelligence report from inside Sindia's camp, which clarified things greatly. The large *campoo*, earlier reported as headed for Burhanpur, belonged to Begum Sumru and it was not destined for garrison duty there after all.[178] The British breathed a collective sigh of relief because the Begum's troops were now no longer accorded a high level of threat status. In fact, the Begum was a potential strong ally, having ceased to be an ardent public supporter of Sindia's cause. General Lake, working earlier with intelligence from Lieutenant-Colonel Collins, advised the government that the Begum had 'long evinced a desire to be taken under the Protection of the English'. And that if a rupture with Sindia occurred in the Deccan, she 'might be contrived to enable those battalions to join General Wellesley'. Governor-General Richard Wellesley had acknowledged Lake's report and sent it to the Deccan as early as July.[179] Confident in the knowledge that the 'Burhanpur battalion' was the Begum's and not Sindia's, Colonel Stevenson advanced and occupied Burhanpur without opposition on 15 October 1803.[180] The remaining small body of Sindia's infantry at Burhanpur retreated in disorder towards the Nerbudda River.[181]

After the bloodless capture of Burhanpur, fourteen of Sindia's European mercenaries surrendered to Colonel Stevenson. The group was composed of nine officers, four sergeants and one *matross*.[182] The senior-most officer was Captain John James Dupont of Holland. Despite the assertions that the French dominated Sindia's mercenary corps, this band of men contained only one Frenchman, that being Captain Mercier. The only Englishman in the group was Ensign Alexander Marrs. The remaining twelve individuals were Portuguese.[183] The fourteen mercenaries, along with 'English' gunners John Roach and 'George Black', who had wandered into Stevenson's camp on 1 September, were initially listed as 'Prisoners of War'.[184] But Colonel Stevenson, invoking

Governor-General Richard Wellesley's Proclamation of 29 August to encourage the defection of mercenaries in Maratha service, issued formal
orders that these men be taken into British pay.

The European Officers & Sergeants who have this day been received from the
Service of Dowlut Rao Scindiah, and all Europeans who may in future come in
from the service of that Chief, or any power confederated with him, are to be
under the charge of the Deputy Adjutant General, & who will draw pay for them
agreeable to rates which will be hereafter determined.[185]

With Asirghar deprived of potential infantry support from Burhanpur,
Stevenson was encouraged and he pressed on with plans for an assault on
the fort. His men began constructing siege batteries and followed up this
move with an attack on Asirghar's *pettah*. Having taken that objective,
Stevenson offered terms to the fort's *killadar*. These were refused and
Stevenson threw more men into the *pettah* that was as close as 150 yards
to the fort in some places. Stevenson was putting the majority of his faith
in two siege batteries and not in negotiation. There was always the chance
that Sindia's infantry might return to relieve the siege and so great effort
was taken to make the batteries serviceable on 19 October. In a final bid
to spare a costly assault, word was sent out that they would commence
firing the next day. Apparently there was the basis for some dialogue but
several sticking points remained. The siege batteries had only been in play
for one hour on the 20th when a white flag appeared and arrangements
were made for the delivery of hostages to Stevenson in agreement with
the terms he had stipulated.

On 21 October the enemy garrison walked out with their personal
property as well as a cash bonus. In writing about the settlement Bird said
that the garrison accepted a 'bribe' of Rs. 20,000 to leave, but William
Thorn noted the money represented the troop's arrears in pay.[186] The
settlement was quite in keeping with the interpretation of contractual
obligation within the South Asian military economy. The garrison had
not been paid and Sindia was therefore delinquent in servicing his debts
to contracting parties within the defence establishment. The garrison,
for their part, could walk out proudly. They had not let the fortress fall;
they had merely relinquished their agreement to fight when it was clear
the British would assume Sindia's option of paying them. It later became
popular to refer to this process as the 'golden key', which was used to
unlock a fortress. It was part of a cultural continuum that the British did
not fully understand but were quite happy to take advantage of. With the
capture of Asirghar, Stevenson had taken the last of Sindia's strongholds
in the south. Locally the fort was known to be essential for controlling
access to the central corridor that linked the Deccan to North India.[187]

But the fortress also proved to be a valuable addition to the prize fund of Wellesley's army. Stevenson reported that, in terms of specie, there was only about Rs.11,000 in gold and silver coins in Asirghar. However, there was such an abundance of other disposable prize property that Stevenson was compelled to have four prize fund property 'managers': one to oversee the interests of the King's troops and three for the EIC's men. Aside from military stores, there were many luxury goods and much jewellery that had been liberated by Sindia's men. 'One ornament which was shewn to a shroff was valued at half a Lac of rupees, and a string of pearls at 12,000 rups. Among the Jewels is an emerald ring, with Shah Alum Padsha engraved on it.'[188]

With British troops in possession of Burhanpur and Asirghar, Stevenson's mission was subject to change. The ailing Colonel divided his time between watching out for Sindia and preparing for a siege of Bhonsle's fortress of Gawilgarh.* He wrote to Arthur Wellesley: 'My principal objects will be either the attack of Gyal Ghur or attacking Scindiah's Force and my proceeding to Nagpoor or going up the Ghaut will depend on the understanding I may receive from you.'[189] The old Colonel's illness was frustrating but so was Arthur Wellesley's approach to coordinating the next combined operation. Stevenson, however, remained a gentleman and resisted any temptation to criticize the young Major-General's slowness to act. Wellesley contemplated his next move. Sindia's loss of infantry and artillery at Assaye were combined with the loss of his logistical centres at Ahmadnagar, Burhanpur and Asirghar. That also meant few places to shelter the surviving elements of his 'regular corps' in the Deccan. His remaining regulars were believed to be in disarray since the defection of senior European mercenary officers at Burhanpur. When combined, the events suggested that Sindia's power in the Deccan was greatly diminished and that might present an opportunity to negotiate a separate peace with him – dividing the Sindia–Bhonsle Maratha alliance.

While Wellesley opened negotiations with Sindia's representatives, Stevenson's men assumed most of the drudgery associated with routine military patrolling. This consisted of efforts aimed at tracking Bhonsle's forces to deprive them of any opportunity to surprise the British. Raghuji Bhonsle's diplomatic representatives remained aloof; holding out for greater concessions, they foolishly refused to cooperate with the British in the ritual of envoy verification. Authors are divided on whether it was aggressive patrolling or Bhonsle's paranoia, but the Maratha leader felt obliged to move his camp five times in three days. For Wellesley's cavalry

* Aka Gyal Ghur.

there was also convoy duty to occupy one's time, but by and large this was a period of waiting to see how things would turn out. Sindia's *vakeel* was open to the idea of a political dialogue and he was received with great pomp and ceremony as recorded by Welsh. The arrival of an accredited *vakeel* from just Sindia's camp and none from Bhonsle was a point which Wellesley used to justify a difference in his treatment of the two Maratha leaders; in effect driving a further wedge between them.[190] Sindia was granted a peculiar form of 'armistice' on 23 November 1803, but Daulat Rao did not take advantage of the opportunity. By remaining active in the field, rather than withdrawing to a specific neutral area, Sindia garnered greater distrust and suspicion.[191] With his men rested and the 74th Regiment 'restored' to operational strength, Arthur Wellesley turned his attention to Raghuji Bhonsle, targetting his infantry for destruction. In the meantime, Stevenson had completed the preparations for a future siege at Gawilgarh but that could be put on hold. Arthur Wellesley ordered the Colonel forward to effect a rendezvous that would unite the two British battle forces. The Maratha confederates moved off on 28 November, just as Colonel Stevenson's force was approaching to assist the Major-General.

The Battle of Argaum*

While the British column was marching down the road on the morning of 29 November, a Maratha envoy approached. He rode up and asked Major-General Wellesley if he would meet with his master Raghuji Bhonsle at a position about ten miles ahead. Wellesley uttered an answer without the slightest motion to rein in his horse, treating the Maratha with deliberate indifference. The determination of Wellesley to ride on must have conveyed not only determination but also an ominous foreboding and the envoy's next question tipped his hand as to the real reason for enquiring about a meeting with Wellesley. Captain Welsh understood Bhonsle's man to ask Wellesley if he would indeed attack the Marathas if he just happened upon them; to which Wellesley replied, 'Most undoubtedly'.[192] The British column continued on towards their rendezvous with Colonel Stevenson's force, also thought to be some ten miles distant. After a long hot march they arrived at the campground selected by the quartermaster.

Wellesley's men were settling in to their camp routine and the Major-General had no original intention of engaging the Marathas that day. However, after climbing an old brickwork tower to reconnoitre the

* 21°+, 76°+. Aka Argaon.

countryside, Wellesley did catch a glimpse of the enemy in the distance.[193] Then around 2 pm, a detachment of Maratha irregular horse approached and engaged the allied contingent of Mysore horse that were riding forward outpost duty for the British. Instructing his men to grab a meal while 'resting on their arms', Wellesley and the pickets pursued the Marathas through the thick fields of *jowari*. They galloped through the tall grain to find the Marathas in force. The enemy formation consisted of Bhonsle's infantry with mounted support apparently being provided by Sindia's as well as Bhonsle's horsemen. Raghuji Bhonsle's army was reportedly under the leadership of his 'brother', usually identified as Nana Babah or 'Munnoo Bappoo'. Bhonsle's infantry were drawn up in an extended line on the plain in front of the village of Argaum.[194] This was to be a battle aimed at neutralizing Bhonsle's infantry arm. They were never as powerful as Sindia's 'regular corps' but they were still a force to be reckoned with. As it was late in the day, Wellesley once again decided to attack rather than risk the chance of retiring before such a force.

The Major-General called for his infantry to be brought forward. The infantry had the support of their bullock-drawn battalion guns and Wellesley's cavalry had what under normal conditions would have been considered their 'galloper guns' – horse-drawn 6 pounders raised from the cavalry regiments. But the tremendous loss of horses at Assaye had resulted in the cavalry substituting bullocks for the four horses that each team of 6 pounder galloper guns required. Wellesley had placed the bullock-drawn cannon near the head of the column to provide faster access to the artillery if needed for immediate reaction. British bullock-drawn artillery always seemed to lag behind and having them lead seemed sensible in that there would be no 'waiting for the guns to come up' if combat erupted unexpectedly. It was a modification to the line-of-march dictated by the Marathas' performance at Assaye and would seem to have been sensible. The column was surrounded on either side by the vegetation and compelled to advance along a narrow road bounded by the tall fields of grain. The village of Sirsoni* marked the place where the road spilled out onto the plain of Argaum.

The British had no inkling they were actually playing into an ambush. It seemed a straightforward advance to contact down the road since the enemy had already been spotted. Bhonsle's infantry were indeed there to 'give battle' but they were the bait for the opening Maratha bid to gain the upper hand. The Maratha artillery had pre-registered their guns on the walled village of Sirsoni. It was the only place where Wellesley could bring up his men and bullock-drawn artillery to gain entrance to the field

* Aka Sirsoli.

by way of the road. The Marathas opened fire at 1,000 yards as the British column followed its pickets onto the plain.[195] When being drawn by animals, cannon were 'limbered' – in effect harnessed for travelling with their ammunition tumbrels and caissons. Some larger pieces needed the assistance of detachable road wheels. In order to fire they had to be uncoupled and taken out of this arrangement or – in simple parlance – 'unlimbered' and then deployed. The Marathas stood a good chance of taking out Wellesley's guns before he could unlimber them. The incoming round shot wreaked havoc among the leading artillery bullocks. The drivers could not manage the stampeding animals that bolted. The beasts then panicked upon seeing that the source of their discomfort lay ahead and they turned back through the infantry ranks with their heavy loads. Several soldiers of the column were crushed to death by their own artillery. The gruesome sight of lacerated animals blindly trampling men, combined with the shock of unexpectedly distant but accurate Maratha artillery fire, caused two battalions of sepoys to run for their lives and seek shelter in Sirsoni.[196] Many of the sepoys who had been at Assaye were terror-struck at the idea that they were in another Maratha artillery killing zone, even though the guns were 1,000 yards away.[197] The sepoy veterans possessed something more than a vague idea about their odds for survival, given the intensity of the Maratha barrage that nearly destroyed them at Assaye two months earlier. Welsh said they were 'thrown into temporary disorder', but it is apparent that these men were panicking and running from the field.

Arthur Wellesley shouted out reassurances seeking to inspire these men but they continued to run. Much to the surprise of the Europeans in attendance, Wellesley did not vent his anger on the men but ordered their officers to take them to shelter behind the village where they were to regroup and then be led back onto the field.[198] As the rest of the column pushed past the bodies of the men crushed in the panic, they formed up in front of the village with Stevenson's troops further back off to the left. It was now apparent that Bhonsle's troops were positioned in a line with forty to fifty pieces of supporting artillery. Approximately 2,000 Arabs were posted to Bhonsle's flank while 5,000 men under Beni Singh formed the centre.[199] There is conflicting evidence over the ethnic composition of Beni Singh's force. Welsh listed them as Ghosseins* while others have generically said Rajputs. Elphinstone avoided generalizations but did mention one of Beni Singh's soldiers in particular was a Brahman.[200] The total enemy forces were estimated as high as 40,000 but that included all of the *pindaries*, irregular horse, camp followers and

---

* Aka Gossains.

some of Sindia's infantry who were camped well behind Bhonsle's in the appearance of being non-combatants.[201]

Blakiston admitted that the sepoy incident was the first time he had occasion to view what he considered as Wellesley's genius. After Major-General Wellesley had restored order among the panicked troops, he led them to their place in the lines that were forming and once there he ordered them to lie down. This had a two-fold effect; it reduced their target profile while it immobilized them in the sense that they could not run while lying down. Later he would write:

What do you think of nearly three entire battalions, who behaved so admirably in the battle of Assye, being broke and running off, when the cannonade commenced at Argaum, which was not to be compared to that at Assye? Luckily, I happened to be at no great distance from them, and I was able to rally them and re-establish the battle. If I had not been there, I am convinced we should have lost the day. But as it was, so much time elapsed before I could form them again, that we had not daylight enough for every thing that we should certainly have performed.[202]

Blakiston thought that Wellesley was quite pleased with himself that day, for Arthur commented 'Did you ever see a battle restored like this?' in direct reference to having the sepoys lie down.[203] It should be pointed out, however, that the British in South Asia had known of the technique for some time. Panicked troops were ordered to lie down in 1780 during the Second Anglo-Mysore War.[204]

The British forces remained on-line and waited for several minutes in the hope that their artillery could be salvaged and brought into action to answer Raghuji's guns. Captain Welsh's unit lay directly between the Marathas and the point of their artillery's initial impact. And for that posting they paid substantially: 'in this position, our corps being drawn up exactly in front of the village, on which the enemy's batteries were pointed, as the only entrance to the plain, severely suffered, in having Lieutenant Turner, two Subadars, one Jemadar and forty sepoys knocked down by cannon shot.'[205] By 4:30 pm it was apparent that the British guns could manage nothing but a feeble response and there was a need to advance without them across the open field.

During this battle Colonel Stevenson, despite his ailing condition, directed his troops from a *howdah* on elephant-back. Stevenson's men had to be ordered forward at double time to close the distance and to bring them into line with Wellesley's infantry. Captain Welsh had trouble seeing them at a distance owing to the smoke of the battlefield but as they came clearly into focus it was a scene temporarily detached from reality. 'It was a splendid sight to see such a line advancing, as on a field-day; but the

Argaum

N

Bhonsle's horsemen
& irregular infantry

Sindia's infantry
(non-combatants)

2 Large formations
of Sindia's horse

Raghuji Bhonsle's regular infantry

Shallow Nullah

Stevenson   Wellesley
Combined British infantry

Cavalry   Cavalry

Sirsoni

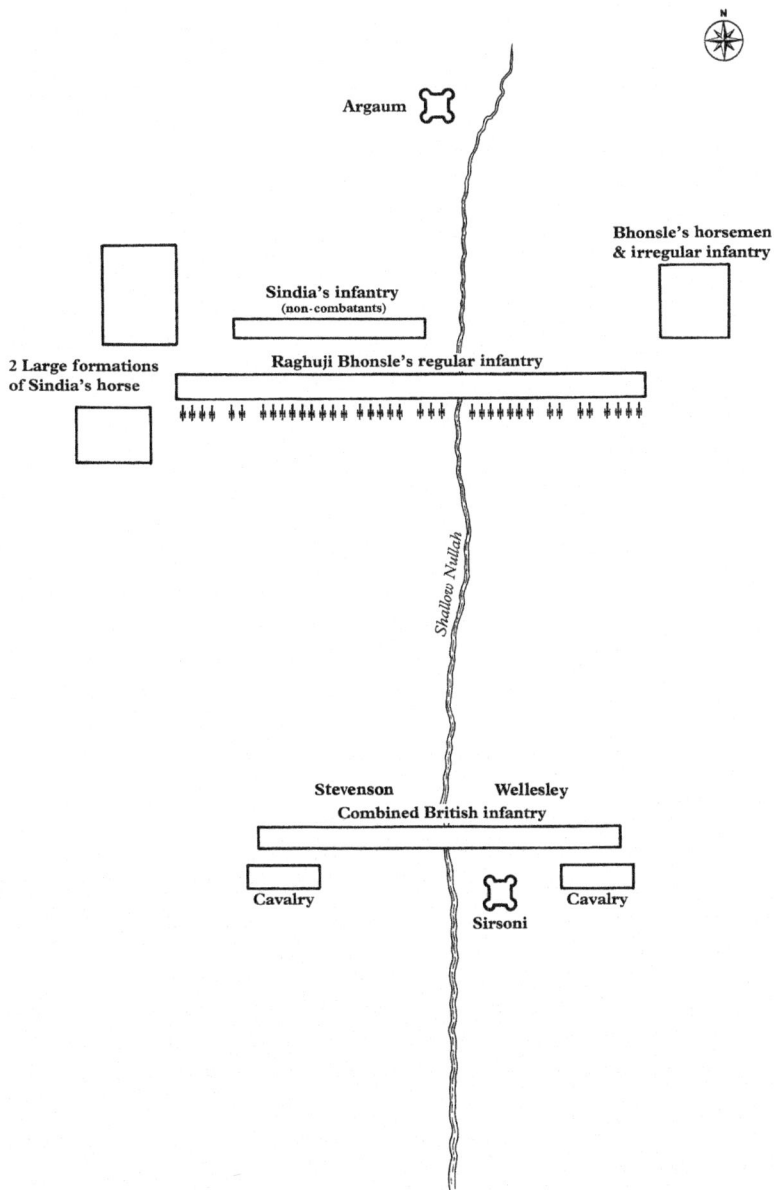

Map 6 The Battle of Argaum 29 November 1803, based on a map by
Emery Walker

pause when the enemy's guns ceased firing, and they advanced in front of them, was an awful one.'[206] The Maratha gunners, with the leading British units now some sixty yards away and closing, fired grape shot. The relatively close range discharge of grape, instead of the usual canister shot, minimized the weapons' ballistic effect, as the heavier pieces of grape could not fully 'open up' their lethal pattern at a distance so close to the muzzle. Welsh noted that these discharges felled about ten men per regiment.

The Marathas then decided to charge His Majesty's 74th and the 78th Regiments. Perhaps the Marathas hoped to inflict a signal victory in front of the previously spooked Madras infantry. About 500–600 Arabs, identified by some authors as Persians, charged the Europeans and were cut down to the last man.[207] The Ghosseins under Beni Singh, and apparently 'dressed like beef-eaters', tried to turn the Madras infantry's flank.[208] The 74th and 78th, having dispatched the Arabs, turned to help the sepoy units and with that move they thwarted the attempt to 'roll up' their line. Bhonsle's infantry received some encouragement from his camel-mounted rocket troops but their desultory fire lacked effective coordination and they soon broke off the action.[209]

Sindia's cavalry under Gopal Rao Bhau charged the Madras 6th Native Infantry but they were repulsed with steady musketry. The British dispatched their cavalry in a charge that broke the Maratha horse and set off a wholesale retreat. Many of the Maratha horsemen were wearing helmets and traditional body armour over stuffed quilted coats that offered protection from British sword cuts. Blakiston observed that the prevalence of helmets and body armour among some Maratha cavalry contributed to a modification in British mounted tactics. British horsemen, in pursuit, used their first sword strike to knock the helmets from their opponents and then dispatched them with a second blow or thrust to the skull.[210] A group of Maratha mounted troops who escaped made it off the field with their gallopers but thirty-eight cannon fell to the British. Bhonsle's infantry faired poorly in their flight from the field. Welsh recorded that upwards of 3,000 fell victim to the British cavalry who rode them down and cut them up well into the night, taking full advantage of the particularly bright moonlight. However, Beni Singh and a sizable number of his men managed to escape and they got to the fort at Gawilgarh.

In hindsight it can be said that the Marathas squandered their tactical opportunity to take advantage of the confusion at Sarsoni. If they had used their cavalry for envelopment and pressed hard on with an offensive infantry assault before the British line could be fully formed and the union with Stevenson's troops completed, then it might have been another story. Imperial historians and Wellingtonians often downplay Argaum as if nothing that followed Assaye could ever live up to the 'drums and trumpets' expectation of British military glory. Ironically the ability of Assaye to cast all else into the shade was a historically contemporary phenomenon. The lingering power of the Maratha artillery to inflict panic in the ranks of the EIC's sepoys at Argaum was not the stuff of polite conversation in Bengal. In speaking about Wellesley's casualties at Argaum, James Mill quoted the *Calcutta Gazette* as saying: 'The British loss, in this battle, if battle it may be called, was trifling; total in killed wounded, and missing, 346.'[211] The British sources stated 5,000 as the total number of Maratha casualties.

Wellesley was forging ahead and although Argaum was not as 'great' a victory in the popular press, it had been vital in diminishing Bhonsle's resources. In his wake Arthur left a trail of broken men who would need to be laid up somewhere until they stabilized, recovered, or died. Argaum had heightened the need for trained medical men. With each new victory came the necessity of establishing yet another field hospital equipped with surgeons because amputations were a by-product of infantry warfare in that era. The Bombay Presidency was beginning to realize the magnitude of the medical crisis. They controlled the rear-area hospitals where the long-term wounded would ultimately be sent. Bombay authorities asked the Major-General if he would allow some of their seconded surgeons to return home to Bombay if he had no particular need for them. The Deccan Campaign was almost over; Sindia's infantry battalions had been pulled out or all but eliminated – only his horsemen remained a threat.[212] The Major-General would now devote his full attention to Bhonsle – the Berar Rajah. Wellesley's next move was to track down those enemy troops who had escaped Argaum and eliminate them. Even though Beni Singh's men had been badly beaten up, Wellesley knew it would be tempting fate to deprive his own men of what feeble medical aid he had been able to scrounge. Arthur Wellesley would write to the Bombay government and tell them that 'upon the whole I do not think I can allow these surgeons to return to Bombay with justice to the troops under my command'.[213]

Gawilgarh

As the first week of December 1803 drew to a close, it was obvious that the British had attained almost total victory in the southern theatre of operations. For the moment they controlled the Deccan by holding the main fortresses that dotted the land, while the larger fortified villages relied on armed irregulars to stand behind the mud walls defending them against *pindaries*. Arthur Wellesley's final objective in the southern campaign of 1803 was the fortress of Gawilgarh. Destroying Bhonsle's military

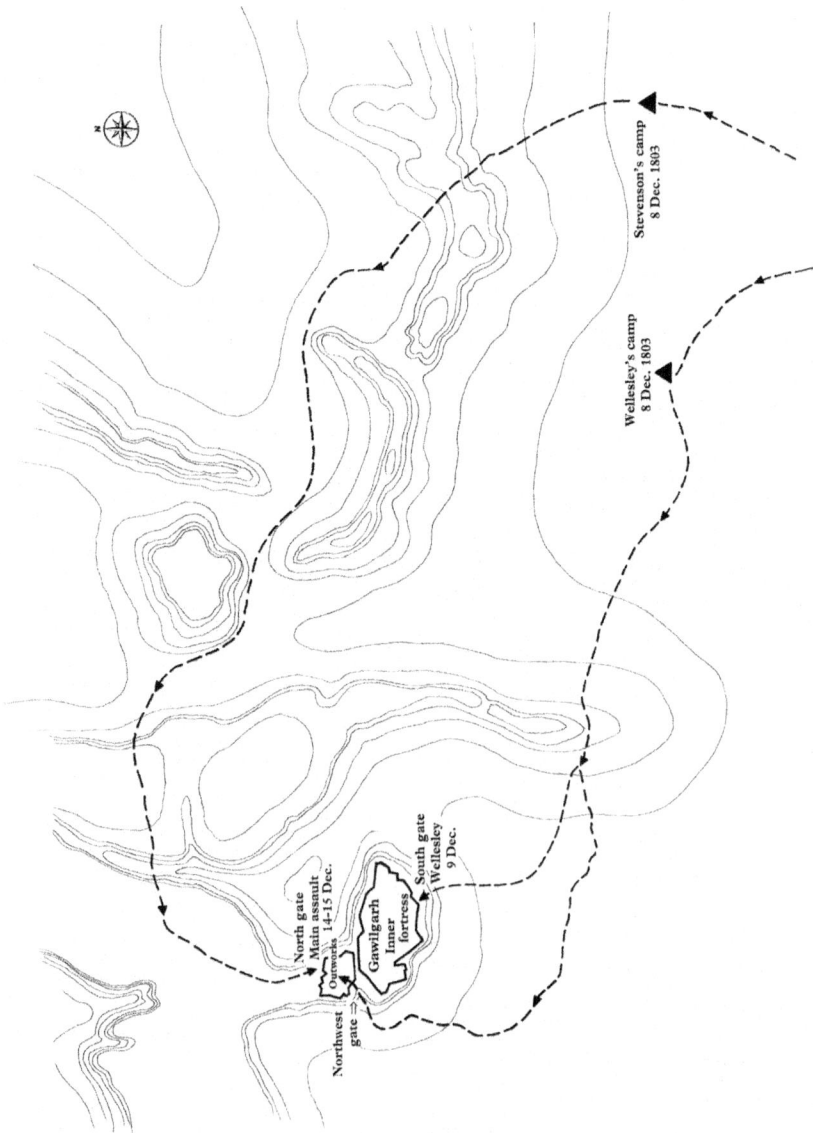

Map 7 The Siege of Gawilgarh December 1803, based on a map in J. W. Fortescue, *A History of the British Army*, 13 vols. London, 1911–20

capability remained part of the Governor-General's directive and so Gawilgarh and the troops sheltered there were both targets. The fort played a role as a fortified sanctuary along one of the main infiltration routes in the transitional zone that linked the Deccan to Hindustan. Along with Burhanpur, Asirghar, Daulatabad and Ahmadnagar, it was as one of the Deccani chain that the British viewed as critical in a bid to disconnect the north from the south. The Maratha corridor was not a single road or path, but rather a series of historic access routes. Although Sindia's Deccani forts had already been taken by Wellesley's army, in effect putting a stopper into the neck of the Deccan as it funnelled north, the capture of Bhonsle's Gawilgarh would further destroy the Marathas' ability to regionally shift troops with impunity. A permanent disconnection in the Maratha ability to project power was essential to the British strategy of containment. In 1803 British troops were, rightly or wrongly, judged to be skilled at taking hill forts.[214] They earned that reputation in the Second Anglo-Mysore War and confirmed it with Popham's capture of Mahadji Sindia's sanctuary at Gwalior during the First Anglo-Maratha War.[215] But Arthur Wellesley knew he could not afford to become reckless. This was not to be a rushed escalade like his effort at the *pettah* of Ahmadnagar. One-third of his dead in that three-hour skirmish were officers and it had been too costly in terms of command and control resources given that fortress's ultimate fall by negotiation.

Unlike many of the great northern forts in Hindustan, Gawilgarh had been built according to an asymmetrical blueprint to incorporate the terrain into the defensive fortifications. Elphinstone provided by far the most scenic description of the fort's location, but then he often tended towards the romantic.[216] An approach from the east or west was ruled out owing to the elevation and an inability to use conventional British siege techniques from those sides. The southern face took advantage of a natural scarp and lack of a suitable approach road. The northern exposure could be reached along an elevated saddle but the fort's architects had recognized that potential deficiency and so that side featured out-lying walls forming a layered defence works. Wellesley's siege plan for Gawilgarh required the opening of breaching batteries from two opposite sides of the fort. Colonel Stevenson began trenching and pushed saps along the northern approach that he oversaw. At night the fort took on a mystical quality as the Marathas fired their so-called 'blue lights' that burned with an eerie glow. These illumination rounds aided Maratha sniping at reconnaissance patrols and British work parties that laboured under cover of darkness.[217]

British light brass 6 pounder field guns were brought forward and aimed at the tops of the outer defence works to suppress harassing

Maratha smallarms and artillery fire. Blakiston indicated that they were constantly engaged in firing to keep the Marathas' heads down so that siege work could progress according to schedule. He wrote that on a particularly hot day the British gun crews had to halt their firing to let their brass cannons cool, thereby reducing the risk of spontaneous combustion when ramming powder charges down the barrel. They feared their cannon barrels were hot enough to ignite the black powder charges as the gun tubes absorbed both the external heat of the sun and the intense internal heat produced from steady firing.[218] Despite delays and the continuation of harassing Maratha fire, the northern battery was finally completed and the heavy British iron guns brought into service. This is also one of the few actions in the Deccan with clear references to Wellesley's men having access to rifles. Ironically the incident involved a case of mistaken identity when Captain Colebrooke called for a rifle to shoot a man at a considerable distance. The man turned out to be Captain Dickson who had gone out to inspect a preliminary breach in the outermost wall of the fort. Fortunately Dickson was identified before the shot was made.[219]

At night the British fired grape shot into the breach every twenty minutes to keep the fort's garrison from repairing the damage.[220] But, owing to the degree to which the guns were pushed back from the main fort by the perimeter walls, Stevenson's batteries could not reach the innermost layer of fortifications with sufficient direct fire from their position.[221] Angled explosive shellfire from howitzers was of no use in trying to create a suitable breach in the inner sanctum. The pulverizing of stonewalls called for concentrated kinetic energy, not exploding shards of shell casing. Yet the heavy guns, firing iron round shot with tightly fitted sabots, could not draw direct aim on the innermost walls owing to protection from the outlying secondary walls. Given the distance, ballistic drop in trajectory and number of walls that had to be penetrated, there was not sufficient firepower to blow a series of holes from the battery straight through into the heart of the fort. Even though the British held the *pettah* for a week from 7–15 December 1803, without direct breaching fire the plan to gain access to the interior of the fort had to be modified.[222]

Progress on the southern side was tortuously slow owing to natural obstacles. Elphinstone watched one day in fascination as a bullock team was hitched to pull a 12 pounder while an elephant was positioned behind to push it. The young man was thoroughly captivated by the drums beating the 'Grenadiers' March' as sepoys toiled and he remarked that 'the whole of the scene was romantic'.[223] Captain Welsh, among the troops tasked with carving out a path for the iron 12 pounders on the southern side, was less enthralled with the labour. The southerly approach necessitated extremely strenuous efforts on the part of several units and the pioneers were driven to their limits in cutting pathways and creating new support roads to facilitate the movement of iron siege guns up the treacherous southern hills. The angle of ascent on that approach was too steep for the 'heavy metal'. The work details, which had to manhandle each piece, experienced several injuries and casualties mounted as men strained their muscles and wrenched their backs levering the guns inches at a time. Thin soil covering and loose footing contributed to numerous accidents. Compression wounds and fractures abounded as limbs were crushed by the unforgiving weight of great iron gun tubes pulled back down the slippery slopes by gravity.

It was just impossible and so the plan was changed. The iron guns on the south side, which could not be brought up and into firing position, were temporarily abandoned. They were pushed off the ascent path and covered with brush so Maratha patrols could not find them and render them useless by spiking them.[224] As Captain Welsh's writing indicates, it must have been obvious to any astute military observer that the southern effort had become merely a feint. He recorded that 'every man of science in our camp could readily foresee that this labour was in vain, further than as a diversion'.[225] Two brass 12 pounders and two 5.5-inch howitzers would be made to serve the purpose of a ruse, creating the sound if not the fury of a fire support role. Problems persisted with what Elphinstone described as an 'ugly' battery, saying it looked 'more like a fold for sheep than anything else'.[226] They were to commence bombardment on 15 December and a small diversionary assault party was to make a feint on the southern flank. According to Welsh it was with surprise and horror that they began firing on the morning of the 15th only to see their iron round shot come rebounding back down from the heights of Gawilgarh, nearly reaching the muzzles from which they had been discharged. The altitude and angle of trajectory were just too much. And to make matters more embarrassing the battery, built mostly of fascines, caught fire.[227] Fortunately there was sufficient water to put out the blaze.

During the attack of 15 December the real push began on the opposite side of the fort after Stevenson's northern batteries had succeeded in knocking a hole in the second wall. That allowed a body of Highlanders inside the perimeter walls where they then hoped to plant their ladders to scale the next innermost wall. The trick to getting into that next defensive zone of the inner fort was to put sufficient men over the top in a group, so that momentum would not be lost as individual soldiers were picked off by the defending garrison. The account in Elphinstone's journal made it clear the undulating ground and variations in the wall construction proved more of an obstacle than the British anticipated.[228] However, a subsequent letter to his friend Strachey revealed even more. For it was inside the second breach near the innermost sanctuary they discovered to their horror that there was a physical divide between their party and the fort.[229] It was within the second defensive perimeter that arguably the bloodiest fighting took place. There was confusion and panic back at the first breach. Some of the troops managed to get to a gate linking the two perimeters and opened it so that the bulk of the British forces could flood in. They were met by stiff resistance as thousands of Maratha troops poured into the mêlée. There was little time for the British infantrymen to reload after initially discharging their muskets. Much of the fighting devolved into a series of individual combats with bayonets and edged weapons. The most ardent defenders were said by some Britons to be those under the direction of Beni Singh, who led a forlorn hope. To a man that group perished in hand-to-hand combat. Many of them were run through with the 16-inch EIC bayonet, a nasty triangular spike of steel tapering to an efficient point that lent it self to puncturing rather than cutting an opponent. Others were simply bludgeoned to death with EIC muskets wielded as clubs in this desperate action. No quarter was asked and none given. Elphinstone wrote that they eventually found a way to the inner sanctuary, but he casually mentioned 'a prisoner was taken who knew the way'.[230] The Rajput contingent in the fort fought with extreme determination but eventually the British prevailed. An undetermined number of the defenders allegedly committed suicide by jumping to their deaths from the hill fort when it was found they could not hope to win; a historic analogy with the death of Rajputs at Chittore would not be out of place.

Wellesley had ordered no plundering but on this occasion he was apparently unable or unwilling to prevent it.[231] Elphinstone wrote, 'All around us, everywhere alike, lay dead and dying, and on one side was an officer calling out for volunteers to hang the Killadar. I saved him by the argument that he knew where the treasure was.'[232] There was virtually no treasure and the *killadar* escaped only to be killed in the pandemonium. In fairness to the British troops who rampaged through the fort, it should be said that it was customary for European armies during that era to allow troops to plunder a fortress that offered ardent resistance after refusing terms of surrender. Large quantities of low-denomination coins, silver plate and clothing were carried out but there was no vast fortune. The fort yielded 52 pieces of artillery, 150 wall guns in calibres from one-half pound to one pound and 2,000 EIC muskets in pristine condition.[233] Maratha losses were estimated as high as 4,000 while those of the British were a mere 126 'Europeans and natives'.[234] The fall of Bhonsle's fortress of Gawilgarh proved to be the final battle of the 1803 Anglo-Maratha Campaigns.

Conclusion

In this chapter we have seen that Arthur Wellesley had to confront some
rather uncomfortable truths in the Deccan. After Assaye it was painfully
apparent that not all Maratha armies were composed of light horsemen
and that Sindia's infantry knew both the theory and the practice of modern infantry warfare. But Arthur also had to confront the truth at having
misread more than Maratha military culture. He had misjudged rather
substantial parts of the South Asian military environment by seeking to
transfer theories from one micro-environment to the subcontinent as a
whole. The failure of the monsoon rains, combined with the truth about
the Marathas' ability to cross rivers, helped heighten his sense of urgency
in August and September 1803 as he rushed to hunt down and destroy
the Maratha units in the Deccan. If unchecked those remaining Maratha
troops could threaten the British Subsidiary Alliance System. An inability
to protect puppets such as the *Nizam* and the *Peshwa* would be a major
blow to brother Richard Wellesley's plan to create manageable buffer
states. As militarily emasculated allies they were forced into dependence
on British organized military muscle for their protection. However, the
British had to do a credible job of protecting their allies or the whole
system ran the risk of unravelling.

In the Deccan of 1803 we see Arthur Wellesley as having to deal with a
foreign military culture which was at odds with his British military culture.
I do not mean 'at odds' in the obvious sense of an adversarial relationship
in combat, but rather that Arthur perceived several aspects of South Asian
military culture as alien to his own British military and political culture.
That he should not understand the businesslike South Asian attitude
towards negotiating military labour contracts on the field of battle is not at
all unusual given the predominance of a 'duty, honour, country' mentality
in his British military background. 'Death before dishonour' has served
as both a behavioural cliché and motto for elite Western military forces,
but how does it translate across the boundaries of competing military
cultures? To the South Asian warriors of 1803 there might be no particular
dishonour involved in negotiating the surrender of a fort to the British. If
you were *killadar* of a fort and your Maratha employer had not paid you
for months, but the British were willing to negotiate an amount of money
that represented your back pay and compensation for loss of the fort as
it affected your future income-earning potential – why not consider their
offer? Did that reflect greed, corruption and the betrayal of one's country?
Not necessarily; what was greedy about wanting what was owed to you
and if you had kept faith that far in the contractual obligation of holding
the fort, where was the corruption? Did the corruption lie above you in

a system that failed to meet its obligation to pay you as promised? And the issue of national betrayal was hardly applicable given the absence of nationhood. If we were to push the notion of a 'warrior's duty' it is worth considering that only survivors can refight their battles and 'live again to fight another day' would seem to have been an axiom that transformed many South Asian sieges into long drawn-out business negotiations.

The difference in warrior values, as found in these competing military cultures, can often be detected in perceptions of how lives were 'spent' for military objectives. In touring the aftermath of the battle for Gawilgarh, John Blakiston and a companion were drawn to a large room in the fort filled with dead and dying Rajput women.[235] Their loving husbands had slit the ladies' throats and the blood-soaked room haunted Blakiston's memory more than the sanguinary field of Assaye. It was an image that jarred his Western sensibilities. He did not really know how to reconcile the voluntary killing of spouses and it seems that Blakiston wished to isolate it as an event separate or detached from the battle proper. Once the option of reasonable negotiation had run out, the Rajput garrison slaughtered their women before engaging Arthur Wellesley's forces because by that point the decision was made that the fight would be to the death of the last man. Theirs was an ancient practice to keep the purity of the women intact by preventing them from being dishonoured upon falling into enemy hands. It would seem that 'death before dishonour' in this case represented a more holistic understanding of both death and dishonour as extended to the South Asian familial identity of these Rajput warriors. We find directly conflicting concepts of chivalry in looking at Rajput vs. Western military culture here. The execution of wives was not the type of loving behaviour that could be easily equated with examples of nineteenth-century Western military culture and therefore it complicated objective military analysis by foreign observers. It is all the more interesting since the Western concept of military honour would dictate an almost opposite code of conduct, in the sense that surrendering to protect female honour would be preferred over risking female deaths as non-combatants. To kill one's own loved ones to prevent capture would be virtually unheard of in Western society in the early nineteenth century. Paying ransom, surrendering all male warriors – these were viable alternatives in a Western military culture where 'death before dishonour' was a motto meant for the macho world of the European soldier who spent his life for king, country or unit pride.

By December 1803, in less than six months of campaigning, Arthur Wellesley had delivered the Deccan. And so he was forgiven the human price he paid for a lesson in modern infantry warfare at Assaye. His fear of command failure apparently motivated him to engage in offensive operations that were beyond the classification of 'doggedly determined'. The concept of one-third casualties at Assaye and marching his men repeatedly into the mouths of cannon appeared as maniacal to a South Asian military culture that saw such behaviour as weighing the cost of life against the bigger business opportunity that was war. By being so drastically at cultural odds with the prevailing South Asian system Arthur Wellesley set himself apart as an enemy to be feared. It is quite human to fear that which we do not understand and combat behaviour is no exception. That is one of the elements that makes terrorism effective. When people operate militarily outside of the norm or at odds with the shared value system of a group they are fighting, then they – as enemies – pose a special threat or point of terror because they are not behaving in a logical and predictable way. Arthur's strategic gains redeemed his tactical indiscretions in the eyes of many Britons who shuddered to think of the alternatives. Retreating under fire was not glorious. And standing still to be cannonaded into oblivion might be glorious to some military cultures, but it was a useless waste if you sought to gain ground and win wars with the lives you spent.

Although Wellesley wrestled with the differences imposed by this alien land there were definite signs that even he was adapting to the South Asian military economy. Arthur saw advantages in exploiting a system in which all military assets, commodities and resources, had a market price. The British had a huge corporate and governmental structure behind them and sufficient leverage to procure a war chest with enough money to buy out virtually all the Marathas' military assets. The pre-war acquisition of southern *jaghirdars* and the allegiance of the *Peshwa*'s retainers – men like General Gokhale – represented British knowledge and use of the South Asian military economy. When it came to buying out forts like Asirghar, the British strategy was made easier by the Marathas' subcontracting of garrison duties. Without sufficient political, national, or religious cause to stand and fight, the loyalty of many garrisons was to survival and a work ethic that went largely unspoken in the military labour market. Powerbrokers bargained for – and hired – soldiers as contracted garrisons. If a better offer came along and the primary retainer was not present to ensure the enforcement of contractual obligations, then it was easy for the contracted employees to change employers and in the process deprive one of additional military assets such as the fort that was being garrisoned.

The British victory in the 1803 Deccan Campaign also signalled the end of an era in Maratha military history. By taking Pune, then Ahmadnagar, Burhanpur, Asirghar and Gawilgarh, Arthur Wellesley had been vital in disrupting a Maratha presence that stretched back in one form or another for two centuries. By temporarily removing the Marathas' hold on the Deccani fortress chain, Arthur had helped to advance his brother Richard's strategy of containment. The British were keen to close the Maratha corridor to North India – a series of military infiltration routes used by Maratha leaders like *Peshwa* Baji Rao I to menace Bengal. Arthur had temporarily closed the valve of the pipeline long enough to allow blocking mechanisms to be put in place. British rapid reaction forces, many of which included irregular native horse units, were ordered by the Governor-General to sit astride the Maratha access routes, further facilitating the containment strategy.[236] Outlying British unit commanders, stationed along roads in the Upper Provinces, received specific orders to make ready the snare intended for Maratha troops moving between the Deccan and Hindustan theatres. The disconnection in 1803 prevented thousands of Maratha light horsemen and *pindaries* funnelling northward to assist Sindia against the British in the Gunga-Yamuna Doab. No longer were the EIC's wars being fought on territories adjacent to her Presidencies or limited to raids into the interior. Richard Wellesley had used his brother Arthur to help push the conflict for control to the very centre of the subcontinent. The Bengal, Bombay and Madras Presidencies had triangulated the map of India; it was only a matter of time until their power would be consolidated.

4   The Hindustan Campaign of 1803

Introduction

This chapter concerns the northern theatre of operations in 1803. It was also known as the Hindustan Campaign, an effort overseen by the Commander-in-Chief of British forces in India General Gerard Lake.* Fortunately Lake ordered his officers to keep journals of the campaign and this narrative would be impossible without that directive. Men like John Pester, William Thorn, Charles Stuart and George Isaac Call left us an amazingly rich and honest collection of their experiences. John Pester was a hard-drinking, hard-fighting man and nothing shocked him for long. On his days off he hunted for the love of the chase and shooting, but his sense of fairness also stemmed from what we would call 'his love of sport'. William Thorn, already a captain, was observant and very intelligent but he later published with political purpose and he comes across here as up-holding the 'official story'. Charles Stuart was a temperate and well-mannered gentleman in the aristocratic sense of the word. As a dedicated British officer in the EIC's Bengal Native Cavalry, he felt a tremendous amount of compassion for the South Asian officers and men that served with him. On many occasions Charles Stuart did not agree with Lake's decisions and Stuart saw tremendous evil in this war. As for George Call, he too was observant and intelligent. But in the writings of the wide-eyed Call we can also see a young man who was being steeped in an elitist military culture – that of His Majesty's Dragoons. We do not see the cliché 'loss of innocence' in Call, but rather we detect the insidious armour of racial callousness growing week by week.

This story of 'Hindustan 1803' begins almost a year before the campaign against Sindia in an effort to show the reader that the struggle for North India was already well under way. As we shall see, all was not well and rebellion was rife. This posed a problem for Richard Wellesley's administration since the EIC's accountants had urged further military

* Later known as Lord Lake.

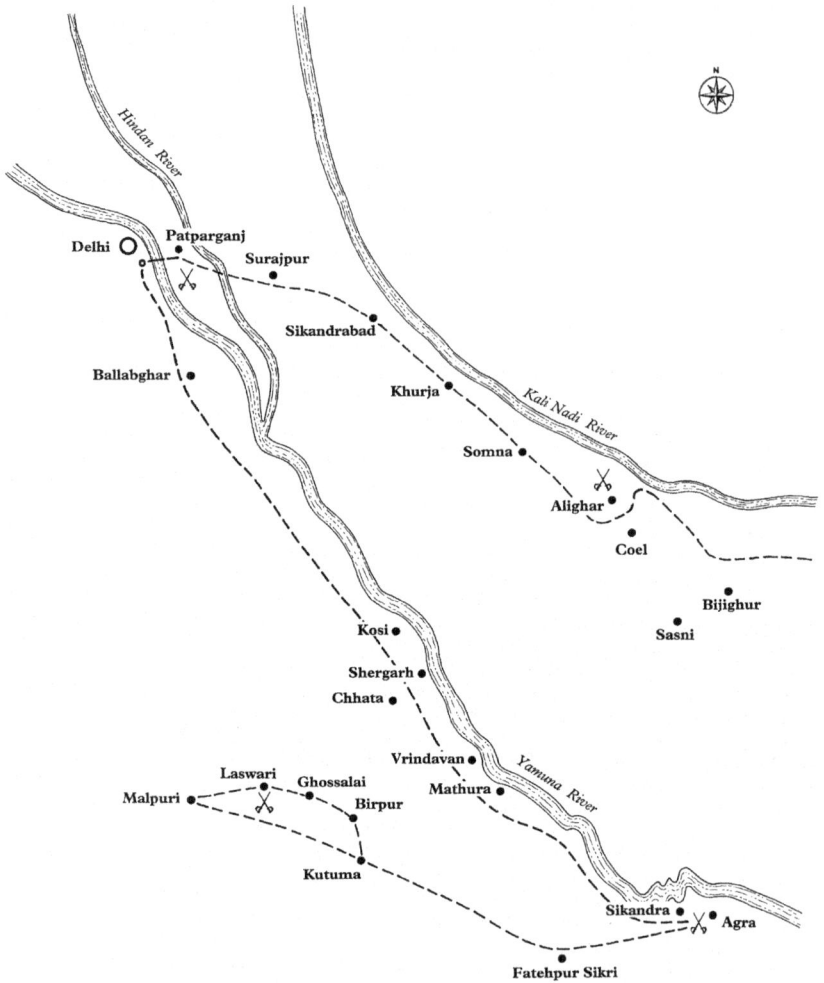

Map 8   An overview of the 1803 Hindustan Campaign: from a map (not to scale) by George Isaac Call, His Majesty's 27th Light Dragoons

cutbacks. They had desperately hoped that their ledgers might show the benefits of peace. Lake, as one of England's senior-most commanders in the field – having fought in Europe and America – knew better than to lie about military preparedness and the projected cost of victory. He realized time and money were both in short supply; his troops were simply not ready for a war with Daulat Rao Sindia in 1802.

The Maratha military tradition of seeking pragmatic cost-effective solutions to military problems seemed to leave Sindia in an ideal position in 1802. He could easily dwarf the EIC's sepoy army in North India by dominating the military labour market through spending his accumulated wealth and extending his indigenous credit lines. Failing that, he could rally hundreds of thousands of *pindaries* if he so decided. But that made little sense to him, as it was an option akin to opening Pandora's box. Once loosed in vast numbers the *pindaries* would strip all the land like locusts and that included Sindia's personal holdings. Why risk such a dangerous policy in a land as revenue-rich as the Doab? Every time the cost-cutting Bengal Presidency demobilized a company or even a platoon of sepoys it meant fresh assets were there to be snapped up. As for firepower, Perron had consolidated the improvements made in defence procurement under de Boigne. Artillery and smallarms were stockpiled in Sindia's arsenals and depots. The stage was set for a monumental showdown in Hindustan. What the British needed was some means of gaining a competitive advantage and, failing that, they needed some way to compromise Sindia's powerful army.

The 'Mud War'

In 1802 the Bengal Presidency authorized troop reductions in a number of EIC military units.[1] In his diary Lieutenant Charles Stuart of the 3rd Bengal Native Cavalry* criticized the expansion of the EIC's territory just as demobilization had swollen the Marathas' infantry ranks with British sepoys. The free flow of trained men to the Maratha 'regular corps' was worsened by the fact that Sindia's troops had no shortage of smallarms, ammunition, or artillery. General Perron, Sindia's C-in-C, saved training costs and mobilization time by 'plugging in' these valuable free agents from the military labour market. Charles Stuart wrote critically:

With the increase of dominion no increase of our force had taken place; on the contrary a reduction of nearly 8,000 men had considerably weakened the Army all the branches of which, had been placed on the peace establishment. On the other hand General Perron was actively employed in adding a new Brigade to his army & many of the men lately discharged from our Service flocked to his standard, while the whole of his force was amply supplied with every military equipment.[2]

The troop reductions put pressure on General Lake to try and get the best performance out of his men. Lake and Governor-General Wellesley

---

* Aka 3rd Native Light Cavalry or simply the 3rd Bengal NC.

had been shocked to see the disastrous outcome of an 1802 experiment in training men for the Bengal Native Cavalry. The concept was to have new recruits 'break' and train their horses as part of their own introduction to cavalry service. Lake thought it a miserable failure and he believed many of the problems stemmed from using inexperienced officers who were not combat veterans. Generally speaking, the Bengal Native Cavalry were poorly prepared and the Governor-General was mortified to find the King's 29th Regiment of Light Dragoons were in a 'nearly equal state of inefficiency'.[3] The only bright spot in the review of troops was the 3rd Bengal NC Regiment and the credit went to Major Middleton. The remedy for addressing this shocking state of affairs was an intensive retraining programme. Lake's plan would serve the dual purpose of up-grading his cavalry troops as well as preparing them for the possibility of all-out war against the Marathas.

Lake's British and native cavalry, with the exception of large numbers of the 3rd Bengal NC, were sent to Kannauj* during the winter months of 1802–3.[4] Lake had secured Kannauj, as well as some other territory in the Gunga-Yamuna Doab and Rohillakund, from the Nawab of Awadh in lieu of outstanding debts.[5] As territories ceded to the EIC, these became the latest in a long string of acquisitions, which the Company called the 'Ceded Districts'.[6] It was an astute move on the eve of a war with Sindia because the territory bordered his and that would reduce the length of future marches and resupply routes. The training ground near the Maratha frontier was ideal 'cover' for explaining the arrival of troops from Khanpur† and Bengal.

At Kannauj the horse soldiers were to study the so-called new uniform system of cavalry movements as taught in person by Colonel (later Major-General) William St Leger of His Majesty's 27th Light Dragoons.[7] The idea was to transform Lake's cavalry into a well-oiled military machine, one that was acclimatized and trained in a setting identical to the area of operations in which it was to be deployed. Perhaps more importantly, the men were destined to fight under combat leaders who had instructed them, an element that must have contributed greatly to unit cohesion and esprit de corps. William Thorn, who participated in the exercises, noted: 'nothing could exceed the celerity and exactness of the manoeuvres made with them by this large body of cavalry'.[8] Most of St Leger's drill was to be applied to old-established tactics but these men were also being drilled in the new doctrine of mobile firepower. St Leger had the men

---

* 27°+, 79°+. Aka Kanauj, Kanouge, Canouge. It had been the seat of a great and powerful empire under Harsha in the seventh century CE.

† 25°+, 83°+. Aka Cawnpore, Caunpoor, Cawnpoor.

train exhaustively with the 6 pounder galloper gun, two of which were to be attached to each cavalry regiment. The galloper was a horse-drawn gun that many cavalry officers knew as 'flying artillery'. In time separate 'horse artillery' units would emerge, but this was mobile firepower raised within the cavalry regiments. EIC officers had studied the successful use of horse artillery by Frederick the Great during the Seven Years' War (1756–63).[9]

Horse-drawn artillery had been known in South Asia since Babur's use of *rakhale* at the First Battle of Panipat in 1526.[10] Maratha armies had possessed one form or another of horse-drawn artillery throughout the eighteenth century and even Jeswunt Rao Holkar's irregular cavalry used gallopers. But the smaller *pindari* bands that plagued the British lacked resources and they had no answer to a 6 pounder laying down a well-patterned curtain of canister shot. There was little incentive for local *pindari* groups to acquire artillery as it was a logistical liability and its mere possession could not be explained in anything less than threatening terms that risked challenge from the British. The 6 pounder 'gallopers' were a major breakthrough in terms of the speed and mobility of firepower when dealing with *pindaries*. The British could pursue a *pindari* unit and engage it at ranges that were impossible to achieve with muskets. The 3 pounder was too small and the 12 pounder too heavy for practical use as a galloper.[11] As a force multiplier, or weapon that increased one's firepower exponentially, the 6 pounder galloper was ideal against a numerically superior force brandishing nothing larger than smallarms. St Leger's training at Kannauj was evidence that British forces were adapting their existing form of warfare to cope with changing requirements in the South Asian military environment. But only now was the British military acquiring parity with the Sindia clan's fire mobility, for they had possessed gallopers for a considerable length of time and de Boigne had standardized their calibres and dimensions.

While most of Lake's horsemen were busily engaged at Kannauj, Charles Stuart and elements of the 3rd Bengal NC were out in the field supporting a number of operations that would come to be known collectively as the 'Mud War'.[12] This series of nasty sieges was directed against the so-called 'mud forts' of the Doab that habitually offered resistance to central authority. It did not matter if the central authority was the Mughals, the Nawab of Awadh, or the British. The landlords who ensconced themselves in these forts were reluctant to pay their taxes. Their 'rude' forts were brown or dung coloured, owing to the use of simple earthen building materials drawn from the immediate environment. But the term 'mud fort' was a bit of a misnomer. Although that name summarized the demeaning contempt in which Bengal's tax authorities held them, it obscured the fact that simple construction techniques made the walls difficult to breach and easy to repair.[13] These defiant or 'Refractory Zamindars' resorted to armed insurgency with cyclical regularity. It was something they had done since before the establishment of Mughal authority although never as part of a coordinated resistance.[14]

Lieutenant John Pester of the Bengal army became a veteran of siege warfare, serving in the British trenches at Sasni,* Bijighur† and Kachaura.‡[15] Two of the 'mud forts', Sasni and Bijighur, belonged to Rajah Bhagwant Singh and figured prominently in a threatened anti-British 'Jat alliance' with Bharatpur. Coincidentally or not, Sasni just happened to be only a day's ride from Coel,§ Headquarters for Sindia's C-in-C General Perron. Charles Stuart wrote in his diary, 'General Perron then at Coel was heard to say that the force we had assembled was not merely for the purpose of subduing a refractory Zumeendar and on our going away he observed, if the English imagined everything was settled with the Mahrattas they were much mistaken.'[16] Counterinsurgency operations in the 'Mud War' provided an ideal opportunity to gain intelligence on the Maratha forces and the subjugation of Sasni as well as Bijighur provided the EIC with depots for future use.[17]

As for Kachaura, it foreshadowed the darker side of war that lay ahead in the struggle for North India. After being surrounded by the British, the garrison had sought terms of capitulation. Lake sent a small force under Captain Müller, the son of a Prussian general, to accept its surrender. Müller and his troops were admitted to the lower portion of Kachaura but not the upper. Supposedly, arrangements were still being made for the evacuation of the *sardar*'s wives and children.[18] In the morning it became evident that Müller and his men were in the sights of the garrison who told them to withdraw or face death. Müller's people came under fire from Kachaura's garrison after making an orderly retreat. Lake saw this as the worst violation of bargaining in good faith and sought to make an example of the garrison. Pester's company and another from his battalion were ordered to seize the *pettah* (close to the fort's *glacis*) and put it to the torch – which they did. The thatched roofs and stored quantities of animal fodder burned intensely. Kachaura's walls were eventually holed with artillery fire and great numbers of the defenders were bayoneted as they sought to escape through the breach. Inside the walls, the remaining

---

* 27°+, 78°+. Aka Sarsney, Sassney.
† 27°+, 78°+. Aka Bidgie Ghur, Bechey Ghur, Bidijaghur, Bidjeeghur, Beejeghur, Beejaghar, Bidgighar.
‡ 27°+, 78°+. Aka Cachoura, Cutcherra, Kouchera, Kachoura, Kachura.
§ 27°+, 78°+. Aka Kol, Koil, Koel.

garrison was 'cut up'.[19] The cavalry were then employed in riding down those few who had got beyond the perimeter:

The enemy, who were extremely numerous, were almost to a man put to death. The ground in every direction near the fort was strewed with their dead, and they very dearly paid for their unaccountable conduct in turning out our troops . . . Hardly any of the garrison escaped the carnage, which by their own dishonourable conduct they entirely brought on themselves, and which they most richly merited.[20]

With the 'Mud War' concluded, the British troops returned to their cantonments. Charles Stuart recalled, 'but scarcely had we settled ourselves comfortably in our Bungalows when our promised tranquillity was disturbed by reports of an approaching Mahratta war'.[21]

General Lake's 'Grand Army' of Hindustan

Final preparations for war against the Marathas in the northern theatre were made during the first week of August 1803.[22] Lieutenant Pester took his grenadiers for target practice, each man in the section firing six rounds. Battalion exercises were conducted in the cool hours of the morning and included the manoeuvres thought most likely to be used in massed firing deployments.[23] He recorded the battalion exercises as 'the most useful and likely manoeuvres to be practised with an army'; they had been designed 'to make the men steady in their firings'. These were preparations for large-scale regular operations. The plan was to be battle ready for the march to Etah, which would serve as the staging area for the assembly of Lake's Grand Army.

General Lake marched from Khanpur on 7 August 1803 with the infantry under Major-General St John. The next day Colonel St Leger followed at the head of the cavalry. It was only sensible that the infantry, with its large baggage train and camp followers, be given a head start on the line of march. Major-General St John, an advocate of light infantry doctrine, had recommended a reorganization of the troops in accordance with the new light infantry tactics he championed. He had proposed that ten men from each company be specially trained in marksmanship and related light infantry skills. There were ten companies in each battalion, so when needed, the specially trained men could be called forward to the front of the battalion to form a special 'eleventh company'. They held potential for flank duty as well since it was believed their expertise could be utilized to hold the enemy at a distance. According to Colonel Pearse's study of Lake, the C-in-C approved of the arrangement as a 'temporary measure'.[24] There were strong opponents to St John's plan who saw it not simply as a recurring trend, but as dangerously shortsighted from the company commander's point of view. At that time infantry companies were only averaging a strength of eighty to ninety men, rather than one hundred. And the removal of ten more men when combat was imminent, especially those selected for their marksmanship and skill, robbed a company of its best men when they were needed the most.

Despite all that has been written about the supposed 'lessons of the rifle' during the American Revolution, General Lake, as a combat veteran of that war, saw no substitute for strength in numbers or the orthodoxy of the musket and bayonet against so strong an infantry as that possessed by Sindia. Although not formally studied it would seem that Lake, like Cornwallis his superior at Yorktown, actually profited by applying his American War experience to India. Lake had been pounded close to senseless by days of heavy French artillery bombardment at Yorktown and defiantly replied by leading a spirited grenadier trench raid against his enemy. There was no substitute for going on the offensive when an infantryman's morale sagged under the weight of siege-induced inactivity. Lake had also seen that the elaborate theories of seizing cities, as a means of crushing enemy resistance through urban and commercial control, did not work in the American Revolution; there was no reason to think they would work here. To him the cities of Agra and Delhi were merely life-support systems for the great fortresses that lay around them. It was the troops inside that counted. The operational objectives had to be kept simple – hunt down the enemy's regular army, engage it, and destroy it. The EIC apparently had to accomplish that or there could be no victory and certainly no peace. If a handful of 'mud forts' caused so much grief, then Sindia's professional standing army could spell disaster for any plans to bring the Doab under the EIC's control.

St Leger's cavalry and St John's infantry were reunited about five days later at a spot where the Kali Nadi empties into the Gunga.[25] It was an area familiar to the cavalry who had trained in that vicinity. Now they looked back fondly on those times. Life at Kannauj had been relatively comfortable, 'many officers had not only glass doors to their tents, but chimneys of brick'.[26] Several wives had accompanied their officer husbands and dining was luxuriant as the evening's claret was supplemented with bottles of shiraz and port on a regular basis. From Kannauj, Lake and his forces worked their way west–northwest along the Grand Trunk Road to the staging area at Etah* where they would rendezvous with additional units such as Lieutenant Pester's grenadiers.

On 19 August 1803 John Pester wrote that the so-called 'half-cast officers' in Maratha service who 'had left in consequence of a proclamation

* 27°+, 78°+.

declaring that all people of that description, born of British parentage, who remained in the service of the native powers, with whom war was now declared, would be considered as traitors to their country, and treated accordingly should the fate of the war throw them into our hands'.[27] The timing of Pester's entry is interesting because it was a full ten days before official public release of the information on 29 August, indicating that the contents of Governor-General Richard Wellesley's Proclamation had been available for some time.[28] The Proclamation clearly played upon perceptions of allegiance and British nationalism in a manner intended to transcend racial differences. While the thought of soon being labelled a 'traitor' conveyed the message that the days of the gentleman's understanding were rapidly drawing to a close. Things had taken a much more serious turn.

The next day, 20 August, General Lake received a letter from Lieutenant-Colonel Collins, the British Resident at Sindia's court. Collins' note said he was making his own way towards Aurangabad, which meant that peaceful negotiations had failed and that war was imminent. According to William Thorn of the dragoons, it was six days later that Lake officially received word from the Governor-General to commence war against Sindia's forces unless a message of a settlement arrived in the meantime from Arthur Wellesley.[29] In reading the memoirs and correspondence of British officers it is apparent that the British high command in North India realized the serious military threat posed by the Maratha artillery. They also knew the prize status that had been affixed to it. John Pester wrote: 'everything bade fair for a bloody campaign, as the enemy are certainly very numerous, supported by a formidable train of artillery, of which we hope soon to have an opportunity of relieving them'.[30]

Pester was a very sociable man and received intelligence from a great circle of friends that included 'Mrs C' at Mainpuri.* She was the wife of Mr Cunynghame,† the District Collector for the Ceded Districts.[31] Mrs Cunynghame wrote, with apparent disregard for security, about her husband's orders to build two bridges across the Isan River.‡ One bridge was to be reinforced to carry the weight of the British siege train destined to pass over it; while the other was to be dedicated strictly for troops. River crossings were going to be important in these operations and while smaller rivers could be bridged, if necessary, the Gunga and Yamuna would have to be crossed repeatedly with boats. The boatmen, as independent entrepreneurs in the South Asian military economy, stood to make a great deal of money in this campaign. Mr Cunynghame received

* 27°+, 78°+. Aka Mynpoorie.
† Aka Cuningham or Cunyngham. The spelling used in the above text was transcribed from the 'Collectors Letter' to Henry Wellesley.
‡ Aka Esah River. It runs parallel to the Gunga and Yamuna between Kannauj and Etawah.

orders to hire more *sebundies* to guard the ferry crossings. Many of the local *sebundies* had been seconded to irregular provincial battalions and now the race was on to find more just to guard the river crossings, since the Marathas could easily move a brigade over a river in a good night's work. Captain Hodgson was dispatched as far as Khanpur to bring in more *sebundies* in a futile effort to seal the riverbank.[32] From more conventional intelligence sources, Pester heard that a great number of Sikhs had joined Perron's forces. Perron, as Sindia's C-in-C, had reportedly sworn all 'the Chiefs of his army' to 'conquer or die with him'.[33] Further reports came in about the strength of Daulat Rao Sindia's brigades: 'they amount to nearly twenty times our number, both in troops and guns, he is said to be confident of success. His artillery reported to be equal to any in the world, and in that he chiefly depends for our destruction.'[34]

General Lake's party arrived at Etah on 23 August and found Pester as well as other troops from the Shikohabad cantonments waiting. Lake lost no time in briefing Colonel Blair, Pester's commanding officer, one of the few field officers in the campaign to have served in the earlier or First Anglo-Maratha War. Colonel Blair had been second in command of one of the corps, which formed the detachment under Goddard, made famous for the daring capture of Mahadji Sindia's cliff-top fortress of Gwalior in 1780.[35] Pester recorded: 'Colonel Blair had distinguished himself on many occasions and was thought to be also the first drill officer in the service.'[36] The men under his command would have had some of the finest drill instruction available in the country. It is not clear, however, to what degree Blair would have served as a repository for 'in-country' knowledge or previous experience against the Marathas. Most units tended to modify official drill and integrate elements of field craft or local knowledge based on weapons and tactics.[37] In some cases the modifications were adopted officially, army-wide.

The Grand Army moved on, picking up General Ware's combined infantry and cavalry force, the so-called Fatehgarh Division,* at Sikandra Rao.†[38] From there they pressed on and camped beside the south wall of the fort at Bijighur along the road that led north to Alighar. News came in that General Perron was preparing to confront the Grand Army at Coel adjacent to the fortress of Alighar.‡ The 'irregular horse' he had gathered were said to be the finest 'ever seen in Hindustan'.[39] And as the British approached, Pester and his men could hear the distant cannon fire of General Perron's gallopers on their daily exercise. Although Perron's mounted units could be seen at a distance, they avoided contact and

---

* Aka Futtyghur, Futty Ghur.
† 27°+, 78°+. Aka Sikanderao, Sikander Row, Sicunder Rao.
‡ 27°+, 79°+. Aka Alighur, Ally Ghur, Ali Ghur, Aligar.

remained aloof. There was no rush to war on the part of these Marathas. An undisclosed number of British, as opposed to 'mixed-race', officers from Sindia's and Holkar's armies were reported by Pester to have taken advantage of the Proclamation and 'come in' on 27 August, still a full two days before the official announcement. The incoming officers from Sindia's service confirmed that Lake's Grand Army faced its most serious challenge in the 'regular corps' and Sindia's artillery, which supported it. 'The British officers who gave us those particulars of the enemy may be supposed to be well acquainted with the real state of their forces, as they were the very men who had disciplined their troops and artillery, and a few weeks since only were with them, some in command of Brigades, others Majors, Captains and Subalterns.'[40] That evening Pester and his good friend Wemyss rode around the fort at Bijighur reflecting on their experiences: 'we went to the spot on which our batteries were erected when we besieged the place last year'. Wemyss had been promoted to serve as one of General Lake's staff officers and he kept his friend Pester well informed of events at headquarters.[41] Wemyss confided to Pester that Lake's staff knew full well that the campaign they were about to embark upon was bound to be bloody. Lake made sure his staff officers had taken seriously the intelligence reports that the Maratha regulars were equal to any sepoy force possessed by the EIC.

The sick from all the brigades were collected and deposited in Bijighur which had also been converted into a British grain depot for the ease of resupplying Lake's army.[42] Charles Stuart along with his comrades in the 3rd Bengal NC also made use of the fort to store their cavalry carbines since the stubby flintlocks 'encumbered both man and horse'.[43] Presumably Stuart and his fellow horsemen preferred to use their EIC-issued swords and cavalry pistols. But Stuart wrote of another larger problem in his diary that night. Battalion strengths had previously been cut to 800 and yet now, on the very eve of this Maratha war, an order arrived to expand his battalion by 100 men; Charles knew that would be difficult to do. Obviously the Marathas drew men from the same pool of military labour and, as noted previously, they had quickly taken up the unemployed EIC troops who had been discharged earlier. Stuart griped that changes in doctrine had hit the cavalry particularly hard. They lost good riders as well as horses in the drive to form new cavalry-based galloper gun crews.[44] And it had been only a little over a year since men of the Bengal and Madras Native Cavalry had been seconded to the Governor-General's bodyguard at a rate of twenty per regiment in accordance with the requirements of a 'peace establishment'.[45] The Governor-General's bodyguard was begrudgingly seen as having skimmed-off the best trained men and a somewhat bitter Stuart recorded that it had depleted the affected unit's mounts at the rate of one troop per regiment.[46] The recent arrival of His Majesty's 8th Light Dragoons from the South African Cape was of help but the shortage of horses meant that they were delayed in leaving Khanpur and had to follow Lake at a later date.[47] Charles Stuart would have to accompany Colonel Clarke to Anupsharh and collect more EIC cavalrymen in a bid to bring the mounted arm up to strength. Stuart's diary makes it clear that he saw long-term problems with a balance-sheet approach to warfare in which potential enemies like the Marathas could play the military labour market to their advantage every time the EIC experienced a demobilization as the result of a cost-cutting drive. This organizational infighting, over detaching men for galloper crews and the Governor-General's bodyguard, exacerbated the shortages brought about by the EIC's corporate cost-cutting moves.

As August 1803 drew to a close there was uneasiness in the British camp about keeping everything under control until the army was unleashed on the Marathas. The British were operating at the end of their logistical umbilical cord and battle was near. They could not afford to alienate those logistical support personnel that might go over to the enemy. Two of General St John's elephant attendants, a mahout and a coolie, were apprehended for plundering a 'bunga' who was bringing supplies into the camp.[48] A sentry had ordered them to halt their attack, but the assailants pretended not to understand the order so the sentry seized them at bayonet point and brought them in for formal charges. Their mistake in judgement was in thinking no one would bother to take action against them for preying on such a lowly member of South Asian society. After all, they worked for a British general, they had jobs and status. However, they were not being judged in the framework of South Asian society but rather within the context of British military culture and its legalistic structure. That meant martial law as applied to an army on the march. The British saw the attack on the unfortunate man as threatening the security of the operation. What if the 'bunga' had run to Sindia's men and reported the food and fodder levels of Lake's army? It was fortunate he had not and that his assailants were caught. Orders were given to summarily hang the General's men as a lesson to others who might jeopardize relations with civilian support personnel who formed part of the army's greater retinue.

Rumours had persisted for several months that Sindia's General Perron would leave Maratha service just like his predecessor de Boigne had done. Others reported Perron actively leading cavalry incursions across the Yamuna River.[49] Lake had written a series of letters to ascertain Perron's sentiments but apparently no answer had been forthcoming. Now with his army perched on Sindia's Hindustani border, it looked as if Lake might finally get some answers. Mr Beckett, aide-de-camp (ADC) and personal secretary to General Perron, rode out on 31 August 1803 to meet the advancing British.[50] Pester maintains that it was a mere one hour before hostilities were scheduled to begin that day.[51] Beckett spoke with Lake regarding the subject of Governor-General Richard Wellesley's Proclamation made public two days earlier. General Lake gave Beckett a copy of the document and sent him back to his master. Lake was not impressed with the defiantly slow pace that Beckett used in retracing his steps. Stuart reported that the old general was so ill pleased that he dispatched his son Major Lake to ride after Beckett. He was alleged to have repeated his father (the general's) threat that if Beckett did not get a move on then the Grand Army would reach Coel before him. With that warning Beckett hurried off. Although Perron had been out on parade with his troops earlier in the day, he rode to meet Beckett when the latter returned with the Proclamation. As Beckett and Perron passed through Coel the locals reportedly hissed at them.

Coel was inconsequential as a town but it was the site of Perron's headquarters as well as one of his more opulent estates. Surprisingly it was not contested, the Marathas withdrawing to the fortress of Alighar. Nevertheless Coel had to be secured and checked out on the off-chance that Maratha troops had been positioned there to ambush any British forces headed to Alighar. But to a large degree the occupation of Coel was a symbolic act. Coel was well known to a number of older British veterans who had travelled there to meet with de Boigne when it was his headquarters and personal residence, which had passed to Perron when de Boigne resigned. The British main column marched past Coel (on their left) as they headed towards Alighar. Their baggage as well as bazaars were detached and sent to a safe location. John Pester was in an ideal situation to observe the opening moves of the battle that was about to begin. He was posted to the advanced guard, which had been ordered to seize an enemy village on the high ground adjacent to the Maratha camp.[52] The fact that a morass flanked the Maratha campsite figured in everyone's memoirs. From his slightly elevated position, Pester saw the precision with which Sindia's troops broke camp and formed into columns. Their skilful use of military geography, in camping within the protection offered by the swampy ground, was not lost on Pester. Charles Stuart observed that the recent monsoon rain had ensured the morass was an obstacle to be reckoned with.[53] General St Leger formed the British cavalry into six columns, with the most advanced taking the high ground between the opposing forces. The enemy's right was covered by the protective artillery fire from the fortress of Alighar while the Maratha mounted units had fanned out to front the extensive swamp. The left of the Maratha line took advantage of the marsh-like irregularities in the ground and the shelter of an intervening hamlet. It was the enemy's left flank, according to Captain William Thorn, which was designated by Lake to be the focal point of the British attack.[54]

In order to engage the Maratha horse, on the periphery of the swamp, the British cavalry had to form a single column and advance to contact.[55] As the British closed on the Marathas they could clearly see Perron's bodyguard dressed in their bright red uniforms. Stuart was able to view Monsieur Fleury in charge of another well-dressed body of horsemen; he was reported to be the only European on the field for the Marathas, which made his identification all the more easy. Thorn wrote that there were 5,000 regular Maratha cavalry in attendance and approximately 15,000 irregular light horse. The Marathas had cleverly lured the British cavalry vanguard around the watercourse, thereby elongating St Leger's formation and denying him a chance to charge on-line. Once the British cavalry were drawn into an over-extended position, the Marathas raked the British horsemen with smallarms fire. Maratha skirmishers killed several British horses and that spread confusion through the cavalry ranks. The British cavalry drew their pistols and prepared to return fire but Lake saw the danger of the lure and recalled the men immediately to avoid a trap. This opening round of the battle did not speak well for the British cavalry since they had been forced to change formation twice without gaining ground or significant position in their dalliance. Their manoeuvres were well executed in terms of tactical command and control, but the necessity of having to change formation appeared as an overall inability to correctly 'read' the enemy and the field.

Lake was a fighting general of the old order who stressed personal combat leadership, a point which should not be forgotten when considering the success of his troops. To have one's C-in-C take an active role in combat leadership – to lay his life on the line – provided tremendous inspiration to the average soldier. General Lake placed himself at the head of his cavalry that was reformed in columns of regiments. The training lessons of Kannauj were to be put to the test. The opening salvo of the cavalry's gallopers was heard by all, serving more than the purpose of suppressive covering fire. That discharge signalled the commencement of the main attack.[56] The British infantry then swept the enemy pickets from their ephemeral posting on the left end of the Maratha line.[57] They pushed on under a continuous covering fire from the gallopers. The galloper crews then began to alternate their fire between crowds of Maratha irregular horsemen, who had appeared on the periphery, and the pockets of Maratha skirmishers who consolidated and bunched up as they fell back.[58] In later passing over the ground, Pester noted the high number of casualties which the British gallopers inflicted. He attributed this to target density and he thought each round 'did execution' among the concentrated enemy infantry ranks.[59]

As the Maratha left was being 'rolled-up', Lake ordered his cavalry vanguard into two lines and advanced with the infantry supporting in a third and fourth line as the morass permitted.[60] The Maratha forces fell back further as the British advanced; but the Marathas had positioned matchlock sharpshooters in the hamlet, which was by now on the British right. Lake was determined to minimize his casualties from this aimed fire and maintain his momentum by sending an entire battalion of sepoys to clear the hamlet; something they did in short order owing to their numerical superiority. Another large body of Maratha horse then arrived. It was an old-style Maratha tactical formation in the sense that it consisted of a large column of irregular cavalry spearheaded by regular cavalry. The regular cavalry, in the lead, were intended to pierce a hole in the British formation. If it had gone according to plan, the light horse would then ride through to the inside of the British formation cutting up opponents as they presented themselves. Maratha light horse often carried a variety of edged weapons, which included *tulwars*, lances, the Maratha spear and the so-called 'Rajput pike' or 'half pike' swung in much the same manner as a battle-axe.[61] But things did not go according to the Maratha plan. The British response was swift and decisive. The 'gallopers' swung into action and their blanketing fire caused loss of control in the Maratha formation, which was put to flight. Lake's cavalry pursued them but Maratha artillery covering-fire from Alighar proved a distraction. William Thorn noted that the fire from the fort continued during the rest of the flanking movement but it had little effect on the British advance with the majority of Maratha rounds being aimed too high and passing over the heads of the British.[62] The sustained forward drive of the gallopers pushed the outlying Maratha light horsemen back further until they abandoned the field.

The remaining enemy infantry now fell back to the walls of Alighar's fortress. This was a well-orchestrated move with the fort's artillery crews adjusting their range, loosing an effective covering barrage to deny the British an opportunity to cut off the Maratha infantry before it pulled back inside the fort.[63] Informants and spies had been been able to get inside Alighar as late as July 1803.[64] The British knew from their intelligence reports that Alighar held plentiful supplies of artillery ammunition to feed the Maratha guns and a siege could prove costly. Lake curtailed the day's action and withdrew back towards the town of Coel where camp was pitched by 2 pm. Casualties were light for the British. The infantry had escaped virtually unscathed while in the cavalry there was only one man killed and four wounded with twenty-one horses listed as casualties.[65] General Lake's son Major Lake had a horse shot from under him that day.[66] Pester's horse, named 'Collector', was wounded when shot through the neck. The heat had proven particularly troublesome for both men and beasts.[67] Charles Stuart summarized the day as not particularly meaningful with regard to the severity of battle. Rather he viewed the real significance in terms of the impact that it would have on Maratha political and logistical support. Stuart implied that the Marathas needed a victory in the opening battle to reinforce their credibility and hegemony in the region. Since the EIC was just beginning to extend its power into the Doab it did not bode well for the Marathas to be driven from the field. Stuart reckoned there were 12,000–15,000 enemy irregular horse opposing them including recently arrived Sikh units from the Punjab. The Marathas' performance that day would give cause for local *zamindars* and petty rajahs to question the wisdom of supplying fresh remounts to the Marathas as opposed to the EIC. It was never a good idea to extend credit to losers in a war of this dimension and if you had to live with the 'victors' it was unwise to be seen as a former 'enemy' supporter.

The day had ended with the British in firm possession of Coel, which contained Perron's infantry barracks as well as his personal mansion. Perron's chief European officers had also constructed homes there.[68] The rumour was that Perron had supposedly withdrawn towards Agra with select members of his personal bodyguard but, before departure, he left Colonel Pedron in charge of the Maratha forces at Alighar. Knowing the fortress was well provisioned to withstand a major siege, Lake opened negotiations with Pedron who sought to hold out until the arrival of a relief force. According to Stuart and Pester, the British believed Pedron was seriously contemplating surrendering Alighar but Sindia's *sardars* would have no part of it.[69] George Isaac Call of the 27th Light Dragoons noted the garrison's command structure was fractured between the French officers under Pedron and three particular Maratha *sardars*.[70] The *sardars* had more influence over the Maratha troops than the elder but newly appointed Pedron. Many of the remaining Maratha officers were unimpressed by General Perron's departure and talk of betrayal by the European and Euro-Asian officers was common. During negotiations with Alighar's garrison, the Marathas practised their gunnery with characteristic skill. Pester, MacGregor and Shairpe had arranged to inspect General Perron's captured house and garden at Coel. They rode back to camp under well-aimed Maratha artillery fire. 'Several guns were laid for us, and one very heavy shot was near clearing the whole of us, it completely covered us with sand.'[71]

Lake's patrols picked up a French officer as he tried to escape from the besieged garrison on 1 September 1803. The prisoner informed them that the garrison had opted to stand and fight rather than accept surrender terms.[72] The ethnic backlash had become quite evident in the flat refusal of the *sardars* to surrender and this was literally a house divided.[73] Colonel Pedron, upon having decided to surrender Alighar, was driven to the upper part of the fort where he remained isolated. Dragoon George Call noted that they considered the story of the garrison's division as a ploy of war and did not take it seriously despite the fact that the Governor-General's Proclamation had now been made public. Charles Stuart was alone, among the diarists, in stating that it was also on 1 September that General Lake received a letter from General Perron asking permission to 'come in' with his wife, family and bodyguard. Stuart wrote that things seemed rather confusing because Perron's reputation was one of personal bravery, yet he had not seemed willing to show any sign of that character trait since the war started. Cryptically, Stuart indicated that he knew more about Perron's surrender than he was willing to commit to paper. Apparently Perron had been under suspicion or was 'discovered' by Sindia some time earlier. There were also rumours in camp that Perron was a spy for the British and as a result Sindia had sent agents to take control of Perron's *jaghir*. Further complicating matters, Perron could not trust or cooperate with Bourquin who was said to be his designated successor as Daulat Rao Sindia's C-in-C. In his diary, Charles Stuart posed some rhetorical questions, which pondered whether Perron was really dedicated to serving the Marathas.

The most rational way of accounting for his conduct is to suppose him unprepared for the event and that undetermined what to do he was at the last decided by the events which ensued. He certainly was treated by us as an enemy for his garden House at Coel was completely plundered of its furniture and wines while this reflects the disgrace on our Army.[74]

The Maratha command structure was now divided along both its horizontal and vertical axis. The Maratha clansmen were at odds with their remaining foreign and Euro-Asian officers. Among Sindia's European mercenaries, the French suspected the British mercenaries would not fight the EIC and Lake's army; while Perron and his successor Bourquin – a fellow Frenchman – were deeply suspicious of each other's motives. Some of the English observers perceived a deeper class division between the French. The latter theory held that Bourquin and his cronies were not so much working-class heroes as vulgar men from the lower reaches of French society – journeyman butchers, apprentice bakers, itinerant salt sellers – supposed candidates for French revolutionary indoctrination if the story was true. They contrasted some of the tired old gentlemen like Pedron, a French officer of the *ancien régime* – men who had seen it all

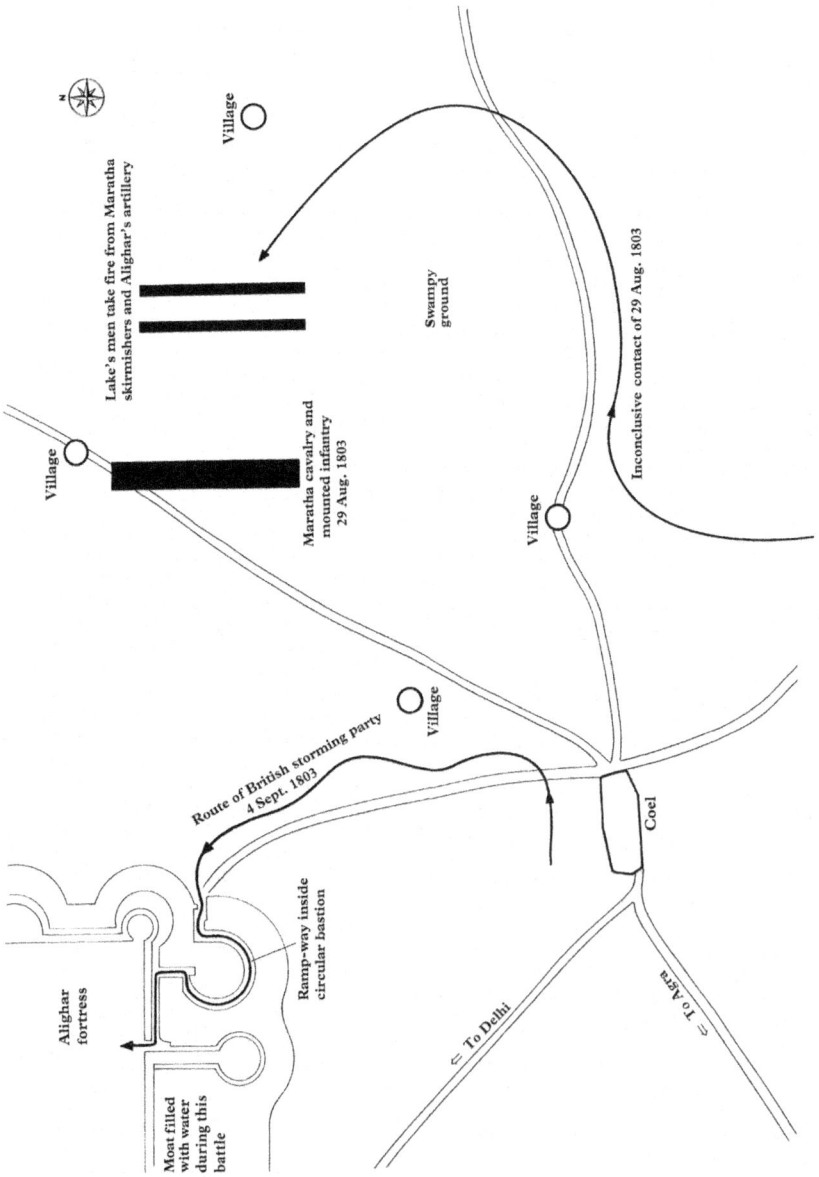

Map 9 Alighar 29 August 1803 and 4 September 1803, based on a map that appeared in Major W. Thorn, *Memoir of the War in India*, 1818

before and only sought to survive long enough to retire with a pension. Alighar's garrison reflected the greater divisions in Sindia's command structure and now it was hopelessly further subdivided on the issue of surrender. All this time the British continued to reap a windfall of intelligence as more ranking mercenary defectors came in from Sindia's units in the Delhi region. Not all were soldiers; some were European civil administrators who worked in the unique government of the Mughal–Maratha state that had Shah Alam II as its figurehead and Daulat Rao Sindia as its helmsman.[75]

The failure to negotiate a surrender of Alighar meant that the assault would be a sanguinary affair. Lake granted no quarter to an enemy that refused his surrender offer. Time was critical for the British if this battle was to be concluded before a Maratha relief force arrived. One rumour had it that Perron was out collecting all his remaining Maratha forces for a timely return. Although there was a particular strategy for taking the fort, ultimately luck, timing and contingency planning did more to win the day. Rather than building extensive trench systems for a conventional siege, Lake was determined to push an artillery piece up to the entrance of Alighar and 'blow the gates' to enable his European grenadiers a chance to rush in. That was no simple plan given the moat that protected the fort, but a young Irishman named Lucan had taken advantage of the Proclamation and just 'come in' from Maratha service with valuable intelligence on Alighar. Prior to his arrival in the British camp, Lucan had completed a trip from Delhi – the heart of Sindia's North Indian 'empire'.[76] Lucan was able to tell Lake that Sindia's battalions in Delhi were to be used in either a march on Coel, to relieve a siege of Alighar, or in ambush against the Grand Army at locations yet further afield. While the proximity of Maratha infantry reserves was somewhat distressing, Lucan also clarified another matter, which did provide some additional breathing room. British intelligence believed that Sindia was ordering troops north from the Deccan to defend Hindustan. Lucan reckoned that the nearest Maratha Deccani units were still 200 miles from the Yamuna River and therefore a pending but not an immediate threat.

At midnight on 1 September 1803, a violent earthquake struck. Thorn placed its duration at an exceedingly lengthy two minutes and he noted several buildings were destroyed.[77] Stuart said it was composed of three sharp shocks. Pester was roused from a sound sleep and had difficulty believing that he was not dreaming until the commotion of the camp around him convinced him that the event was real. A number of the old-timers assured him it was the severest quake in memory. 'The motion was very like that of a boat in a moderate sea.'[78] For some of the Marathas' followers this 'omen', coupled with the loss of Coel by the supposedly unbeatable Perron, was too much. There was a lingering feeling that the unfolding political and military events were of cosmic proportion.

By 3 September the British were hard at work in the garden or 'Sahib Bagh' of Perron's summer home, 'making ready' for the assault on Alighar.[79] They prepared the materials they would need to reinforce the hasty redoubts that would be built to house the British 18 pounder siege guns, which were Lake's biggest serviceable ordnance.[80] Although most often associated with the use of sabot-fitted iron round shot against stone fortress walls, they would be employed here with grape shot to clear Maratha artillerymen from Alighar's walls. The Maratha artillery now fired periodically into Perron's garden as they caught sight of worthwhile targets. Pester thought the activity in the garden was intended to mislead the enemy's appraisal of British tactics. 'The pioneers employed as yesterday, cutting materials for batteries, which we imagined to be a deception, as it was pretty certain from Wemyss' account that we should attempt the place by a coup de main.'[81] There were also 12 pounder crews assembling in the garden under the watchful eye of Shipton of the artillery. But no one at that point seems to have realized how valuable those guns would prove be in the coming fight. The 18 pounders were too heavy to move under fire but a 12 pounder, while a brute to manhandle, could be forced along by burly men aided with hand-spikes. Shipton told Pester in confidence that the assault would be the next morning and that he, meaning Shipton, had been designated 'Captain of the Artillery' to lead the dangerous positioning of the guns needed to 'blow the gates'. Standard practice was to use a 6 pounder when blowing gates, but it was dangerous in that you had to run it up to the very door you were trying to blow open. As if by way of encouragement, newsletters from the Deccan arrived that night which informed Lake's camp that Arthur Wellesley had taken the *pettah* at Ahmadnagar.

Time ran out for Colonel Pedron after five days of negotiation and Lake decided on 4 September that there was no use delaying the assault. According to Pearse's *Memoir of the Life and Military Services of Viscount Lake*, Pedron had favoured capitulating but he was prevented from doing so by his second in command, who had the old colonel arrested.[82] The Maratha forces inside Alighar knew their time was short and they had already begun to run a mine under the main gateway for fear that the British would indeed risk a *coup de main* by wheeling up a gun to blow the gates in.[83] A moat or ditch, which varied from 100 to 200 feet across and with a depth of 30 feet, surrounded the fort. At any given time there was a minimum of 10–12 feet of water in the ditch. The garrison had almost completed work on a drawbridge and if the assault was delayed the passage across the ditch could be cut at will by the enemy's raising of that single bridge.[84]

Lucan, having recently defected from the Marathas, warned C-in-C General Lake that the fort, like many others in India, was protected by an intricate series of inner gates to thwart a simple assault. And that if Lake intended to take the fort via a *coup de main* he would have to get his men over the causeway across the ditch, through the main entrance and then navigate a maze-like circular bastion once inside. The circular bastion was a death trap with flanking loopholes where an unseen enemy could fire smallarms at their leisure. Stopping for a moment or hesitating under fire to try a phoney door could cost multiple lives. Without decisive leadership any such action was doomed to failure because men, regardless of whether they are enlisted men or officers, will falter in the face of a withering fire from an unseen enemy. Theoretically one should have selected an officer or NCO who could motivate and inspire troops and make the impossible an attainable goal. The British regimental system also required a known and trusted leader who would not retreat and bring disgrace on the unit. A natural leader, a madman and a father figure would just about sum up the desired characteristics. Although this action was officially led by Colonel William Monson, the recently defected Mr Lucan – to the surprise of many – would serve to guide the troops. The assault party was composed of members of His Majesty's 76th Regiment under Major McLeod, the 1st Battalion of the 4th Regiment of Bengal Native Infantry (NI) commanded by Lieutenant-Colonel Browne, four companies of the 17th Regiment Bengal NI under Captain Bagshaw, and ultimately the 2nd Battalion of the 4th Regiment of Bengal NI. Colonel Horsford of the artillery had completed the two batteries of four 18 pounders each to provide the needed covering fire.[85] The British forces would carry ladders to escalade the walls adjacent to the gates in an effort to neutralize any remaining defenders there, while a 6 pounder was wheeled up to 'blow the gates'. The apparent last-minute decision to use a 6 pounder seems to have been based on concerns for mobility on the approach to the innermost doors.

In the first half of the eighteenth century it was more common to use a petard to blow gates open but the petard had fallen out of favour and few soldiers were trained in its use. As a field-expedient alternative means of 'blowing in' the gates, it became common practice to use a 6 pounder with the muzzle placed against the doors.[86] The military men of the day often argued about the precise technique. But the most common way was to load a powder charge only, without cannonball or other projectile, in the cannon. Then place the muzzle of the gun directly against planks pressed against the door so that the tremendous pressure generated by the muzzle-blast of the black powder charge buckled the doors open. Some wooden doors were reduced to a mass of splinters. If the cannon was loaded with shot it had a tendency to merely blow a hole in the door the size of the cannon ball, which was much too small to admit a man or be of any practical use.

Under cover of darkness Monson and his men approached the causeway, which led to the gates. To their surprise they found a collection of sixty or seventy Maratha troops out 'smoking [sic] under a tree in front of the gateway'.[87] After a brisk skirmish the Marathas, to the last man, were killed. The enemy in the fort apparently took the opinion that this was a failed British attack and did nothing to challenge the position. Fearing an alarm, the British withdrew four hundred yards back from the walls of Alighar. However, with the sounding of the morning gun the assault was renewed. By timing their advance to coincide with a heavy barrage by the 18 pounders, Monson's men were able to cross the open ground and the causeway. Pester had been appointed by General Lake to accompany the storming party for the purposes of reporting on their progress. He noted that they got as far as the entrance to the first gate before the Maratha troops on the walls noticed them. It had been hoped that some of Monson's men could follow guards, posted outside the gate, into the fortress as they retreated. But a fight ensued and the element of surprise was lost; at that point the entire area erupted in fire. In addition to extensive smallarms fire from the wall, the storming party was hit on both flanks by Maratha artillery fire from guns placed in half-moon batteries on either side of the main gateway.

A recently constructed Maratha addition to the fort entrance offered Monson the false hope of easy access. Unfortunately the entrance was shut and to worsen matters the Marathas had trained two or three guns to fire grape shot on it. Monson lost time and men in the effort to take advantage of what seemed an opportunity that could not be passed up. Mounting British casualties forced another modification to the plan. Two ladders were brought forward to assist in an escalade of the gate and Major McLeod of the 76th with a number of grenadiers tried to climb over the top. Within seconds a bank of Maratha pikemen appeared and drove the British soldiers back down their ladders. Things were looking pretty grim but spirits rallied for a moment as the 6 pounder was brought forward to use on the doors. Amazingly its blast had no effect and so word was desperately sent for a 12 pounder.[88] As Monson watched helplessly and waited for the larger gun, the Marathas poured down artillery as well as smallarms fire on his position. The shattered bodies of British soldiers were now layered two to three men deep before the gateway.

The British 18 pounders were not the only guns used in an attempt to sweep the Marathas off Alighar's walls. George Call revealed that the cavalry's galloper guns were used specifically to try and neutralize Maratha snipers.[89] The enemy marksmen favoured their long-barrelled matchlocks for accurate aimed fire. The British galloper crews, who attempted to 'take them out', in turn drew the attention of Maratha swivel guns and light artillery posted to the walls. A 4 pounder ball hit the horse beside Call and the beast dropped on the spot as if it had been poleaxed. Pester, who was mounted in order to ride back to Lake with a situation report, had his horse wounded twice although he himself managed to survive unscathed. As a cavalry officer, Call also noted, in the midst of this battle, that an estimated 50,000 Maratha horse and *pindaries* had fled the general vicinity along with the remainder of Perron's 700-man mounted bodyguard.

The greatly anticipated British 12 pounder was painstakingly wheeled forward under an intense fire, but its heavy weight caused it to crash through a concealed mine gallery which the enemy had skilfully tunnelled underneath the area in front of the main gate. Exploding mine galleries filled with black powder had been the cause of numerous British casualties during the 'Mud War'; fortunately for Lake's men this mine collapsed before it could be charged. With the temporary delay of the 12 pounder's carriage sticking in the hole caused by the tunnel's collapse, the British incurred a dramatic increase in casualties as the Marathas dropped in fire from two heavy mortars that they had held in readiness to defend this point. They loaded grape shot instead of explosive shell into the preregistered mortars and the metal rained lethally down from its high angle of trajectory. 'This misfortune detained us considerably, and at this time it was that we lost so many of our officers and men. Never did I witness such a scene before the second gun could be hauled up; the sortie was become a perfect slaughter-house, and it was the greatest difficulty that we dragged the gun over our killed and wounded.'[90] Despite the considerable size of the 12 pounder it took four to five blasts at point-blank range to shatter the door.[91] This process was reported to have taken a full twenty minutes, which was excruciatingly long due to the continuing fire from the fort's walls. By now the Marathas had seized the assault ladders abandoned by the retreating grenadiers, sending their own men down to engage Monson's force in hand-to-hand combat.[92] The losses in this grim struggle were high and Monson himself was lucky to survive with a deep-slashing pike wound.[93] The combat leadership role of the British officer corps was once again reflected in their casualty rates. Six British officers were killed in the Maratha counterattack at that first gate. Four grenadier officers, the adjutant of the 76th Regiment, as well as Lieutenant Turton of the 4th Bengal NI, were dead and Pester rode back through heavy smallarms fire to inform Lake of the successful breaching of the outer gate, but at a frightful cost.

British casualties were mounting elsewhere as well and distance from the fort's walls was no guarantee of safety. George Call watched as Maratha sharpshooters continued to take their toll of British officers. He saw their matchlock snipers kill four lieutenants outright while wounding another seriously in addition to wounding a captain and a major. Call also witnessed the damage that heavy Maratha artillery could do to infantry formations at a distance. These men did not die alone. They were felled in groups. Incoming rounds were large enough to smash trees yet retain sufficient kinetic energy to still do lethal damage. In one case, cited by Call, up to seventy men became casualties while clustered beneath a large tree where they sought concealment on their approach to the fort. A devastating round from one of the Marathas' massive 32 pounder guns hit them.[94] Those impacted directly were obliterated. Tree splinters pierced some, still others were struck down by flying smallarms and accoutrements turned into secondary projectiles. Pieces of human bone and equipment radiated out, from the point of impact, inflicting more casualties among the densely packed group of men who had formed up tightly behind the tree to reduce their target profile. Charles Stuart confirmed the devastating power of a 32 pounder, which was ensconced in a well-defended bastion. It was a 'pucka' position with breast works and loopholes so that Maratha infantry could defend the gun crew. The scene proved uncomfortable for the British diarists. It evoked thoughts about dehumanization and the mechanical aspects of death in combat. A seemingly anonymous enemy behind the parapet walls had killed these men.

Alighar's exterior gate had been taken; however, that was just the beginning of the action to seize the fort's inner tower that led from the gateway to the upper fort. The closest modern analogy to this inner roadway travelling up an inclined plain might be one of those circular or corkscrew ramps that one finds in multiple-level automobile parking garages. The bastion's interior circular roadway had walls containing 'loopholes' or apertures through which the enemy could fire in relative safety. The Maratha infantry kept up a bristling gauntlet of smallarms fire while their artillerymen positioned a stubby short-barrelled carronade to fire grape shot into the passageway. A second gate in the passageway was forced open with difficulty and the British quickly pursued the defenders as they retreated, slipping through a third gateway. All this time the 12 pounder had to be pushed and pulled up the circular rampart and many men fell wounded or dead from the flanking smallarms fire. Often the bronze brute slipped backward crushing feet beneath its iron-rimmed wheels. Captain Shipton of the artillery was wounded in the process and although unwilling to give up his command of the piece, he could not execute his duty.[95] His bravery was beyond question but in such a wounded state he had become an impediment to effective leadership under fire.

Ironically, after failing to open a final set of inner gates that blocked the rampway upward, Major McLeod found a way through a small wicket door into the bastion tower and troops were summoned to follow him. Extensive hand-to-hand combat ensued up the stairs and along the parapets with a number of the Marathas being pushed or jumping from the walls into the moat. The attackers found themselves in a square where they faced Maratha infantry directed by the fort's senior Maratha *sardar*, identified by Charles Stuart as Bali Rao.* This valiant Maratha officer was reported to have led a forlorn charge into British bayonets rather than face surrender.[96] Major McLeod then turned his attention to the fort's prison cells. He had to intervene to save the incarcerated Pedron, who was in danger of being bayoneted behind bars by a British 'Light Infantryman'. McLeod offered the Frenchman quarter and in return received the old man's sword.[97]

The diaries of John Pester and Charles Stuart both indicate that revenge was responsible for the deaths of hundreds of Maratha troops killed after the battle. Pester was less than sympathetic in his explanation for not taking prisoners during the actual assault of Alighar, 'They were told what they might expect if they waited the result of the storm.'[98] Yet Pester also realized that the orgy of killing had extended well beyond any reasonable limit.

Many of the enemy were shot in attempting to escape by swimming the ditch after we got in, and I remarked an Artilleryman to snap his piece at a man who at the same instant dived to save himself. The soldier cooly waited his coming up, and shot him through the head. As the heat of the business was over, I remonstrated with him on putting them to death at that time, but the man declared that he had lost some of his oldest comrades that morning, and that he wished to be revenged, reminding me also that we had received orders to spare none.[99]

Some Maratha troops were shot in the water and some drowned, but reaching the other side was no guarantee of escape. Many were 'run-through' by George Call and his brethren in the 27th Dragoons as they emerged on the adjacent plain. Call said his unit 'was ordered to cut-up all who attempted to run-off'. As if a participant offering an explanation for some sort of juvenile schoolyard brawl, he added that the enemy had provoked them with insulting names.[100]

---

* Aka Balle Row, Baji Rao.

John Pester, who had been responsible for carrying the news of the exterior gateway's fall to the C-in-C, returned with the General so that Lake could see for himself the frightful cost. Pester had many friends among the dead and dying on the ground and the scene shocked Lake even though the crusty old man had done his personal share of killing in the trenches of Virginia. Pester remarked of Lake: 'I never saw anyone more distressed for a moment when he entered the sortie, and saw officers and men heaped on one another.'[101] As if being accosted by a ghost from his own unresolved past, a wounded grenadier approached Lake. The man who 'hoped his honour was well' said he had served in Lake's company at Yorktown twenty-two years earlier. Lake, although often noted for his gracious treatment of the wounded, invited the injured man to headquarters to see what could be done for him.[102] However, Lake's fury and desire for revenge contrasted the compassion shown to his old comrade-in-arms. He was so enraged that he ordered the cavalry pickets to ride out, travelling further afield than during the battle and track down those who had escaped from the fort earlier in the day. Once the escapees were located they were to be summarily executed. The nearby Maratha survivors, many of whom had managed to avoid being shot or drowned in the moat and now wandered the field in a dazed condition, were also rounded up and executed. Charles Stuart's criticism was bitter with remorse:

An officer saw a Riding Master of one of the regiment of cavalry go up to a poor unarmed creature and strike his head off and on demanding the authority for so cruel an action was told the Commander-in-Chief had ordered every man to be killed. In the moment of indignation he might have over-looked the cruelty and injustice of this order but I lose all patience when I hear such barbarity cooly defended on principles of policy. The place was strong and the garrison had every reason to suppose that their defence would have proved successful. They behaved as good soldiers and faithful servants and God forbid we should be reduced so low as to have recourse to deeds of cruelty to ensure our success.[103]

Ironically Governor-General Richard Wellesley, in later commemorating the victory, cited the British surrender proposals as proof of British humanity.[104]

Lieutenant Charles Stuart indicated Lake was still in the vicinity of the gateway as Pedron appeared, a very tired old man dressed in a green jacket with gold lace and epaulettes; he was brought out as a prisoner. Lake, through an interpreter, said to Pedron, 'See here sir what you have to answer for?', pointing to the dead and wounded.[105] Stuart did not think that Pedron was to blame since he had no authority over the garrison. He had been a prisoner of the Marathas before the British attacked. It would seem, with regard to the costly final stand made by the garrison, that it was Bali Rao who acted against Pedron's wishes and urged a stalwart defence that cost Lake's men so dearly. Although British casualty figures vary greatly among authors, ranging from 223 to 278, there is agreement that the British officer ranks did suffer in accordance with their tradition of leadership under fire as seventeen officers were rendered as casualties.[106] George Call said about 2,000 of the enemy were killed, with the only prisoners listed as Pedron and two Maratha 'chiefs'.[107] Within a day of the action the sun bloated many of the bodies in Alighar's moat. Human features became distorted beyond easy recognition. Large numbers of bodies drifted on the surface and the putrefying flesh created an unbearable stench that sickened all. Stuart saw detachments ordered to dig mass graves and the pioneers then drained the ditch that surrounded the fort.[108]

John Pester noted three hours of plundering were allowed after the victory but Charles Stuart, literally a much more sober critic of Lake, told a different story. Stuart recounted how the plundering led to an orgy of drinking and that lasted for three full days. The *arrack** flowed freely and many troops were to be seen openly roaming the town with bottles in their hands. Often they would stagger into the private quarters of citizens who protested vociferously only to have a handful of coins tossed in their direction as a silencing gesture.[109] Stuart's observations on the exploitation of the soldiers by local merchants, over the price of *arrack*, proved to be an interesting treatise on the redistribution of wealth. The Marathas had collected the treasure, the British had liberated it and the soldiers were reported to have spent almost all of it, on the hyper-inflated prices charged by the merchants for their alcohol.

Alighar had been a vital objective for Lake and its logistical significance tempts one to say that the victory shortened the war.[110] Among the spoils of Alighar were large quantities of blue uniform jackets with red facings as well as new arms and accoutrements of the type used by Sindia's regular battalions.[111] Powder, shot and entire tumbrels of Spanish dollars fell to the British. Alighar was the primary depot for Perron's battalions and 281 pieces of artillery were captured. This included 33 brass guns and 60 iron guns of different calibres, 4 brass howitzers, 2 mortars and 182 wall pieces.[112] General Lake's son and Wemyss toured the fortress with John Pester who recorded in his diary, 'The guns and tumbrils (of French model) uncommonly fine'.[113] With the battle over, George Call

---

* The term was used generically by the British as alcohol applied for the purposes of water purification. In earlier British–Indian wars it specifically meant the spirits distilled from the fermented juice collected from palms. Several Middle-Eastern varieties of arrack continue to be distilled from a variety of sources; commonly with anise flavouring added, they can often average upwards of 40 per cent alcohol.

wondered what the bazaar people thought about the British taking Alighar in one morning rather than the ten months that the Maratha garrison had bragged it would take. Call directed his question to Ram Narrain, a *shroff* or we might say 'monetary agent and currency dealer' who travelled with the Grand Army.[114] The reply was that the camp followers had come to expect it since the earthquake of 1 September, as it had been interpreted as a sign that the British General would triumph over the Marathas.

The elusive Monsieur Fleury

Timing was everything if Lake intended to 'take out' the main forts in the Doab before Sindia's remaining officers could coordinate offensive and defensive actions. Charles Stuart believed the Marathas knew the strategy that would hurt the British most was a combination of offensive light horse raiding coupled with defensive positional warfare. In an abstract sense we can see that the concept, if it had been properly followed, was not far removed from what Shivaji had done when confronted with Mughal incursions into the Deccan: mobile raiding based from a chain of fortresses garrisoned by capable infantry. The victory at Alighar purchased neither peace nor rest for Lake's men and word arrived that on 2 September 1803 a group of 5,000 Maratha horse had attacked the British detachment at Shikohabad. It was a moderate-sized outpost where Lieutenant-Colonel Coningham's five companies of the 1st Battalion, 11th Bengal Native Infantry were stationed. The attack on Shikohabad was not a chance encounter with a *pindari* band. The Maratha officer commanding this bold attack was reported to be Fleury, the same European mercenary who had been noted on the field during the opening skirmish near Alighar on 31 August. Fleury's tactical appreciation of Sindia's cavalry was quite evident. After having withdrawn from Alighar he had staged a surprise attack within forty-eight hours, selecting Shikohabad as a post deemed by the British to be secure. Shikohabad, on the frontier of Etawah, was open country and Coningham had only one artillery piece to hold the Maratha horse at a distance. Since the raiding party contained men from Sindia's regular cavalry they rode with horse artillery. They put their gallopers into action quickly, using their specialized howitzers to fire over the walls and hit the defenders who fell back on the cantonment's square. Charles Stuart noted that Coningham's single piece of artillery was rendered useless because its ammunition supply was destroyed when the Maratha howitzers dropped an explosive shell on Shikohabad's magazine.

Despite the nature of the raid, Charles Stuart spoke favourably of Fleury's gentlemanly behaviour in taking time to negotiate with the wounded Colonel Coningham. In view of Lake's execution of prisoners at Alighar this was quite chivalrous – unless Fleury was playing a different and more politically complex game. He allowed the British to march off with their single cannon as well as their smallarms.[115] It was extremely thoughtful of Fleury at the very least because the loss of a battalion gun was second only to the loss of a regiment's colours in bringing disgrace on a British military unit. The fact that the gun's ammunition had been destroyed was immaterial when considering regimental honour. In agreement for safe passage and the right to keep their weapons, Coningham and his troops agreed not to serve against Sindia for the remainder of the war. They were permitted to cross the Gunga River thereby re-entering the EIC's formally recognized territory. Fleury did retain a hostage in the form of Mrs Wilson, wife of EIC Captain Wilson. The intrepid Mrs Wilson was apparently more outraged by the plundering of her wardrobe than by her captivity. She took great umbrage at having scores of uncouth horseman delve into her baggage trunks and parade her personal clothing for all to see. John Pester, still back at Alighar, heard from secondary reports that he had lost all the furniture and possessions which he had stored in his bungalow at Shikohabad.[116] Luckily for the hard-drinking Pester, he had surmised as early as July 1803 that Shikohabad was too remote for safety's sake and he had refrained from storing large amounts of his European wine there.[117]

It was 2 am on 5 September by the time the British could dispatch Colonel Macan with the 3rd Cavalry Brigade (His Majesty's 29th Regiment of Dragoons and the 1st and 4th Bengal NC) to relieve the cantonment.[118] Given Fleury's speed and head start, it was doubtful Macan could catch the Maratha officer. The loss at Shikohabad was a blow to British pride as the cantonment and market were completely pillaged before being burnt. Yet the raid held potentially greater significance as a Maratha signal to local *zamindars*. It was clear that the British foothold in North India was tenuous and any signs of a major Maratha success were bound to trigger a new round of *zamindar* rebellions.[119]

The ability of large bodies of Maratha horse to avoid detection and pursuit underscored the need for the British to develop a mobile or *Indianized* doctrine. Young Cornet George Call had tried to use news of Fleury's success as part of a lecture he gave to a number of South Asian troops.[120] He stressed that the attack showed the need for the British forces to stick together and fight in a supportive and combined fashion. Call confided in his diary that his young age worked against his credibility as an authority figure, when he spoke to the well-seasoned EIC veterans. They had not only heard it all before, in contrast to the young gentleman, they had done it all before.

As Macan's brigade of cavalry advanced they occupied the Maratha possessions that lay in their path and deposited token members of the force before pushing on. It was an attempt to capitalize on contiguous territorial growth, but it was a dangerous strategy owing to the fact that every time men were left to occupy a location it depleted the brigade. By 7 September, the effective remainder of Macan's force reached Firozabad, steadily tracking Fleury and the Marathas towards Agra. The Marathas had recrossed the Yamuna River to escape their pursuers by less than a day. Macan prevented them from doubling back by seizing the mud-walled fortification at Firozabad, which they had abandoned. Located only twenty-four miles from Agra it had served as a Maratha depot and when it was taken it was found to contain nine pieces of artillery, some cattle and a considerable store of grain.[121] Macan pressed on the following day but his Maratha opponents had melted away with great effect. Perhaps they had taken shelter within the great Red Fort at Agra. Dragoon William Thorn accompanied Macan's group and he commented on seeing the Taj Mahal at a distance as the brigade passed within 11 miles of Agra.[122] The British were not yet prepared to tackle Agra and the remains of Macan's detachment changed direction so that they could rendezvous with the 8th Regiment of Light Dragoons and three battalions of sepoys under Colonel Clarke. Later as a reformed brigade, under the command of Colonel Vandeleur, they rejoined Lake.

Lake's camp at Alighar

Lieutenant Pester and his friend Wemyss went out foraging with members of Charles Stuart's unit, the 3rd Bengal NC.[123] Fresh fodder was scarce since the monsoon had failed and grasscutters needed to travel further afield.[124] Their specialist talents were indispensable to the functioning of an army in this environment. William Thorn wrote that the grasscutter 'gathers forage, consisting of the roots of grass, which he digs up with an iron instrument resembling a mason's trowel. These roots, being carefully washed, constitute an excellent food; and in fact no other could well be obtained in a climate which, during the season when hot wind prevails, is so completely bare of vegetation, that not a single blade can be discerned above ground.'[125] Maratha mounted units dogged the forage party but with the judicious firing of their galloper the British kept the enemy from venturing too close. The new British doctrine of mobile firepower was allowing a greater economy of force and operational freedom. In previous times it would have meant assigning more cavalry or slowing the foraging party down with infantry or bullock-drawn artillery to support it in hostile territory. But with the assignment of a 6 pounder galloper the foraging party could move at ease around the country and it usually provided sufficient firepower to dissuade Maratha light horsemen.

On 6 September, word was received that Sindia was preparing to engage Lake's army as it crossed the Yamuna River. Sindia's advanced units had already confiscated the boats at the ferry crossings and were reported to be raising batteries at the most likely fords.[126] Reconnaissance patrols brought word that Sikh horsemen were to be seen in greater numbers and rumour had it that the Maratha regulars were drawn up waiting for the Grand Army.[127] General Lake, after securing Alighar with a small detachment, departed with the remainder of the army. When he halted, on the first day of the march to Delhi, he received a letter from General Perron requesting permission to proceed with his family, possessions and escort to Lucknow.[128] He had quit Sindia's service and, much like his predecessor de Boigne, it was his intention to take his earnings and leave not only Maratha service but also India. Perron was to be accompanied by two of his trusted officers, ADC Beckett and the elusive Monsieur Fleury. Lake quickly agreed to allow the General to keep his bodyguard, receiving 'his opposite number' with all the ceremony and honour accorded a C-in-C. Sindia had lost one of his most experienced mercenary leaders, inflicting an incalculable blow to command, control and communication in his army. Worse than that, with one defection Daulat Rao Sindia lost his greatest source of military and financial intelligence in the Doab. General Lake would have the first chance to debrief Perron, but eventually Governor-General Richard Wellesley's staff would have an opportunity to interview the man who Sindia had entrusted with his military assets.

The Grand Army continued its push towards Delhi, taking the abandoned fort at Khurja 50 miles northeast of Alighar. The garrison was reported by many to have fled at the approach of the army that had taken Alighar. Charles Stuart, however, suggested that the fort was virtually handed over intact by the *killadar* who was reported to be the son of Alighar's unfortunate Colonel Pedron.[129] The Maratha garrison had departed without removing the great quantity of grain which was stored there. While their withdrawal might have saved the lives of Maratha soldiers in the short term, the long-term prospects for adequate supply were being diminished by the trend towards depot abandonment in the face of British advances. Places like Firozabad and Khurja could not hold out against a force the size of Lake's. And, although no single such cache of grain could supply all of Sindia's needs, the systematic loss of depots that formed a logistical chain was bound to hurt his future options. Both Stuart and Call noted that Lake's army passed Sikandrabad on 10 September. Although usually described as a small mud fort, Stuart recorded that it was also the location of an extensive cantonment for Perron's troops.[130] The Maratha regulars were quickly being deprived of their command structure, logistical support, peripheral outposts and now the cantonments where they usually sought relief from the rigours of the field. It was reported that Louis Bourquin had taken Perron's place as senior field leader of Sindia's army and the new commander was under immediate pressure to halt Lake's advance. The British received word that Bourquin had crossed the Yamuna with sixteen battalions of regular infantry, 6,000 cavalry and a large train of ordnance. John Pester got a note from his companion Wemyss at dinner that night which said it was the opinion of all those at headquarters 'that we should have an action certainly tomorrow or the next day'.[131]

At 2 am on 11 September Lake struck camp and by 3 am began the march towards Delhi. When the lead units of his army halted at 10 am they had marched 18 miles under a very hot sun. They emerged from a thick patch of jungle at Patparganj* southeast of Delhi and began pitching their tents.[132] Lake was pleased with the army's swift progress and as a reward he issued a dram of liquor. This rather tranquil and happy camp scene lasted about an hour before all hell broke loose.[133] At that point the cavalry pickets informed Lake that Maratha horse were approaching. In response to this the 27th Light Dragoons and the 2nd and 3rd Bengal NC were ordered out. Lake personally took command of his three regiments of cavalry and led them on. 'By degrees' the Maratha horse retired slowly, luring the British horsemen within the range of their masked batteries. General Lake sent a number of his men forward to probe the determination of the Maratha horse. Survivors spoke of the 'elephant grass' being so high that it obscured any chance they might have had to see the semicircle of artillery that surrounded the thicket.

The blinding yellow–orange flash and roar of the opening salvos shattered not only the tranquillity of the day but the eardrums of men closest to the guns who reeled in shocked reaction that added to the confusion. The accompanying pressure wave generated by the explosive muzzle-blast, which flattened the obstructing grass, was immediately followed by other, unnatural, and far more eerie auditory sensations that played upon deafened ears. Grape shot tore and chain shot scythed through the grass with a shearing sound which was followed by either a metallic clatter or muffled thuds depending on whether the individual projectiles struck equipment or the flesh of men and horses. None of the diarists commented on having consciously heard any screams until after the battle. Shock and disorientation were hallmarks of a Maratha artillery ambush.

* 28°+, 77°+. Aka Putpergunj, Putpergunge, Putper Gunge.

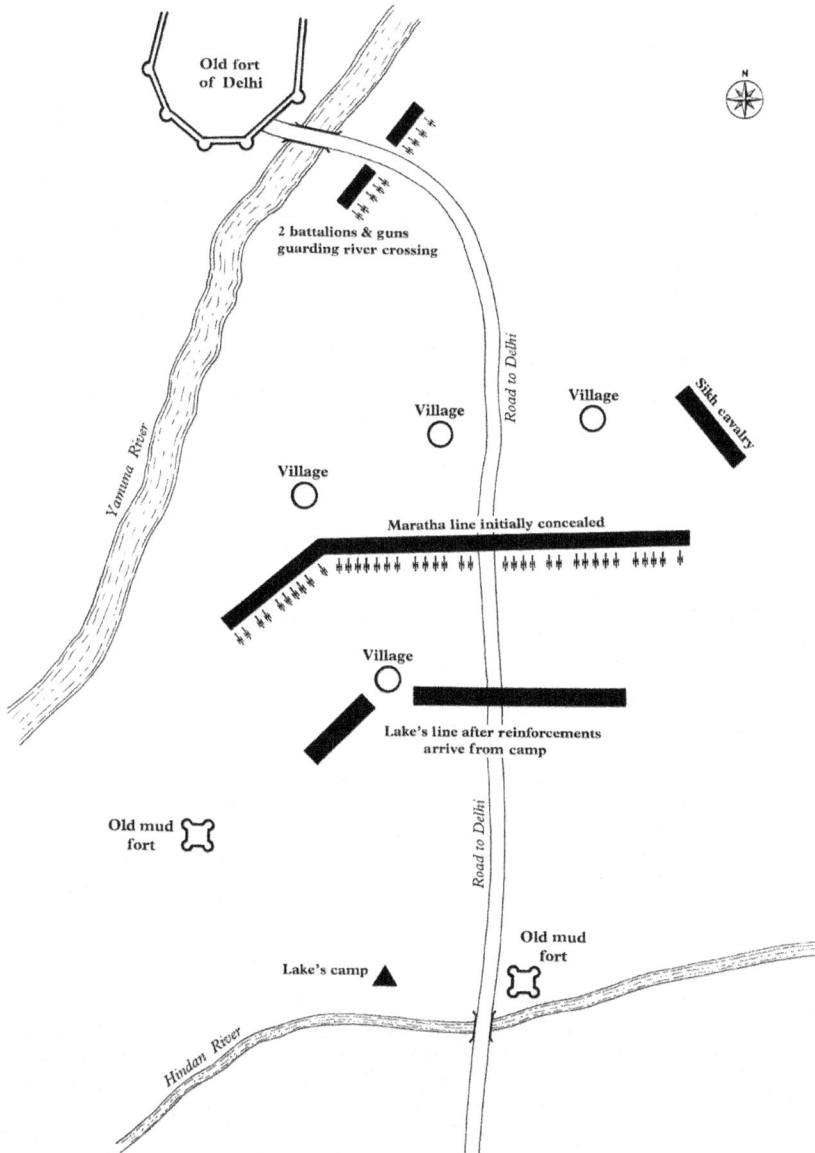

Map 10 The Battle of Delhi 11 September 1803, based on a map that appeared in Major W. Thorn, *Memoir of the War in India*, 1818

Although men were being literally punctured and sliced as slabs of meat, the survivors were too preoccupied with their own fate to register immediate concern for their brothers; evidently there was a strange, survival-oriented filtering-out of fellow soldiers' anguish, a reversion to the more basic instincts and the internal battle posed by the desire to run vs. the training and indoctrination of staying to fight.

The accounts conjure surrealistic images: the shredding of environment and an army, a scene soon enveloped in the white, acrid haze of gun smoke which clung closely to the vegetation in the heat and humidity of that September afternoon. Some of the survivors used the analogy of a storm, others accepted it as 'a rain of heavy metal', unleashed by the Marathas. The opening roar of Maratha guns brought the sudden realization that this was to be an infantry battle and not the skirmish of horse. In reflecting on the magnitude of the ambush George Isaac Call later noted in his diary:

the guns were planted semi-circularly, of course the shot came but to one point. They were very well laid and the firing well served. The enemy must have been certain of victory, otherwise they would not have brought such large unwieldy cannon and so great number into the Field, from the opposite side of the river; how could they suppose in the event of a defeat, they could transport them to the river side (which was 7 miles from the scene of action) with such slow animals as Bullocks. No doubt remains, from the force brought into the field, of both horse and foot, and the well laid snare into which we were drawn, and the number of Guns, so well arranged, that a decisive victory over us, would be the reward for their late losses.[134]

Lake's 'reconnaissance in force' had revealed a well-planned deployment, which demonstrated Bourquin's appreciation of topography. The Maratha forces were said by William Thorn to be drawn up in order of battle, with infantry having the benefit of cavalry support and artillery cover from numerous guns.[135] The Maratha infantry were on rising ground with entrenched positions flanked by swamps to channel the British into a frontal assault through the designated *beaten zone* for their artillery. The volume of fire in this killing zone would have been impossible for Lake to predict as high jungle grass screened and camouflaged many of the additional guns. The site was on a small doab bounded by the Hindan River* to the east and the Yamuna River to the west; it was also referred to as the Dharderi Plain.† The battle, which was about to take place, would be referred to by some as the Battle of the Dharderi Plain. Others called it the Battle of Patparganj, but later Lake's efforts here would eventually become known as the Battle of Delhi.

---

* Aka Hindon River, Hindun River.   † 28°+, 77°+. Aka Plain of Darderry.

Lake's cavalry remained the focus of the Marathas' attention as the result of a delay in bringing forward the infantry. A 'great loss' in men and horses was incurred during the hour which it took to unite the infantry and cavalry.[136] Call, as one of the 27th Dragoons who had moved forward with Lake, found himself in the thick of the action. The young officer wrote:

their guns, which opened on us most warmly and being kept up in salvos, at length, forced us to retire a short distance, where we halted in line – this caused their guns, which had been hidden by the high grass, to be advanced and for an hour and a half played on us so much, that both men and horses were mowed down like grass; our gallopers were of little use and the tumbril of the 27th blew up early in the action.[137]

The entrenched Maratha infantry proved capable of withstanding what feeble fire the depleted British cavalry ranks could produce. It was during this delay that General Lake had his horse shot from under him. His son, Major Lake, serving as the General's ADC, dismounted and gave his horse over to his father. Major Lake took the mount of a fallen trooper and kept it until one of his own horses was brought to him. Unfortunately the Major's remount was shot and killed beneath him. George Call saw the General's second mount, meaning the horse his son had given up, shot from under him as well and the search began for a third horse for the C-in-C.[138] Under a blazing sun, with no chance to stop and refresh, men were dropping from more than Maratha fire, as they were rendered motionless on the ground by sunstroke. Major Middleton and Cornet Sanguine of the 3rd Bengal NC were both listed as having died of excessive heat.[139] Middleton was an excellent horse soldier as well as a particular friend of Charles Stuart and his death was mourned for several pages in that officer's diary. Remember, it was Middleton alone who was singled-out for praise by Lake in the 1802 review that revealed the deficiencies in the British cavalry.

The combat force under Lake was thought to have numbered 4,500 while Maratha troop strength was listed as 19,000 of which 6,000 were cavalry.[140] General Lake was said to have feared a direct assault on the Maratha trenches might yield horrendous casualties. Imperial histories have painted what followed as an amazing picture of Lake's infantry and cavalry executing a carefully choreographed cavalry lure with an infantry follow-through. The standard battle account has it that in order to draw the enemy out onto the field, Lake ordered his cavalry to feign retreat. As the General's horsemen slowly fell back his infantry rushed forward screened from the Marathas by the retiring British cavalry. However, George Call's diary hints that luck had more to do with it. The bulk of the infantry had not even reached camp when the initial cannonade took place and they had to race to the sound of the guns. Grenadier John Pester's account of the battle had Lake opting for the use of 'cold steel' – meaning the bayonet – as the result of not being able to get the bullock-drawn British artillery up in time. 'The rapidity of our advance was so great that our Brigade guns and field pieces were obliged to drop in the rear, and we soon found that it was the General's intention to close with the bayonet.'[141] Call's account had Lake's cavalrymen already beginning to fall back before his infantry could get out of the camp area; 'by this time, the right wing had just reached the encampment and marched under General Ware, directly to our support, we, retiring towards them, and drawing the enemy's guns from out of the hollow and long grass, nearer to ours'.[142]

The Maratha infantry, seeing Lake fall back with his cavalry, presumed that they were driving the British General from the field. So they sallied forth from their trenches with their light artillery being pushed forward to ensure victory. Their exhilaration soon turned to shock as the suppos-edly retreating British horsemen divided their formation in the centre, wheeling left and right respectively, to reveal Lake's infantry drawn up into one line followed by a second line of additional cavalry. This second cavalry rank was then dispatched to the flank with their gallopers to hold off a body of Sikh horse that Bourquin had ordered to attack on the British right. The left flank was protected by the 1st Battalion of the 2nd Bengal NI Regiment, which was strengthened by four guns under Artillery Colonel Horsford's personal direction. General Lake posted himself at the head of his prized infantry unit, His Majesty's 76th Regi-ment, and acting in unison the entire British formation moved forward.

The infantry were being asked to advance into a deadly storm of heavy metal with little chance of survival. Once again the role of officer leader-ship should not be underestimated. This was more than an advance by a senior officer at the head of a body of infantry. This was the personal leadership role of the C-in-C. A great man willing to put his own life on the line as he urged his men forward in a manoeuvre which evoked the spirit of parade-ground precision but under the most adverse of survival conditions. Lake's men demonstrated extraordinary discipline under fire, keeping their muskets on their shoulders until ordered into firing position at a distance estimated to be only one hundred yards from the Maratha line. The Maratha artillery, by way of reply to the challenge, renewed their trademark maelstrom of metal. Round, grape and chain shot were all unleashed on the steadily advancing infantry. Lieutenant Pester and Colonel Blair were mounted and on the left of the line. As Blair conveyed his orders to Pester, a cannon ball grazed between them plunging into the troops behind them and killing a great many of the grenadiers of the 14th Bengal NI. Seconds later another cannon ball grazed Pester's horse. But the third incoming round was grape shot and its pattern easily took down Pester's mount.

As Pester recalled: 'A grape shot passed through the housings of my pistols, and shattered the stock of one of them, and I felt my horse staggering under me; another grape had grazed his side, and lodged under the skin; a third went through him. It entered at his near quarter and passed out at the other.'[143] Pester's horse fell mortally wounded but the lieutenant escaped with just severe bruising. He then took the offer of one of Blair's remounts and dashed forward of the line. In his anxious bid to rally the infantry, Pester noted the left of the line falter and at full gallop rushed to offer words of encouragement. But General St John at that moment gave the order to fire and, according to Pester, 'I was actually within twelve yards of the front rank men, at full speed, when the whole gave their fire.'[144] By the luck of timing Pester escaped the fate of being cut down by his own infantry. Despite the continuing barrage of anti-personnel rounds the single long line of troops held, fired that one volley and then charged with fixed bayonets. Pester's account made it sound as if it was with great energy and zeal that 'they rushed on with an ardour nothing could resist' and closed with the bayonet.[145] Dragoon George Call may not have been watching Pester's infantry but the foot soldiers he observed were near dropping from exhaustion. Call, noting that they had little energy left after being on the march since 2 am, wrote: 'Many were so exhausted at the end of the charge they had not strength to bayonet.'[146]

The Maratha line broke but as they ran the British infantry formed into columns of companies allowing the entire cavalry force, both native and European, to pass through their ranks and chase down the enemy. The advance of the British cavalry caused the Marathas to lose cohesion and panic. The cavalry pushed forward through the high grass. A number of the wounded British infantry raised their hands to alert the cavalry to their presence on the ground in the deep grass. Confusion in the thick of the battle was such that their own cavalry trampled an unknown number of the British infantry, as they lay wounded or dying. Stuart's brother horsemen reported being urged on by the wounded shouting 'Cut them up, Cut them up!' The cavalry advanced to the Maratha infantry on the banks of the Yamuna and broke their resistance. The British gallopers were used to rake the river with grape shot as the Marathas scrambled into the water. Many swam across the watercourse and struggled to pull themselves up on to a sand bar and sanctuary, only to be felled by the gallopers trained upon them. Some cavalrymen pushed as far as two miles along the bank, riding down enemy fugitives. Eventually they were held in check by Maratha artillery on the far bank around Delhi.

Lake reformed his infantry, having them wheel left to push remaining enemy stragglers, and those who had sought refuge in the broken ground, to the banks of the Yamuna River. William Thorn noted that on this occasion the gallopers 'did great execution upon the fugitives'.[147] Call wrote, 'they plunged in and endeavoured to swim across; our gallopers raked with grape some hundred and many were drowned attempting to get across the rapid stream which represented a red river'.[148] John Pester agreed, 'the river appeared boiling by the fire of grape kept up on those of the enemy who had taken to the river. It was literally, for a time, a stream of blood, and presented such a scene as at another period would freeze a man's very soul.'[149]

It is important to note that Lieutenant Charles Stuart did not fight in the action on the Dharderi Plain or as some referred to it the 'doab in the Jumna'. He had been detailed to serve with the baggage escort. Stuart admitted that the account in his diary was brief because it represented information gleaned from other members of the 3rd Bengal NC who were present. Stuart was told the unit had been lured into a thick grassy area by their pursuit of some Maratha horsemen. As the Maratha cavalry melted back by degrees the British cavalry pushed forward with their gallopers. The survivors said that what happened after that took them by complete surprise and the shock was overwhelming. The Maratha massed batteries began a deadly cannonade from positions concealed by the grass and delivered a classic textbook artillery ambush with technical precision. C-in-C Lake, at the head of his troops, sent for his infantry while ordering his cavalry to wheel off and withdraw slowly at the walk. Stuart's interviews indicated to him that this was the point at which Lake's experience was crucial to winning the battle. In 'slow walking' his cavalry retreat, Lake prevented a panicked flight from the field that would have resulted from a faster pace. The well-executed movement, although not without severe casualties, allowed the infantry time to come forward which would not have been otherwise possible. The Maratha line was then seen advancing out of nowhere. The Maratha officers were riding slowly beside their infantry, as opposed to riding at its lead, urging their men on and delivering orders of direction from the saddle. Lake was heard to shout 'Bring me the 76th' several times, but his prized Europeans were delayed by the fact that several had been in the process of bathing at the campsite after their early morning march.[150] They had to scramble for their clothes and kit when word of the ambush was received.

According to accounts given to Charles Stuart, when the British cavalry wheeled into column and the infantry passed into the intervals, the

Marathas could not recover full command and control. They were caught out in an uneven or broken line of advance, many of their infantry having rushed ahead thinking the British enemy was in full retreat. The Marathas did not have time to carefully relay their guns after rapidly pushing them forward in pursuit of Lake's retreating cavalry. As a result, many of the Maratha rounds, especially the grape shot, flew over the infantry's heads, but the trajectory was such that the rounds did great damage to the British cavalry who were behind the infantry as they advanced. General St John, on the left of the British line, led a bayonet charge into the Maratha right that was estimated to be their strongest point and therefore the portion of the line that demanded priority. Although St John's men broke the Maratha line, Stuart's informants commented that the Marathas fell back in a very orderly fashion until they reached the banks of the Yamuna River.

While the battle was over for Lake and his men, Charles Stuart and the baggage guard were still struggling miles away. The baggage train had been specifically ordered not to march before the cavalry or infantry because it posed the potential hazard of tying up the roads. Despite the intentional delay in their departure, things began to run terribly late for the baggage train that day. The ponies, camels, elephants and thousands of bullocks were so bothered by the heat and 'myriads of flies' that the baggage train could not be controlled. Men and beasts were strewn across an ever-lengthening line that extended for at least 10 miles.[151] The bullock-drawn hackeries that carried grain were among the greatest impediments to progress and the desire of the bullocks to stray off in search of water was a constant annoyance. The lumbering beasts slowed everything down to a 'creeping pace'. It was fortunate that the Marathas had withdrawn their light horse and *pindari* units to support their infantry closer to Delhi. But Stuart's guard remained vigilant. Most of his men had been awake since well before 4 am – forced to eat their breakfast in the saddle. By noon the men had long since exhausted the supply of water they carried. Upon finally reaching some small *nullahs* they found that the water had become badly contaminated by the large numbers of cavalry that had preceded them, muddying and fouling the water as they passed. Stuart remarked: 'Many died from fatigue and want of water. Several officers told me they were glad for what dirty water they could get from the nullah.'

Captain Doveton was observed collecting dirty water in bottles and giving it to a group of three Europeans who had dropped by the road. One of them lay on his back foaming at the mouth, which according to Stuart was a common sight on that hot September day. The sepoys fared better than the Europeans on the march, but there were several riotous scenes when the army came to a well. No reverence for rank or discipline was paid by the mobs of British troops who clambered over one another to get a chance to drink from the few sparse wells. In the afternoon Stuart heard a distant cannonade but had no way of knowing of the tremendous battle in progress. As night began to fall the officer commanding the rear guard decided to split the baggage train. Those closest to the camp were to proceed along the road in darkness and discern their route from the firelight of surrounding hamlets. Stuart made it clear that if it were not for the light of nearby villages the men and beasts would have been lost in the darkness. The infantry of the baggage train's rear guard were ordered to go back down the road they had just travelled and round up as many stragglers as possible; then they were to establish a secure perimeter for the night.

When Charles Stuart finally caught up with his unit he learned a great deal from one of his friends who was a regular guest at General Lake's dinner table. Stuart's confidential source relayed the information that Lake was particularly pleased with the capture of the Maratha guns and the brave steadiness that the cavalry had shown while advancing into the enemy's fire. The Grand Army had fought exceedingly well in consideration of the fact that they had been under arms for seventeen hours despite the intense heat. According to Stuart's friend, who was on very good terms with Lake, His Majesty's 76th Regiment was not the first infantry unit to attack the Maratha guns as reported. Stuart recorded that since the heat affected the sepoys less than the Europeans, it was indeed a sepoy unit that first reached the Maratha artillery during the final push. Stuart, very much the English gentleman in service to a native unit, insisted that the sepoys should have received credit for having been the first to assault the murderous Maratha artillery. But Lake's dispatches made no mention of his dinner-table admission.

Historically contemporary British casualty reports listed Lake's loss as 478 killed, wounded or missing, while Duff would later state them as 585, of whom fifteen were European officers.[152] Maratha losses were not actually recorded but estimated as 3,000 by Thorn.[153] Call recorded 4,500 enemy casualties of a presumed enemy force totalling 24,000 including 5,000 mounted Sikhs and 7,000 Maratha horse. Part of his calculation was based on the reports of twelve prisoners.[154] Charles Stuart's informant in Lake's entourage proved very helpful in trying to come to terms with the reality of casualty figures. He had ridden past the Maratha artillery positions as the enemy were driven from their guns towards the river. He later told Stuart that many of the Maratha gunners must have feigned death for their bodies were conspicuously absent when he rode past the same area later. There were 'far fewer on the ground than there had been before'. This is crucial in understanding how inaccurate the body-count figures were. In actual fact the number of Maratha troops listed as killed was not a body count but a victor's biased estimation based on ego rather than physical evidence in the form of corpses.

George Call admitted that a great many of the body-count numbers were derived by sending representatives to regiments and asking the men how many of the enemy they had killed.[155] Call went on to explain how enemy casualties for the day were calculated. He wrote, almost apologetically, that it had been seven years since a cavalry action of that magnitude had been fought in India and that as the result of the lapse, coupled with liberal amounts of bravado, many of the men may have inflated the figures. The cavalry brigade theoretically consisted of 800 men committed to battle and the men were claiming they killed three enemy soldiers each on average ($800 \times 3 = 2,400$). Call wrote that his cavalry brigade's tally was reported through their adjutant-general Colonel Clinton as 2,400 Marathas. Note the fate of the wounded in Call's chart below. Consider also that this cavalry tally minimized the impact of the British infantry; particularly if you consider the grape shot 'kills' in the river to have been inflicted by cavalry gallopers and that those 'ridden over' were victims of the British horsemen as well.

Killed by Cavalry	2,400
Killed by Infantry or Guns	700
Killed in the River by Grape shot or Drowned	800
Badly Wounded and then 'Ridden Over'	600
Total Enemy Killed	4,500[156]

What is one to make of this great disparity in casualty figures? If this Maratha artillery trap was so well laid why are the Maratha casualties up to nine times greater than the attacking British? Aside from tremendous exaggeration, there is a very simple answer in Charles Stuart's diary. The British counted the women, children and other non-combatant camp followers they killed as 'Maratha casualties'. Stuart of course could not have seen this because at the time he was miles away trying to sort out the problem of how to contain the baggage train. He had learned of many a 'savage act' from his brother officers in the 3rd Bengal NC who had personally but unsuccessfully sought to intervene in these needless deaths. Stuart's comrades were mortified at the barbarous treatment handed out to the enemy by George Call's unit – His Majesty's 27th Dragoons. Stuart was outraged at the King's horsemen. Within the confines of his diary he levelled charges of cruelty against them.[157] Stuart was obviously frustrated; the accounts of civilian slaughter by Anglo-Saxon troops of King

George's cavalry shook his moral foundations. But he could not bring formal charges against them since he was not present when the killings took place and the stories would be easily dismissed as hearsay evidence. Eyewitnesses from the 3rd Bengal NC did not want to testify against the Dragoons – interservice rivalry was bad enough between the EIC's troops and His Majesty's soldiers. And this was wartime. Inviting division among British troops might be construed as unpatriotic at the very least and, in a worst-case scenario, treasonous.

Stuart tells us that a number of the King's Dragoons used their pistols extensively in the bloodletting along the riverbank. The ferociously angry accusation of cruelty concerned the shooting of non-combatants at that location. As was the case in most major Indian armies of that era, the Maratha regular army travelled with a large number of camp followers and bazaar people. 'Many attempted to make their escape by swimming but whenever an unhappy wretch showed his head above the water a dozen pistols were levelled at it. An old man with a child on his shoulders stood in the water afraid to proceed. A dragoon, untouched by remorse, shot him and the child, I suppose, must have drowned.'[158]

The implication is that Stuart, as a white Anglo-Saxon officer serving with the South Asian officers and men of the Native Cavalry, developed a warm sense of humanity and compassion. He rode with and fought along-side South Asians and saw them as equals. It seems that if he ever har-boured racial prejudice he drew a distinction of human equality around those who soldiered with him. Although speculative, it is tempting to be-lieve that in his accusation of cruelty we see not only another aspect of interservice rivalry, between the EIC's and King's troops, but also a facet of dehumanization. If correct, the assumption is that the King's Dragoons were more racist and elitist because of their military culture and racial isolation within the South Asian military environment. These exclusively white officers of the King's troops were a minority and would no doubt have felt threatened by the abilities of South Asian cavalry officers who could ride or shoot as well, if not better, than any of them. Consider also that Charles Stuart's comments of cruelty were uttered in relation to the death of a South Asian enemy civilian. This could be construed as indi-cating that Stuart's care and concern was not just for his fellow soldiers but also for non-combatants, even when they were in service to the oppo-sition. As for interservice rivalry, between the King's and EIC's 'native' units, there had always been rifts. Some cases were basically economic, in terms of how much *batta* one received while on campaign. Other inci-dents of rivalry were put in terms of the moral and professional character of soldiers who served the EIC as a business entity rather than the army of George III. The racial element was always present in the slur that EIC officers, like Stuart, preferred the companionship of 'blacks' to that of soldiers of their own race.

Lake's army had fought to the point of exhaustion but they had been victorious in driving the Marathas from the field. Lake's baggage train was so far behind, his troops slept without cover on the battlefield. However, there were more pressing worries as they awoke on the morning of 12 September 1803. There were so many dead littering the field that an entire battalion was needed for the burial detail since the huge numbers of bodies posed a severe health threat in the intense heat. That was one of the considerations in moving the army's campsite 2 *coss* from the battlefield to Patpargunj. The dooley bearers of Lake's army were contract employees, like those who served Arthur Wellesley's army, and it was difficult to order them about in the fashion you would expect of military stretcher-bearers. The 'doolies' had proven extremely reluctant to risk their lives under fire. The net result was that most of the wounded had lain on the field for hours in the heat before the bearers would consider going out to pick them up. Stuart said that the dooley bearers 'always keep out of danger' and he bemoaned the fact that the hospital tents were so far away that many lives were lost for want of immediate aid. When the baggage troops had finally arrived they had set the medical tents up at the place originally designated as the army's campsite, a point just west of the Hindan River's bank and miles from the actual engagement. That posed a large logistical problem because it meant the wounded had to be moved back from the battlefield, travelling up to seven miles to get medical attention. The unity of the Grand Army was lost again. The fragility of the arriving wounded meant the hospital tents could not be moved forward to a more advantageous location. Carrying stretcher cases for miles meant more casualties as many men bled to death from wounds inflicted by Maratha anti-personnel rounds. It would be several days before a complete evacuation could be performed. Stuart also conceded that the field was no place for performing an operation. Huge numbers of amputations had to be performed as limbs were pierced, mangled and left useless by canister as well as grape shot. There was a shortage of surgeons. The volume of cases, the heat in the tents and the plague of flies that swarmed around the hellacious scene overwhelmed those who were on duty.

John Pester provided a gruesome description that seemed more reminiscent of an abattoir than a hospital.

About thirty surgeons were absolutely covered with blood, performing operations on the unfortunate soldiers who had their legs and arms shattered in the action, and death in every shape seemed to preside in this assemblage of human misery . . . In one corner of the tent stood a pile of legs and arms, from which the boots and clothes of many were not yet stripped off.[159]

Charles Stuart wrote with pity of Cornet Crowe, a promising young officer. The bones in one foot were crushed by the impact of a projectile. The unfortunate Crowe was compelled to travel to several different aid stations before he could find one where his foot could be amputated. Stuart knew that the amputation meant Crowe was 'rendered useless as a cavalry officer', since he could no longer use both stirrups. Stuart hoped that provision would be made for young Crowe on Lake's staff. It was a form of military welfare that had been applied in several existing cases among the C-in-C's staff. Crowe would not be automatically 'invalided out' since the Grand Army was engaged in a serious campaign where all talented Europeans were put to use whenever possible. He would, however, have to find a different military occupational speciality. Although maimed, he had survived, which was more than could be said for many wounded who died of heat, haemorrhaging or shock, as they lay for hours and in some cases days before the dooley bearers picked them up and took them to the aid stations. At least they were better off than the Maratha wounded abandoned on the field only to be plundered and killed by Lake's camp followers. Lieutenant Pester came across two of the naked Maratha sepoys along the riverbank still clinging to life five days after the battle. 'We knew them by their hair, which is never cut like our soldiers. They were a fine soldier-like looking cast of men; their clothes were stripped off them [plundered], but the most friendly office of the two was left undone – putting them out of their misery.'[160] Despite the lack of hope for recovery, Pester sent the two wounded men off to a field hospital.

Before leaving the field, another task that was given priority was securing the Maratha artillery that had been abandoned. There were several very sound reasons for treating this issue with some urgency. Among the guns captured by the British at the Battle of the Dharderi Plain were some unique gun tubes that were laminated in a manner that the British had not seen before. According to Charles Stuart, 'to render them more durable than if made of brass alone they were lined with iron or rather an iron cylinder being first prepared it appears as if the brass were cast over it which has been done so well that there is not the smallest separation between the two metals'. Stuart also noted that several of the mortars and howitzers of a large size were fitted with a sophisticated elevating screw, which allowed them to be used with a certain amount of interchangeability. The tumbrels were in poor condition but the draught bullocks were particularly fine and superior to those the British possessed, so they were taken to pull Lake's artillery.[161]

Since the Grand Army was going to continue its march it was important not to leave the Marathas any weapons they could salvage for reuse. In addition there were still large forts such as Agra to tackle and one could not have too many pieces of artillery for use against a fortress of that dimension. But the superb Maratha artillery was also sought for its 'prize value'. A gun in fine condition could be resold to native allies or even European shipping interests at high value. One that was damaged, even irreparably, brought rupees on the scrap market. Once articles were disposed of, their cash value was credited to the prize fund. It was common in campaigns such as Lake's for everyone from the C-in-C to the *puckallies* (water carriers) to draw proportional shares on the prize fund to which they were entitled. A cavalry detail was formed into search teams to locate and mark the whereabouts of Maratha cannon. The Maratha guns had been so well hidden that their exact number and whereabouts were not known for days. The search details must have eaten particularly well on that occasion for Charles Stuart tells us they were also tasked with destroying the large number of wounded cattle which still roamed the battlefield.[162] John Pester told of the men in his unit who were sent to retrieve other Maratha gun bullocks illegally plundered by British soldiers and camp followers alike. 'In the course of an hour our Non-Commissioned, who were to collect those of our Corps, cleared from the lines of the 2nd Regiment only one hundred and twenty-three fine gun bullocks; they were sent to the Prize Agents to be disposed of on account of the army at large.'[163]

The official report of captured ordnance noted that 68 artillery pieces were taken from the Marathas. These guns were complete, meaning mounted on fine field carriages with traces and limbers. The majority (23) were light 4 pounders used as mobile field guns that could be pushed forward with the infantry when it advanced, but heavier guns included an 8-inch howitzer, two brass 20 pounders and five stubby 18 pounder carronades, the remainder being a very diverse assortment. William Thorn, like most British officers who saw the Maratha artillery, was fascinated by its modernity and in particular the so-called 'French' pattern elevating screws.[164] This addition allowed a number of guns to serve more than one purpose. By adding an elongated elevation screw a number of cannon could be used as howitzers to deliver 'plunging fire' at a higher angle of trajectory. 'And to the mortars and howitzers the same kind of elevating screws are, by a simple and ingenious adjustment, made to elevate the piece to any angle, and give either of them the double capacity of mortar and howitzer.'[165] The brass guns cast in India were primarily from Sindia's foundries at Ujjain* and Mathura. Ironically, the only guns that burst were two iron pieces cast in Europe. None of the Indian cast-iron guns suffered that fate.

---

* 23°+, 75°+. Aka Ougein, Oujein, Ougene.

The Marathas had come into the field with more than enough artillery ammunition to finish Lake's force. There had been 61 tumbrels laden with ammunition but 24 were destroyed in battle and the remaining 37 captured. In addition, the cavalry found two tumbrels laden with treasure, which presumably was to pay Sindia's troops and *banjaras* in the field. Normally the treasure would have been turned straight over to the prize agents for inventorying as part of the prize fund. But as of that moment, no agents had been designated. This resulted in a rather strange meeting that came to Stuart's attention.[166] Several cavalry officers met to discuss a proposal to keep the tumbrels of treasure and divide it only among the cavalry since they had brought it in. It was the type of decision that could affect interservice rivalry, increasing tension between His Majesty's troops and those of the EIC. The King's troops often lorded their royal commissions over the Company's officers. Status and motivation featured regularly in taunts and jeers and bravado took its toll on truth. Although many of His Majesty's officers made derogatory remarks about the 'money-grabbing' values of those who served the EIC, rare was the King's soldier who refused a share of the prize fund. They were willing to play up the perception of elite status, with regard to their combat ranking, but they also thought that India – the country that robbed them of their health and seemed such an alien environment to die in – owed them something. And if that something was cash, that could be interpreted as compensation for having to serve under such conditions. So why not take the money and be done with it? Fortunately, the clandestine meeting broke up with the horsemen having made the decision to follow the standard procedures for the disposal of treasure through the prize agents.

Charles Stuart's diary makes it quite clear that there was a tremendous cultural difference between the King's mounted troops and those of the EIC. But not culture in the context of British social class. In that respect Stuart outranked virtually all the Dragoons as he was the son of a baron. The difference was that of military culture. Stuart painted the King's Dragoons as leading a softer life on routine marches. They did not have to ride exterior picket duty; rather, they were 'let off' with the easier job of interior picket postings. One could argue, on the Dragoons' behalf, that Lake probably wanted them 'in close' so that they could respond at a moment's notice in case there was an emergency situation that needed an immediate reaction force well versed in the latest galloper movements.[167] Others no doubt believe the King's Dragoons earned the softer daily postings. For during pitched battle the General was more likely to use them as his primary cavalry troops and therefore they were prone to incur higher casualty rates in larger battles. They were told they were the best – the problem was they believed it implicitly. It was their inculcation as an elite force that tainted their military culture with an insufferable air of superiority and that directly bred interservice rivalry. At that time the rivalry in the cavalry may have been more clearly defined than that found between infantry units. One finds examples in European military history of the cliquish nature of cavalry units and the elite status was occasionally underscored with enlistment from the upper reaches of society. But as we can see in the case of Charles Stuart, one could find a member of the aristocracy in the EIC's Native Cavalry. That may in some small way explain the underpinning of liberal racial attitudes in the EIC's units at that time. It was not a case of who was better, King's Dragoons or EIC Native Cavalry. That was impossible to quantify on either an individual or 'service' basis. The EIC's regiments often had more years of Indian experience because they were constantly being called out. Yet the overall quality of cavalry, the King's and the Company's, was so low that Lake had seen fit to initiate large-scale retraining at Kannauj. The rivalry remained very real.

Five battalions of Maratha sepoys had managed to escape from the battle outside Delhi and two of those had remained virtually unscathed. Lieutenant Charles Stuart was the most honest diarist in stating that in fact the majority of the Maratha infantry at the battle site were not killed but rather 'dispersed'. Although these were Sindia's enlisted men, as opposed to his European or Euro-Asian mercenary officers, terms of employment were offered to them under the Governor-General's Proclamation. Stuart reported that many of the Marathas' sepoys had 'now taken service with us'.[168] Their fate was similar to that of many of the other Maratha military resources found on the battlefield. These enemy infantry were not destroyed, but rather recycled and incorporated into the EIC's war machine.

The Battle of Delhi was a ferocious combat but it proved more politically than militarily important in the ultimate effort to push Sindia out of North India. In gaining Delhi, Lake secured access to the ageing Mughal monarch in whose name Sindia controlled the Doab. The British were about to achieve the all-important political objective of 'freeing' the Mughal Emperor. Lake's quartermaster, named Campbell, entered Delhi on 14 September 1803 to make arrangements for the General to visit the Mughal Emperor Shah Alam II. Members of the 6th Bengal NC had reportedly seen hundreds of Maratha wounded along the roads and a great many had died of their wounds since the battle.[169] While Campbell encountered no resistance from Maratha stragglers, he did have a problem locating sufficient boats for Lake's army to cross the river.[170] Maratha forces had commandeered many craft to make their escape towards Agra. However, Lake's staff were encouraged that day as Louis Bourquin, Sindia's highest-ranking European field commander, surrendered. Four other officers, identified by William Thorn as Gessin, Guerinnier, Del Perron and Jean Pierre, also surrendered with Bourquin.[171] Apparently this party had to be taken into protective custody to save them from what was termed 'popular resentment' and they were eventually dispatched to Fatehpur en route to the Bengal Presidency. Most of the British diarists, who witnessed Bourquin's surrender, tell of him speaking low French and they cast a very negative picture of him. George Call threw in the indignity that the Frenchman had been a baker and sold salt beef in Calcutta.[172] Call also noted that Bourquin was only followed by 100 enlisted men at the time of his surrender – supposedly indicating the extent of the Maratha defeat.[173] What Call did not know was that Bourquin had been Perron's protégé. Although never a renowned soldier, he had gained extensive combat experience in the campaign Perron had waged against the army of Irish mercenary George Thomas. As for the 100 Maratha sepoys, they were only a fraction of the thousands who remained in service under Ambaji Inglia's command.

Lake crossed over to Delhi by boat on 15 September followed by his infantry. The next day the British General visited Shah Alam who reportedly called Lake his 'friend and deliverer'.[174] The elderly Mughal was often described as a tragic figure. He had been a virtual puppet of the Marathas since Mahadji Sindia became Regent in 1784 and the Afghan adventurer Gulam Kadir* had blinded the Emperor in 1788. 'Possession' of the Mughal Emperor became a bizarre ritual for the powers that competed for control of the Doab or, at the very least, its riches. Although both the Marathas and the British campaigned and collected taxes in his name, he was more than a figurehead. He was a status symbol for those who sought to control North India. 'Liberating' and then supposedly 'caring for' the Emperor had become one in a series of rituals by which the contending powers tried to legitimate their bids to control North India. While Lake exercised his charm and showed great kindness to the Emperor-poet, the men of the Grand Army struggled to move the tons of captured Maratha artillery across the river. In their flight from the battlefield the Marathas had not taken time to destroy their barges or 'platform boats on which they crossed their own troops and Artillery'.[175] The British now made good use of the captured Maratha barges. Like Shah Alam II's ancestor Babur, Lake would use boats to float his 'new' artillery down to Agra for siege purposes.

---

* Aka Gollaum Cawdor.

On 23 September 1803, the same day Arthur Wellesley met the Marathas at Assaye, Lake said good-bye to Shah Alam. Lake appointed Lieutenant-Colonel Ochterlony, deputy adjutant-general of the EIC's troops, as the official British Resident to the Mughal court at Delhi. Lake also arranged to have an old palace residence fitted out as a more permanent hospital and a proportion of the army's medical staff were designated to remain for the care of the wounded who were still in a delicate condition. The field hospital, with its massive backlog of amputations and wounds to treat, had been forced to stay back near the battle site for twelve days. Lake had left a battalion there to guard it – knowing full well that Sindia might easily exploit a target that would cause Lake to divide his forces and double-back.[176] General Lake could not afford to keep leaving great numbers of his men to garrison each of his conquests as he had done at Alighar. Lake's fighting strength was rumoured to have been as little as 5,000 men on the eve of the battle at Delhi. Therefore, in order to preserve his resources, he set about employing a number of the officers who had served Sindia's army. An unspecified number of British mercenary officers, who had 'come in', were recruited to raise a corps of Mewati troops.[177] The Mewati corps along with a battalion and four companies of native infantry were left at Delhi with Ochterlony.

Agra had to be neutralized as a garrison and weapons production centre and it was so much easier to control the Doab by controlling its flanking waterways. The rivers defined the territory, made it fertile and ultimately provided the corridor by which to annex it more effectively. Lake intended to use a number of the captured Maratha guns for the siege of Agra's fortress. Those artillery pieces, along with other 'heavy metal', were sent by boat down the Yamuna as the army marched along the west bank.[178] The plan was to take the 'high road' to Agra because it was so well marked with *cossminars* and the route carried Lake's army through the rich territory of the Bellum Rajah.[179] Its high degree of cultivation held the promise of making life easier while on the march. The initial portion of the march was relatively uneventful in military terms but the absence of Maratha regulars did not guarantee tranquillity. Everyone complained about the Gujars* and the problems they caused. The Gujars had served extensively in the Jat armies of the region. They were well acquainted with the manner in which armies operated and were quite inclined to use that knowledge in the commission of theft.

Charles Stuart once again found himself posted to the rear guard; perhaps this was the price for not having trained at Kannauj. As the tail unit

---

* Aka Goojers.

had passed the last of the *nullah*-cut roads, on the outskirts of Delhi, Stuart became aware they were being followed by Gujars. They were attempting to make off with the hackeries that could not keep up. Problems grew progressively worse as the column passed Tughluqabad on its right.[180] The old fort had become a nest for these plunderers who turned Stuart's ten miles of rear-guard duty into an anxious task. The Gujars were a security threat as they murdered stragglers after having robbed them. And we know that Gujars had threatened North Indian campaigns since the time of Babur, who made a rather brief reference to his method of dealing with them in 1520: 'Sayyid-Qasim Eshik-aqa was in command of the rear guard. He caught some Gujars coming up on our rear, cut off their heads, and brought them in.'[181] Lieutenant Pester's diary revealed there was historic continuity in counterinsurgency tactics. Small parties of infantry were sent out and they shot thirty Gujars on 24 September; on the following day a *Jemadar*'s patrol from the 2nd Bengal NC 'cut up' twenty more.[182] Lake's men felt no sympathy for the Gujars who, aside from thieving, had cut the *dawk* and killed some *harkarrahs*. The Gujars cruelly mutilated their captives by cutting off their right hands as well as noses.[183] Constant vigilance was needed and Charles Stuart's exasperation was evident in his comment, 'The patience of Job, would not hold out on a rear guard during a long march in a hot climate.'[184] When the rear guard finally made it up to the camp at Faridabad* it was assumed that the last of the hackeries could make it in on their own since no one would dare to abscond with the supplies in full view of an armed British camp. However, the assumption proved to be quite wrong. To Stuart's chagrin three or four carts were plundered in sight of the encampment as the rear guard came in. Here were a group of brigands who were seeking to exploit one of the biggest opportunities that the South Asian military environment could offer. One did not need the power of Sindia's battalions to steal from Lake's army, merely an incredible amount of nerve and the luck not to get caught.

On 26 September George Call of the 27th Light Dragoons realized that something very important was underway. Two Maratha *vakils*, representing the 'Commandant' of five Maratha 'regular corps' battalions, had arrived unexpectedly in camp. Ever since Lake's victory at Delhi there were persistent stories that the General had been lucky to escape the ambush alive because there were at least two battalions of Maratha troops who opted not to fight. But those battalions had 'come over' following the battle and were now in British service. As negotiations progressed with the *vakils* it became evident these five complete battalions

---

* 28°+, 77°+.

had remained on the Delhi side of the Yamuna River, having never been committed to the battle.[185] How many Maratha battalions were still out there and why had they chosen not to fight thus far? That these were intact battalions and not straggler or composite units was supported by George Call's observation that their total battalion allotment of 25 pieces of artillery (5 guns per battalion × 5 battalions = 25 guns) was also up for negotiation on condition of employment with Lake. This astounding negotiating process was an aspect of the South Asian military economy that flabbergasted many of the Britons. Was it audacity that led these men to believe they could travel the countryside as virtual free agents in the military labour market seeking the best offer? Here was an army for hire but terms could not be agreed on at that time. The meeting broke off but the negotiating would resume at a later date.

By 29 September, it had begun to rain heavily and Lake's army halted to round up the baggage train. Delays resulted from the slowness of the carriages and patrolling against Gujars. When the C-in-C learned how far the baggage train was strung out and the degree of difficulty that the rear guard was experiencing, he exploded in anger. Lake issued a stern ultimatum to the officer in charge of the rear guard; if the hackeries with the public stores were lost then the gentleman in question would be forced to pay for them.[186] The rear guard got into camp at 1 am on the first day of October. They had been in the saddle for thirty hours and Lake's words still rang in their ears. Stuart commented they had been awake for forty-eight hours and that neither men nor animals had sufficient food because they were in almost constant motion trying to keep up with the progress of the infantry. Yet the relentless march of Lake's army meant they had to get up at 5.30 am – less than five hours away – for the last leg of the journey to Mathura. The extraordinarily long hours in the saddle also took their toll on the horses ridden by the Native Cavalry. This helps to explain why so many more of the Native Cavalry's horses were 'knocked-up' compared to those of the King's Dragoons. Quite simply the horses of the Native Cavalry were abused by prolonged service without proper rotation or care in the field. This also meant that Native Cavalry officers in the field were often forced to spend more on remounts than other officers because of the attrition inflicted upon their mounts.

At Mathura on 2 October 1803, Lake's infantry rendezvoused with Lieutenant-Colonel Vandeleur.[187] A month earlier Vandeleur had gone south with elements of the 8th Light Dragoons and a brigade of Native Cavalry in hot pursuit of Monsieur Fleury and his Maratha horse. Their chase through the Doab and then south of the Yamuna River was a separate action of the war in one sense and it was very much a glimpse of the future 'predatory' war that broke out against Holkar in 1804. Fleury had

long since 'come in' with Beckett and Perron, but his horsemen had conducted an extensive incendiary campaign and their scorched earth tactics
were very effective in a land that had been left parched by a deficient and
rather late monsoon. The destruction of fodder was their specific objective. If you could eliminate the feed needed for the British horses then
their cavalry could not pursue you without accompanying *banjara* convoys
which, in turn, could be preyed upon.

Colonel Vandeleur had reached Mathura two days before General
Lake. There Vandeleur had overseen the surrender of Sindia's Colonel
Dudrenec and two other European mercenary officers in Maratha service, identified as Smith and Lapenet.[188] Chevalier Dudrenec had been
sent from the Deccan with some of Sindia's regular battalions in July to
reinforce Perron in Hindustan. But like other mercenaries, induced by
the Governor-General's Proclamation, or simply unwilling to fight fellow
Europeans, they had left Maratha service to seek safety and a pension with
the British. The holy city of Mathura was of great military significance to
Perron as a manufacturing centre, but he had defected already. It was not
only part of his *jaghir* and vital to revenue collection; it was also the site
of one of Sindia's major cannon foundries.[189] The British force apparently encountered no resistance from the townspeople of Mathura who
seemed quite willing to receive the British troops and they traded with
them as well. This appears to have been noteworthy to Charles Stuart,
as he could discern no particular reason why certain towns welcomed
the British and others rejected them. Less objective diarists painted the
towns that welcomed the British as those who were trying to 'throw
off' Sindia's 'oppressive yoke', or – if you followed Governor-General
Richard Wellesley's lead – seek freedom from Perron's 'French state on the
Jumna'.

Lake's Hindu sepoys were anxious to enter Mathura and have an opportunity to make *puja*, performing purification rights in the Yamuna River.
They had requested that the C-in-C halt for one full day to allow them
this opportunity.[190] However, the reports of enemy personnel in the area
compelled Lake to keep the army on a high state of alert. Lake was reluctant to let entire units or large groups of men go at any one time for fear
of weakening his strength or giving the enemy a chance to take hostages.
But Lake was also conscious of the morale of his Hindu infantry. As a
result, Lake transmitted special orders to selected officers to allow small
numbers of Hindu sepoys to visit Mathura but they were to stick together
and return before nightfall. Lake's concern over the spiritual well being
of his Hindu troops remains under-reported.[191]

While at Mathura Lake took the opportunity to personally interview
Chevalier Dudrenec. The Colonel was one of Sindia's senior European

mercenary commanders and he willingly provided data on his men, pieces of artillery and logistical arrangements. Dudrenec had 'come in' but where were his battalions? The Chevalier also gave reports on Sindia's disposition in the Deccan and the state of revenue collection. But Dudrenec was far from the only source of intelligence as several of his French staff and British brigade officers had by now 'come in'. Oddly enough, Stuart spoke in mixed terms of Dudrenec who had enjoyed an excellent military reputation as an officer in Maratha service during the dark days of the Holkar–Sindia interclan warfare that preceded the 1803 campaign. But Charles Stuart saw the surrender of Dudrenec and 'several other officers who followed his example' as shameful in having occurred this late in the war. Stuart perceived Chevalier Dudrenec as having kindled some form of false hope in the hearts of his men – then smashing their confidence in a contemptuous moment of surrender. Stuart knew that the British would profit not only from the intelligence but also from the blow to the morale of the Maratha sepoys. Yet Stuart expressed a note of scorn or shame that implied that a true gentleman, if he were going to defect, would have done so at the beginning of the war. Stuart saw Dudrenec as waiting and starting on campaign with his loyal men, as if he were a Judas steer, leading his men on, not only in terms of confidence, but also on to their doom. 'The ill conduct of all the French officers, while it reflects dishonour in them, will be of great service to us for it must have a tendency to dishearten the troops they have so shamefully abandoned. The Brigades [Dudrenec's] are supposed to be near Agra.'[192]

Thanks to reports from the Maratha mercenaries who had 'come in', Lake's intelligence on a strategic level was magnificent with regard to how many men, guns and horses Sindia had remaining. But, as we have seen in Arthur Wellesley's Deccan Campaign, that did not preclude intelligence shortcomings on a tactical level. Dudrenec confirmed the strength and physical disposition of the Deccani brigades that had travelled north with him. However, Dudrenec could not give Lake their precise and immediate whereabouts since his departure. Lake desperately wanted to avoid another ambush. Cornet Call was sent out as part of a mounted reconnaissance party on 4 October, and they found themselves in the suburbs of south Agra angling for a look at the Red Fort. The tension was definitely growing. The young dragoon realized they were now within range of the 'Great Gun of Agra', a monstrous piece manned by the Marathas and firing a ball measuring 23 inches in diameter.[193] From their final position the British party could see seven enemy battalions were encamped on the glacis outside the fort. But it appeared by their baggage and impromptu camping arrangements that they had not been allowed to enter the *Madaghur* – the fort's outer work. It was a most unusual situation and the meaning of it was far from clear.

By the next day Charles Stuart and the bulk of the army were marching past the main part of the city of Agra. Their intention was to move against the fortress from an eastern approach.[194] Stuart was aware they were taking occasional fire from very large-calibre guns as they passed but he had no idea of the specific range. William Thorn tells us that, having encountered virtually no resistance, Lake's deployment continued, his cavalry having managed to surround and isolate Agra on 8 October.[195] During that time a treaty with Ranjit Singh, the Jat Rajah of Bharatpur,* was formalized and as part of the mutual defence clause he had sent 5,000 horsemen to Agra to help tighten the British noose around the city by cutting Maratha logistical lines. Lake attempted to negotiate with the occupants of Agra's fort but confusion reigned supreme in the Maratha stronghold and no answer was forthcoming.

The British strategy for taking Agra unfolded with a peculiar rhythm that suggests the plan was modified several times to incorporate late-breaking information.[196] Intelligence was being gained from defectors of Sindia's Deccani brigades, some of whom were found wandering near Agra in small groups. Brigadier-General Clarke was dispatched with his brigade, back over the same route he had just covered, to a position west of Agra. Early explanations of his retrograde movement suggest a blocking or anvil force to smash any remaining Maratha troops or those who sought to escape the assault on the fortress from the east of Agra. However, the diaries of Stuart and Call indicate Clarke's mission was revised with a much more specific target in mind. Rather than serve as an anvil for an eastern assault or as a western mop-up detachment intended to pick up escapees, it is clear that Clarke's men were ultimately tasked with taking out a large body of Maratha infantry who were ensconced in the Jami' Masjid Mosque.[197]

According to Charles Stuart, on the morning of 10 October, there was to be a British main force advance from the east with the immediate aim of clearing Maratha light infantry units and snipers from the ravines found along the approach to the fort. This bold thrust was supposed to coincide with Brigadier Clarke's advance from the west and a flanking cavalry movement. Those flanking horsemen were to secure the area in the vicinity of the Taj Mahal, thereby cutting off an obvious escape route for the Marathas when the fortress of Agra fell.[198] Brigadier Clarke's brigade was supposed to penetrate the city by dividing into two columns: one under his personal command, the other under Colonel White. Each column

---

* Not to be confused with Ranjit Singh the Sikh leader.

was to travel a separate route through the city, with the intention of converging on their target, effecting a pincer movement against the Maratha infantry in the Jami' Masjid Mosque to the west of the fort.[199] As the day wore on Lake's main force to the east saw columns of smoke rising from the city and heard loud peals of musket fire indicating Brigadier Clarke's men had run into very determined opposition. Accounts are confused but it seems that the Brigadier's party got lost in the inner city and then came under intense musketry from the Marathas firing from windows and rooftops. Brigadier Clarke's second column, under Colonel White, had more luck in finding their way to the target. But unexpectedly stiff resistance and the danger of running out of ammunition forced Colonel White into a holding action near the walls of the fort. It was at that point that the Brigadier's column took an unacceptable number of casualties. They were hopelessly trapped, so Clarke ordered a retreat. One of Brigadier Clarke's runners managed to find the other column, giving Colonel White the order to retreat as well. Pester's account states that Colonel White thought that it would be 'disgraceful' for British troops to retreat and he refused to abandon his position as long as he had a single officer left alive. White sent a letter under fire to Lake stating what had happened. The C-in-C responded to White and 'paid him a handsome compliment in Orders, [but] ordered Brigadier Clarke to return immediately to camp'.[200]

Back on the eastern front with Lake's main force, the ravine clearing operations were hampered by the Maratha infantry's expert utilization of terrain. The ravines contained numerous crevices and their weather-worn surfaces made these natural trenches much less perceptible than any man-made example of field engineering. The presence of freshly turned soil or the distinct unnaturalness of a man-made berm often draws attention to newly constructed trenches. But in the sun-bleached terrain these natural features did not betray any glaringly obvious indication that the undulating ground held anything more than a few centipedes or snakes. And from these imperceptible slits, Maratha sharpshooters proved particularly effective. South Asian snipers were rarely acknowledged for their historic ability to demoralize their enemies. But they were every bit as effective as their American counterparts in bringing grief to the British. As noted earlier in this book, designated marksmen in Maratha battalions did not commonly use the general-issue flintlock musket and most Maratha snipers in this battle used much more accurate long-barrelled matchlocks.

Maratha casualties were very low in this initial round of fighting as Sindia's men skilfully exploited the natural cover and concealment, leaving Lake's men scrambling to try and locate the source of fire. Pockets of resistance had to be winkled out and Lake's men suffered a number of casualties just as they thought they had finished their sweep of the outlying ravines.[201] Sporadic fire continued but the British had enough men through the ravines to think about forming up for the next objective. They hurriedly reassembled; their objective was now to neutralize, or if possible seize, the lethal artillery that the Marathas had posted outside the fort in the well-fortified support battery. As the British troops advanced in broken formation to consolidate their apparent gain, the area erupted in fire. Much to the embarrassment of the British they were now attacked by a large number of Maratha troops who had maintained their fire discipline and lay nestled among the rocks in silence, waiting for the British. The ambuscade was delivered with rapidity and before Lake's forward numbers could react they again suffered numerous casualties. Fortunately for the beleaguered advance party, Major General St John – Lake's chief advocate of light infantry tactics – saw the tide of events turning and hurried support to the dangerously isolated men.

The Maratha ambush party, having apparently delivered their volley of smallarms fire, then appeared to break and run for the safety of the *glacis*, which lay in front of the fort. The British, invigorated by what they perceived as a positive change in their luck, dashed forward having been convinced that they could now capitalize on what seemed to be a reversal of fortune for their Maratha opponents. But the attack and this feigned retreat of the Marathas were part of a greater defensive strategy. The British were 'under' the range of the Marathas' wall-mounted artillery and the elevation of the heavy guns could not be depressed to take advantage of the closing target. However, as St John's relief group sprinted forward they entered within the range of the flintlock-firing Maratha infantry placed on the fort's wall above the *glacis*. Severe fighting ensued in the vicinity of the *glacis* and in the defensive outworks of the fort as more British troops were pushed into the fray. According to Thorn, despite considerable casualties amounting to 228 killed or wounded, the British assault was successful in capturing 26 Maratha cannon complete with field carriages, limbers and 29 ammunition carriages, while inflicting 600 casualties on the defenders. For British enlisted men and sepoys this hotly disputed area outside the fortress proper did not seem to make much sense. Why had all these guns, tumbrels and equipment been left outside of the fort? Within forty-eight hours of this victory, 2,500 Maratha troops, who had also remained outside the fort's walls, 'came over' and marched into the British camp after brief negotiation.[202]

Lake's men then began the arduous task of laying formal siege to the inner fort. In extreme heat and under backbreaking loads they carried supplies up the ravines. They commenced excavation of support trenches.

Being careful to keep their heads down to avoid sniping, they started construction of the British grand siege battery. Labouring in the intense heat brought accompanying dehydration which made life miserable, but annoyance and disease were also made worse by swarms of flies. The Maratha gun bullocks in the fort were starving and so the garrison turned them out to fend for themselves. Pester and his friends took the liberty of slaughtering some of the animals for fresh meat. But they got little pleasure from the meal owing to their dining table's location adjacent to the unburied dead along the trenches, men who had fallen four days earlier. He lamented, 'It was hardly possible to eat anything, the stench was so terrible.'[203]

A secondary battery was built three hundred and fifty yards from the southeast side of the fort towards the river. The rapid sequence of events compelled the Maratha garrison to begin surrender negotiations and to expedite their case they used two of their European mercenary officers as their intermediaries. This was all somewhat bizarre because the two officers, Colonel Hugh Sutherland and Colonel George William Hessing, had been under house arrest in the fort after being ousted in a coup by the Marathas.[204] Sutherland and Hessing wrote to Lake, after explaining to their subordinates-turned-captors that if Lake stormed the inner fort and they offered armed resistance then the British General would be exasperated and 'exterminate the whole of them'.[205] Hessing and Sutherland urged the garrison to surrender their arms and abandon all treasure in return for their freedom. Captain Salkeld, from Lake's army, arrived in the fort to conduct further negotiations among the politically divided defenders. Several Maratha *sardars* were for continuing the fight but many of the enlisted men saw little chance of a breakout. The truce soon broke down and Captain Salkeld narrowly missed being killed by a cannon shot as he returned to his lines and renewed heavy firing broke out on both sides.

For several days there were limited exchanges of fire as sharpshooters from the fort tried to inflict as many casualties as possible on the British troops constructing the grand siege battery. During the day Maratha gunners on the fort's walls fired grape shot sweeping the trench tops to slow enemy progress. The labourers in the grand battery were given constant reminders to keep their heads down. But night was particularly dangerous as Maratha trench-raiding parties crawled to the edge of the grand battery and fired in at point-blank range. Lake's artillerymen, despite such tempting targets, maintained rigid fire control and refrained from using anything heavier than musket fire to drive off the nocturnal infiltrators that probed the position.[206] Lake's men were well supplied with 'heavy metal'; having transported the best pieces of Maratha artillery previously captured at Delhi, they now sought to employ the guns against their former owners. On the morning of 17 October, the grand battery's eight 18 pounders, four 12 pounders, six howitzers and six 10-inch mortars opened fire with the assistance of an additional supporting mortar battery. The guns commenced firing on the fort's southeast bastion, which had been selected for its apparent weakness. Secondary British artillery fire, to enfilade the defenders, was provided by four 12 pounders to the left of the battery and on the right were two 12 pounders situated by the river.[207]

Despite the heavy volume of suppressive fire that the British grand battery achieved, the Maratha artillery on the fort's walls was well aimed and still operational. Maratha gunners repeatedly demonstrated their accuracy with stunning results and perhaps the reinforced walls of the British battery gave men a false sense of security. Pester, who had gone into the British grand battery as an observer, was shocked to see 'the enemy's shot every moment came into the embrasures and killed the men. One shot carried away five men, and we were completely covered with the blood and the brains of the poor fellows.'[208] However, by the end of the first day, firing from the grand battery was pronounced as having produced adequate results. The mortars' explosive shells had done considerable mischief among the garrison troops, while the 18 pounder solid round shot had caused the red 'sandstone' walls of Agra's fort to come off in huge flakes. The next morning the garrison sent word that they would capitulate without further resistance. Between 5,000 and 6,000 Maratha troops marched out of the great fort with their personal possessions. Lieutenant Pester claims to have taken Sindia's flag from the ramparts, 'a red silk ground with a white snake diagonally', and replaced it with the British Standard.[209]

The vast storehouses of Agra's great fortress fell to Lake's army. Tremendous quantities of grain, weapons and war materiel were obtained including Rs. 22 *lakhs* apparently earmarked for paying the Maratha troops who were months in arrears. The prize agents amassed a fund of Rs. 24 *lakhs* and 40,000.[210] Grant Duff noted 162 pieces of artillery fell to Lake that day as well. Among the more interesting pieces of booty was the 'Great Gun of Agra'. The fort had been a cannon-casting centre for decades and the mammoth piece of artillery was more than just a showpiece, it was fully functional despite its massive size.[211] The *shroffs* of Agra reportedly offered the British a *lakh* of rupees for the gun so that the metal could be melted down and recovered.[212] Like many other 'special' South Asian guns, it was said to have had an alloy content of five metals including significant quantities of silver and gold for ritual purity as well as metallurgical qualities acknowledged in historic Indian gun-casting lore.

The British had intended to take the gun as a trophy by river to Calcutta but the raft that was constructed could not support the weight and the gun sank into the river bed where it remained for some years until reportedly being broken-up and recovered. In addition to the 'Great Gun' there were 86 iron cannon of various descriptions and 76 brass guns including carronades, howitzers, mortars and gallopers. Several guns proved to have laminated barrels. These were like the previously noted 'bar guns'. The name was derived from the longitudinal iron bars that formed the barrel's hexagonally shaped inner bore cylinder.* The hexagonal-bore cylinders were, in turn, encased within conventional-looking bronze outer-barrels to give the exterior appearance of a normal cannon barrel. The Marathas also left large numbers of muskets in the Jami' Masjid Mosque and some were taken into service for EIC recruits whose arms had become worn.[213]

The capture of the fort, known as the 'key to Hindustan', had far-reaching strategic significance for the EIC. Agra did of course hold the potential of being an ideal garrison town owing to the extent of urbanization. There was also the aspect of revenue collection in an area where revenue records were well documented and sophisticated enough to allow reasonable annual projections to be made. But perhaps the greatest strategic aspect of the victory was the opening of river traffic on a scale previously unknown to the British. During the preceding century British incursions into North India had been somewhat ephemeral and British penetration of the interior tended to follow slow, predictable patterns of annexation. Men, material and communications could now travel with greater speed between Calcutta, the British capital of India, and Delhi, the new symbol of British political legitimacy in North India. From a strategist's standpoint the logistical gain was paramount. Heavy artillery, tons of ammunition and thousands of bushels of grain could move with much less effort on water. There would continue to be threats along the riverbank, but having control of the river meant access to a greater hinterland supporting the South Asian military economy.

Despite the stunning string of successful battles by Lake in the Doab and Arthur Wellesley in the Deccan, Sindia's ability to resist with a 'regular corps' of infantry had not yet been crushed. Intelligence sources reported Sindia had thirteen battalions, or up to 12,000 men, and they still had 100 pieces of artillery.[214] Now there was some fear that Sindia would counterattack Delhi to take back the Mughal seat of power, the symbol that had formerly signified his claim on the region. Seven battalions from the Deccan, formerly under Dudrenec, were thought to be

---

* Meaning they appeared as hexagonal when viewed in cross-section or looking down the interior of the barrel from the muzzle towards the vent.

'linking up' with three of Bourquin's battalions who had not been committed to battle at Delhi in September.[215] Many of Dudrenec's sepoys had refused to change allegiance when their commanding officer surrendered himself to Colonel Vandeleur. There were also stragglers from Agra and the action at Patpurganj who were still in the vicinity. These combined Maratha groups, now in service to Sindia's new C-in-C Ambaji Inglia, had chosen not to defend the city of Agra as one might have expected. By remaining to the rear of Lake's army and leaving the burden of defence to the garrison force, they had preserved themselves for a bid to recapture Delhi. If the Marathas hoped to retain their economic and political power among the North Indian tributary states it was imperative to regain the appearance of controlling Delhi and the Emperor.[216] General Lake also realized that the Maratha battalions still possessed enough artillery power to pose a clear and present danger. He had no choice but to continue his campaign as a search and destroy operation specifically targeting those remaining Maratha infantry troops, their cavalry escort and their 'very superior artillery'.[217]

General Lake established a large hospital at Agra. He had Sindia's marble-walled *durbar* hall converted for that purpose.[218] It was spacious and accommodated the European and South Asian wounded as well as many of those from Delhi who were still recovering from the battle there a month earlier. Lake and his troops left Agra on 27 October 1803 and began the pursuit of Sindia's remaining battalions. Unexpectedly heavy rain slowed Lake's progress and so he opted to leave the most cumbersome baggage and siege guns at Fatehpur Sikri* to permit greater speed on the march. Late on 31 October, after a day's march, the Marathas were located at a distance judged to be within effective striking range. Lake ordered all of his cavalry forces into the saddle and they set out at eleven o'clock at night to bring the Maratha battalions to battle. Lake's plan was to use the mobility of his cavalry troopers to close the final distance and then attack the Marathas in a fashion intended to force them to go to ground and defend themselves. This tactic was to provide time for Lake's infantry to march up and engage the remaining Maratha battalions. The forced night ride covered the twenty-five miles in six hours. At sunrise on the morning of 1 November 1803, Lake and his eight regiments of cavalry found the Marathas now totalling about 9,000 regular infantry, with 72 artillery pieces augmented by 4,000 to 5,000 cavalry.

The bulk of the Maratha infantry at Laswari were thought to be composed of battalions who had travelled up from the Deccan to defend the revenue-rich Doab at Sindia's request. Although a number of their

---

* 27°+, 77°+. Aka Futtypur Sikri, Futtypore Seekrie.

mercenary officers had 'gone over' to the British, like Dudrenec at Mathura, these men at Laswari* had opted to fight on under Ambaji Inglia. As British reconnaissance parties at Laswari reported in, it appeared as if the Maratha left flank was in confusion and the speculation was that they were actually in retreat. Lake judged it appropriate to attack at that point, as he wanted to take advantage of the apparent chaos on the Maratha left wing. The C-in-C hoped that his infantry could come up in time to save the day. However, the Marathas had been given sufficient notice of Lake's approach to take at least some defensive measures. The Maratha right flank rested on the village of Laswari and their left on the village of Mahalpur.† The Maratha right flank was partially covered by a ravine and to their rear was an awkward ditch described as a 'deep rivulet' or 'deep nullah'.

An overview of Laswari

In trying to reconstruct the Battle at Laswari it is very difficult to reconcile the conflicting first-person British accounts. A great deal of the problem has to do with fitting the jigsaw puzzle of personal stories and anecdotes together. The lack of a unified perception reflects the limited vision of individuals who were denied an overview of the tactical manoeuvring that took place that day. Some groups of men, for example an undetermined number posted with Major General St John, never left their assigned positions and therefore could only contribute views from their stationary perspective of events. However, after having studied several accounts, I believe that a great deal of the confusion results from assumptions made about the Marathas' position as it was encountered, as opposed to what it would have looked like if the British had arrived later in the day. It seems that the Marathas were caught unprepared for the rapidity of Lake's advance over the final 25 miles that had separated them on 31 October. If the Marathas had been able to complete their preparations this battlefield would have been tactically quite similar to Delhi or Argaum in using a lure to draw the British into a formally prepared *beaten zone* or point where their artillery-fire was focused. Reports of initial confusion in the Maratha ranks and broken formations on their left suggest that Ambaji was in the process of arranging his men in a form of 'L' ambush. In a classic 'L' ambush, the line extending as perpendicular to the horizon is hidden and is the 'surprise' portion of the ambush intended to deliver unexpected flanking fire to an enemy attacking the visible lure that

---

* 27°+, 76°+. Aka Laswarri, Laswarry, Lassuary, Laswaree.
† Aka Mohaulpoor, Mohaulpore, Malpoorah.

is the line parallel to the horizon. In this case, the eastward extending per-pendicular portion of the line consisted of firmly fixed artillery. This ar-tillery dominant section of the deployment was intended to cannonade the British after they had crossed the rivulet at the southeast entrance to the battlefield as they proceeded westward to attack the Maratha infantry line. The highly visible portion of Ambaji's infantry line that ran north to south, the lure that invited Lake's attack, rested on the village of Mahalpur. As it turned out, the Maratha plan nearly worked. The long line of Maratha guns stretching east towards Laswari was obscured by tall grass and that explains a great deal of their initial effectiveness. Later, as the battle progressed, the Marathas reformed their lines. That these move-ments were accomplished in the heat of battle under sustained British fire further testifies to the proficiency and skill of the Maratha sepoys.

The opening movements

The Marathas had cut the embankment of a large reservoir; this action inundated the roadway that Lake sought to use for the approach of his cavalry to the battlefield. The right wing of the Maratha infantry lay in an adjacent position concealed in the tall grass behind their strong line of artillery. Several British accounts confirm the manoeuvring of Lake's cavalry around the periphery of the boggy roadway caused problems. The horsemen swung wide of the muddy area to seek the dry ground that was deemed less dangerous for the horses and the gallopers. As the lead units skirted the area the Marathas had flooded, considerable clouds of dust rose from the sun-baked fields. The extremely dry topsoil beside the roadway soon filled the air as the rushed movements of the densely packed British cavalry formation threw tremendous quantities of fine, choking dust into suspension and the cloud hung about the men. William Thorn, who was with a galloper crew, suggested that the advance was only saved by Lake's superior powers of perception as many junior officers and their men were totally disoriented in the dust cloud. Lake ordered his advanced cavalry guard to head towards a position he had predetermined would intersect any possible Maratha retreat. He intended these horsemen to halt the escape of the Maratha guns should they try and withdraw in this confusion.

The initial group of British horsemen in the advancing wave contained members of His Majesty's 29th as well as 8th Regiment of Light Dragoons. The dragoons penetrated the battle area on a tangent and took a small number of artillery pieces but they had, in actual fact, ridden past a large number of the guns they were seeking to intercept.[219]

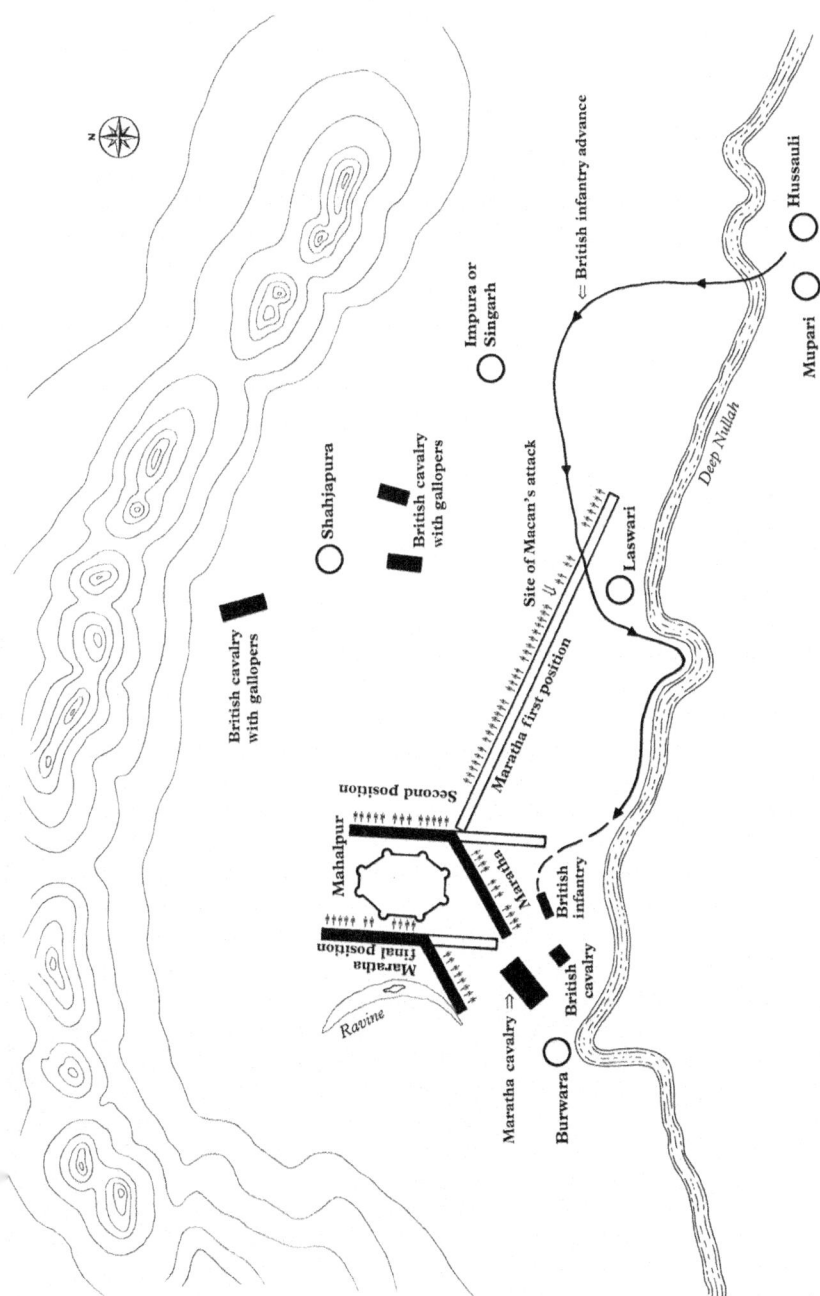

Map 11 The Battle of Laswari 1 November 1803, based on a map that appeared in Major W. Thorn, *Memoir of the War in India*, 1818, as well as a rendition by Emery Walker

The British cavalry pressed on through the area lying in front of them, covering the ground east of Mahalpur without noticing the long line of guns parallel to them on their left. However, the Marathas did not have any intention or, for that matter, any means of moving their guns at that moment. The Marathas were using the old Mughal-era tactic of anchored artillery. Heavy chains were strung between the gaps in the artillery line and anchored to hackeries, carts and caissons.[220] These obstacles, combined with field-expedient embrasures, lay behind the tall grass waiting to thwart any British attempt to ride through the guns. The British cavalry would be forced to jump these chain barriers and that would greatly reduced the dragoons' ability to sabre Maratha gunners.

In order to attack this line Colonel Macan and horsemen from His Majesty's 29th Light Dragoons combined with troopers of the 4th Bengal NC Regiment. They charged the right of the Maratha artillery line. The Maratha infantry delivered substantial musket fire from behind their hastily erected field-expedient fortifications. But the Maratha artillery held their fire until Macan's men were only 20 yards from the muzzles of the guns.[221] With an evil roar the grass erupted in a blast that quickly revealed that the horsemen had misjudged the size and extent of their target. In dragoon William Thorn's words: 'a frightful discharge of grape and double-headed shot mowed down whole divisions, as the sweeping storm of hail levels the growing crop of grain to the earth'.[222] The casualties were severe among Macan's men but the horsemen rode through the emplacement and reformed on the other side to ride back again.

Many of the Maratha gun crews had sought shelter under their artillery pieces on the first pass and they re-emerged to try and turn their guns, load and fire. The chains must have greatly impaired their efforts and with each succeeding pass their numbers grew fewer. George Call of the 27th Light Dragoons remarked: 'Our loss of Cavalry which was made in the first attack, was very great, owing to the enemy's well served fire of grape and our incapability of breaking the chains.'[223] Despite a devastating fire that included chain shot, Macan's men rode back and forth three times before they were ordered to withdraw under fire. Owing to the anchor chains, a lack of bullock teams and the galling fire, the British cavalry managed to bring away only two small field pieces as they retreated.

By noon the British infantry arrived, His Majesty's 76th Regiment and six battalions of Bengal NI, after marching all night and into the heat of the mid-day sun. It was judged prudent to let them rest shortly and take refreshment before entering the battle, but a messenger arrived from the Marathas requesting a disengagement. Grant Duff's history of the Marathas states that one of the Maratha officers, on seeing the approach of British infantry, offered to surrender his artillery if terms were granted.[224]

Although conditions were agreed upon, the passing of one hour's time saw no change in the Marathas' military disposition and so the decision was made to resume the battle. The British infantry formed into two columns and moved between the *nullah* to their left and the Maratha line to their right. They were to try and turn the Maratha infantry's right flank which had changed front with precision, pivoting back in the first of its two ninety-degree increments to form a straight line stretching from north to south on the perimeter of Mahalpur.

The Maratha infantry were well posted, having manhandled many of their artillery pieces back with them. Drag ropes and handspikes were used to pull the guns back to the reformed line. The Marathas should, at that point, have had the benefit of cavalry support, which was hovering in the rear. But the Maratha horsemen held back waiting to see where the British dragoons were headed. Major-General Ware's men were in the first British infantry column and Major-General St John's men were in the second infantry column. The British cavalry forces were now divided. Some of the horsemen under Macan were posted to support the infantry while another body of cavalry operated as a diversionary force approaching the Maratha left on the opposite side of the battlefield. This cavalry feint to the north (Maratha left) was made all the more credible by advancing in a leapfrog manner which allowed the gallopers to lay suppressive fire. One regiment's galloper crews would open fire, targeting Maratha artillery and distracting their attention from the British infantry advance. Another set of British regimental galloper crews advanced past the first crews a few hundred yards before going to ground. There they would unlimber their gun and deploy covering fire while the first crews limbered-up and rode past a few hundred yards to the next position where they deployed and repeated the leapfrog process of fire and movement.

The British infantry tried unsuccessfully to use the tall grass to conceal their movements. The Marathas poured down showers of grape shot from large artillery pieces delivering high-angle fire. The 76th Regiment in their customary vanguard role bore the brunt of this barrage and casualties mounted quickly. Lake felt a great deal of attachment to his prized '76th'. And rather than have them linger in the shower of grape shot he had them move forward with the 2nd Battalion of the 12th Bengal NI and five companies of the 16th Bengal NI, without waiting for the rest of the column to form.[225] As the men advanced things got worse. The Maratha gunners dropped the elevation of their guns and loaded canister shot packed full of musket-ball-sized projectiles. These rounds inflicted severe casualties and to compound matters the Maratha horse formed in an apparent bid to charge the 76th Regiment.[226] In order to keep the Maratha horse from riding through the ranks and breaking up the formation Lake ordered His Majesty's 29th Light Dragoons to prepare to charge them.

While awaiting Lake's orders to charge the Maratha cavalry, a number of dragoons halted in a hollow behind a galloper where William Thorn was positioned. Normally this might have seemed a safe and logical spot to post them while waiting to charge but it proved exceedingly dangerous as the result of Maratha counter-battery fire. As Thorn's gun crew laid down covering fire for the 76th infantry, the Marathas found his range and proceeded to deliver round shot on his position in an attempt to knock out his 6 pounder and deprive the advancing 76th of the galloper's fire support. The dragoons, partially concealed by the grass behind Thorn's galloper, were patiently waiting for the word to attack but according to Thorn, 'the shot rolled and ploughed up the ground in every direction among our ranks, with the most mischievous effect'.[227] One round killed the 29th Light Dragoons' commanding officer Major Griffiths, just as they were about to be committed to action. Captain Wade quickly took the dragoons' command and the order was given to charge. The Maratha horse were repulsed before they could reach the British infantry.

As with other battle accounts of this war, it soon becomes apparent that the British placed a great emphasis on combat leadership roles among their senior officers. A corresponding level of officer leadership was not possible in Sindia's army, which had been gutted by officer defections. British officer leadership did make for high casualty figures among the senior ranks but it must also have spoken to the sepoys and enlisted European ranks about the concept of human equality under fire. Colonel T. P. Vandeleur of His Majesty's 8th Light Dragoons had fallen mortally wounded and the corpulent General Ware had been decapitated by a cannon ball in a gruesome fashion that carried his skull clean away from his falling body. The British sepoys could plainly see their senior leaders placing their own lives on the line. This would not always be true of infantry warfare in the twentieth century, as 'resource management' approaches to combat leadership tended to remove men above the rank of colonel from the assault line. These British officers at Laswari, up to and including the rank of general, were willing to do just what they asked their men to do rather than remain safely out of the line of fire 'directing' the battle. C-in-C General Lake had a horse shot from under him and his son Major George Lake once again stopped to lend his mount. The young Lake mounted another trooper's horse but before he could move out he fell wounded by Maratha artillery fire, cut down before his father's eyes.[228] 'This touching incident had a sympathetic effect upon the minds of all that witnessed it, and diffused an enthusiastic fervour among the troops, who appeared to be inspired by it with more than the ordinary portion of heroic ardour.'[229]

To divert Maratha smallarms fire the 29th Light Dragoons charged the Maratha infantry, braving musket volleys and grape shot, as they wedged their way through two standing lines and sabred the enemy sepoys. This permitted Lake to lead his infantry forward and seize the Maratha batteries that were their immediate objective.[230] The dragoons then wheeled left and attacked the Maratha horse that had remained menacingly near. The Maratha horsemen tried to escape through an opening but the rear of their main formation had its escape route cut and the momentum of battle now favoured the British. Lake's infantry continued to press forward past the guns and drove the Marathas back towards a small village and mosque. However, Sindia's retreating infantry were then exposed to numerous elements of Lake's cavalry, which intersected the area choosing targets of opportunity and producing confusion in the Marathas' line of retreat. The Maratha left wing did manage to fall back in reasonably good order contesting 'every point inch by inch'. But Lieutenant-Colonel John Vandeleur's dragoons of the 2nd Cavalry Brigade broke into their column cutting many down and capturing the rest with their baggage intact. Once again George Call provided the details of what followed: 'the cavalry now revenged themselves for their former loss by cutting them up in great numbers – out of 16 Battalions, 2,000 men only were saved'.[231] Additional numbers of Marathas were then killed by the explosion of their own ammunition tumbrels that were hit by fire from the British cavalry's gallopers.

Lake's army lost about 800 men killed or wounded and several were men of rank. As already noted, Major-General Ware, infantry leader of the right wing of the Grand Army, was killed in the last stage of the battle by Maratha cannon fire.[232] The burial parties laid thirteen British officers to rest and 'All the surgeons of the line [were] ordered to the General Hospital to assist in dressing the wounded'.[233] Lake had two horses shot from beneath him that day and at one point a Maratha soldier, attempting to kill the British C-in-C, placed the muzzle of his weapon so close to the General that powder burns were left on Lake's coat.[234] Man and beast had suffered privation to make the victory possible. The cavalry had been forced to cover 42 miles in less than twenty-four hours and then pressed into service from morning to evening without food or water. Lake's infantry had suffered similarly, making forced marches of 65 miles in forty-eight hours and then, after only an hour to rest, they were thrown into a battle that was already in progress.

The 'Report of Ordnance . . . Captured at Laswaree' indicated 71 pieces of artillery and 64 ammunition-laden tumbrels were taken by the British.[235] A number of additional tumbrels, as noted, had exploded on the battlefield, indicating Maratha artillery ammunition had been plentiful. While Maratha armies had previously carried huge guns into the field, the artillery captured at Laswari suggested that the remaining battalions had felt their resources being squeezed. Of the 71 pieces, 7 were carronades, but the short-barrelled brass pieces gave nothing up in firepower since they were 18 pounders and 16 pounders and there were also 9 howitzers. The howitzers, ranging from 8 inches to 12 inches in bore diameter, were ideal as multipurpose guns. Although they could be used to lob explosive shell at high angles deep into troop formations, the Marathas made use of them at Laswari for delivering indirect fire with grape shot.[236] The largest percentage of guns captured at Laswari were 6 pounder brass field cannon, which numbered 26 of the 71 pieces. William Thorn noted that of the brass 6 pounders captured at Laswari, all were cast in India with the exception of one, which was cast in Holland.[237] Also captured intact were 4 Maratha galloper guns. The gallopers were all iron and cast in Europe but they were quite small, 2 were 2.5 pounders and 2 were 1.5 pounders, not at all common for Maratha galloper crews, which usually relied on 6 pounder brass guns or howitzers which they cast themselves.

The Battle of Laswari was expensive for Lake but the Maratha battalions suffered more. The Marathas lost 2,000 men as prisoners of war but the dead and the dying on the ground were said to number 7,000.[238] Lake later ordered the release of the prisoners with the exception of some forty-eight principal officers.[239] The seventeen battalions had been smashed beyond recognition and the Maratha commander Ambaji Inglia was barely able to escape.[240] The entire Maratha bazaar fell into the victor's hands complete with supplies, camp equipage and 1,600 bullocks. Three tumbrils full of cash were also seized. Among the military booty taken from the field there were 5,000 stands of smallarms and an impressive 44 sets of colours.[241] The colours were indicative of the *esprit de corps* enjoyed by these Maratha troops since they were largely composed of Sindia's 'Deccani Invincibles'.[242] Grant Duff, writing years later after interviewing survivors, could not deny the tenacity with which these men fought but he refused to give the Marathas credit per se. 'Few if any of those men were natives of Maharashtra, they were chiefly from Oude, Rohilcund and the Dooab, for except Sivajee's Mawulees, and men trained in the ranks of the Bombay sepoys, the native Mahrattas have never made good infantry.'[243]

A week after the Battle of Laswari, on 8 November 1803, Lake's army marched away from the killing field that had become quite rank from the number of unburied human and animal corpses.[244] It took almost another week before the wounded and captured artillery were sent off to Agra. The Grand Army had completed its immediate objective and during December it celebrated the victory and reformed its units. The Mughal Emperor sent a *khilat* or ceremonial robe of honour to General Lake,[245] while the officers 'of the British Indian army' presented their C-in-C with a silver service valued at £4,000.[246] Lake's victories had altered the balance of power in North India and several rulers entered the mutual defence pact, which Governor-General Wellesley had advocated. Much ceremony was made over the commitments of the Maharajah of Jaipur, the Rajah of Macherry and Ranjit Singh, the Jat Rajah of Bharatpur, who would later side with the Maratha leader Holkar in his 1804–5 campaign against the British.

Conclusion

Lake's Hindustan Campaign of 1803 represented the attainment of two policy goals as outlined by British India's expansionist-minded Governor-General Richard Wellesley:
• penetration and possession of the Gunga-Yamuna Doab;
• displacement of the Sindia clan's political claim on North India.
In achieving these points it seems, at first glance, that Lake's campaign achieved the desired military objectives and war aims of 1803 as outlined in Richard Wellesley's dispatches. The Mughal Emperor Shah Alam II was safely in British hands, the Doab had been conquered and Sindia's standing army was eliminated. But a closer look would reveal that possession of the Emperor did not constitute definitive control of the revenue system or the complete allegiance and cooperation of subordinate powers.

Owing to administrative complexity and alienation, the nuances of the economic as well as political system of North India were very difficult for the British to master in 1803. The possession of Shah Alam did give the British the appearance of being Sindia's political successors in relation to their guardianship of the Emperor and their claim on the Doab. But the British, like the Marathas, were considered foreign masters by many of North India's indigenous elites. The Mughal throne had long since lost its real power but its administrative infrastructure lingered on, for now. This would seem to have set the stage for a prolonged period of conversion during which time indigenous elites, such as the Jats and Rajputs, would use political recognition, subordination and allegiance as bargaining chips in a greater power play of subcontinental proportion. These bids to carve out claims of autonomy from decaying political infrastructures eventually came full circle in outlasting the British Raj. And it could be said the process did not reach a form of historical conclusion until the 1970s when the Republic of India formally terminated the political power of all 'princely states'.

Although the British had 'conquered' the Doab it was far from pacified or territorially secure. The British had focused their military might in such a way that it allowed them to push on with primary targets, but in doing so they had to by-pass several areas which were now potential points of insurgency behind their lines. The British were learning the tactical aspects of rapid reaction in North India and that was of benefit to their imperial control mechanism. The 1803 campaign showed that overwhelming firepower was not a guarantor of victory; if it was then the Marathas would have won the war. But the British advances made in the use of firepower mobility (i.e. gallopers) were recognized almost immediately as particularly well suited to counterinsurgency in North India. An immediate reaction capability was very useful in minimizing or containing insurgency before it got out of hand. This could be seen in Macan's reaction following Fleury's raid on Shikohabad. Much agricultural land was burned, but the driving and relentless pursuit of Maratha cavalry denied them opportunities to prey upon serious military or civilian targets. If kept on the agricultural periphery of society, bands of *pindaries* and marauding horsemen were a security problem but not a military threat to the British armies who set about the serious business of eliminating more conventional infantry-based threats.

Sindia's battalions had been dispersed and therefore directly removed from Maratha service, but the majority of his sepoys had not been physically destroyed. Charles Stuart of the Bengal Cavalry had it right when he indicated that Sindia's battalions were not so much destroyed as they were 'dispersed'. More than 2,000 Maratha sepoys at Agra literally walked into British service as many of their comrades had done before them at Delhi. The EIC was acquiring and incorporating the human and physical assets found within the South Asian military environment: fortresses, cantonments, transport animals and soldiers. They were all coming into the EIC's control. Lake's 1803 Hindustan Campaign was instrumental in making the EIC the dominant player in the South Asian military economy. And with that domination, particular emphasis was placed on the EIC's control of the North Indian military labour market. The elimination of Sindia as a rival bidder for the services of sepoys, or as an alternative employer for mercenary officers as well as horsemen, meant that more men, and in particular South Asian men, saw their profession as soldiers tied to EIC ascendance. What many did not seem to see or realize was that their careers would then be tied to British corporate interests. And that, as such, they would also be subject to the corporate practices of the EIC. Deficit spending to win wars meant a boom in hiring cycles. Conversely, budget cutting meant demobilization. As far as fixed physical assets, the British were still learning about the intangible costs associated with their victories over North Indian fortresses.

The fortresses of Alighar, Delhi and Agra had come into British possession but the British may have misinterpreted the manner in which they had 'fallen'. There has been a great tendency for Western authors to classify all South Asian defensive fortifications under one heading as *Indian forts*. But there was a tremendous difference in architectural design, physical construction techniques, intended military usage patterns and the political purpose of construction. Shah Alam's fort at Delhi had not been contested. It was very much a political prop for the Emperor although, as we shall see in the next chapter, it was the scene of an unusual siege attempt in 1803. The battle for Delhi against Bourquin had been conducted miles away from the city. And so the British acquired Delhi's fortress without direct attack. Lake's 1803 campaign did, however, involve attacks on two of India's great fortresses. Suffice to say that Alighar and Agra were similar in both being large North Indian fortresses, but they differed greatly in the context of design, construction and purpose. However, both fell to Lake's army rather quickly.

Many sources say that Alighar was taken by a *coup de main* and Agra fell after fire was opened from a grand siege battery forcing the garrison's surrender. But I believe that was a superficial and incomplete analysis, which may have misled Lake and others into thinking that 'taking an Indian fort' was in effect simple. Alighar's drawbridge was under construction at the time of the attack. If it had been completed before the attack, then a *coup de main* would have been impossible and how would the British have coped with the great water-filled moat? But history is riddled with 'what ifs'. Of greater relevance to the fall of Alighar and for that matter Agra was Richard Wellesley's Proclamation of 29 August 1803. The Proclamation had led to devastating defections among officers as well as sepoys but it did more than that in terms of *command, control, communication and intelligence.** The successful attack through Alighar's main gate was made possible by the intelligence provided by the defector known as Mr Lucan. He not only provided General Lake with recent intelligence on the disposition of the Marathas' divided forces, he also served as

---

* Known by the military abbreviation C^3I.

personal guide for Monson's storming party. Lucan's knowledge of the fort was instrumental in getting the assault party through the interior rampway of the fort allowing the entrance of men to the inner fort via the wicket gate.

At Alighar and Agra, the garrison's command and control were dysfunctional as the result of having been severely divided on the issues of trust and loyalty. Both garrisons had imprisoned their European senior officers, as we saw in the cases of Pedron at Alighar and the unfortunate Hessing as well as Sutherland at Agra. Both garrisons suffered internal division over the issue of what course of action to follow. Relinquishing Agra for the settlement of wages in arrears was a real option and the situation was not unlike that which Arthur Wellesley faced at Ahmadnagar. But Lake's victories at Alighar and Agra had pre-empted the necessity of prolonged sieges that could last for months and produce nothing but casualties and expense. A strong case can be made that the conditioning of the British in 1803 – meaning the British were conditioned to expect rapid victory against Indian forts – laid the basis for Lake's future disaster at Bharatpur in the 1804–5 Holkar Campaign when that Jat fortress held out and forced a humiliating British withdrawal. That reversal of Lake's fortune in 1805 may also explain why his brilliant performance in the 1803 Hindustan Campaign is often overshadowed by stories of Arthur Wellesley in the Deccan.

5   'Coming in'

Introduction

Mercenary employment practices underpinned the military labour markets that helped to drive the South Asian military economy. Although never culturally exclusive, mercenary hiring lingered historically in South Asia because it met corresponding military, economic and social needs. This helps to explain the ease with which the EIC, a credit-rich foreign commercial entity, could build three huge sepoy armies that would become the foundation of a later million-man British Indian Army. Despite embracing the mercenary essence of this cornerstone, British attitudes towards Europeans in South Asian military service were rapidly changing at the dawn of the nineteenth century.

This chapter specifically concerns European mercenaries in Maratha service but it also demonstrates some of the specific techniques used in British manipulation of the South Asian military economy. When combined with diplomacy, these control efforts held the promise of delivering regional security as well as greater political reward on an imperial basis. The story concerning the Marathas' European and Euro-Asian mercenaries has never been clearly placed against the backdrop of covert intelligence, financial negotiations and espionage, as found in the EIC's records. As this chapter develops, it also presents a tale of intrigue that calls into question the accepted reasons for British military victory over the Marathas in 1803.

In search of the pagoda tree

George Carnegie, a native of Scotland, reached India in 1799 seeking to make his fortune. A cousin had bragged about his Calcutta business prospects but they evaporated before George arrived. To his chagrin George Carnegie found that he could not apply for the EIC's civil service from inside India as appointments had to be obtained before leaving England. After soliciting advice from various sources and weighing his decisions against the disgrace of returning home, he came to believe that he had two options. The first possibility was to borrow some money and purchase a commission in His Majesty's Light Dragoons. The second option was to join the army of Daulat Rao Sindia. The latter course of action was recommended by many Britons in India as a sure way to make an exceedingly large amount of money. If he were willing to serve twelve years it was reckoned he would have enough to retire to Britain and purchase a business as well. George Carnegie learned that 'most of the Great Men' in Sindia's army were 'French and Scotch'. As a result it was comparatively easy to get a recommendation to join the brigade of Scottish Major Sutherland in Sindia's service.

George's letter to his mother Susan Carnegie, in October 1799, specifically notes that there was open talk of what would become of Sindia's mercenaries.[1] 'At present it is generally believed that our Government will insist on Sindeah's turning all the French and perhaps British Officers out of his Service.' George expressed the desire that the situation be resolved, but that did not stop him from joining. And he eventually went so far as to recommend this adventurous life of mercenary service to his younger brother Thomas. The Carnegies were optimistic but not frivolous in their consideration of political events. They did weigh the dangers of joining Maratha service and their letters explored questions about duty and honour in the context of service and war. George noted of the Marathas: 'At present they are our good allies, and so long as they continue so, there can be no disgrace in serving them.' Susan Carnegie wrote to a nephew in Calcutta:

I have a letter from George himself some days ago, who represents the India Co. [sic] as encouraging of late English Officers of character to enter the service of the Country powers, their interest evidently being to preserve Peace with the Co., while that of the French and other foreigners was to stir up dissention in hope of sharing the Spoil.[2]

Was there really a possibility of war with the British? Writing to his mother while en route to India, Thomas noted: 'I do not see that in the Service of Sindeah a man runs any risk of being obliged to serve against his country.'[3] There was hope that the British would be allowed to resign their commissions honourably if war between Sindia and the EIC became inevitable. Upon arrival in India, Thomas was a guest of his mother's old friend, Bombay Governor Jonathan Duncan. 'The Governor still is of opinion that Scindia's Service is much better than the Company's whilst he is on good Terms with them, which he has been for a long time and likely to be!!'[4] A week later young Thomas, who was destined to die of disease and not war wounds, made the observation, 'The most of

Scindia's Officers are British, and was anything to happen, he would lose all of them.'[5]

An honourable profession

Governor-General Richard Wellesley had surrounded himself with bright young men like John Malcolm.* Yet it would be a mistake to think that the Governor-General rejected all the wisdom and talent of age. Richard could not say that the eccentric John Ulrich Collins was his friend. Nevertheless, Richard showed more respect for the older man's judgement than his brother Arthur who dismissed 'King Collins' as a buffoon. Richard knew better; he had read the service records. Collins had soldiered in the Bengal Infantry, making ensign in 1770 before Arthur had reached his first birthday. Noted in despatches for his personal bravery, Collins had also survived that very unpopular and controversial conflict known as the Rohilla War. But more importantly, he had served as military secretary to Richard's predecessor, Governor-General Sir John Shore.† That meant Collins had extraordinary background knowledge of not only military matters and military men, but also the underlying political issues of the day.[6] Collins did not like Richard Wellesley – he did not have to in order to do his job well – nevertheless he served his master admirably in the great game of intelligence. The Lieutenant-Colonel, known in the customary military fashion of abbreviation as Colonel Collins, was appointed as official British Resident to the court of Daulat Rao Sindia in 1798. But he was more than a passive observer. He ran one of the most effective intelligence networks that the British had in 1803. Collins was able to provide Richard Wellesley with information that extended to the inner workings of Sindia's clan as well as his *durbar*. Collins' informants ranged from humble 'newswriters' to Sindia's most senior advisors. Occasionally Collins was fed misinformation, but such was the nature of the game and it did not diminish his personal strengths as a covert information analyst. Collins was often astute enough to accurately weigh his sources against each other to maximize the value of his 'Marhatta reports' to Richard Wellesley. Collins' analytical ability, combined with his skill at cultivating informants, was to pay great dividends.

During 1801, the reports from Collins were heavily focused on the factionalism in Sindia's camp. There were significant differences of opinion among the *sardars* on a number of issues, which included relations with Holkar, the *Peshwa* and the British. In addition, there were serious objections being raised over Sindia's handling of Perron and the foreign

---

* Aka 'Boy Malcolm'.   † Aka Lord Teignmouth.

mercenaries. A certain degree of the dissent was clan-related but much was generated by personal bids for power and ranking. It had been common knowledge in Sindia's *durbar* since mid-1799 that Perron as well as Dupont were contemplating retirement.[7] It was widely assumed by the British that Scotsman Hugh Sutherland would fill Perron's leadership role; he was looked to by many as the acting head of the British officers within Sindia's mercenary corps. Soldiers of fortune from Scotland were by far the biggest component of the British contingent and Sutherland played a pivotal role in what developed as the Scottish connection. However, the French mercenaries seemed to enjoy preferential treatment under Perron and several were keen to follow him up the promotional ladder.

Information moved freely between Sindia's mercenaries and the British. After all, there was no immediate threat of an Anglo-Maratha war in 1801 as Holkar and Sindia were still locked in their own fratricidal conflict. Social as well as professional visits were commonplace between British officers and their countrymen in Sindia's service. The liaisons were of great use in the formation of a rather elaborate, yet at that time still informal, intelligence network. This ultimately developed into a rather cellular system of operatives that complemented Collins' list of contacts with very few overlaps in personnel. Agra became a popular meeting place as many British visitors arrived with the publicly expressed desire of seeing the Taj Mahal. Of course the Taj was not the only impressive piece of Mughal architecture in Agra and the great Red Fort drew many visitors as well. It was still being repaired in places during 1801 as it had sustained damage in 1799. The fort's *killadar* had rebelled that year and Perron was sent to recover it. The siege lasted fifty-eight days and the garrison capitulated after a sustained bombardment that wreaked havoc with the sandstone walls in one specific location that would later be of interest to General Lake for the purposes of opening a grand battery.[8] In 1800, Sindia had placed the ailing veteran Colonel John Hessing in command of the great fort at Agra.[9] The old Dutchman was an unhappy-looking soul but he surrounded himself with a loving extended kinship network that was also linked to the family's mercenary profession.[10]

Among the visitors to the fort was Lord Metcalfe, 'then a young civilian and Assistant Resident at Daulat Rao Sindhia's camp'.[11] Metcalfe noted, 'I breakfasted by invitation with the Dutch Commandant, Colonel J. Hessing. I found him with his son [George Hessing], who commanded in the engagement at Oojein,* where his battalions were defeated; a Mr. Marshall, an Englishman, and two other officers, whose names I have not learnt.'[12] Young Metcalfe, an obvious gourmand, devoted several lines to describing the elaborate breakfast that Hessing had provided.

---

* 23°+, 75°+. Aka Ujjain.

But in the rather stereotypical fashion of the innocent abroad, Metcalfe added: 'The Dutchman was as polite as a Dutchman could be, and very well meaning, I am certain.'[13] The cordial visits by British visitors were extremely useful for relaxed intelligence networking free from the cloak-and-dagger aspects of military spying. One reason was that Agra had become a family centre for Sindia's mercenaries. The extent of intermarriage – but more importantly, interracial marriage – among the mercenaries at Agra gave it a unique atmosphere compared to a number of British settlements where interracial relationships were becoming taboo and formal marriages of that description were practically impossible.[14] Coel had been popular in de Boigne's day but by 1801 Agra contained the cross-cultural legacy of a European mercenary settlement.

Agra's Colonel J. Hessing was General Perron's brother-in-law, having married a sister of Major Derridon and Madame Perron.[15] John Hessing's daughter married Captain Robert Sutherland.[16] That meant that Robert Sutherland became George Hessing's brother-in-law. Sindia's British, Dutch and French mercenaries were related by intermarriage in a fashion not unlike the unions found in the intermarriages of families within the Sindia clan. There were apparently three more significant intermarriages of mercenary families, two of which were by way of South Asian sisters. Unfortunately our understanding of the exact extent of intermarriage is placed in question by the fact that there are apparent errors in Herbert Compton's *A Particular Account of the European Military Adventurers of Hindustan*, which has been accepted for years as the standard biographical source on many of the mercenaries. Compton listed only one Sutherland, that being Robert. But in reading Compton's biographical sketch of Robert Sutherland it is apparent that he produced a work that unknowingly merged the lives of Robert and Hugh Sutherland.[17] Despite the resulting errors, Compton was apparently correct when he noted that Perron, Hessing, Sutherland, Derridon, Filoze and Bourquin were not only linked by their mercenary service to Sindia, but also all related by marriage.[18] Compton followed much of the traditional British-based anti-French rhetoric and painted the intermarriage connections as being divided on national grounds between Britain and France. However, that assertion is more than suspect. The Hessings were from Holland and the Sutherland correspondence file clearly indicates that several French officers were extremely close confidants of Hugh Sutherland. Major L. Derridon was Hugh Sutherland's comrade in arms during the war and as late as 1806 he served as Major Sutherland's business representative in Mathura.[19] Hugh Sutherland's post-war correspondence indicated he was still friendly with the mercenary Pedron, presumably Pedron the younger who surrendered Khurja to Lake.[20]

In the relaxed family setting of Agra the European mercenaries in Sindia's service talked about their problems concerning the Maratha *sardars* who tested Daulat Rao's power. The mercenaries felt pressure to follow de Boigne's old advice about remaining loyal to Daulat Rao Sindia above all else. But times had changed. Luckwa Dada, Sargi Rao Ghatke* and Ambaji Inglia were now among those powerful members of the inner circle pressuring Sindia to reduce the mercenary faction. On 20 August 1801 Sindia presided over a private Maratha ministers' conference where his father-in-law Sargi Rao Ghatke raised serious doubts about the wisdom of entrusting so much power to the foreigners. Ghatke 'strongly remonstrated on the impolicy of intrusting the important fortresses' of Ajmer, Agra and Alighar to Monsieur Perron. As well as 'the imprudence of assigning to him a Jeydad,[†] so large a proportion of his [Sindia's] territories', he specifically pointed out that Sindia might, at some future period, be placed in a critical situation if it became necessary to deprive Perron of command. Sindia's father-in-law felt that Perron would not willingly surrender the forts in his possession and that Sindia should wait until things had settled down somewhat and then seek an opportunity to retake them by force if necessary. But Delhi was a different matter. Ghatke told the Maharajah to waste not a moment in seizing Delhi from Perron's administrative control. He thought that Perron's hold on Delhi would have a disastrous effect in exciting the jealousies of other powers – presumably the British. In reply to this rebuke Sindia was reported to have said: 'I have the Highest confidence in the fidelity of Mr. Perron.'[21] In his report on the meeting, Collins confided his observation that Sargi Rao Ghatke would lend his support to any plan that would reduce the power and influence of Perron.[22] This meant that Maratha factionalism, over the role of the mercenaries, was obvious in Sindia's *durbar* two full years before the hostilities of 1803. Perhaps more important to the British intelligence community than the two-year lead were the statements of Ghatke as one of Sindia's kinsmen. His pronouncements suggested potential wedges to further split the tenuous bonds of Sindia's *durbar*.

A suitable pretext

Additional intelligence began to filter through to Richard Wellesley on the conflict emerging between Ambaji Inglia and the foreign mercenaries. In particular, offence was said to have been taken at the degree to which

* Aka Sharzarao Ghatge, Surja Rao Ghatke.
† Note that Perron was expected to use the revenue income from his *jaidad* to pay for the battalions of regulars under his command.

Perron personally ran Daulat Rao Sindia's political and military affairs in North India. Ambaji had been one of Mahadji Sindia's pillars of strength in the darkest days of the First Anglo-Maratha War. He had been present when Goddard attacked Gwalior and Ambaji managed to walk out alive, bringing with him members of Mahadji's family – an act of heroism duly rewarded. In those days Ambaji had no quarrel with the European mercenaries, showing no particular animosity towards men like de Boigne. Why should there be a problem? Monsieur de Boigne's stature was not particularly great until the 1790s with his authorization to expand beyond two brigades. Perron's profile was different than that of de Boigne and that rankled men like Ambaji.

Daulat Rao Sindia did not owe a personal debt of gratitude to Ambaji as his great uncle Mahadji Sindia had done. But Daulat Rao's ability was only a fraction of Mahadji's and so Ambaji bided his time waiting for a chance to reassert his own lost power. Within the nether world of Sindia's *durbar*, Ambaji found himself forming relationships of convenience to further his cause. This pattern became more pronounced with increased pressure from the British to support the *Peshwa* and the Treaty of Bassein. On several occasions Ambaji was allied with Sargi Rao Ghatke and those periodic liaisons of convenience lasted, in one form or another, until 1805, despite Ambaji's later courtship of the British.[23] But the price of a transitory alliance with Ghatke was often too much for Ambaji and it produced a series of spats. And the rivalry between Ambaji and Ghatke, when mutually opposed to Daulat Rao Sindia, rarely did anyone any good as the young Maharaja was more prone to sulk or become reactionary when the veneer of manipulation became too thin to avoid transparency. This was not always an easy situation for the British to read, but Collins struggled on in trying to decipher the often ambiguous and contradictory signals from the *durbar*. One option contemplated by Collins, in a contingency plan to weaken the Marathas, was to offer Ambaji asylum in EIC territory so as to better exploit the growing hostility amongst the Marathas. Collins realized there was an opportunity to play one Maratha off against the other in such a fashion that it would accentuate factionalism.

In short it was sufficiently apparent from the discourse of Khamgaur Khan,* although he met with no encouragement to disclose his sentiments from me, that Ambajee is anxious to obtain the protection of the British Government, in return for which he would I believe, feel no hesitation in entering into any engagements which could, in reason, be expected from him; and even consent to send his family, either to Britain, or Benares, there to remain as hostages for his good faith.[24]

---

* 'Khamgaur Khan' served both Ambaji and Sindia.

Sindia's distinguished veteran artillery commander Gunput Rao* was another of Collins' intelligence sources. He provided information that Perron was urging Sindia to seek additional alliances to bolster the Maratha position against British expansion. Richard Wellesley, at that time, was entertaining plans for a first-hand inspection trip in the Upper Provinces and Perron was sensitive to the rapidly changing diplomatic events. Perron was not pushing for a 'French state' with South Asian diplomatic recognition, as some authors have implied. His proposals for Maratha alliances with the *Peshwa*, the *Gaikwad* of Baroda and eventually Holkar suggest Perron had devoted considerable thought to his survival strategy. A balance of power or external multilateral treaty system also held the potential of defusing the situation via an armed truce. General Perron had already opened a dialogue with the Sikhs on Sindia's behalf.[25] Perron went so far as to meet with Sikh leader Ranjit Singh's uncle to lay the groundwork for further negotiation. Perron knew well de Boigne's advice to avoid war with the British. And Perron had no incentive to seek war with them; indeed that would be counterproductive to his mercenary game. The object was to live a very long time – or at least long enough to retire in comfort as de Boigne had done. For his part, Governor-General Wellesley had no objection to Perron retiring a rich man. He just wanted Perron and the other French mercenaries out of Maratha service as soon as possible. The remaining British mercenaries would be easy enough to control. Perron's position should theoretically have been rendered more secure by his efforts to form a North Indian alliance system but his safety was increasingly jeopardized by his Maratha rivals. Collins assessed Perron as politically aware and adept at survival, stating the General 'is by no means deficient in the acts of management and conciliation'.[26]

Collins was given specific instructions to try and obtain Sindia's agreement to enter a separate Subsidiary Alliance Treaty in January 1802.[27] It was part of a multi-track approach by the British to divide the Marathas up via alliance packages to pre-empt the need for an all-out war. It was hoped that, at the very least, the *Gaikwad*, the *Peshwa* and Sindia could be split off from Holkar, which would make victory a much more attainable goal. Collins was warned that he should present this offer of alliance as a move designed to strengthen Sindia and under no conditions should he portray it as vital or necessary to the safety of the EIC.[28] The proposal held out the supposed benefits of alliance; but not without a corresponding price. Richard Wellesley wanted to render Sindia's battalions more controllable or responsive to British overtures. If the British mercenaries

* Aka Gunpat Rao, Gunpat Row.

were to replace Perron then command and control of the 'regular corps' would be on a distinctly different footing as far as the EIC was concerned. The secret instructions to Collins particularly stressed the removal of the French officer cadre from Sindia's service.[29]

The Governor-General knew there was little chance of Sindia agreeing to all the preconditions of the proposed Subsidiary Treaty Alliance. Giving up Agra and Delhi were difficult demands, but relinquishing command of the battalions that guaranteed his safety from internal Maratha squabbles was asking too much. Collins could not make the argument that Sindia needed British forces for his defence; a point which did appeal to lesser figures like the *Peshwa*. And the feeble supposition that this alliance 'is for Sindia's own good' was wearing thin. Richard Wellesley pushed his own political agenda by writing that he saw the 'Troops in the service of Scindhia under the command of French Officers' as the biggest threat that the British faced. Ironically, the Governor-General was far from convinced by conspiracy theories, which alleged that Sindia was actively participating in an alliance with Napoleon or his agents in the then current War of the Second Coalition. Rather than blow with the more popular winds of conspiracy, Wellesley, perhaps in seeking greater credibility with London, tacked hard against the conspiracy fear in arguing that a well-intentioned Sindia might not be able to control the actions of the French mercenaries if they sought to stage a coup at some future date and then delivered the 'regular corps' over to Napoleon's agents for political or monetary reward. (This greatly contrasted Wellesley's stance on Tipu Sultan who he actively portrayed in 1799 as working with Napoleon.) But Wellesley's personal doubts about Sindia were expressed in negotiating instructions forwarded to Collins through Bengal government secretary Edmonstone. Imperial risk was embodied in the concern that 'Could it even be supposed that Scindhia possessed that power of control over the French Officers commanding Troops in his service no considerations could require the British Government to expose the security of its interests to a precarious dependence upon the fluctuating views and doubtful disposition of Scindhia.'[30] But mere supposition would not suffice. Wellesley needed a solid legal argument to present to the EIC in order to justify pushing Sindia any further on this mercenary issue.

Richard Wellesley was forced to admit privately to his Bengal advisors that legally he had no 'right to the dismission' of the French mercenaries based on potential future events and there was also the nagging fact that Perron and his associates 'were entertained previously to the commencement of the War [with France]'.[31] But timing and presentation were not to be undervalued and a European peace still had not been signed, although it was in the foreseeable future.* The Governor-General continued to see the War of the Second Coalition as a suitable pretext to push Sindia for dismissal of the French officers. Wellesley reasoned:

> after the conclusion of a general peace in Europe, the importance therefore of excluding the power and influence of France both now and hereafter from the Government and Counsels of Scindhia is proportionatley encreased, and as the existence of the War with France furnishes a just ground of argument for demanding the exclusion of French subjects from the Service of Scindhia, which would not exist under a different situation of affairs, the early prosecution of that measure cannot be neglected without hazarding the loss of the most favourable opportunity which can be expected to occur for its successful accomplishment.[32]

Wellesley already knew, from his intelligence sources, that the British as well as the 'country born' Euro-Asian mercenaries in Sindia's service had been largely subverted to the EIC's cause. Is it any wonder that the Governor-General offered Sindia a substitution of English troops for the French officers under Perron as part of his treaty offer? Or that he promised not to raise any objection if Sindia wanted to appoint British mercenaries to replace any French officers removed from the regular battalions?[33] John Collins had to continue playing the game and let all the options run out on Sindia's chance to bargain. But these talking points that the Governor-General sent through secretary Edmonstone's correspondence came dangerously close to tipping off the Marathas that the sell-out of the British mercenaries was in fact a foregone conclusion. Collins was being urged to raise points that could only be predicated on secret intelligence and on more than one occasion he had to exercise his professional judgement on whether to risk pursuing the Governor-General's attempt to 'assist' in Sindia's decision-making process. Wellesley and secretary Edmonstone had considered that there was also a possibility that Sindia might face an explosive situation if he agreed to evict the French officers and they refused to go or – worse than that – what if they seized the army by force? As a result of this wild speculation Collins was authorized to use deadly force against the French if Sindia sanctioned their removal and they refused to go. Beyond that Richard Wellesley was willing to intervene directly. 'If the movement of any body of troops should hereafter be necessary to aid Scindhia in securing either the total or partial dismission of the French, and especially the removal of M. Perron the Lt Gov. of the Ceded Provinces and the officers in Command of the forces will be ready to afford the requisite forces at the proper season.'[34] It was an interesting offer and one whose covert linkages were facilitated

---

* The Peace of Amiens, signed on 27 March 1802, brought a fourteen-month lull to the Napoleonic Wars.

by the fact that the Lieutenant-Governor of the Ceded Provinces was Henry Wellesley, another one of Governor-General Richard Wellesley's younger brothers.

The problem with Richard Wellesley's plan was that it was too clever by half. Wellesley was seeing Perron in Francophobic terms of the 'French threat' to India. Sindia's French officers were indeed mercenaries and descriptions of that occupational speciality might, under some circumstances, imply selling out to the highest bidder. But there was no immediate danger of that among the older veterans. Perron was not a stupid man nor was Chevalier Dudrenec or Monsieur Fleury. They had seen the system reach its zenith under de Boigne and knew that the route to prosperity as well as retirement lay through British territory. The hypothetical scenario of their receiving a better offer from Napoleon was a British fiction. The traditional value system of the South Asian military economy had suited them perfectly well. These men were not revolutionaries. They were Sindia's military contract employees. They wanted to work for Daulat Rao as long as he did not jeopardize their lives unnecessarily by engaging in war with the British. Many of the French officers had aspirations matching their British mercenary counterparts. If the EIC was willing to take the British mercenaries under their wing why should the French not be allowed that same opportunity? Even if the EIC did not want to hire them, given the historically contemporary European political climate, there was no need to assume that they were all potentially rabid Bonapartists. The assumption was an insult. Why should they not also be entitled to extend their employment contracts or at the very least be offered the opportunity to convert them to EIC service or simply retire on an EIC pension? France, the Revolution and Napoleon – those things were all very far away. Many of the French mercenaries had lived in India for most of their adult lives and several had directly invested in the EIC or held its commercial paper as a preferred investment vehicle offering both security and liquidity. These men had homes and extended families spanning the country from Pondicherry to Lucknow to Agra. Several believed their military service was just as well tied to the EIC. Regardless of motive or intention, at least it seemed the EIC was a committed stakeholder and would not soon leave the subcontinent.

Changing plans

There were a growing number of Maratha *sardars* who resented Sindia as a weakling needing foreigners to maintain power, and hatred began to centre on Perron. The C-in-C realized in 1802 that his fortune and his personal security were in danger. He was the leading foreigner who wielded the most power in the shape of the 'regular corps' and real riches in the form of land-based revenue as well as trade concessions such as the salt monopoly in Rajasthan. There were several *sardars* who wished Perron dead. And that is one of the major reasons why the General had cautiously delayed going to meetings of Sindia's *durbar* in the Deccan the preceding year; he feared a *coup d'état* organized by jealous Maratha rivals. Perron had felt safe in the Gunga-Yamuna Doab with never less than a brigade to cover his actions. But by the spring of 1802 it was apparent Perron had to do something. If he remained aloof from formal meetings for much longer there was a danger that he could jeopardize his own credibility and that might mean his removal or replacement by another mercenary such as Sutherland. Reluctantly, not being able to delay a personal appearance any longer, Perron took his troops off to Ujjain for a rather confrontational meeting with Sindia's *durbar*.[35]

Sindia was also in a vulnerable position with regard to his own growing marginalization in the *durbar*. His father-in-law Sargi Rao Ghatke and his so-called uncles were openly urging courses of action that paid little to no regard for Sindia's future. James Skinner's memoirs make much of Sindia trying to orchestrate an armed showdown at dinner between Perron and a Pathan contingent. But this story does not figure to any great degree in the records of the other mercenaries or Collins. Sindia was forced into a position of having to decide between the support of his clansmen on one hand and, on the other, the physical protection offered by Perron and his brigades. Collins learned of Sindia's plan to escape from Perron's influence in a bid to establish his independence and safeguard his well-being. The break with Perron would present the British with an opportunity of sorts. Collins wrote, 'I have also been told, in confidence, that whenever the disturbances in this quarter are composed, so far as to admit of Sindia's repairing to Agra, it is the intention of the Maharage to deprive the General of the command of those fortresses which he now possess's in Hindustan.'[36] Perron's fate may not have been sealed but it was evident to Collins that as of April 1802 the General would be wise to have a contingency plan if he wanted to follow in his predecessor de Boigne's footsteps and live long enough to enjoy retirement from Maratha service. It was an ideal time to extend the hand of friendship to Perron and it behoved the British to do everything possible to expedite his safe retirement should the opportunity present itself.

Perron had backed Sindia's initial opposition to Ghatke's bid for the post of Chief Director of Public Affairs. Sindia favoured Ambaji Inglia for the directorship and Perron, as Sindia's servant, was supposed to swing full support behind Ambaji's candidacy. However, in trying to do so,

Perron alienated Colonel Hugh Sutherland who had apparently already promised his personal backing to Ghatke's effort to attain the extremely powerful administrative post. The reasons for Sutherland's support of Ghatke's candidacy remain obscure, but there is no evidence of a French–British power struggle as suggested by some British imperial historians. Collins' reports expressed fear of a slaughter if Perron and Sutherland squared off in the factional dispute that was threatened. Members of the *durbar* who supported Ambaji Inglia's bid for the post had to accept Perron as their defender, even if they hated the General. His battalions were needed to counterbalance the corresponding units Sutherland controlled in support of Ghatke. The confrontation was postponed, as opposed to being resolved, when Ambaji Inglia withdrew his name owing to the strength of influence Ghatke wielded as Daulat Rao Sindia's father-in-law.[37] Collins had witnessed firsthand the degree of division amongst the mercenaries in Maratha service.[38] But these differences had often previously centred on military or money matters as opposed to nationality. Their obvious points of difference were on more than one occasion bridged by a common bond of suffering imposed by the deteriorating civil–military relations with Sindia's chief ministers. The familial relationships of the European mercenaries, in this Maratha family drama, were somewhat ironic given the extent of the previously mentioned inter-marriages that involved the Sutherland clan and Perron.[39] Hugh Sutherland had also by this time become linked to George Carnegie's family network. Hugh had feared for the welfare of his young daughter Isabella whom he sent to Scotland and placed under the care of Susan Carnegie – the mother of the previously mentioned George Carnegie. Hugh Sutherland had also invested £4,000 in two shares of an East Indiaman to be captained by George Carnegie's brother James.[40]

George Carnegie, writing in July 1802, noted that he had been present at Ujjain when Perron came to face the *durbar*. Carnegie did not like the idea of Sutherland being subsequently transferred to command of the 2nd Brigade in the north – but Perron had his reasons. Carnegie wrote:

We are sometimes alarmed by reports that the Company wish to get from Scindea a rich tract of Country which runs up the east bank of the Jumnah river from below Coel nearly to Delhi; honest John Company wants this country because it is rich and will be a vast addition to his fine estate . . . it is too well known this Honble [sic] Society set no bounds to their avarice, and altho' unable to devour their own loaf, they greedily grasp at that of their neighbours.[41]

The letter is also important in shedding light on a different interpretation of Perron's character and his actions. Carnegie said: 'The French Officers are suspicious (without cause) of the British [mercenaries] forming

Cabals against their interest; and the General, being a weak man, lends an open Ear to the false insinuations of designing interested men.'

Carnegie drew a line between the French officers and Perron, insisting Perron 'will soon resign the Command of the Army to Colonel Sutherland'. The letter directly contradicts later secondary accounts such as Compton's that have a decidedly *anti-Perron as head of the French clique* tone. Perron did not fail to promote his British officers as was so often charged by imperial historians. Rather, it was that personal alliances were allowed to prevail despite lucrative promotional offers. George Carnegie admitted that Perron offered to make him Major of Brigade to the 1st Brigade. However, Carnegie put his trust in Hugh Sutherland as one of those larger-than-life mercenary figures who seemed to have the one commodity that a soldier of fortune could not live without – good luck. Hugh had survived many a fray, but beyond that he was a shrewd businessman who knew how to make money from this type of soldiering. George Carnegie had seen other business opportunities disappear and that partially explains his willingness to readily integrate Sutherland into his extended clan. He offered peace of mind to Hugh Sutherland by having his mother Susan Carnegie look after Isabella Sutherland in Scotland; the young girl was to be physically safe from the turmoil in South Asia. The Carnegies welcomed Sutherland's investment in the family's shipping venture. That suited Hugh because it served to diversify his investment portfolio, hedging his property-dominant South Asian asset risk with internationally traded shares specifically linked to the EIC's shipping interests. This mercenary networking was not that far removed from the alliance systems found in Maratha military culture, both seeking to profit from the South Asian military economy. George Carnegie wrote: 'Colonel Sutherland must be my Main Stay in the end.' Carnegie then declined the offer to be made Major of Brigade extended personally by Perron, in favour of being transferred north under Sutherland. That continuation of Carnegie's role in Sutherland's brigade compelled him to retain his subordinate rank of captain. An optimistic Carnegie wrote: 'When the General retires, he will be succeeded by my friend and Country Man Colonel Sutherland. From that hour French influence will cease, and the British Nation may ever after depend on the most powerful Prince in India as their trusty friend and Ally.'[42]

From the Ujjain *durbar* meeting, which featured a meeting of northern and southern brigades, Collins received detailed information on the status of the units personally under Perron as of April 1802.[43] Each brigade was composed of ten battalions, with 716 flintlock-armed regulars to a battalion and an extremely lean officer complement of two to three officers per battalion – a possible consideration in any combat leadership analysis.

Each battalion had five guns, consisting of four standard field pieces and one howitzer, each with a complement of *golundauze* and *lascars* counted separately from the sepoys. As noted previously, this battalion-based artillery firepower was more than double that of British battalions that had two guns each. Maratha firepower was further augmented at the brigade level with each brigade having its own artillery park consisting of sixteen large-calibre guns and a *rissalah* of regular cavalry equipped with their own gallopers. Perron's *rissalahs* were said to total 4,000 regular cavalry matching the quality and description of those that the EIC had previously purchased from de Boigne when he left Maratha service.

Perhaps of greater long-term strategic value as intelligence was the complete pay scale of Perron's forces, which Collins obtained from the General himself. This data could be used to calculate the cost of buying out the mercenaries and other troops in Sindia's service. There were two pay scales because Perron reduced the pay of troops serving in Hindustan where there were far more opportunities for a Maratha sepoy to make money in service. Prize funds alone proved a compensatory advantage but shares in lucrative northern-based brigade 'business' activities were also a factor.

What follows below is the European pay scale, also known as 'Deccany pay':[44]

Majors	per month	Rs 1800
Captains	per month	Rs 600
Lieutenants	per month	Rs 300
Privates	per month	Rs 9

'Native officers' in proportion to the above.

The Deccany pay scale was reduced to so-called 'Hindustanny pay' as follows:

Majors	per month	Rs 1200
Captains	per month	Rs 400
Lieutenants	per month	Rs 200
Privates	per month	Rs 7

The Marathas remained much more egalitarian than the British in their attitude of equal pay for equal work in what was considered the hardship posting – the Deccan. The stipulation was made in Sindia's army that both Europeans and 'Natives' serving south of the Nerbuddah were to be paid according to the Deccany scale.

This intelligence was of tremendous value for those who wanted to calculate the projected cost of an offensive war and weigh it against the cost of purchasing these human assets in the context of the military labour market. The accuracy of these figures was never in doubt as Collins was by this time negotiating directly with Perron about the cost of his army. Collins seemed somewhat surprised at the degree of sophistication used to retain Sindia's investment in human resources – namely his pension provisions. But then Sindia and the EIC were competitors within the South Asian military economy and the free-market forces of 1802 meant it was in his best interests to lock in his battalion strength through competitive employment packages. Any EIC demobilization was a windfall. Collins duly reported, 'General Perron, with whom I live on terms of civility, informed me, a few days ago, that Sindia had assigned lands for the maintenance of Invalid sepoys on a plan somewhat similar to that of the Invalid Establishment in the Honourable Company's Service.'[45]

Early in 1802 the major reason for seeking the removal of French mercenaries was the existence of a state of war between Great Britain and France. Therefore, Richard Wellesley did not welcome news of the Peace of Amiens in the sense that it robbed him of greater immediate support for interfering in Maratha affairs. Collins wrote: 'your Lordship was fully aware that the continuance of the war with France afforded a just ground of argument for demanding the exclusion of French subjects from the service of Scindia which would not exist under a different situation of affairs'.[46] For their part the Marathas were quite aware that the perception of a French threat had led to a coup against the mercenary corps in Hyderabad as the *Nizam* succumbed to pressure to remove Monsieur Raymond and his associates. Maratha 'newswriters' in Hyderabad had reported the process and its memory was evoked in Sindia's *durbar* during 1802. Peace or not, the Governor-General's position on removal of the French mercenaries was to remain firm but Collins, like other sceptics, realized that Governor-General Wellesley had been pushing the legal interpretation of the EIC's mandate for involvement. Collins mindfully wrote:

I can only express my regret, that the sudden intelligence of peace, should tend so materially, to weaken the grounds which your Lordship furnished me with for obtaining the most important object of my present mission, as I hurriedly conceive, it solely remains for me, under existing circumstances, the endeavour to attain by conciliation, that which I might, otherwise have claimed on the plea of right.[47]

Frustrated by what he saw as the loss of the War of the Second Coalition as a clever piece of cover, albeit a somewhat thin political rationalization, Richard Wellesley needed to change his plans. Over the next six months he reviewed his options and thought about how best to neutralize Sindia's military strength. The effectiveness of decapitating the command capability of Hyderabad's sepoy battalions, by removing their European officers, seemed well documented by the trusted John Malcolm;[48] and the lesson of buying out mercenaries was part of the historic record in the Anglo-Mysore Wars.[49] But it still came as a surprise, when early in 1803 the Governor-General did an abrupt *volte-face*. He apparently no longer believed there was a 'more certain, as well as a more honourable mode of effectually destroying this French party and its adherents', as he had written to Henry Dundas five years earlier.[50] It was time to start laying the groundwork for subverting Sindia's army by way of pension and employment offers. Richard Wellesley no longer saw any dishonour in buying out his enemy's army.

Charles Stuart's diary exemplified the criticism within British ranks over the manner in which EIC sepoy demobilization had fed Sindia's expanding army. Some observers realized that true professional soldiers would prefer service in any army as opposed to retirement and loss of their warrior identity. Richard Wellesley now considered how to formulate a policy that would allow all mercenary officers, as well as the sepoys in Maratha service, to exchange their rank and pension privileges for corresponding remuneration with the EIC or its allies. The exact method of formulating a policy to buy out soldiers from Maratha service remains somewhat obscure. Like many policies it seems to have evolved in stages and no one was able to foresee its eventual magnitude or ultimate cost. It is clear that Governor-General Wellesley had asked for opinions in the opening months of 1803 which set off rumours among the intelligence operatives that a deal was in the works to let the mercenaries resign Maratha service and either retire on an EIC pension or, if they were lucky, opt for EIC employment. A formal framework did exist by the end of June 1803 when Wellesley authorized his C-in-C General Lake to take the first steps necessary to put this plan into action.[51] But as we shall see in the case of the most important 'retiree', the spring of 1803 was when negotiations for defection began.

The basic conditions of the offer to convert pensions and employment contracts would later be referred to generically as 'the terms of the Proclamation', a reference to the guidelines established in the formally published document, which for obvious security reasons, was not scheduled for public release until 29 August 1803 when the British field armies would be in position. The Proclamation was intended to serve multiple purposes. Not only would it weaken the Maratha war machine but it would also supply the additional necessary troops to conquer and garrison Sindia's northern territories. By playing the military labour market as a dominant investor, the EIC under Richard Wellesley could increase the probability of its victory. Wellesley saw a means of siphoning sepoys from

Sindia's service while at the same time eliminating Sindia's command and control capability by gutting his officer corps. The Governor-General was not going to waste time pulling men out in ones and twos – he was prepared to authorize new battalions. Richard Wellesley noted:

A considerable number of the sepoys, who were discharged from the British army at the late reduction, are said to have entered into M. Perron's service, and it is supposed that if any new corps were raised in the vicinity of their station, many would return to the service and that Scindiah's European officers might be induced to resign the service of Scindiah by offers of a present subsistence, and of a future establishment in the service of some of the allies, or tributaries of the British Government.[52]

The Commander-in-Chief's retirement

Perron had benefited from de Boigne's departure in more ways than one. There was the promotion to C-in-C, but de Boigne's flight had also demonstrated to Perron that the British could indeed be trusted to cooperate in any plan that weakened the Marathas. Before departing service de Boigne was said to have advised the remaining European mercenaries never to seek open conflict with the British. He was also alleged to have warned that it might be necessary to dismantle the 'regular corps' – better to demobilize the sepoy battalions than to have them become the object of a rivalry that precipitated war. The British appreciated de Boigne's judgement and went out of their way to assist in his departure, including help with the sale of his military assets. The Bengal government went so far as to 'purchase' the General's personal Mughal bodyguard.[53] The British also helped with the conversion of significant amounts of cash into an interest-bearing EIC 'shares' account. By assisting de Boigne to leave Maratha service, the British also advanced their own cause by gaining a bounty of intelligence from him. The lesson was not wasted on Perron.

During March of 1803 Perron opened correspondence with General Lake on the issue of quitting Maratha service and entering British territory for the purposes of leaving India. In some ways it was no surprise, as we have seen, that there had been rumours of Perron's departure ever since the tumultuous power-sharing squabbles of 1799. Richard Wellesley dictated Lake's role in coordinating Perron's departure directly on 27 March.[54] Wellesley authorized Perron to proceed to Lucknow, accompanied by a body of infantry and cavalry, which would later be exchanged for British military protection. He was to receive a passport upon entry into British territory. 'The passport must describe the number and rank of his suite civil and military; and copies of it must be transmitted to the officers of the several stations, and to the civil magistrates of the districts through which he shall pass.' Perron would become a guest of the British Resident at Lucknow who would be ordered 'to treat him with every mark of respect and consideration' before his departure for Calcutta. Once there the Governor-General promised: 'I will receive him in a manner conformable to his wishes, and will use every means to facilitate his voyage to Europe.'[55]

As can be seen by the preceding reference to Perron's retinue, once this operation to detach Sindia's C-in-C commenced it could hardly be considered a covert movement. The eighth clause of Lake's instructions revealed that Perron's departure was to be accompanied by an application to Sindia for permission to leave service. The cover story, when and if needed, was to be that Wellesley was handling this matter in a supportive fashion denoting assistance to Sindia by aiding his personnel. 'I shall inform Scindiah of Mr. Perron's request, and I shall state that I have complied with it from motives of respect for Scindiah.' We must remember that in March 1803 many optimists believed there was still a chance to avert war. If Perron's departure could be used to pre-empt the need for hostilities, then British action had to be portrayed as friendly cooperation in expediting the removal of another retired mercenary from Daulat Rao's service. But, peace or war, this move was shrewd on the part of the Governor-General, who left no doubt of its importance in his letter to Lake. 'You will observe that I am strongly disposed to accelerate Mr. Perron's departure from Scindiah's service; conceiving it to be an event which promises much advantage to our interests in India.'

It is apparent, in reading the secret correspondence generated by this ploy, that Richard Wellesley's patience was greater than Lake's during what was to become a long and very complex period of negotiation. There were, however, potentially great rewards to be gleaned from assisting Perron in his bid to leave Maratha service. Wellesley wrote to Lake: 'I wish you to afford every encouragement and facility to General Perron in his return to Europe through our provinces.' The matter was to be kept classified but it was also necessary to keep this channelled properly within the EIC. It was a delicate covert matter but it also held tremendous potential for collateral political damage to Wellesley if it went wrong. The memories of Governor-General Warren Hastings' involvement in the affairs of Awadh and the subsequent gut-wrenching seven-year impeachment process had inflicted a toll on the transparency of Wellesley's government in India. It was important that the Perron negotiations remain covert yet generate a sufficient paper trail in the Secret Department of the EIC so as to be of use in refuting any charges of political intrigue or inappropriate procedure on the part of the Governor-General's office. Wellesley contacted Lake in a less than subtle, but successful, conspiratorial bid to seed the trail with deflective official documentation. As Governor-General, he was acutely aware of the issue of culpability in the Hastings trial and so he needed to cover his tracks in what amounted to interference in Sindia's internal affairs. In a 'Private' letter to Lake, Wellesley wrote: 'I think it would be proper that General Perron's application to you, with your opinion upon it, should be converted into an official form in the Secret Department; in which I will also record my answer, with the regulations proposed for the conduct of General Perron during his journey from Coel to Calcutta.'[56] In the quoted passage also note in particular the identification of Coel as the starting point of Perron's 'coming in' – a full five months before Lake's campaign began there.

The level of anticipation among Wellesley's intelligence men was growing. These were high stakes. If Perron's departure had Sindia's blessing, in ridding him of a point of division within his *durbar*, all the better. But the important thing was to get him out of Maratha service as C-in-C. Many of the mercenaries were now speculating that it would pave the way for Sutherland's ascendancy as C-in-C, something that had been denied when de Boigne departed. British mercenaries like George Carnegie hoped that if Sutherland was made C-in-C he could discourage the talk of war with the EIC, thereby permitting the continuance of a lucrative mercenary lifestyle. Holkar was still on the periphery and there was money to be made in waging war on him. However, the members of Sindia's *durbar* were leery of making a Briton head of the army when war with the EIC looked increasingly inevitable over British demands to recognize the political *fait accompli* embodied in *Peshwa* Baji Rao II's agreement to the Treaty of Bassein.

By the end of April 1803 the covert intelligence network showed the strain. John Collins was placed in an awkward position as a number of his more dependable intelligence sources had fallen under suspicion. Spying had been routine but the war party in Sindia's *durbar* was increasingly critical of the European presence and intelligence leaks. Collins wrote to Richard Wellesley advising him that 'War with the English is publicly talked of in the Marratta camp at this place. This occasions some alarm to the British subjects in the service of Sindia. Colonel Doddernigue, Major Brownrigg and Major Smith have assured me that they would have visited me long ago, but for the prohibition of the ministers of this court.'[57]

At that point the negotiations with Perron ground to a halt. The reasons for the cessation were not at all clear at that time and Perron's motivation was subsequently questioned by biased imperial historians of the nineteenth century. There has been a tendency to portray the failure of Perron to 'come in' in April 1803 as an indication of his greed. Supporters of that theory paint Perron as not wanting to give up the chance of gaining *lakhs* of additional rupees in Sindia's service during a war with the British. However, that speculation is not substantiated in the intelligence-related files covering the period from April to September 1803. Richard Wellesley's plans contained some practical points for defection, but life is rarely so simple and in this case it was much more complicated from Perron's viewpoint. His life was in danger with regard to the extent of his covert dealings and he had to have a contingency plan to extract his wife, family and fortune from Agra if he was discovered during the process of moving everything to Coel. Four months was not an unreasonable amount of time to make Wellesley wait in the context of Hindustan 1803. Simple mail delivery could take weeks and it might easily be compromised in a land where one's allies as well as employers were accomplished spies.

Perron, in a polite style that refutes nineteenth-century portrayals of him as a less than cultured man, wrote to Lake in April 1803:

The propositions which the most noble the Governor General has been pleased to make, and which your Excellency has had the kindness of forwarding to me, are evidently intended as the means rather of facilitating and rendering agreeable my journey through the Honourable Company's Territories than as the smallest restriction upon any desire which I could possibly form; and as I cannot help looking upon some of the propositions as extremely flattering proofs of the Most Noble the governor general's kind disposition towards me, I should most certainly show myself unworthy of such consideration, were I in the slightest circumstances to deviate from the wishes that may be communicated to me during my march either by the gentlemen resident at particular stations or by those whom your Excellency may think proper to send agreeably to the intentions which your Excellency has done me the honour of making known to me.[58]

But Perron went on to indicate that he was experiencing more difficulty than he had originally expected. Perron specifically noted setbacks concerning his 'own private arrangements'. Sindia also demanded his immediate attention and he warned Lake to tell the Governor-General that it would be at least May before he could safely put his escape plan into action.

The Governor-General's Proclamation would not be read in public for another four months but its contents had already begun to impact on Sindia's officers. The two issues that weighed most heavily on the minds of the British mercenaries were: (1) the threat of being charged with treason if a war should break out before they defected; and (2) the financial extent of pensions and chances for further employment with the EIC. The second issue was by far the most important as most British mercenaries had already stated that they would not serve against their fellow countrymen. The discussions with EIC personnel were encouraging but would the promises of equal rank and pension provisions be honoured once they 'came over'? Maratha prize funds regularly exceeded those of the EIC. Did one really expect compensatory pay supplements or merely the EIC's rank equivalent in wages? And if the mercenaries banded together in an act of collective bargaining, could they exert enough pressure on the system to extract more money? These issues divided them and there might not be a chance for renegotiation once a deal had been struck. And what if a peaceful solution to this tension with the Marathas could still be found? George Carnegie was aware of the historic precedent for buy-outs and he bravely played the waiting game. He saw hope and merit in deferring his defection until events dictated a more definite course of action. Carnegie later admitted that as of May 1803 there was discussion of resigning, which was the prerequisite to accepting British pension and employment terms.[59]

The covert negotiations were kept up while Perron publicly maintained the line that military preparations were continuing for a possible war against the British. Collins wrote: 'at this Durbar it is confidently asserted that General Perron is extremely active and busy at Coel, in levying new troops, and in making other preparations with a view to Hostilities'.[60] For their part the Marathas continued to try and limit the damage Collins could do as an intelligence operative. Active steps were taken to cut his access to basic information channels in the camp. In addition to denying direct meetings with the mercenaries in Sindia's service, there were moves to halt intelligence outflows from other sources, including Sindia's enemies. Collins notified the Governor-General: 'Several of Holkar's Sardars have advised him to turn my Newswriter out of his camp, and I should not be in the least surprised were this event to happen since a more brutal person, than Jeswunt Rao Holkar never, I believe, existed.'[61] Within Sindia's *durbar* the pressure was rapidly rising to do something about the European mercenaries and the British troop mobilization. Ambaji Inglia was putting events in motion to bring the situation to a head.

The diary of Charles Stuart (3rd Bengal NC) confirmed that by June 1803 it was common knowledge among British field officers that Perron had already negotiated his safe passage through British territory.

About this time too, an idea was entertained that General Perron having amassed a large fortune was anxious to quit the Mahratta Service and it was supposed he would take refuge in the Company's dominions having previously obtained the necessary permission. That this event was expected is beyond a doubt for the Officers Commanding at Sassnee & Bidjeghur were ordered to salute him should he arrive at either of those places, which from their vicinity to Coel, it was probable he would.[62]

By mid-July 1803 Wellesley, acting in conjunction with the council that formed Bengal's 'Supreme Government', authorized British field officers to prepare to accept as many of Perron's defecting officers as came forward to offer their services.[63] While still more than a month from official public disclosure, the large-scale process of undermining the Maratha military machine had begun. Wellesley had generously endorsed buy-out pay levels that exceeded recommendations established in a preparatory study filed by Major Frith.[64] As it turned out, Frith was a keen advocate of acquiring a number of very specific Maratha irregular horse units. He had researched buying out those Maratha *rissalahs* that served as heavy cavalry under mercenary officers. They represented a military asset that the British needed in the Doab. These were elite mounted troops associated with men like James Skinner. They were troopers selected for their horsemanship and skill with smallarms. Able to fight as cavalry or mounted infantry, they used both Maratha and European drill interchangeably as needed in the execution of their specialized tactics. Under de Boigne such *rissalahs* had been issued gallopers, some of which were 4.5-inch howitzers. The howitzer gallopers were very effective against refractory *zamindars* because their explosive shells could be rained down over perimeter walls as plunging fire. The shellfire killed exposed infantry and ignited stockpiled fodder held in anticipation of long sieges. These 'irregular horse' units identified by Major Frith, not to be confused with light irregular or *pindari* horse, were serious players to be reckoned with in the military labour market and they had no direct equivalent in the British order of battle prior to 1803.

There is substantial evidence that General Lake was one of the more ardent supporters of the plan to buy out the human resources of the Maratha armies and Lake monitored North India closely. But despite his being based in Hindustan, the northern theatre, Lake saw great merit in focusing on defections in the Deccan first. There were fewer of Sindia's troops in the Deccan – further away from the central front in the Doab. Perhaps they could be detached more easily, which would leave the thinly guarded chain of fortresses in the northern Deccan vulnerable to Arthur Wellesley's troops. Many of the mercenaries travelling to the Deccan from the Doab had carried advanced word to their comrades. But the fate of the sepoys had taken second place to officers and the guarantee of pension funding had been slow to materialize. Edmonstone, in his capacity as secretary to the Bengal Supreme Government, sent a letter in mid-July 1803 to Colonel Close, the British Resident at Pune.[65] Edmonstone informed Close that the Governor-General in Council had approved the plan known as 'Lake's offer system', and it was going to be extended to other parts of India beyond the Gunga-Yamuna Doab. It was official. The

covert efforts to subvert and recruit some of the finest human resources
in Maratha service had received the official stamp of approval. In point
of fact the programme was well known in the intelligence community but
in the meantime it had to remain covert for security reasons. From the
standpoint of those British personnel trying to attract and recruit defectors, the key element of the official word was that the plan now had full
government financial backing. That meant the mercenary recruiting drive
was fully bankrolled by the Honourable East India Company. There was
no holding back; it was time to buy as many men out of service as possible.
And those who could not be purchased with cash or pension-and-rank
conversion schemes might be *turned* for intelligence purposes, accepting
large amounts of cash to stay in position and relay secret information to
help the British war effort.

The EIC was particularly interested in immediately removing the
British, European and American mercenary officers from Maratha
service.[66] But it was important that the mercenaries be allowed to try
and resign or extricate themselves before the policy was made public. The
British command knew that there was a more than reasonable chance that
those who could not resign or escape would face imprisonment or death
at the hands of the Marathas once it was known that an official welcome
policy awaited them in the British territories. The public announcement
of the Proclamation would be proof of the British strategy of coopting and
subverting the Maratha officer corps. Bengal was ignorant of traditional
Maratha attitudes to mercenary service. It had to be assumed that those
men – some of them loyal to the Marathas, some of them British spies
and all potential soldiers for the EIC – would be placed in immediate
danger once the Proclamation became public. Edmonstone specifically
told Close to issue the appropriately signed Proclamations only *after* he
had received official word that formal hostilities had commenced.[67]

In July 1803 Sindia's Deccani officers were requested to swear an
oath, 'that they would serve in or out of the Maratta Territory'.[68] But
according to the young Scot named Daniel Stuart, 'The officers however declined taking this oath which was not noticed particularly by
Colonel Pohlman.'[69] As we shall see there was good reason for Pohlman
to ignore the request to swear allegiance. However, few historians seem
to have examined this critical junction in events. John Pemble's analysis
of events stated that Sindia ordered his officers to take an oath of allegiance requiring them to fight all Daulat Rao's enemies regardless of
nationality.[70] Pemble, among others, tended to view this as the water-
shed point where officers such as Skinner came over. Pemble wrote, 'The
governor-general's proclamation played no part in their action, since they
had not heard of it at this time. They quitted Sindia's service because they

were kicked out, not because they sold themselves to the British.'[71] But once again, let us be clear in our understanding of the events. Sindia never issued a blanket expulsion order and a very small minority of European mercenaries risked their lives in serving him beyond 1803. The public reading of the Proclamation and knowledge of its content were two different things. As for being 'kicked out', few ambitious mercenaries could afford that. To get the best employment package on offer it was essential to resign one's commission in Maratha service. An honourable resignation was a demonstrable sign of good conduct which was the preferred criterion for joining one of the EIC's armies.

It also became apparent during that eventful month of July 1803 that Sindia's *durbar* was not about to commit to any political agreement that might be construed as a British diplomatic victory. The Treaty of Bassein was not going to bind Sindia, Bhonsle or for that matter Holkar. The British rationalized that if there was no political option to be had, then war was inevitable. This triggered a deluge of secret correspondence and covert message traffic. Lieutenant-Colonel David Ochterlony had evolved as General Lake's chief personal intermediary with the British mercenaries in Sindia's service. Ochterlony's junior officers had been regular visitors to Coel until mid-July.[72] And Ochterlony was a personal friend of Hugh and Robert Sutherland. That friendship also provided Ochterlony with a chance to establish a personal relationship with Robert Sutherland's in-laws, the Hessings.

Ochterlony had visited the Sutherlands and Hessings at their homes in Agra. They shared many happy moments and mutual concerns. Ochterlony, like Sutherland, had children as the result of his intimate relationship with at least one South Asian woman. Ochterlony was worried about the future of his Euro-Asian daughters if they stayed in the hierarchical structure of British Indian society. The European mercenaries with South Asian spouses lived openly in the cosmopolitan setting of the Agra cantonment. This atmosphere of integrated racial harmony was particularly healthy for the children who were raised with a positive self-image and enjoyed the best aspects of both their South Asian and European cultural identity. This integrated cultural experience was so special that Ochterlony referred to it in his correspondence as the 'Agra system'. The uniqueness of the 'Agra system' was contrasted by a repressive attitude towards interracial marriage in British controlled cities like Calcutta.[73] There it was customary to send interracial offspring to boarding school and in some cases to keep South Asian spouses in a separate or 'black' part of town.

Ochterlony knew that it was time to get his friends out. Some of the younger British mercenaries in Maratha service could afford to play the waiting game and see if the cash offers, to be bought out of service, might grow higher. But the Sutherlands and the Hessings had too much at stake and they were already quite wealthy from years of service. Ochterlony wrote to General Lake from Bijighur to up-date him on the latest intelligence from Agra.[74] Sindia's preparations for war were being accelerated. The *harkarrahs* reported cannon shot of different calibres was arriving at Agra from Gwalior. And Sindia's men had begun to collect all the available boats along the Yamuna River.

Despite his closeness to Sutherland, Ochterlony was not yet willing to divulge the extent of their friendship to General Lake. There had been a certain logic to thoughts of having Sutherland assume command on Perron's departure. But some members of Sindia's *durbar* were now actively advocating expulsion or death for the Europeans as potential traitors. In language reflecting professional detachment, Ochterlony urged the General to proceed carefully, extending plans particularly towards removing Hugh Sutherland from service. Ochterlony was convinced Sutherland could be of special service owing to his role as *de facto* leader of the British mercenaries. Lake was cautious, however; he was worried about Perron and realized that key informants in Maratha service should be used to *turn* as many potential defectors as possible. He thought it unusual that Perron had not written lately but if in the meantime he could also engineer the removal of Sutherland, as Perron's logical successor, there was an increased chance of also removing the mercenary officer cadre intact. That would reduce any remaining European mercenary officers to a handful of renegades and outlaws. Ochterlony adopted an attitude of nonchalance and told Lake he did not really know Sutherland well enough to judge how he would react to an offer not merely to 'come in' but to betray the entire war effort by bringing over as many officers and sepoys as possible. 'I have not had a sufficient intercourse to pretend to appreciate his Character & tho he might be carefully communicative to a powerless acquaintance I do not know how far he might revolt at any direct offer that might seem to imply a suspicion of his acting with Duplicity Treachery or a Breach of Hospitality.'[75] But Ochterlony also prodded Lake to make the best use of an opportunity that had presented itself by way of Sutherland's previous letters. Or, in simple terms, Lake should make any further offer sound like the gracious acceptance of Sutherland's previous offer to help. Ochterlony urged Lake to write to Sutherland and tell him 'that in case of misfortune you hope he will render that service to his King & country which must be the result of his communication'.[76]

Within two days Ochterlony wrote again to Lake regarding events inside Agra. Ochterlony had been such a friend to the extended

Sutherland–Hessing family that he had sent them the latest editions of the *Calcutta Gazette* and medicine for the treatment of Hessing senior who was gravely ill.[77] Ochterlony was also in personal contact with Hessing's doctor whom he used as another intelligence source. On 23 July 1803 Ochterlony wrote to Lake: 'The Old commandant of Agra is dead & I hear from the Dr. that there were he thinks still many men employed in the fort.'[78] He also used the unidentified doctor to confirm details on the fort's artillery including the number of guns, their composition and origin. Ochterlony's intelligence on Agra was coming almost exclusively from the mercenary colony, and its accuracy gave him cause to doubt the value of more traditional sources. He declared he no longer had faith in newswriters as a source and, fearing their role in disseminating misinformation, he put greater emphasis on the role of the mercenaries as the best source of intelligence open to Lake.

Ochterlony's mercenary contacts also apprised him of the fact that the Marathas had tried to regain control of the brigades in the 'regular corps' by imposing a new parallel command structure. Various Maratha factions tried to exploit the latest civil–military split in the *durbar* that revolved around control. Maratha *sardars* with civil rank in Sindia's *durbar* were trying to foment unrest in the brigades aimed at the overthrow and expulsion of the European mercenaries. This had reached crisis proportion in a revolt that occurred as some battalions were shifted out of Hindustan and towards the Deccan. This revolt would seem to have coincided with the previously mentioned oath of allegiance incident that Pohlman had ignored. Ochterlony wrote to General Lake to tell him how the backlash manifested itself. 'In each corps there is a Subedar Adjutant who has a sort of interior management which gives them a very great influence & it is supposed that the late Mutiny on the March to the Deccan was secretly instigated & encouraged by them.'[79] Soon after this, countermarching orders were received from Sindia who feared the Doab was too exposed.

Things were getting chaotic. General Lake was surprised to hear that Perron had apparently sacked Hugh Sutherland from Sindia's service. This was a real problem as Lake had not heard from Perron recently and there was massive confusion over who had dismissed Sutherland and why – what had become of him? This information vacuum touched off a frantic search to locate Sutherland and take advantage of his vast knowledge. General Lake was concerned Perron might be reneging on his agreement and opting to stay in Sindia's service. Lake wrote to Governor-General Wellesley, 'Colonel Sutherland, lately dismissed from the command of a brigade by Perron, might be able to give much valuable information, and be instrumental in drawing over other officers from Perron.' Governor-General Wellesley did not doubt that Perron was 'coming in' but he did think the crisis over Sutherland's and Perron's whereabouts might indicate a worsening of affairs. But in crisis there is often opportunity and Wellesley authorized Lake to grant more concessions and/or cash if he thought they might win over more mercenaries, *sardars* or even placate Sindia's chief ministers whom he identified as 'confederates' in the context of the Maratha Confederacy. Richard Wellesley returned Lake's communication having pencilled in:

I do not know where Colonel Sutherland is to be found; if the Commander-in-Chief should know, his Excellency will be so good as to take immediate means for securing Mr. Sutherland's assistance. In general I wish the Commander-in-Chief to understand, that I shall cheerfully sanction any obligations or expense incurred for the purpose of conciliating the officers or ministers of the confederates. – Wellesley.[80]

A crisis in command and control

During the first week of August, about the time Collins was withdrawing from Sindia's camp in the Deccan as British Resident, the Maharajah's war machine started to come apart at the seams. A tumultuous power struggle had begun and there were three main factions pulling in different directions. First there were the mercenaries and sepoys who had no intention of fighting the British. These ranged from General Perron down to individual enlisted men. Most of this group were interested in British pensions and employment. The second faction was composed of major Maratha powerbrokers and leading *sardars* of Sindia's *durbar*. They had watched and waited long enough. Sindia had allowed the mercenaries to build an unprecedented level of power because he needed them to guarantee his leadership role, his revenues and some would say his life. This second group – the Maratha *sardars* – was predominantly pro-war although even amongst that majority there was no consensus of opinion on political or military strategy. They ranged from independent light horse advocates of predatory warfare to those who urged a more conventional united front with Holkar. The third faction was a small but determined group of renegade mercenary officers who had no intention of 'going over' to the British. Some of this third group saw this as their time of opportunity; a chance to rise rapidly to senior command if the Marathas would still trust them. They were men who saw that the mass defections would mean the Marathas would have to fill gaps in their command structure from the level of platoon sergeant to brigade commander. There was still a chance, in their minds, that the Marathas might opt for one of them to fill the position of C-in-C. Other members of the third element, with no intention of 'coming in', included those men wanted by the British as deserters and outlaws. The latter category included various races, religions, and nationalities. It would be tempting to think they were all criminals but that too would be an over-simplification; times had changed. The EIC was about to institute legal mechanisms to intervene in what had been a relatively free-flowing military labour market. However, registering for a passport to cross in and out of EIC territories was just not on as far as the renegades were concerned. It deprived them of the freedom and anonymity needed if they were going to make a living at 'freelancing'.

Many of the mercenaries, among those waiting to come over to British service, initially accepted their fate with the sense of adventure they were known for. A few felt betrayed by both sides – seeing the war as unnecessary. And a very small number changed allegiance with a zeal that spoke of an opportunity for revenge. This was a peculiar time when it came to examining the transitory nature of loyalty within military culture. Did one remain loyal to one's countrymen, one's race, one's rank or to oneself? On 10 August 1803 Captain Beckett* (Perron's ADC) wrote a letter to Major R.W. Rotton, a friend and fellow mercenary in Sindia's service. Beckett was almost apologetic in telling his comrade that the disastrous political situation had not been foreseen when Rotton was accepted for service; he wrote that, in essence, no one thought the seriousness of affairs would last but that it had indeed escalated to the point that there was no turning back.

Beckett forwarded Rotton's discharge and wished him luck in the hope they would meet safely after the crisis had passed. Then Beckett wrote that he knew Rotton was a loyal Briton who could not serve against his fellow countrymen. In other words, as chief secretary and ADC to Sindia's C-in-C, Beckett was urging the now discharged Rotton to do the right thing, to 'go over'. But as we shall see in the latter portion of this chapter, Beckett need not have worried about his friend's course of action. As for Beckett's next move, he was at Coel, which was very close to the body of British forces at Bidjighur. Therefore he would have to be clever about how he extricated himself. He was obviously under intense scrutiny by the Marathas owing to his rank and responsibilities but he continued helping his countrymen get out before the shooting started. He and Perron would stall the war as long as they could but when it came time to leave they would have to make a run for it.

---

* Aka Becket, and in some EIC pension lists as Becketts.

The European and Euro-Asian mercenaries who were waiting to defect received word that the Maratha *sardars* of Sindia's *durbar* were going to take back what they saw as their rightful and traditional leadership roles. They were not backing down on the issue of military as well as economic control and, distrusting the mercenary corps, they locked Sindia into a position of confrontation with the British that meant one thing – total war. The course of action for the defecting soldiers was now very clear. They had to get out immediately but do so in a way that guaranteed their physical safety while maximizing their chances for a pension with the British. Some calmly and coolly ignored Maratha mobilization orders while moving to the periphery of Maratha territory under the pretext of going out to put down local uprisings that were fabricated as part of their cover story. Others were more concerned that they live up to what they saw as the letter of the gentleman's agreement. They wanted to go through the formal resignation process from Maratha service so that they could, presumably, take maximum advantage of the EIC's offers of pensions and subsequent employment. Their military professionalism was reflected in their bravery. Not, as usual, bravery before the enemy on the field of battle. Rather, bravery in staying until the discharges were authorized and signed. A number of discharges went smoothly thanks to the help of higher-ranking mercenaries like Beckett who took on the role of out-processing as many of his comrades as possible. However, there were those who waited for discharges that never arrived and a minority found themselves imprisoned by Maratha *sardars* who saw the mercenaries as traitors who had worked for the downfall of the Maratha cause. There were also those who managed to get their discharges and then tried to get out of Maratha territory with the money and riches they had accumulated legally in Sindia's service. Several were plundered and released with nothing to show for their years of dangerous work. Inevitably, others simply disappeared on their way towards the British lines, victims of revenge killings or victims of highway robbery at the hands of local criminals. This was yet another reason to consider the British offer to 'come in' with sepoy defectors. There was no guarantee of safety but if you could bring over a section, a company or in some cases an entire battalion, then your chances of making it to the British lines were much better.

As Sindia's command structure imploded there was a series of contradictory orders from the competing Maratha factions. In the absence of a comprehensive policy decision from Daulat Rao, local and regional *sardars* began to take what they saw as appropriate action. In Alighar Pedron, the French commander, found himself under house arrest and in Agra Sutherland and Hessing suffered a similar fate. With the British massed on the eastern front adjacent to Coel and Alighar there was a suspicion that it might be too late to save that sector. Stationed at Sikandrabad* was a brigade that should have been deployed at Alighar or at the very least been held in reserve to come to the relief of that fortress when Lake attacked. Sikandrabad was southeast of Delhi on the Grand Trunk Road that led straight to Alighar and that proximity was to prove a life saver when the brigade received instructions to fall back on Delhi rather than move forward to the eastern front where Lake waited. Amidst the uncertainty, mercenary George Carnegie fired off a situation report to David Ochterlony as the man most likely to have Lake's ear on the issue of defection and its timing. Carnegie had put all his faith in Ochterlony's judgement based on Hugh Sutherland's advice. Now Sutherland was imprisoned in Agra and Carnegie needed to make up his mind. Without his mentor Sutherland, Carnegie pondered how best to 'come in'. If they left without their discharges, before Sindia's command and control had been sufficiently crippled, there was a danger they could be hunted down as deserters and shot. Carnegie briefed Ochterlony on the latest intelligence with regard to his unit's position and the probability that the British component of the mercenary officer corps of his brigade could be extracted as one group. He also added his observations on what would happen to the combat performance level of Sindia's 'regular corps' once the majority of its officers had defected. The letter also holds particular interest because it suggests that Carnegie and Sutherland's faction of mercenaries did not know that Perron and Beckett had already secured their terms. Carnegie seems, in the absence of direct up-to-the-minute intelligence, to have believed Perron's cover story that he was preparing to meet the British in battle at Coel.

There are 8 Battalions in this Brigade all commanded by Britons or the sons of Britons – most of them thinking their fortunes desperate were resolved to try their fortunes with the *General* [Perron] and his Secretary [Beckett], but I have succeeded in convincing most if not all of the shame they would bring on themselves and families – and I am now confident the whole will resign when I do – but we must stick to our posts, till all hope of accommodation is fled. I am confident the Brigades would fight to the last man, but when deserted by two-thirds of their officers, I am of opinion they will be disheartened and make a poor stand. The Brigade marches toward Delhi tomorrow . . .[81]

Carnegie's letter to Ochterlony also confirmed that key players in Sindia's Deccani brigades were ready to 'come in' and that the counter-marching towards Hindustan had begun.

---

* 28°+, 77°+. Aka Sikandarabad, Secundra; not to be confused with Sikandra Rao (27°+, 78°+).

I have a letter from Dan'l Stewart 1st Brigade dated at Adjunta Ghaut the 30th July – all remained quiet at that time. Twelve Battalions are on the march to Hindostan under Dundernaick [Dudrenec] and Brownrigg – I am confident the latter and his three officers Captains Marshall, Harricott [Harriott] and Atkins, will resign, as I know all their sentiments on that head.[82]

What kind of situation would greet them on the British side of the lines? There was no point in having waited to get the best pension settlement if one were going to be shot by a British sentry who did not know that these Maratha personnel were 'coming in'. Among Carnegie's concerns was the issue of whether all the British field forces knew about the impending defection. There were still ten days until the Proclamation was read publicly. Many assumed that would be the point of no return, but what if advanced elements of Lake's army started operations before 29 August? Carnegie hoped Lake's forces would follow the same procedure as used when the mercenaries were tipped off before the offensive in Gujarat. 'Pray do not you think that before the blow is struck, your government will give us notice – it is customary, and was done by the Bombay Government when they made war on the Giqwah [*Gaikwad*] in Guzerat.'[83]

British mercenary Lewis Ferdinand Smith provided one of the better explanations as to the sequence of events in late August that brought about the complete destruction of Sindia's command structure at the highest level, that of the C-in-C. Ambaji Inglia had orchestrated a bloodless coup against Perron that brought the General's world crashing down around him. By paying a 'tribute' to Sindia of Rs. 25 *lakhs*, Ambaji purchased the title and rank of *Soubha of Hindustan** from Sindia. Having acquired that exalted Mughal rank, Ambaji surpassed Perron in civil as well as military rank. Smith obviously saw the move as the settling of old scores on more than one count: 'by appointing Umbajee to the Soubah of Hindostan, Scindea delivered Perron over to his most implacable enemy; he dreaded the rapid arrival of Umbajee who would have assuredly drained his purse, if he had spared his life'.[84] Following Perron's sacking, Bourquin 'was the first to revolt against Perron'. The General survived with the help of many personal bodyguards whose loyalty he ensured with liberal amounts of cash. However, Perron was now in a real bind. His immediate retinue was loyal and many troops in outlying areas could be gathered because they had not heard of events. But much of Perron's cash as well as his family were still in Agra. He began to make attempts to gather his family and as much money as possible

---

* Here taken in the Mughal rather than Maratha context of *Soubha* to mean Viceroy of North India.

before heading towards Coel to defect. Bourquin, seeking to exploit the turn of events, convinced Perron's second brigade to mutiny. Bourquin then wrote to several of Perron's cavalry units informing them of Ambaji's ascendancy and offering large cash rewards for Perron's imprisonment.[85] It was to be a race against time for Perron to gather as much as he could before going over. Needless to say, he was a dead man if he fell into the hands of Ambaji or bounty hunters seeking to take up Bourquin's offer.

Fortunately for Perron, the temptation of greater rewards intervened to distract Bourquin. After having incited mutiny in Major Gesslin's second brigade, Bourquin had the audacity to lay siege to the Mughal Emperor's quarters. It was a bizarre scene as relayed in Smith's account. Bourquin declared himself for the Emperor, no doubt hoping for a coveted Mughal rank of authority as saviour of his imperial highness. But Bourquin was denied access to the old sovereign apparently because Monsieur Drugeon, the *killadar* of Delhi, refused to recognize Bourquin's authority.[86] The forces at Delhi were badly divided in terms of allegiance. Bourquin had sought to align himself with Amabaji but he could not overcome those still loyal to Perron who were left fighting a holding action until they could slip away. There was no French state on the Yamuna, there was no French unity, and the scene that followed showed there was no longer any control over the European and Euro-Asian mercenaries left in Maratha service. Bourquin was furious at Drugeon's defiance. In retaliation for refusing to obey his 'orders to release' the old Mughal, Bourquin wheeled up his artillery and commenced siege operations against the fort at Delhi. 'The two brigades consisting of eighteen battalions with one hundred and ten pieces of cannon were in this state of mutinous confusion and anarchy, when Bourquin heard that General Lake, with rapid marches, was approaching Dhailee [Delhi].'[87]

The siege of the fort in Delhi was halted and the guns quickly shipped across the river to meet the British on the Dharderi Plain in what we now know as the Battle of Delhi (see chapter 4). When Lieutenant Pester entered Delhi almost three weeks later, he noted the vacant remains of one of Bourquin's six gun heavy batteries near the Emperor's quarters.[88] Bourquin went from laying siege to Delhi's fort to defending the city. He had failed in his bid to secure the Emperor who might yet serve as Sindia's or Ambaji's puppet. Perhaps worse, from a tactician's point of view, he had allowed himself to be distracted from preparations to defend Delhi against Lake's advance. There was time for little more than a hasty ambush. More battalions were melting off on the periphery; without their usual officers to instil faith, lead them steadily into battle and maintain *esprit de corps*, there was little reason to stay. The offer to go over to the

British was a serious option and many Maratha sepoys questioned what was the best course of action. Was it worth staying in Sindia's service now? Or, was it better to forget one's arrears in pay and find a leader or spokesman to negotiate terms with the British? Indecision led to a number of hesitant moves and the British were aware that at the so-called Battle of Delhi in 1803, there were a number of uncommitted Maratha sepoy units hovering near the battleground. Some just waited to see the outcome of events in the hope of throwing in their lot with the winners. This was, after all, another aspect of the traditional South Asian military labour market. Soldiering for a living was dangerous at the best of times and only survivors could take their pay back home.

Lake had written to Governor-General Wellesley on 25 August 1803 prior to the official release of the Proclamation. He informed Richard that he was unable to contact Perron who inexplicably failed to answer letters asking for a meeting in person or via a trusted confidant. Despite his disappointment at not getting Perron to come in yet, Lake was happy to report, 'some of the European Officers in the Service of Perron has [sic] already come over, and I have no doubt that with the exception of one or two individuals the whole will'.[89] Two days later, Lake received a rather cryptic letter from Perron. It made some odd apology about not being empowered by the Marathas, under civil–military distinctions, to negotiate a settlement to the impending war. Lake did not know at that time that Ambaji was the *Soubha* of Hindustan. Perron's letter hinted that dangerous developments had slowed his plans. But Perron was stalling for time, still hoping to extricate his funds from Agra. He closed the letter by assuring the British General that he (Lake) had 'the happiness of serving a Government whose character is superior to such paltry doubts and suspicions . . .'.[90]

At that point Lake lost his temper. Not being able to read between the lines of Perron's bid for more time, the British General fired back a reply stating he had no reason to believe Perron was empowered to negotiate on behalf of Sindia. Then in a rather blatantly unguarded moment, Lake wrote several lines that held the potential to jeopardize Perron's life if they fell into the wrong hands: 'considerations of regard inclined me to submit to your attention some propositions which concerned yourself alone, and which would have enabled you to carry into effect the intentions you formerly expressed of quitting this country with ease and safety'.[91] Perron was within a day's ride and messages arrived quickly so he wasted no time now that Lake, his only remaining saviour, had compromised his life. Perron replied the next day (28 August) that he would send his ADC Captain Beckett to talk to Lake straightaway.[92] The clandestine nature of the meeting was also underscored by specific orders that Ochterlony was to personally see that Beckett got through the lines.[93] As recorded in the diaries of Call, Stuart and Pester, the meeting with Beckett did take place but secrecy was such that none knew the real issues. They assumed Perron was negotiating a surrender of Alighar. We do know from Beckett's correspondence that Perron continued to ask for more time to get his affairs in order – he was still trying to get massive cash reserves out of Agra.

On the eve of the Proclamation

About midday on 28 August 1803, a group of high-ranking mercenaries arrived in Lake's camp at Bijighur where preparations were under way for the advance on Alighar.[94] Among those who 'came in' that day were Captain Kenneth Bruce Stuart and Captain George Carnegie. They were bound to attract special attention owing to their family connections. Kenneth Bruce Stuart was said to be the country-born or 'half-cast' son of General Robert Stuart. While George Carnegie's mother Susan was the cousin of EIC Chairman David Scott.[95] George Carnegie's stepbrother also happened to be Lieutenant-Colonel Nicholas Carnegie of the Bengal Artillery. Nicholas had been a cadet in 1777 with David Ochterlony and was eventually promoted to Major-General, officer commanding the Bengal Artillery.[96] Whether or not Nicholas was instrumental in strengthening the relationship between his stepbrother and Ochterlony, as soldier turned spymaster, is not clear. At the very least it would provide a familial reference point for Ochterlony to assess loyalties. George Carnegie was content to build the covert relationship with Ochterlony on the grounds that it was the path recommended by his mentor Hugh Sutherland. Although Ochterlony was a loyalist born in America, he was accepted as a vibrant part of what we might term 'the Scottish connection' in this war. But Ochterlony was emerging as a far more versatile intelligence man. He had networked with other mercenaries in Sindia's service, as we have seen in the connections to the Hessings at Agra. Ochterlony was Lake's most trusted personal intermediary in dealing with the mercenaries defecting from Maratha service. Lake, in reference to Ochterlony's intelligence role, wrote to the Governor-General, 'He is the only man that I have seen since I came to India that I could repose much confidence in, but I feel perfectly secure of him, which you may easily imagine from knowing how I am situated in regard to confidential people.'[97]

In a later letter to his mother, Captain George Carnegie acknowledged his personal strategy for playing the benefit system under the provisions of the Proclamation. He did not seek to take unnecessary advantage of the EIC's situation, he was merely following the latest twists and turns in the evolution of the military labour market that the EIC sought to dominate. He felt real sympathy and compassion for Daulat Rao Sindia but there was nothing he could do once the Maharajah was locked on the road to war. Carnegie might gloat over his skill in playing the game of brinkmanship, holding out for the best pension offer, but he took absolutely no pride or happiness in Daulat Rao's dilemma. He wrote to his mother, 'I was advised by some of my friends to resign Scindea's service seven months ago [May 1803], but as matters have turned out, it is fortunate for me that I stuck to my Post till a few days before the Sword was drawn, and consequently enjoy the benefits of the Proclamation, a copy of which you will see in the newspapers.'[98] If he had panicked and left Maratha service before the pension plan was approved or voided his eligibility options by not obtaining a discharge, it would have been different. But the Captain kept his nerve and was rewarded by an outstretched hand from General Lake. George Carnegie went on to explain to his mother that he had left with Perron's blessing and by choice according to provisions embodied in Sindia's employment contract: 'having obtained General Perron's permission to resign the Service, a privilege all British Subjects could claim, as on these terms we entered into Scindea's Service'. When Carnegie crossed over to the British lines, he was surprised to find the British already in possession of Coel but he was more impressed by the friendly reception General Lake gave them. 'We immediately waited on the Commander in Chief who showed us the Proclamation and gave us much Credit for joining him before it had been made public.'[99]

Captain Kenneth Bruce Stuart, who came in with Carnegie, later filed a deposition in support of his pension claim and that document confirmed that the evaporation of Sindia's command structure had occurred in earnest during the second week of August 1803. It would seem that K. B. Stuart, like many others by then, knew that war was unavoidable. On 10 August 1803 he tendered his resignation to C-in-C General Perron. On 11 August, Stuart's resignation was accepted and he began the march to Bijighur. Kenneth Stuart and his comrades set off to see their old contact Ochterlony who had been of so much assistance in coordinating their 'coming in'. As he noted in his account to government, he was warmly greeted by Colonel Ochterlony who conveyed 'to me His Excellency the Commander in Chief's approbation of my conduct and an invitation to his head Quarters'. Captain Stuart and his companions were personally debriefed by Lake and his staff. K. B. Stuart's deposition to the Bengal authorities further stated, 'I remained with the Grand Army four or five days, and with his excellency's permission came down to the Presidency where I have just arrived.'[100]

The European defectors, under the Proclamation, were paid according to very specific criteria. The verification of cases, for civilian accounting purposes within the EIC, had to come through military channels. It was important to establish confirmation of their rank and identity. But later special claims and requests for additional compensation would also bring questions about how their individual cases were graded.[101] Was it a willing defection? Was the individual turning intelligence to the British prior to defection? What was the quality, level and/or amount of intelligence they provided? Provision was made in the grading system to account for the relative value or usefulness of their intelligence. Did they 'bring over' sepoys and/or artillery when they came in? How and when did they receive knowledge of the Proclamation's date and content? This last question was tricky for some to answer. Their reply might adversely affect their entitlement as well as their legal status. In the case of European sergeants and enlisted men in the boondocks, places like the northern Deccan, it was often a case of being the last to know. Others knew well in advance but said they did not know until the last minute because they thought they might be accused of treason for not turning intelligence earlier as some of their comrades had done. The question itself was not that sinister and was often merely aimed at finding out which units might still be outside the communication loop and therefore potential enemies. The fact that the Proclamation was systematically extended to cover the entire 1803 campaign period shows the degree to which the timescale of defection was of secondary importance to the primary task of removing as many men as possible from Maratha service.

Captain Charles Stuart of the 3rd Bengal Native Cavalry, whose diary figured so prominently in chapter 4, observed what he considered to be a peculiar scene on 29 August 1803, the day the Proclamation was made public:

In the course of yesterday and today many English and half caste officers of Perron's army came in – for some time past their situation has been very precarious and uncomfortable. On the 29th of August they shut themselves up in the Garden House [meaning Perron's Garden House at Coel] and were not altogether free from the apprehension that the Maratha troops when retiring might vent their rage on them.[102]

Charles Stuart's diary indicated that these officers were not being held as hostages but rather they were isolated as self-declared non-combatants. The significance of the Garden House scene can only be discerned by comparing it to one of the letters that George Carnegie wrote to his mother Susan. It represents another split or division among the mercenaries that resulted when Lake propositioned them to turn against Daulat

Rao Sindia and fight their brother soldiers in Maratha service. Appeals to patriotism were now balanced by feelings of loyalty to one's non-British comrades who may not have been able to voluntarily leave Sindia's service. George Carnegie wrote:

I had done my duty to my country in leaving poor Dowlet Row when he most wanted my services . . . I did not feel myself authorised (even in this just and necessary War!!!) to assist personally in the destruction of an Army, in which almost every individual was known to me. In short I declined following the Commander-in-Chief, and took lodgings in Coel till I could determin [sic] what course to steer.[103]

Perron as the lynchpin

As we saw earlier, Perron willingly exposed himself to danger by personally providing intelligence to John Ulrich Collins as early as 1802. But there is also reasonable evidence that Perron did a great deal to personally oversee the final dismantling of Sindia's mercenary officer corps and ensure its delivery to the British. Unfortunately Perron's actions have often been misunderstood or misrepresented in British imperial history as the selfish actions of a greed-motivated French revolutionary. Oddly enough the alleged proof for the imperialist indictment revolves around an incomplete examination of James Skinner's memoirs, Skinner being the most famous of Sindia's Euro-Asian mercenaries.

In the mid-nineteenth century, J. Baillie Fraser published two volumes that were sold as the *Military Memoir of Lt. Col. James Skinner*.[104] But the books included a great deal of anecdotal material that Fraser added to notes on the English translation of Skinner's original manuscript. J. Baillie Fraser did not differentiate, for the reader, which material constituted Skinner's original entries and which were his own neatly inserted stories. Among the latter were a few fabrications in accordance with nineteenth-century hero worship for political purpose. The timing was interesting given Skinner's ethnicity and the political backdrop of pre-1857 India. The stories added to the original Skinner papers were often those contributed by Fraser's brother and his cronies who knew James Skinner in later life. The problem was compounded over time when secondary biographical portraits of Skinner relied on Fraser's *Memoir* as Skinner's personal testimony. James Skinner, so long accustomed to speaking South Asian languages with his men, is now believed to have dictated his thoughts in a transitional or regional form of Urdu.[105] The handwritten copy of his diary, which survives in Britain's National Army Museum, was transcribed in English by an uncertain hand apparently unfamiliar with many of the events in Skinner's Maratha service; the writer's peculiar transliteration of names and places further complicates an accurate reconstruction. Although there is no doubt J. Baillie Fraser had access to this manuscript, as evidenced by bibliographic anomalies, a physical comparison reveals that a great deal of Fraser's two volumes was composed of filler from unverified secondary sources. But of greater interest regarding this tome was the historiographic omission of specific information found in the appendix of the original Skinner diary. Perhaps it is sheer coincidence that certain material in the appendix did not meet Fraser's literary or political purpose.

The stories concerning James Skinner's exit from Sindia's service vary in craftsmanship and construction. Many twentieth-century biographies portray a noble and gallant Skinner unwilling to give up the Maratha cause until the last possible moment. Skinner was supposedly dealt an unexpected blow when his loyalty was questioned by the Maratha *sardars*. As a non-white, the British had rejected him at an early age. Then in 1803, because of his father's Scottish ethnicity, or so the story goes, he was considered a security risk to the Marathas. Despite the tales of a valiant Skinner soldiering on until the day of battle at Alighar, the appendix of his diary reveals a different story. Oddly enough, the evidence shows a gap of almost a month from the time Skinner was cut loose from Maratha service until he went over to the British. A copy of his discharge from Maratha service confirms that Skinner had been a free agent since 1 August 1803.

This is to Certify that Capt. James Skinner entered my service in 1796 as an Ensign & throughout his Service he Conducted himself as a good soldier and excellent officer, having on Several occasions distinguished himself & was in my fullest confidence & is now discharged for Political reasons.

Camp at Coel
1st August 1803

Signed C Perron
Commander in Chief of
Maharaj Dowlut Rao
Scindeah[106]

Why Skinner did not go straight to Lake in the first week of August or join Carnegie and Stuart later in going over is a mystery, given that his discharge pre-dated theirs. Perhaps Skinner's actions were influenced by the fact that he did not get along with Sutherland, as recorded in his diary. It may have been that Skinner avoided Carnegie's group because he perceived it as being composed of Sutherland's men. However, Skinner's discharge, issued four weeks before the public reading of the Proclamation, does make three points very clear.

First, leading mercenary officers were actively seeking and obtaining their discharges a full month before the Battle of Alighar and almost two full months before Assaye. The implications were disastrous for Sindia and the Maratha cause as a whole. This was far more than a loss of skilled soldiers; more importantly for the enlisted men under fire, it was the loss of their combat leaders who could not be readily replaced before the war started. The removal did irreversible damage to the morale of Maratha enlisted men as well as the vital command and control capability extending from the platoon to the brigade level. If one systematically studies the battles of the 1803 campaigns, the impact on Sindia's offensive capability is glaringly evident. Fleury's mounted attack on Shikohabad was one of the few offensive actions for Sindia's army. Assaye, Burhanpur, Alighar, Delhi, Agra and Laswari were all a matter of waiting for the British to arrive, not a case of offensive strategies designed to rob the enemy of the initiative. Even those actions that featured baiting the British and then counterattacking (as at the Battle of Delhi) were characterized by a lack of Maratha infantry momentum, which highlighted leadership deficiency. The great reliance on artillery and in particular concealed artillery ambushes was made all the more necessary because of a lack of offensive infantry capability – a direct reflection of the loss in officer leadership.

Second, General Perron willingly granted the discharges at the most strategically critical time. In other words, Perron as Sindia's C-in-C was acting to strengthen British interests, actively reducing the number of veteran combat leaders in Maratha service, at a time when he knew the British were about to commence military operations and were actively welcoming the departure of officers from Sindia's service. Perron knowingly and willingly facilitated Governor-General Richard Wellesley's plans to incapacitate the 'regular corps' that was the Marathas' most lethal weapon against British infantry forces. He acted to help the British while he continued his covert efforts to ensure his own safe extraction as well as that of his family and fortune.

Third, there was no attempt to cover up the fact that the discharges were granted for political reasons, as stated in the final line of the Skinner discharge. This release was issued with wording to expediently facilitate Skinner's re-employment as a mercenary in British service. These men were not discharged for failure to do their duty and their discharges were, what we would term today, honourable. The references to his good conduct did nothing but increase his potential value to the British. If Perron were acting to further the Maratha cause, or his own fortunes as tied to the Marathas, he could have refused the discharges or linked them to dishonourable conditions thereby casting aspersions on the mercenaries' ability and jeopardizing EIC willingness to engage in a policy that might be portrayed as spending good money on technically undesirable human assets. In granting honourable discharges to the mercenary officers

Perron was handing Richard Wellesley further ammunition to counter any internal British opposition to his policy of buying out the enemy. The Governor-General could say these were fine men and justify purchasing their loyalty with pension and loyalty packages. And if the war should fail to materialize, or produce a peace whereby rapprochement with Sindia was necessary, Wellesley could tell the Maharajah that he had done nothing to subvert the war effort – he had merely employed some good men that Sindia had seen fit to discharge; nothing more than Sindia himself had done in hiring EIC sepoys in 1802 when Bengal's budget cuts necessitated their release back into the North Indian military labour market.

During the first days of September 1803, Perron's personal horse guard had been within sight of Lake's outward pickets and although the troops could be heard practice firing their gallopers, no attempt was made by the British to engage them. It had been known by the British for over four months that when Perron came in these men would accompany him with the full blessing of Richard Wellesley. Perron had originally intended to stay in place until Sindia sent out his official replacement. But Bourquin's bounty had proven a surprise and now it was time to make a break. Perron went to the trouble of writing a letter to Pedron, the French commander in Alighar, speaking of the latter's duty to defend the fort. But it seems that the letter was one of the last pieces of Perron's cover story. Pedron was a captive of the Maratha officers who now held command of Alighar. And the letter was meant more for the perusal of Pedron's Maratha captors than the old man himself. It was important to keep the cover-up until the last possible moment: it must appear to Sindia's men that Perron was resigning, not completing the sell-out of the army. On 5 September Perron wrote a letter to Lake that indicated that he was not only exasperated with the charade but ready to complete the final phase of his defection from Sindia's service. The letter to Lake was not filled with the rhetoric of honour and glory; it confessed the bitterness of failure. The defence of Sindia's lands was a 'solemn duty' and he might still have considered it his duty if it were not for 'the treachery and ingratitude of many of my European Officers'. Without naming him he pointed directly to Bourquin: 'especially of some in high situations in the Army, render my attempts fruitless, and my persistence on them a folly . . . The treachery of my European Officers has now left me little influence or command in this unhappy country.'[107] Lake's reply was brief. Perron was told to come in now and to stick to the agreed route to Sasni where Colonel Ball was waiting to receive him. Perron was further instructed to limit his cavalry bodyguard to 400 men and to adhere to the agreed times for entry.[108] H. G. Keene relied on the memoirs of Perron, as related by the latter's great grandson, to show that the General had no faith left in the system. Perron wrote: 'The successive treacheries of Bourquin and Pedron, and the suspicious conduct of almost all the other officers, had inspired the natives with such distrust of Europeans that our lives were in hourly danger . . . For myself, I only saved mine by great sacrifices of money.'[109]

When Perron and Beckett arrived they brought over a great number of documents pertaining to Maratha affairs. The most valuable were revenue return records for North India. These were deposited with the British revenue collector at Moradabad and then forwarded to Richard Wellesley's office. When Wellesley and his staff saw the cache they sent immediate word to Lucknow to have the British Resident there debrief the two mercenaries on the exact extent of the revenues generated by the lands under Perron's control.[110] Wellesley's men could then put that information to direct use in their reports to the Court of Directors. It was important to have high revenue figures to offset the expense of the war and the more revenue gain Wellesley could show, the safer he would be from criticism that his deficit spending was out of control. It should be noted that Beckett is often dismissed in imperial histories as haughty and somewhat villainous. But it would seem he earned the hatred of Lake's staff officers for his refusal to pay what he saw as unearned homage to their military ability. Beckett – although a loyal Briton – told Lake's staff, upon 'coming in', that they could not have beaten Sindia's army if the defections had not taken place and that on a unit-for-unit basis the Marathas' sepoys were better soldiers. The disparaging comments of Lake and his followers have it that Beckett was arrogant and behaved in a most ungentlemanly manner. But this portrayal was the result of Beckett's audacity, or should one say bravery, in having delivered his stinging rebuke while a guest at Lake's dinner table.

Lucan's revenge

The previously mentioned mercenaries in Perron's Garden House, including George Carnegie, were among those who had resigned service and 'come in' on 28 August 1803, the day before the Proclamation was made public. Although they were entitled to the pension benefits of the Proclamation they remained in self-imposed isolation, having opted not to accept Lake's additional offer to take immediate service with him in the field against Sindia's forces. One man, who did accept Lake's offer to fight against his former comrades, was the Irish mercenary Lucan. According to notes in the diary of dragoon George Isaac Call, Lucan did not arrive in Lake's camp until 4 September 1803.[111] That made him a very late and yet rather timely arrival. He had not come in with any others but wandered into Lake's camp alone, fuming after having been plundered and abused by Bourquin. His maltreatment, at Bourquin's hands, allegedly came after the Frenchman refused to accept Lucan's resignation. George Carnegie noted of Lucan: 'He vowed revenge and immediately offered his services to the Commander in Chief.'[112]

Lucan's desire for revenge is supported by Charles Stuart's diary. It states that Lucan volunteered to lead a mission to intercept Bourquin but Lake vetoed the proposal. Lucan, however, once having been denied permission to go after Bourquin, disobeyed and attempted the mission on his own – but failed to get sufficient support.[113] At about that time Lucan learned of Lake's advanced preparations for the assault on Alighar. Lucan volunteered to serve as a guide for Monson's storming party. Lucan's knowledge was deemed to have been absolutely fundamental to the success of the assault. He knew not only the specific number of interior gates that had to be breached but, perhaps more importantly, those inside which were false gates. As the diarists assert, the period under fire in breaching the gates 'did great execution' and a preponderance of the evidence shows Lucan made the difference in the success of this as a *coup de main*. The only way for Monson's party to arrive in strength at the wicket door was via the direct route, which Lucan guided them through.

Monson, whose reputation later suffered greatly because of his retreat before Holkar, was wounded early in the assault. For much of the heavy fighting it was Lucan who took the lead in urging the grenadiers forward. He allegedly shouted: 'after me there are 5 lakhs of rupees inside'. The plundering however got out of hand with the entire British camp joining in; not just soldiers but camp followers and coolies as well. But the storming party was so exhausted it is said they did not get their fair share of plunder. Blanks were fired to disperse the crowd but to no effect and two or three tumbrels of 'dollars' were given up to plunder. Men were scaling the walls and swimming the moat to get in. The British moved a hackery in front of the fort gate to block access but every manner of goods were being carried away, from money, to old tents, to firearms. Charles Stuart was sickened by the fact that British dead were lying there intermixed with those of the enemy, both being crushed in the riot. That Lucan was not mentioned in dispatches was an oversight pointed out to Lake, who said he would rectify the situation. George Call observed that Lucan's citation for meritorious service was actually via the Governor-General in Council by Orders to the Commander-in-Chief.[114] As the result of his gallant behaviour, Lucan was granted a commission in the 76th Regiment of Foot.[115]

The intelligence gained by using Lucan represents one of the few examples in the 1803 campaigns where the British had the upper hand in immediate tactical-level intelligence, as opposed to strategic-level intelligence. The latter had previously been provided by men like Collins who told them of troop movements and the number of guns. But at Delhi, Laswari and Assaye, Lake and Arthur Wellesley had lacked immediate tactical-level intelligence, a deficiency that contributed to making their victories costly. Intelligence on Alighar, gained by Lucan 'coming over', was vital to quick victory as it facilitated movement through the inner fort. Lucan and other defecting mercenaries were also able to confirm the hostile stand-off amongst the garrison divided between Bali Rao and Pedron.

The fact that Lucan was given a field commission in a King's unit as opposed to a commission in an EIC unit, as other mercenaries were, is also indicative of special treatment. It also might suggest there was more to Mr Lucan's story than has met the eye. The 76th was Lake's favourite infantry unit, but he was promoted within weeks to being one of 'Lake's Staff Officers', being clearly listed under that heading as 'Hindostanee Horse Paymaster'.[116] This role and subsequent data indicate Lucan was fluent in South Asian languages. It seems highly unlikely, but possible, that a former Maratha field officer would so readily be given an active high-risk commission – let alone a King's commission – unless something was known about his background. However, it may be that the type of intelligence he revealed was deemed to be persuasive. He was obviously versatile, providing intelligence on existing Maratha fortifications, troop disposition, morale, logistical and financial situations. He proved a valuable link with the defecting irregular horse units since he could speak their languages, knew their *modus operandi* and was trusted by them as a leader. And let us not forget that it was his infantry leadership under fire as the gates were blown in at Alighar that was the most graphic display of his willingness not just to betray the Marathas but to sacrifice his life if necessary to defeat them.

Lucan fought like a man possessed of a higher calling. He led Monson's party at Alighar in 1803 and ironically died defending Monson's retreat during the 1804–5 Holkar Campaign, although the author Compton later questioned Lucan's motives in implying that the Irishman's final charge was motivated by a desire to win a name for himself.[117] With Lucan on his staff as an officer in a King's regiment, Lake could add a new dimension to a European unit. In the past the EIC units had more advantage in dealing with South Asian peoples, gaining local intelligence and coping with the local environment. But several EIC units were viewed with suspicion as potentially dangerous sources of intelligence leaks. The acquisition of Lucan allowed Lake a chance to take advantage of new options that were viewed by some as more secure. The diarists make it apparent that intelligence was readily lost via the EIC's sepoys to the camp followers and from there to the enemy. The acquisition of Lucan meant that Lake could execute direct orders with irregular horse units outside the EIC communications circuit, which had previously allowed information to be disseminated to the Marathas.

Weighing strategic vs. tactical intelligence

John Pester admitted that soon after Alighar the British had learned of Sindia's additional forces being in the vicinity of Delhi from former Maratha officers. They warned that Brownrigg's brigade was heading north and the defectors also promised the British that this would be a severe contest. Pester recorded: 'We expect an action with them shortly, and all the officers lately in Scindiah's employ assure us that they depend entirely on the formidable train of Artillery to defeat us.'[118] But knowing what is going on at the strategic theatre level, such as the shifting of Brownrigg's brigade from the Deccan to Hindustan, did nothing to increase tactical intelligence at the battlefield level. It was advantageous to know how many guns were headed north, but what lay beyond the next thicket? Both George Call and Charles Stuart made it abundantly clear in their diaries that the Battle of Delhi represented an intelligence failure for Lake's army at the tactical level.[119] They laid the blame squarely on the shoulders of one man.

It was common practice in Lake's Grand Army of 1803, as it had been in South Asian armies since Babur's day, to send the quartermasters in advance of the army to select campsites. This often meant finding a location within immediate marching distance that would provide the basic needs for the army's brief stay. On the morning of 11 September 1803 Lake's quartermaster-generals, Nightingale of the King's troops and Campbell of the EIC's troops, had gone out ahead of the army with a small party to reconnoitre the ground and scout probable campsites.[120] With the scouting party that day were the EIC's adjutant-general Colonel Gerrard and Captain Covell of the 27th Light Dragoons. This was of particular significance because Colonel Gerrard, in addition to being the EIC's adjutant-general, was also head of the Spy Department for the Grand Army.

While riding near the Hindan River the scouting party observed small bands of what were thought to be Maratha horsemen in the distance. Shortly thereafter a suspicious man was stopped and questioned in the vicinity of the riverbank. The man said that Bourquin had passed through the area and crossed over the Yamuna River ahead of Lake's army. After the scouting party rejoined the British forces, Captain Covell reported to General Lake that it was his considered opinion that the army should occupy a small height of land near the Hindan–Yamuna plain to give them a better chance to observe the area. Covell then proceeded to tell Lake about the suspicious character and his warning that Bourquin had just passed through the area and crossed the river ahead of Lake's army. Apparently the C-in-C was less than pleased and Charles Stuart's diary directly quoted Lake as saying, 'Gerrard, did you hear this?' Colonel Gerrard admitted having spoken to the same man but also confessed that he had done nothing about it. Charles Stuart, miles from the scene but having benefited from his fellow officers' observations, recorded that Lake and his men had no inkling that the Marathas were on the same side of the Yamuna River as the British. Gerrard had failed to verify the man's story by looking for signs of a major crossing along the riverbank and also failed to follow up the information with secondary patrols or information gathering.

It appears that it was Covell's delayed intelligence that motivated Lake to take to the field with his cavalry forces as a reconnaissance in force and check out the Maratha horse that the scouting party had seen in the distance. Covell's name was later listed in the 27th Dragoon casualty reports as dangerously wounded. He was hit during an assault on a small village, which flanked the battlefield. Ironically it was a position that he had urged EIC deputy quartermaster-general Captain Salkeld to occupy when they had been out scouting for a campground. Later Covell's name was listed among the dead. Dragoon George Call was unusually blunt in saying that Gerrard's greatest shortcomings as head of the Spy Department were his 'inactivity' and his habit of 'offering insufficient reward' which meant that he seldom got good intelligence from the local population.[121] Both Call and Stuart agree that the massive failure of battlefield intelligence that day was Colonel Gerrard's fault.[122] Call went on to say that it was this incident that led to Gerrard being immediately sacked. The point that seems to have escaped previous imperial histories is that the position as head of the Spy Department for the Grand Army was then given to Lieutenant Lucan.[123]

Following the Battle of Delhi, George Call asserted that the suspicious stranger, questioned near the riverbank, was a Maratha agent sent to plant disinformation among Lake's men with the intention of making their ambush complete.[124] Supposedly the news of Bourquin having already crossed the Yamuna was intended to make Lake abandon caution and rush forward with his forces into the well-laid trap anchored by the semi-circular ring of Maratha artillery pieces, concealed by the tall grass and flanked by additional batteries clustered in the small villages nearby.

Since Lucan had used his linguistic ability to secure his position as paymaster of the Hindustani horse it seems logical to speculate further that it must have been that same ability that made him a sound choice for the Spy Department, a job in which he needed to be able to communicate with local people. Perhaps as a former Maratha officer he would have recognized the stranger at the river. However, the great anomaly in this reasoning or rationalization over how Lucan became head of the Spy Department for the Grand Army is his identity as a former Maratha officer. This was a man who had only just 'come over' from the Marathas one week earlier. How wise was it to take an enemy soldier who had defected and put him in charge of a payroll? More importantly, how could you take a sworn enemy turned defector and in one week make him head of covert intelligence for your army? Did Lake believe that Lucan was trustworthier than most men? Did Lake have little other choice because of a shortage of men who could speak South Asian languages? Or might we speculate that Lucan could have been a British intelligence operative for some time prior to his defection and was therefore a 'known quantity' as it were? By bringing Lucan not just into EIC service but placing him in His Majesty's 76th, the intelligence loop was reduced in size and complexity; and, as previously suggested, Lake's reliance on leak-prone local personnel was further reduced. That increased Lake's command and control factor through the process of consolidation, although ironically it negated the carefully crafted parallel structure of the King's and the Company's troops within the Grand Army. As efficient as this system modification may have been for Lake's purposes, it leaves some rather large unanswered questions about trust vs. knowledge. If we were to take an analogous battlefield situation, out of context from any other war, it is inconceivable that an enemy officer would be granted an equivalent battlefield commission in an elite unit, provided a staff officer's position with command over troops in the field, given control of a payroll as well as direct personal access to the C-in-C and then placed in charge of intelligence.

Not everyone viewed Lucan's actions in a favourable light. Regardless of motivation, some saw Lucan as a criminal in having served Sindia only to turn on him. George Carnegie's mother Susan wrote about what she considered to be Lucan's treachery. 'I was scandalized at the first accounts of the Maharatta War – to find a Traitor recommended to notice who deserved to be hanged – an insult on public morals! Would rather have seen his name among the slain than honoured or rather dishonoured. Miss Carnegie [George's cousin] and brother James at London judged he would not act so.'[125] Well after his mother's blistering attack on Lucan, George Carnegie confided his thoughts. 'You express an honest indignation at the treacherous conduct of Lt. Lucan towards his old master. Nor in my opinion can the many who support him, fairly or fully clear his Character.'[126]

Motivational factors

As a former leader of Maratha sepoys, George Carnegie made two specific points pertaining to the performance of Sindia's 'regular corps' in connection with the defection of the mercenary officers:
- It was expected that their resistance would crumble without leadership and adequate payment.
- Their ability was beyond question and the experienced combat leadership of their veteran officers held the potential to sway the outcome of history.

The Brigades thus deserted by their leaders, and without the means of subsistence, were expected to disperse or become an easy pray [sic], but the obstinate resistance made by these troops on the 11th and 23 Septr, and 1st Inst, does them infinit [sic] honor, and shews what they might have been capable of, if supported by those Officers, who had so often led them to Victory!![127]

The diary of George Isaac Call of the 29th Light Dragoons provided another piece of the puzzle with regard to the destructive forces unleashed by the defection of the Europeans and the subsequent attempt by Bourquin to seize power. Call wrote that after the Battle of Delhi (11 September 1803) the surviving French officers, who had served in the action, sent a letter to Lake's HQ in which they begged 'to throw themselves on the mercy of the British Flag'.[128] The French officers were debriefed by Lake with ADC Lieutenant Duval serving as interpreter. Bourquin admitted they 'brought into the field 7 tumbrils of money, by way of plunder to their Army, in case we should have been defeated'.[129] But during the action it was apparent that Bourquin was losing and 4,000 Rohilla horse, allied to Sindia's men, fled with four cash-filled tumbrels. The disintegration of the 'regular corps' following the mercenary defections had set off a scramble for cash and loyalty was scarce.

The string of European defections gave the Marathas little reason to trust the foreign officers and those who could not evacuate soon enough suffered accordingly. The garrison at Agra imprisoned Hugh Sutherland and George Hessing, who became their hostages. In the midst of the chaos command and control were sharply divided along ethnic and professional lines and to add to the mayhem the Marathas had seven battalions with artillery outside the fort's perimeter camping on the *glacis* and occupying the town as well as mosque.[130] The problem was they were not united in motivation or purpose. Five of the battalions were reported to be those of the Irish Major Brownrigg's 5th Brigade.[131] They had been in the Deccan but were ordered north to Hindustan by Sindia for previously elaborated reasons. They arrived too late for the battle at Delhi and marched for Agra where they were greeted with distrust. It seems that owing to the vast treasure inside the fort at Agra there was considerable jealousy and many of the garrison inside suspected that these latecomers might have already gone over to the EIC. Even if they were sincere in their claim to be loyal to Sindia's cause, few of the garrison wanted to let the recent arrivals in because the pay for the fort's defenders was months in arrears and if the defence was successful it would mean the garrison would have to share the treasure.[132] Therefore, the Deccani battalions were kept at safe distance outside but Brownrigg came in to 'join' his fellow Europeans as guests of the garrison. Mercenary Lewis Ferdinand Smith spoke of the vast sums of cash in Agra as having 'caused a fermentation' in the minds of the garrison.[133] The garrison was also uneasy about what their former commanders, who they now held hostage, were saying. Smith remarked: 'I believe they were likewise intimidated by the threats of Scindea's European officers, who were prisoners in the fort: Colonel George Hessing, Majors Sutherland, Brownrigg and Derridoven,* and Captains Harriott, Marshall and Atkins, who were told by the garrison, that if the money was lost, their lives would answer for it to the British Government.'[134] Later a serious debate erupted in Lake's army about what was to happen to the 25 *lakhs* of money captured at Agra. Perron had not managed to get his money out of the fortress and from Bengal he tried to file claim to all the treasure. Ultimately EIC officials relied on the testimony of Colonel Hessing that the funds were intended, at least in part, to pay troops in Sindia's service.[135] Later, in a footnote to his diary, George Call reported that he learned that the battalions in service to the Marathas at Delhi had only offered to defend Agra on the condition that their back pay was granted.[136]

As we saw in chapter 4, Lake's assault inflicted heavy casualties on the Maratha sepoys in the half-moon battery at Agra. But there were still the better part of three battalions in and around the outerworks after the British were driven back from their attempt. A man identified as 'Afzul the Tindal' began to speak with the defenders on Lake's behalf. After concluding a basic agreement for their defection to the British, Afzul asked what else they could bring over.

* Presumably this is Derridon.

In the afternoon, the 3 Battalions which had been driven into the half moon battery (more properly called the Madaghur), sent into the General, intimating a wish to surrender and to bring over the Battery of 32 Guns: they marched into our camp a short time after; and also a party of lascars (with 200 Bullocks), who had been brought over by the Rhetoric of one of our Tindals for which service, he was promoted to a higher rank the next day.[137]

Lake was somewhat amazed at this battlefield bartering for loyalty but he had begun to learn the finer points of the South Asian military economy. The Proclamation had been part of a lengthy and well-executed plan with carefully concocted wording subject to review by the EIC's judicial representatives.[138] But this bartering of men and guns was nothing like that; it jarred the old man's sensibilities. With enough cash it seemed Lake could hedge his military bets for success – if not purchase outright victory.

Lieutenant Pester's diary entry for 14 October 1803 speaks of yet another large defection at Agra that exemplified the impact of the Proclamation on the immediate military labour market.

This evening seven Battalions of the enemy surrendered themselves, with their guns, tumbrils, colours and ammunition to Colonel White, and marched up and encamped on the left of our line, at a distance of about one mile. The terms given them were to employ them in our service with the same pay as they received in Scindiah's, the officers to continue with them, with the rank they formerly held. Colonel White, with five Companies of the 16th [Regt. Bengal NI], took possession of the Jumna Musjeed in which they had been posted . . . Many of the enemy who came over this day are very smart fine fellows, and soldier-like looking men.[139]

Note that Pester uses the phrase 'surrendered themselves'. But these men were negotiating their employment with the EIC. These were the men who just four days earlier had staunchly defended the Jami' Masjid Mosque and disgraced Brigadier-General Clarke. Now here they were 'coming in' to be received by their trusted opponent Colonel White. The Maratha sepoys had wisely negotiated for and received permission allowing their 'officers to continue with them'. This may not have been exactly as Richard Wellesley had planned it, but the Proclamation system worked. It delivered not just mercenary officers from Maratha service but crack battalions of sepoys that the EIC could readily employ. The defecting battalions bargained in good faith within this spot market. The mercenary officers had raised these battalions, led them in spirited defence against Clarke's assault and now shepherded them into EIC service. There was surprising continuity as Britain became the dominant player in the South Asian military economy.

Although George Carnegie rejected the path that Lucan had taken, he did eventually end up in EIC service. He stayed on his EIC pension until the fall of Agra. But when he heard that Sutherland was free and that Brownrigg as well as several companions had been hired into service, he volunteered for active duty. Carnegie had not wanted to 'take up service' until he knew there was no chance of having to face his old comrades in battle. Colonel Ochterlony, who had played such a vital role in bringing men in, was by then Resident at the court of Delhi and he offered Carnegie a choice position.[140] Eventually Carnegie served in the 1804 defence of Delhi and was mentioned in dispatches for his courage in personally leading a party into Holkar's siege batteries where they spiked the Maratha guns before Delhi.[141] Holkar's insurgency in 1804–5 tested Richard Wellesley's grip on Hindustan, but it never rivalled the geographic dimension or conventional threat level of the combined 1803 Anglo-Maratha Campaigns. Regrettably George Carnegie died of hepatitis in Delhi during 1805. But before his death he foresaw that peace would bring cutbacks in military spending. That meant there was a strong possibility that the EIC would reverse its policies and terminate the service of the mercenary officers that it had recruited out of Sindia's service. Carnegie's letter humorously played on his new identity as a 'Red Coat', but it also reflected the cross-cultural identity and bonding of a special band of British mercenaries in EIC service who were proud to call themselves Marathas. 'Much, certainly, depends on the liberality of the Company to us Mahrattas, after peace is restored. But, as we have fought hard and bled in their Service (6 of our number killed last year), they cannot decently turn us adrift, or, (Being a Red Coat), I should say more properly, to the right about.'[142]

Within the narrow band of books that deal with the mercenaries who 'came in', there is lingering confusion over the eligibility period of the Proclamation process. Several authors have assumed that the offer ended in October of 1803. And the issue is clouded by the status of several men like Bourquin who were classified as prisoner of war (POW) when they surrendered. In Bourquin's case, it was his renegade action against the Mughal Emperor and resistance at the Battle of Delhi that earned him POW status. However, the terms of the Proclamation proved extremely flexible and as time passed the Governor-General realized there was not only a chance to subvert the Maratha 'regular corps' but, in effect, commandeer the entire North Indian military labour market. Colonel J. R. Sheppherd was Ambaji Inglia's 'Irregular Corps' commander in Bundlekund having served that senior *sardar* for seven years.[143] Colonel Sheppherd remained an intelligence operative during the worst months of the 1803 campaigns and worked behind the scenes to try and influence

Ambaji's decisions regarding the war. Ambaji had begun negotiations with the British in October 1803 about the possibility of not only deserting Sindia's cause but of entering a direct British alliance.[144] After having successfully helped to steer the outcome of events in Bundlekund, Sheppherd negotiated terms under the Proclamation. Sheppherd had not only been in contact with Richard Wellesley's men during the course of the campaign but he had also monitored the treatment of mercenaries who had defected earlier. He accomplished this by maintaining correspondence with the defectors Lewis and Armstrong, carefully inquiring about what terms they had settled for. When Sheppherd finally 'came over' on 11 December 1803, he brought in his entire brigade complete with gallopers.[145] His brigade was small with only 634 men but they made up for their size in firepower, tactical repertoire and *esprit de corps*. Sheppherd's gift to the EIC under the Proclamation included fourteen European mercenaries ranging in rank from sergeant to the colonel himself. Among the infantry, 600 men were armed with muskets and bayonets while it seems at least thirty were issued matchlocks and apparently served as designated marksmen. There were thirteen cannon listed as 6 pounders and they were complete with limbers, grape shot and drag ropes. Sheppherd's men left little to chance in the field as they served as a self-contained army. Mechanics, carpenters and even rope makers accompanied the guns to keep them serviceable under the most adverse conditions.[146] The willingness of the British to accept the brigade into service in the final month of the war seems to indicate that the Proclamation remained open as long as it was of service to the war effort. Advance knowledge of its terms had helped ease complicated covert negotiations. In effect the Proclamation formally started in July 1803, a month before the first battle of the Anglo-Maratha Campaigns, and ran right up until the final peace negotiations in December 1803.

The southern theatre

Earlier in this chapter there was a reference to General Perron's ADC Captain Beckett having urged Major R. W. Rotton to do the honourable thing and 'go over' to the British. But he was bound to have done the right thing. As it turned out, Rotton, who served as brigade commander for the mercurial Sargi Rao Ghatke, was one of Richard Wellesley's most highly prized spies in Maratha service. Like many of the British mercenaries, Rotton had originally joined the Maratha service out of economic necessity. He had applied unsuccessfully to Governor-General Richard Wellesley in July 1801 for assistance in finding placement in the EIC's service. Having failed to win a position with the EIC, Rotton joined Ghatke's retinue and was readily promoted owing to his military and administrative talents. Ghatke was keen to build a personal army able to compete with any other. Daulat Rao Sindia owed his leverage over the *durbar* to his 'regular corps' and Rotton's application for Maratha service in 1801 coincided with that competitive period when leading members of the *durbar* were manoeuvring for position. Sargi Rao had a voice in the *durbar* as Daulat Rao's father-in-law; but with a regular brigade he also had real power for whatever military, political or economic purpose he chose to apply it.

In January 1802 Major Rotton wrote a letter requesting a personal meeting with Richard Wellesley, which was granted. The resulting assignation guaranteed that he would have a special role to play as a spy reporting directly to the Governor-General and his staff. Rotton was extremely valuable to Wellesley for two reasons. He was a chief source of covert intelligence for what was to become the southern theatre of operations and he had access to Ghatke's military plans as well as insight into his political decisions inside the *durbar*. The Major was stationed at Burhanpur in the Deccan, which made him pivotal in encouraging early defections in that sector. The fact that Rotton had personal correspondence with Richard Wellesley prior to Maratha service worked to his advantage in communications with the Governor-General's staff. He was to be considered a known entity in Maratha service, which helped in his vetting as an intelligence source. Reports on Maratha troop distribution and vulnerabilities quickly found their mark and little escaped his observation. Rotton's letters to EIC officials were written with a precise command of the English language and his use of judicial logic, found in his later pension review claim, was nothing short of lawyerly. Major Rotton's appeal for a pension review made it explicitly clear that although his intelligence flow was directed via Richard Wellesley's staff it was General Lake in particular who apprised him of the official plan to undermine the officer corps via the Proclamation, and Lake was instrumental in engineering Rotton's extraction.[147] This was a great relief to Rotton because he had been planning his escape from Maratha service for what had seemed like ages.

The Governor-General made sure that both the civil and military secretaries to the Bengal government knew that Rotton was an intelligence source and that he was to be encouraged in his efforts. Rotton sent intelligence directly from the Deccan to military secretary Merrick Shawe.[148] As a spy for the Governor-General he did not report to theatre commander Arthur Wellesley and that explains the absence of his reports in

Arthur Wellesley's papers. Rotton's pivotal position enabled him to confidently circulate word of the 'buy-out' to his comrades at Burhanpur and in the Deccan, well in advance of the Proclamation's public disclosure on 29 August 1803. The lead-time also allowed Rotton to plan his escape route and prepare a cover story for his departure. Unfortunately, Richard Wellesley's men designated Lucknow as the safe station where Rotton was to be received. Eventually, as the Governor-General's plan gained momentum and the benefits of the Proclamation were opened to all officers as well as sepoys, more reception areas were prepared, especially in the Deccan. But Rotton made his escape according to his earlier instructions and he endured enormous difficulty and expense in getting up to Lucknow. The Major had not only sent up-to-the-moment military intelligence and facilitated mass defections, but he also served as a direct intelligence pipeline between Sindia's *durbar* and the Supreme Government of British India in Bengal. By correlating that information from Ghatke's perspective with data from Colonel Collins the intelligence circuit was virtually complete.

Through informal meetings, troop transfers and correspondence the mercenaries had known for months that war was approaching. As of April 1803 they had been actively encouraged to leave service. They had discussed among themselves what amounted to a bail-out clause in their contracts. For several years it had been customary among British mercenaries joining Sindia's forces to make their service contingent upon the proviso that they would not fight against their countrymen in the form of the EIC's armies or His Majesty King George III's troops. Fighting each other as British mercenaries in service to 'Indian princes' was, however, fair game; although not to be actively encouraged. This produced some rather interesting moments for the *feringhis* when Sindia and Holkar fought each other in what amounted to the Maratha civil war of 1802.

The British mercenaries' proviso may have indirectly increased the division between the British and a minority of French mercenaries in 1803 because it was apparent the British were to be specifically propositioned for defection under the provisions of the offer incorporated into the Proclamation. Initial formulation of the document was thought to be directly tied to those individuals who had already established a pre-existing condition for leaving Maratha service. They could resign honourably and accept conversion of their pensions and re-employment. Some of the other Europeans were not initially clear on where they stood and had to wait and see what official offer for non-Britons was embodied in the Proclamation. The final authorization for across-the-board officer purchase in July 1803 meant many did not hear until well into August, when combat deployments had been ordered and communications were further disrupted. This accounted for some rather cautious negotiations on the eve of major battles. As we have seen in the dilemma faced by George Carnegie, the uncertainty of waiting from April until August 1803 was one of *bail out* vs. lucrative *buy out*.

It was clear that the disintegration of the Maratha officer corps was well under way by August 1803. But it was important to keep the pressure up in the hopes of accelerating the process and winning over more defectors; there was a new emphasis that same month in delivering intact sepoy battalions. The British invitation to leave Maratha service was covered in broad terms under the Proclamation and it was hoped that the document could be physically distributed well in advance to British officials so that it might be simultaneously released across the countryside on 29 August 1803. Regrettably, late revisions and distribution problems meant that many mercenaries would have to be reassured by word-of-mouth. A month before the Battle of Assaye, John Collins wrote to Arthur Wellesley thanking him for his assistance in facilitating the Proclamation process. Collins wanted to selectively locate and extract specific mercenary leaders in a bid to further weaken the Marathas by taking out key men. Some were natural leaders who could serve as broker/dealers bringing in entire battalions of Maratha sepoys. There were also a few men who did not fit that description but held Collins' interest for other specific reasons.

Last-minute troop shifts by Sindia made it difficult to ascertain where all these men could be physically *brought in*, and a few guardian shepherds would not be amiss in guiding those members of the flock that were strung out along the Malwa and Bundlekund corridors. Special attention was devoted to Captain Grant on the grounds that he was vital to Pohlman's Brigade (the old 1st Brigade cherished by de Boigne), which was rated by the mercenaries themselves as the finest. Collins wrote to Arthur Wellesley:

I received your letter of the 19th instant [August], with its enclosures, yesterday evening. I think it will be practicable to convey copies of the proclamation of his Excellency the most noble the Governor General, to Major Brownrigg, Major Smith, and Captain Grant. The first commands a corps consisting of four Battalions of Infantry. The second is attached to Colonel Dudrenec's Brigade, and in fact commands it, he being a very active good soldier, whereas the Colonel has not a military idea. Captain Grant is Brigade Major to, and in great favour with, Colonel Pohlman, an Hanoverian. The latter commands one of Sindia's best Brigades, and would, I know, be glad of an opportunity of leaving his present situation, at least so he assured me, for I met him, in my morning rides, several times while I was encamped at Julgong.[149]

Collins had apparently made very good use of his morning rides, exercising his intelligence network as well as his horse. Note that Brownrigg had already been ordered back to Hindustan where he intended to eventually link up with Sutherland at Agra. While the Major Smith that Collins referred to was Major Ferdinand Lewis Smith whose *Sketch of the Regular Corps* proved such a valuable source on Bourquin's rebellion at Delhi and the failed attempt to seize the Mughal Emperor.

Lieutenant Colonel Collins also wrote to Arthur Wellesley about the developments in the northern theatre, as he understood them. 'Colonel Sutherland has already resigned the service of the Maharage.'[150] Arthur Wellesley knew who Sutherland was, having had personal correspondence with Brownrigg and Sutherland dating back to 1800 when they were his Maratha allies. The mercenary duo had volunteered to help Wellesley in the campaign against the 'freebooter' Dhundia Waugh.[151] With Sutherland taken out of the running as Perron's replacement in August 1803, there was reason to believe the entire mercenary command structure was on its last legs. However, it would be a mistake to think that Collins wanted all the mercenaries to defect. The old man was swayed by reports from 1802 that George Hessing was a dreadful officer, often described in derogatory terms. But much of that negative association can be traced to those who blamed Hessing for Sindia's 1802 loss at Ujjain against Holkar. Collins, dutifully keeping Arthur abreast of developments, sarcastically noted: 'Colonel Hessing the elder, a Dutchman who had seen much service, died a short time ago. His son, a half caste, also commands a Brigade, and as he is a wretched officer, I hope he will continue where he is.'[152] John Collins was opinionated on the combat leadership abilities of specific mercenary officers.

In short, with an exception of Frenchmen, I believe Pohlman, Brownrigg and Smith, are the only experienced officers in the service of Scindhia. There are several English gentlemen attached to the corps commanded by the two former, who, of course, would be glad to accept of the benefit held forth in His Excellency's proclamation. As to the natives I am doubtful, though it is likely I think, that many of them will follow their European Officers.[153]

European mercenaries below officer rank had more to lose if they resigned Maratha service early and missed an opportunity for a pension conversion scheme or alternative employment. For some, particularly the lowest ranks or those that had military occupational specialities that put them at greater risk in combat (i.e. artillery crew), survival was their first and foremost thought. This was evident in the 'Deposition of John Roach Englishman and George Blake Scotchman lately manning each a gun in the service of the Begum'. In their statement about leaving the service of the Begum Sumru* at the end of August, it was apparent they had to make their decision about leaving service in something of a vacuum. Officers with large incomes were jealously guarding their opportunities, there was talk of spies everywhere and the Marathas were reportedly jailing men for failure to serve. The lower ranks, such as Roach and Blake, did not know whom to confide in for fear of incarceration or worse. 'Hearing the detachment was to march to oppose the English', they had 'left camp by permission' according to their contractual understanding of not having to face their countrymen in combat. They claimed they did not know about their options under the Proclamation and suffered doubly as the result of being 'robbed of everything' immediately after having left the camp.[154] Roach and Blake seemed unaware that their patron, Begum Sumru, was herself negotiating with Lieutenant Colonel Ochterlony for the delivery of their entire brigade into British hands.[155]

As a young lady the Begum had learned to campaign in the field with her husband, the notorious Walter Reinhard, who was said to wear a 'sombre' scowl. That this Sombre Sahib became *Sumru Sahib*† is part of the legend. But the Begum was a willing apprentice in the art of military command as well as a shrewd diplomat, excelling where poor Walter could never hope to compete. Perhaps the Begum was the most mercenary of military entrepreneurs in that she bargained for the best individual as well as collective defection terms.[156] She went so far as to broker a transfer of her *jaidad* lands under the provision that the British also officially transfer her revenue-collection rights to a new territory. The British coveted her land to gain a better hold on the Doab. She accepted a new *jaidad* strategically located adjacent to EIC territory on the condition that her income rights to the land be officially recognized by the British government in Bengal. In her meetings with Ochterlony she established a strong bargaining position. The Begum had led Sindia to believe she would support him but she carefully kept manoeuvring her battalions to keep them out of combat at places like Assaye. Later, when British appeals to her brigade commander Saleur fell on deaf ears, she wrote to him in French and addressed a separate 'Persian *Hookumnama*' to her sepoys indicating the deal was done and it was time to enter British service.[157] For their part the British had also driven a hard bargain. They arranged to exchange the Begum's *jaidad* on the western banks of the Yamuna River for an alternative piece of land bordering what was to become their new frontier. For the British, the deal with the Begum meant recognizing her perpetual land ownership and revenue-collection rights. What they

---

* Aka Zeib oo Nissa Begum, Begum Samru.
† Aka Samru, Somroo, Sumroo.

gained, aside from a potential western buffer state, was a contiguous piece of property along the Yamuna River, thereby achieving their own goal of a strategic waterway to Delhi.

Begum Sumru had ordered four battalions, totalling some 2,000 men, to march from the Deccan to Rajasthan in order that they might 'go over' to General Lake.[158] The arduous march was difficult but it proved successful in keeping the battalions intact. While some might question her South Asian or Indian loyalty, she was following a much older or pre-national code of conduct. The Begum was an independent broker/dealer in the greater South Asian military economy. Historically India had seen many women warriors, from the Maratha women, who fought in tandem on horseback with their husbands as noted by famed traveller Ibn Battuta, to the *Nizam* of Hyderabad's elite female detachment of the 'regular corps'.[159] But the Begum's status in South Asian military culture owed nothing to her gender as a military leader; it was her prestigious ranking as a military entrepreneur that elevated her standing. She was an accomplished powerbroker within the military economy. Her infantry brigade gave her the credentials and the presence that meant both Sindia and the British courted her. We do know from the memoirs of several British officers, as well as mercenary George Thomas, that many men sought her sexually well into her middle years. But her attractiveness to the British in 1803 was more an issue of power politics and their bid to control the South Asian military economy. She was a capable military commander having won many battles. Yet we can see by the experiences of many male mercenaries that being a successful battlefield commander was no guarantee of being made a part of the British administrative process after 'coming in'. The Begum was no fool and she knew the difference between flattery and business. It was her 'regular corps' that proved to be the ticket to retaining power, wealth and status. That also explains why she would not let Sindia squander her troops at Assaye. She moved her men to Rajasthan as part of the pragmatic pattern of survival within the South Asian military economy. Death could be honourable; but survival with such highly sought-after assets as a regular infantry brigade was far more profitable.

Of kin and clan

Following the Battle of Assaye, Arthur Wellesley wrote to Collins with news of captured enemy intelligence that took the form of a number of sheets of paper from a Maratha 'orderly book'.[160] There was a keen interest on Wellesley's part to find out which British mercenaries had failed to 'come in'. It was growing increasingly apparent that the officers of one brigade were not responding in the same manner to the Governor-General's generous offer. Had they not heard of the terms? Or were these men renegades? The latter was a distinct possibility in Arthur Wellesley's mind. He had reason to hunt down some English-speaking officers from Sindia's service for having contravened the unspoken gentleman's code of conduct on the battlefield at Assaye.

I have some reason to complain of Scindiah's English officers, and I shall bring the subject forward publicly as soon as I can ascertain the matter more completely. My soldiers say that after they were knocked down by cannon or grape shot, they were cut and piked by the horse belonging to the campoos [brigade], which indeed is perfectly true, and that horse was cut to pieces by the British cavalry. But they say besides that they heard one English officer with a battalion say to another, *You understand the language better than I do: desire the jemidar of that body of horse to go and cut up those wounded European soldiers.* The other did as he was desired, and the horse obeyed the orders they received.[161]

Arthur Wellesley's letter made it apparent that he was tracing the fate of Pohlman's brigade in which the previously mentioned Captain Grant had been said to serve as brigade major. The dispatch included a list of mercenaries, drawn from the orderly book, but it was not clear at that time which men were on the field and which had already left service. The paperwork was probably an earlier pay record or roster, which would not include all the departures precipitated by the Proclamation. The Major-General noted the list of mercenaries was 'a matter of curiosity' and he wrote to Collins that 'I shall be glad to hear from you upon the subject'.

By October the Deccan contained what seemed to be an increasingly large number of transient bands of defectors who had left service at various times. The battle lines of this war were shifting and many small groups of defectors were uncertain where to find Wellesley or where to report. Numerous accounts testify that they were unable to fend off the plundering raids of local *pindaries*. The British would be of greater help to the former soldiers of Sindia if there were a means of collecting them more quickly. Keeping track of all the incoming personnel and processing them was a bit of a headache but it was to be given priority to further the cause of reducing Maratha power. It no longer made sense to have one or two places for their reception. Two weeks after the Battle of Assaye Arthur Wellesley tried to expedite the process by designating three locations at which Maratha 'officers and troops' could report if they 'may be desirous of availing themselves, to the benefits held out to them' in the Governor-General's Proclamation. These consisted of Major-General Wellesley's 'British Army' HQ, the divisional HQ at Pune and the District Collector's Office in the fort at Ahmadnagar. Beyond simply receiving them, Arthur Wellesley ordered his men to 'treat them as friends . . . and to act respecting their payment in the manner pointed out in the proclamation'.[162] The Proclamation was extended and the British wanted to *turn* as many remaining Maratha troops as possible before Sindia could withdraw them to the north in a bid to regroup. White or brown, officers or enlisted men, Sindia's or Bhonsle's – it mattered not; the understanding was that all personnel considered as belonging to the enemy were to be welcomed in or retained by whatever suitable means in conformity with British war aims. Their conversion would weaken the Maratha war effort and they might be needed if Holkar should intervene at this crucial point. Barry Close, the British Resident with the *Peshwa* in Pune, wrote to confirm that a Marathi-language version of the Proclamation was drawn up in the hopes of recruiting the wayward members of the *Peshwa*'s personal retinue who had sided with Sindia when their master turned puppet by accepting so-called British protection under the Treaty of Bassein.[163]

During the second week of October John Collins replied to Arthur Wellesley's inquiry about Captain Grant. It was a nervous time owing to the lingering threat that any outstanding or renegade officers would be hanged for treason; as it turned out the threat was idle and never carried out.

I am shocked and surprized at what you relate respecting the savage cruelty of the English Officer in Scindiah's Service. I thought our countrymen were incapable of action with so much inhumanity. But it is evident those fellows must be the scum of the nation. With an exception of Grant, and Beckett, I know none of the English Officers whose names are inserted in the return. The former is nephew to Mr. Charles Grant, one of the directors, the latter is Secretary to General Perron, and now attends the General I believe, You will perceive by the enclosed Gazette that all British subjects bearing arms in the service of Scindhia, of the Bhonsleh, now that these Chieftans are at open war with our Government, have incurred the guilt of High Treason, and I should heartily rejoice to see every one of them executed excepting such as may be detained by Force. In this latter description I hope Mr. Grant may be included not only on account of his family, but because I should be sorry to find that he had attempted to impose on me by a false confidence. In fact this Officer positively assured me, on the day that I left Scindhia's camp, it was his decided intention to leave the Mahrattas and repair to some English settlement the first moment that a favourable opportunity presented itself. He also affirmed that Lieutenants Stuart, and some other Englishmen, entertained the same design. I have not the least doubt but that W. Grant received a copy of the Proclamation issued by the Governor General on the 29th of August last, as my Native agent at Burhampour informed me, by means of a cossid [messenger], that it had been delivered to him. It is possible however, that he had not been able to effect his escape, and I sincerely hope this may prove to be the case.[164]

Lieutenant Colonel Collins had made special efforts to make sure Captain Grant understood his options under the Proclamation. He was not only one of Collins' intelligence network but nephew to one of the few men who could bring down Richard Wellesley's administration in India. Captain Grant's uncle – Charles Grant – was a leading member of the EIC's Court of Directors but also the man destined to be Chairman of the EIC.[165] Charles Grant led a powerful lobby among the directors that ultimately proved instrumental in opposing the Maratha policies of Governor-General Richard Wellesley.

A month after Assaye, on 23 October 1803, Arthur Wellesley wrote to Collins:

Mr. Grant and Mr. Stuart were not in the action of the 23rd September. The latter is arrived at Poonah, and says that he and Mr. Grant, and Mr. McCulloch, quitted Scindiah's camp on the – [blank in manuscript], and went to Burhampoor: from thence Mr. Stuart went to Poonah, but he does not say in his report what became of the other gentleman. He says that they had not heard of the Governor-General's proclamation of the 29th August, and regrets that they had not, as he is convinced that many Natives would have come away with him. The motives with these gentlemen for coming away were their reluctance to serve against their country, and the fact that the English officers in Hindustan had gone to the British settlements, of which Colonel Pohlman received intelligence from General Perron on the 12th September. The brigades engaged were Pohlman's, Dupont's and Begum Sumroo's; in the whole sixteen battalions.[166]

The letter shows that Arthur Wellesley's Deccani intelligence system had yet to come to terms with all the information that was filtering in. As we can see, this deposition stated Begum Sumru's battalions as engaged at Assaye but we know the Begum kept them out of combat. As for Grant and Stuart, they said they had left Sindia's service for two reasons. They were not going to fight against their fellow countrymen and they had already heard that their counterparts in the northern theatre had gone over 'to the British settlements'. They had not heard of their specific opportunities under the Proclamation to bring over as many sepoys as they could. That, after all, had been a fairly late amendment to the offer and they had been engaged in extensive countermarching for months. The despatch also revealed that Pohlman had received word from Perron as late as 12 September 1803. As we now know, Perron's departure was complete by then. There was a vital need to reach Pohlman and ensure that he understood final instructions. Sindia had more to lose in Hindustan than in the Deccan and the degree to which he shifted troops underscores the fact that the Deccan was not the primary focus of the war. During July Sindia ordered 'the 4th and 5th Brigades under Brownrigg and Dodernaigue' up to Hindustan.[167] Defecting mercenary Major Smith said that in the pre-existing war plan, offered to Sindia by Perron but declined by the Maharajah, the interior of the Deccan was to be the scene of light horse raiding and *pindari* action.[168] Perron wanted to keep the 'regular corps' out of Deccan for several reasons. Pohlman's brigade had to march under Sindia's direct orders as the blocking force in the Deccan. But Perron's letter reached Pohlman somewhere near Adjanta and it set off the final chain of events for the collapse of command and control of the 1st Brigade.[169]

Who was this Stuart whose name kept cropping up? Why was southern theatre commander Major-General Wellesley interested in him to the point where Stuart was included in despatches inquiring about Captain Grant, the nephew of Charles Grant of the Court of Directors? This question leads back to 'Mr Stuart's' deposition upon arrival at a reception station.[170] If we look at various lists of mercenaries such as that provided by Major Ferdinand Lewis Smith or the later EIC pension records, the Stuart most often listed, as a captain in Sindia's service, is Captain Kenneth Bruce Stuart who served in Hindustan and arrived in General Lake's camp with George Carnegie.[171] It would seem that this 'Mr Stuart' of the Deccan, who made his deposition at Pune, had his name transcribed incorrectly. He was a proud son of Scotland whose name should have been recorded as Mr Daniel Stewart not as Mr Stuart. George and Thomas Carnegie left three letters that indicated why concern for 'Mr Stuart' ranked with that for Captain Grant.

As early as 1801 Thomas Carnegie referred to Daniel Stewart being promoted to lieutenant and noting that he was a 'very good Officer'.[172] A month before the war broke out, George Carnegie wrote to his mother, 'Lieut D. Stewart is 600 miles from me.' At that time Carnegie was writing from near Delhi when 'Mr Stuart' had just been posted one month earlier to the 'Burhanpour Battalion' in the Deccan, a point that was approximately 600 miles away.[173] Later, when George wrote about his departure from Maratha service, he mentioned receiving news from Daniel Stewart's brother who served in the EIC's army. 'I am very happy to learn from Lt. Charles Stewart that Daniel has got safe to Bombay, but after great hardships and plundered of everything.'[174] Daniel Stewart ('Mr Stuart') was another key player in the Scottish connection that cross-linked the EIC and the officer corps of Sindia's army.

Daniel Stewart was a part of Carnegie's circle, the Scottish inner circle if you will. The clan and extended kin networks of the mercenaries mirrored those of the Marathas in terms of cross-cultural conflict analysis. The Carnegie clan connections ranged from Maratha field officers in Sutherland's brigade to the Board of Directors of the EIC. Susan Carnegie served as guardian for Isabella Sutherland in Scotland. And the

Sutherlands were intermarried with the Hessings as well as the French connection in the form of Perron and Derridon. Susan Carnegie, the widowed matriarch, was the cousin of EIC Chairman David Scott as well as being an old friend of Bombay Governor Jonathan Duncan.[175] Her letters brought a bounty of introductions for her sons who were openly encouraged by these powerful EIC executives to join Sindia's army. As for networking, we must also remember the extent of fraternal connections and alliances. George Carnegie's stepbrother was Lieutenant Colonel Nicholas Carnegie of the Bengal Artillery, commanding artillery at Fatehgarh; a very helpful reference given Nicholas' connection to his old classmate Lieutenant Colonel Ochterlony since their cadetship in 1777.[176] These connections were of crucial importance to British intelligence during the 1803 Anglo-Maratha Campaigns. But the linkages of defecting personnel to the upper echelons of EIC power were also worrisome. Arthur Wellesley and John Collins both breathed a sigh of relief to find that Grant and 'Stuart', or more appropriately Daniel Stewart, were among the mercenaries who did not wait to see how events turned out but rather left service late on 12 September 1803 after having encountered difficulties in extricating themselves from a tense situation referred to in some circles as the 'mutiny' of the 1st Brigade. They were late in their escape but the main thing was they got out eleven days before the bloodbath at Assaye.[177] Grant would have made a particularly interesting Maratha hostage had he stayed and refused to fight, and no doubt the thought of disgrace to his family was a major consideration in his decision to resign Sindia's service.

The road to Assaye

We know from numerous EIC reports of battles observed prior to 1803 that Sindia's mercenary officer corps, both European and South Asian, led at the head of the formation. Their leadership style had figured largely in accounts of the battles against Holkar in 1801–2, particularly the Battle of Hadespar. Any officer leading one of Sindia's sepoy units at Assaye was bound to have faced an extraordinarily great chance of being killed or wounded, if not by British artillery or musketry then by the infantry's bayonets or the cavalry's sabres. However, British after-action reports stated only one of the Marathas' European officers was found dead on the field.[178] Some imperial historians insist that the European officers made good their escape on horseback following the fight. But given that the number of Maratha casualties was at least 1,200 it seemed odd that there were not more European mercenaries killed or wounded. South Asian leaders of rank who subsequently died of their wounds, such as

Jadun Singh, were reported. Where were Sindia's European killed and wounded? Or was the assumption of European casualties a faulty deduction based on Wellingtonian claims that Arthur Wellesley fought the 'French-led' Maratha infantry at Assaye? How many European mercenaries were at Assaye?

In 1986 I wrote a paper about 'Wellington and the Marathas' indicating that the Marathas were not a technologically backward force but an artillery-savvy lot who made the Battle of Assaye a 'near-run thing' for the future Duke of Wellington.[179] Like many before me, I made reference to Pohlman's performance on the field as exemplified by his ability to quickly order Sindia's sepoys to 'change front'. I was delinquent in my research and accepted Pohlman's presence despite the fact that none of the eyewitness reports confirmed his whereabouts. William Thorn, fighting alongside Lake far to the north, stated in his authoritative work that the regular battalions were 'under the command of two French officers, named Pohlman and Dupont'. I did not accept Thorn's French assertion from the standpoint of either ethnicity or military affiliation.[180] Instead, I opted to follow John Pemble's lead in making assumptions about the nationality of the European officers based on their names. Despite the claim by most that Pohlman was a Hanoverian, my research indicated an archaic reference that he was a Hessian. As it turned out, all of that was speculation and I was mistaken. As the years passed I coordinated additional archival research and began to concentrate on the letters of British intelligence operatives and the pension adjustment claims of retired mercenaries. After twelve years of intermittent research it was quite apparent that my Wellington article was incorrect with reference to Pohlman and the story had to be set straight. Pohlman was officially classified as a British subject and he was nowhere near the battlefield of Assaye on 23 September 1803.

The deposition of 'Mr Stuart' was important in linking several sources that shed light on the disintegration of command and control in Sindia's 1st Brigade just eleven days before the Battle of Assaye. It stated:

When Col. Collins left Camp, Mr. Stuart asked Col. Pohlman to give him his discharge. Colonel Pohlman said that he could have his Discharge from General Perron only. Mr. Stuart continued to press Col. Pohlman for his discharge, telling him that Hostilities might commence before he could hear from General Perron, and that if Col. Pohlman persisted in refusing him his discharge, that he was determined to go away with the best manner he could – at length on the 12th of September, Col. Pohlman was induced to give him his discharge and Mr. Stuart immediately quitted Camp, and went to Burhaunpore. Mr. Grant and Mr. MacCullough obtained their discharge at the same time, and went back with Mr. Stuart to Burhaunpore.[181]

That the departure of the men was part of a larger effort coordinated with Perron is without doubt.

When Mr. Stuart was near Adjuntee, he saw a letter from General Perron to Colonel Pohlman in which General Perron mentioned that in consequence of the apprehensions which were entertained of a war breaking out his English officers had taken their discharge. That he therefore wished Colonel Pohlman would send him three French officers namely Honore, Mercier and Perrin. Mr. Stuart says that Honore and Mercier positively refused to go but he thinks Perrin would go willingly. It was in consequence of this letter that Col. Pohlman was induced at the urgent request of Mr. Stuart and the other English officers to give them their discharge.[182]

Pohlman was obviously present in camp on 12 September having refused until that point to act independently and issue a discharge under the resignation or bail-out provision of enlistment. Pohlman waited until he received word from Perron. Note that Pohlman did not tell his subordinates they could not resign, rather the issue was that the authorization had to come from Perron as C-in-C in accordance with the chain of command and what we also know now to be the efforts of Perron to coordinate a larger plan. The testimony of 'Mr Stuart', as taken down by Lieutenant-Colonel Close, went on to say that:

Colonel Pohlman had insisted on having his discharge, and used every endeavour to procure it from Sindiah, but to no purpose. That though Colonel Pohlman was induced to march, Mr. Stuart knows that it was not Colonel Pohlman's intention to fight, and that he had sent his resignation to General Perron. Mr. Stuart also heard a report that Colonel Pohlman was not in the action.[183]

There was considerable division among the other European officers in the 1st Brigade. The testimony had elaborated: 'Gautier, Honore and Mercier were very averse to the war, and would have got their discharge if they could. Dupont too was very anxious to go away and was in fact for some time a prisoner in his own camp.'[184] The French officer who opposed them was the man who had the most to gain from the resignations. Major Dorson was second in command and he was instrumental in ensuring Pohlman would not be on the field at Assaye. 'Dorson was very inemical to the English, and he was too apprehensive of Pohlman's leaving camp from his unwillingness to serve against the English, that he surrounded his tent at night with Marathas and other people to prevent his getting away.'[185]

The whereabouts of Pohlman during the Battle of Assaye are covered in a thirty-four-page file entitled 'Rejection of a Claim preferred by Lieut. Colonel Pohlman formerly in the Service of Dowlut Row Scindia, to be allowed Pay for a period of 5 Months antecedent to his receipt of

Allowances from the British Government'.[186] The report includes letters of testimony from General Lake, official minutes from the Bengal government's Secret Committee and references to decisions rendered by leading members of Richard Wellesley's administration including John Malcolm. General Lake reported that the Marathas seized Pohlman as a prisoner after he publicly declared to Sindia 'that he would not act against the British Government and demanded permission to retire from the Maharajah's Service'.[187] Pohlman was sent off to prison in Ujjain and his possessions plundered. Despite speculation about Pohlman being Hanoverian or Hessian, his petitions and Bengal's reciprocal rulings all uphold his nationality as being that of a 'British subject', which explained his entitlement to retire from Maratha service. It would seem in signing discharges in Perron's absence he committed an act that the Marathas construed as criminal. Pohlman knew men like Stewart could not obtain Perron's release owing to the timing of events. So he acted on Perron's behalf and signed the discharges thereby facilitating the departure of the 1st Brigade's officers. The Marathas finally released Pohlman from prison in February 1804, a full two months after the formal terms of peace with the British were concluded. However, for a period of five months after that, Pohlman was not seen or heard from. But having recovered himself as a consummate mercenary and broker/dealer in the military labour market, he came into EIC territory in August 1804 bringing with him three battalions of former Maratha sepoys and 200 irregular horse.[188]

The five-month gap, between his release from prison and his formal entry into the EIC's military service at the rank of lieutenant-colonel, was the subject of Pohlman's later appeal. He felt he should have compensation for that period, but the Company ruled against it. The legal wrangling over their decision is quite revealing about the classification of defectors and the issue of 'special allowances' alluded to earlier. An argument was put forward that Pohlman deserved to be ranked within the '1st Class' among those who had 'come in'.[189] This implied that he deserved being rewarded for greater sacrifice and it was an acknowledgement that his life was placed in danger by his actions. The Secret Committee debated the issue of whether '1st Class' personnel should receive what Lake thought they were worth or what the Bengal government calculated they should get on a standard scale for pension/rank equivalency.

The argument changed to focus on pay for the period of the war during which Pohlman was incarcerated. He could have his lieutenant-colonel's pay for his EIC service starting in 1804, but he was not entitled to pay for that unaccountable five-month period before 'coming in'. For the time in jail, Bengal wanted him to receive retirement pay of Rs. 300 per month rather than the field pay and allowances Lake figured was worth Rs. 840 per month. To argue their case, Committee members went to the trouble of preparing boardroom charts complete with projected annual savings and costs to be applied if this case set a precedent for other mercenary claims.[190] Pohlman's pay for the time in jail was knocked down to Rs. 300 per month. To his credit, Pohlman served the EIC well. He won fame during the 1804–5 Holkar Campaign bringing his battalions into service under Lake in the dark days following Monson's retreat.[191]

Conclusion

Some adventurous Europeans went to India seeking their fortunes at the close of the eighteenth century. When they got there many found it was not as easy as just 'shaking the pagoda tree' and coins were not going to fall into their lap. As if to make the economic misery of these would-be entrepreneurs worse, the EIC worked in an exclusionary monopolistic fashion. Unless you were already a part of the patronage system it was difficult to break into the lucrative trade networks that the Honourable Company dominated. Changing course and opting to become an administrator was not a viable alternative either. In those days you could not join the EIC's civil service in India, you had to be 'sent out' with an appointment. But there were always options and the best alternative for earning riches, as recommended by the likes of Bombay Governor Duncan, was the so-called Maratha service of Daulat Rao Sindia. The traditional Maratha practice of hiring various mercenaries led to the channelling of some rather interesting foreign nationals into positions of power within Sindia's military forces. However, beyond money and power there was little reason for them to maintain a lasting commitment to their positions or their employer. And when push ultimately came to shove, between Richard Wellesley and Daulat Rao Sindia, those foreign nationals posed a security risk for Sindia.

The most serious Maratha military threat to Wellesley's plans was that posed by Sindia's 'regular corps'. However, it was ripe for destabilization and subversion. It was going to be much easier for the British to win this struggle if they exploited a series of Maratha weak spots that included the continued Maratha willingness to employ mercenaries at all levels of operations from sepoy to C-in-C. The incorporation of men with divided political, national and racial loyalties predisposed the Maratha brigades to disruption of their command and control. It is not historically accurate to simply say *The Marathas' main military weakness was their use of mercenaries*. Maratha military culture was broad enough to have offered command opportunities to European mercenaries since the reign of Shivaji and it would be a historic mistake to see the European mercenaries as a modern and ruinous experiment introduced by Mahadji Sindia. They were part of a historic continuum in Maratha military culture. Differences in religion, caste and ethnicity had all been suppressed by the pragmatic Maratha approach exemplified by Shivaji. The larger problem for the Marathas was not the use of mercenaries *per se* but the collective absence of Maratha military institutions for indoctrination aimed at instilling a military culture of allegiance to duty. These institutions need not have been bricks and mortar or military schools. They could have been much more basic institutional forms of indoctrination such as a clan-based Maratha regimental system building allegiance to the clan as an integral part of allegiance to the unit.[192] Maratha sepoys followed their officers as part of the chain of command but beyond that they followed as part of a special bonding process and their relationship in the South Asian military labour market. What was missing in this military culture was the next intermediate level of bonding, i.e. that of the military middle managers – the officers and NCOs – to the clan itself.

The South Asian military environment that shaped Maratha military culture was geared much more towards transitory military entrepreneurs who gained money, rank and power by successfully exploiting chances to sell their skills on the military labour market that existed within the South Asian military economy. While the sepoys of Sindia's 'regular corps' were equal on a man-for-man basis to any comparative British army unit, its heterogeneous officer corps was lacking a cohesive employment ethic capable of passing as dedication to duty. The days of sworn officer allegiance to the Maharajah had dwindled with the death of Mahadji Sindia. Few veterans respected Daulat Rao the way they had Mahadji. He was an empathetic figure at times but there was no binding allegiance to serve and protect him. Men such as Carnegie wrote letters expressing remorse at their inability to help their 'prince' Daulat Rao once he committed to war with the British.

During 1801–2 Governor-General Richard Wellesley believed he could subvert control of Sindia's battalions if Perron and other 'French officers' were removed. The War of the Second Coalition gave Wellesley a pretext for attempting to extricate them by coercive means. From the standpoint of legality, this was a precarious ploy. John Collins and others saw obvious parallels with the slippery slope that had resulted in the impeachment of Governor-General Warren Hastings on charges of 'High Crimes and Misdemeanours'. But with the arrival of news about the Peace of Amiens, Wellesley's opportunity of ousting Perron vanished. And so he was forced to consider other means of removing Perron in the hope that the 'regular corps' would come under more easily manipulated British mercenary leadership. What apparently happened next was that Richard Wellesley reconsidered Perron's position in view of intelligence reports and recommendations. There was another option. It would be more efficient and effective to formally adopt Perron as an intelligence source and then assist him in retiring. He could be managed in such a way as to be of greater service to Wellesley's policy objectives. The mercenary's goals remained the same, retire rich and live long enough to enjoy it. Perron had not reinvented himself. On the contrary, it was Richard Wellesley that had changed his way of thinking in allowing himself to see Perron not as the problem but as part of the solution.

Perron could be brought in and EIC pensions used to buy out the other officers. But the further the plan developed the more evident it became that all of Sindia's military assets were for sale if this situation was handled carefully. Although it should be said the stratagem exceeded the wildest expectations of the British and they seemed genuinely surprised that the Marathas would honour contractual relationships that worked against their own interests during war. Elphinstone was amazed to see that Sindia honoured the resignations of mercenary officers, as they were about to face Arthur Wellesley in the Deccan.

I cannot think how I happened not to mention the great pleasure I had in hearing that all the British subjects had come off in so honourable a manner from Sindia's service. I am particularly glad that Stuart is safe in person, and sound in character. The resignations of the British subjects, and their being accepted, showed principle and foresight, on both sides, which make each more respectable.[193]

By using opportunities provided by the defecting mercenary officers, the EIC gained increased access for the subversion of entire Maratha sepoy battalions. They could have their service contracts directly converted to EIC employment. The sepoys were often personally attached to their officers and NCOs with bonds of loyalty that had cultural, organizational and financial explanation. Many officers were more than mere administrators or combat leaders. In several cases these officers had trained their men and spent years with them in the field under canvas developing together with them into a brigade. Unit identity was not just limited to military identity in terms of European military culture; it was also a source of a special warrior identity within South Asian society. And if you were good as a unit, it meant you had additional leverage in the South Asian military economy. If the prince, *sardar* or warlord you fought for as a sepoy was nowhere to be found and owed you months worth of back pay, then chances were you had greater loyalty to your commanding officer – that same officer whose military knowledge and stealth you depended on to keep you alive in combat. An absentee patron with cold detachment, or in a rash moment of trying to seek revenge, might order a battalion to fight to the death. But a serving field officer, who realized the value of his troops, was less likely to squander his men foolishly.

For the Maratha sepoys it was an era of loyalty to one's officer leaders rather than an age of political indoctrination based on higher callings to a country; that was yet to come. The concept of being 'in service' seemed to fall back upon the more traditional kinship patterns common to the military labour markets found in South Asian history. Consequently, after officers learned of their opportunities to 'go over' with their men, many sepoys saw their survival and income as tied to their officers; so they went over with them in an act of trust tempered with opportunism. A mutual symbiosis had developed as the Marathas' mercenary officers and their sepoys lived together in the field during years of clan conflicts, such as the Sindia–Holkar war. They had depended upon each other for survival and in a sense they had each other's best interests in mind. Men like Colonel Sheppherd unknowingly assumed traditional roles in the military labour market that had characterized South Asia for centuries. He was a British mercenary in Maratha service and many of his men were Rohillas, yet in brokering employment for his brigade that had served Ambaji Inglia, Sheppherd became another in a historic line of broker/dealers within the South Asian military economy.[194]

The Wellesley administration proved extremely competitive and deficit spending was an essential military tool in this war. As demonstrated in the next chapter, the espionage and extent of the *buy-outs* from sepoy to C-in-C serve as the greatest explanation of how Sindia's powerful war machine could be defeated in less than 150 days of campaigning. Command and control were shattered on an army-wide basis and experienced combat leadership was denied to the 'regular corps' as defections reached a wholesale level. That the British recruited entire battalions out of Maratha service, such as those at Agra, should indicate that it is entirely inappropriate to talk in terms of the 'destruction' of Sindia's army. The military assets that the Sindia clan had gathered from the South Asian military environment were being recycled and absorbed by the EIC.

As for the Euro-Asian officers referred to as 'half-caste', a peculiarly selective form of patriotism was implied in the appeals for 'British subjects' to accept the terms of the Proclamation. It was used to court men like James Skinner, who owed their existence to the racially oriented gender imbalance of British colonies in that era. There had been a distinct lack of white women and, out on the supposed margins of civilization, these Europeans bedded Indian women who apparently bore them marvellous hybrid off-spring who proved to be such tenacious and skilled warriors – at least that was the later nineteenth-century romantic rationalization. But with the 1803 Proclamation these sons of interracial relationships were being forced to decide between two societies, perhaps not unlike the Indian diaspora in Western society today. Several decided to honour their fathers, a decision which could be underpinned with the teachings of a male-dominated warrior society. After all, most owed their occupational identity to their fathers. Like their fathers, they were warriors; their uniforms provided a warrior-caste identification mark recognized in both British and Indian society. Regardless of religion, they became Britain's politically sanctioned *kshatriyas*. Once the terms of Richard Wellesley's forthcoming Proclamation were known, Skinner resigned Sindia's service and eventually defected to the EIC. It was not an act by a member of an indigenous elite to gain leverage in what was becoming the politically dominant society; that is the spin of Cambridge historians who speak of a 'collaborationist mentality'. The decision was much more basic. These were soldiers – military service was their occupation as well as their lifestyle. The British were now the biggest and most dependable military employers in the land.

6 The anatomy of victory

Introduction

The traditional picture of British victory in 1803 is that of a well-oiled and finely tuned military force marching with automaton precision over the bodies of their brave, but nonetheless dead, opponents. The popular iconography is that of the proverbial 'thin red line' of British infantry, ceaselessly pressing forward through the acrid haze of gun smoke to find death or glory. The following passage concerning Arthur Wellesley's victory at Assaye, from Jac Weller's book *Wellington in India*, indicates the extent to which nineteenth-century drums-and-trumpets-style military history survived intact.

> Every soldier of the 78th was a giant by Marhatta standards. Their kilts were swinging in unison; their weapons and buckles were polished and their belts freshly whitened. Nobody who saw this magnificent regiment deliver its first assault at Assaye ever forgot it. At sixty yards from the enemy guns which continued to fire, they halted as if on parade, presented their muskets, fired, recovered and reloaded. Then they went on again with their bayonets gleaming.[1]

This book, however, has called for a revision of that imagery and the cultural legacy of that iconography. I do not intend to take anything from the bravery of the men who stood before the Maratha guns as their ranks were thinned by merciless clusters of canister shot. Arthur Wellesley's performance under fire ranks as one of the finest displays of personal bravery in combat leadership to be found in British history. But I do question the explanations we have received for victories in the 1803 Anglo-Maratha Campaigns and what they have come to mean.

This chapter argues that the conventional understanding of how Britain attained victory in 1803 is incorrect. The documentation and evidence clearly demonstrate this was not a 'walk over'. Furthermore, victory had more to do with espionage and growing British domination of the South Asian military economy than with superior technology, discipline or drill. Western explanations for victory were constructed for political as well as cultural reasons and they have been used to underpin

284

a cultural domination model for the 'rise of the West'. Consciously or not, Westerners have distorted South Asian military history and in getting the analysis wrong we have prolonged the process of cultural imperialism.

Explaining the British victories of 1803

The evidence laid out in this volume and the subsequent reasoning leads to one inescapable question: if the Marathas were old hands at infantry warfare and their military was potentially powerful enough to thwart British expansion in 1803, how or why were they defeated in two campaigns that lasted less than five months? Over the years a number of explanations have been offered. At one time it was popular among Maratha historians to attribute defeat to deviation from the historic model set by Shivaji. However, as we saw in chapter 1, the understanding of Shivaji's military has been poor at best. The basis for his kingdom was a chain of strong forts that relied heavily on conventional infantry for their ultimate defence. Shivaji's mounted arm, composed of highly skilled horsemen of varying descriptions, was used for mobile operations, reconnaissance and the rapid projection of force to exploit enemy weaknesses. But it was his regular infantry that gave Shivaji the ability to seize territory and hold it, forcing the Mughals to formally acknowledge Maratha military power. The geopolitical reality of having to recognize Shivaji at the negotiating table was the ultimate admission of his conventional military power. Any explanation for Maratha defeat in 1803 as the result of deviation from Shivaji's 'mounted guerrilla strategy' is deficient.

The vast majority of popular explanations for Maratha defeat can be summarized in four points offered by Stewart Gordon.[2] Interestingly enough the four points tend to follow a pattern, not of Maratha weaknesses or failures, but rather of alleged British strengths. This one-sided explanation process suggests that the historic reasoning and evidence have been stacked towards a preferred or predetermined outcome that implies the British were building from strength to strength.[3] It therefore should come as no surprise that the four areas offered as 'explanations for victory' tend to support cultural domination. The four common explanations are:
- The momentum of British victory in the Napoleonic Wars
- Superior British training, discipline and drill
- Superior British artillery
- Superior British credit

To assess the validity of these four points, let us consider them on an individual basis.

The momentum of British victory

This is perhaps the weakest explanation offered from the standpoint of timing and the historic record. The Peace of Amiens was signed between Great Britain, France, Spain and Holland, on 27 March 1802. Therefore, proponents of this theory would have to account for an armed truce or period of military inactivity that was in effect when Richard Wellesley launched his Maratha War of 1803. Does peace not constitute a break or interruption in momentum, if indeed momentum existed at that time? If one broadens the timeframe as well as the geographic scope of the premise to that of the global battlefield there are still problems with this theory. It would be very difficult to prove, in a colonial context or for that matter a European setting, that Britain established self-sustaining military momentum.

Let us consider the possibility of establishing a case for the momentum of British victory in South Asia. Arguably the greatest military threat the British faced before the Marathas in 1803 was that from Mysore. Richard Wellesley had certainly treated it as a priority when he arrived in India. Tipu Sultan fell in 1799 during the Fourth Anglo-Mysore War, the British strategic position having benefited from earlier victories. But the slow-grinding process of the four Anglo-Mysore Wars started with the struggle against Tipu's father Haidar Ali in 1766 and did not end until the final siege of Srirangapatnam some thirty-three years later. Granted the four Anglo-Mysore Wars only consumed a total of thirteen years out of that thirty-three-year span, but the 'on again, off again' nature of four consecutive wars suggests a glacial pace rather than the dynamic momentum of rapidly accumulated British victories. Certainly the defeat of Colonel Baillie's detachment in the Second Anglo-Mysore War was a major British setback to both military operations and morale. As for the Third Anglo-Mysore War 1789–92, chapter 2 of this work noted Major Dirom's description of the horrors encountered by the British during their monsoon retreat – hardly a shining example of victory's momentum in what corresponds to the opening phase of French Revolutionary and Napoleonic warfare 1792–1815. Or do you discount the setbacks in the Second and Third Anglo-Mysore Wars as temporary tactical battlefield reversals irrelevant to the manner in which you count wars as victories?

If the theory of self-sustaining military momentum were valid as an imperial maxim, why did it not hold true in other British wars during the Napoleonic period? Colonial combat theatres in North America during the War of 1812 suggest some rather significant shortcomings in the theory. The invasion and burning of York (Toronto) in 1813 and the collapse of the British-sponsored 'First Nation's Confederacy' in Canada were far from victories. While the British had managed to march on Washington and torch the president's house (1814), that yielded a coat of white paint rather than a decisive military victory as the British withdrew. The crushing defeat and death of Arthur Wellesley's brother-in-law, Major-General Sir Edward Pakenham, at New Orleans in January 1815 also suggests that the generalization is totally inappropriate in a greater colonial context. Or, shall we remove North America from colonial juxtaposition because it featured White vs. White conflict to a much greater degree?

Superior training, discipline and drill

Ever since the incursion of Alexander the Great, there have been ways and means for South Asians to learn first hand about Western military techniques.[4] Once the techniques became known in South Asia, those that were deemed worthy of adoption or replication invariably found some form of assimilation. However, Western military historians have failed to adequately acknowledge the pre-existing South Asian infantry tradition and its ability to learn and grow. The Vedic period abounds with stories exalting the virtue of missile weapons – the role of the bowman being at times synonymous with the power of a god as exemplified by Indra's identity as a bowman. But let us not lose sight of the theoretical as well as precedent-setting value of examples found in classical Indology. One does not deploy bodies of men (numbering in the tens of thousands) to a predetermined point and then release missiles or projectiles on a specific signal without two forms of discipline and drill: that of the individual as weapons' operator and that of the group moving in unison to a designated location and responding to the command of 'release' as ordered. During the early years of the so-called 'Gunpowder Age' the Islamic trade connection to India served to transfer competitive models of military organization and training before the arrival of Europeans. As for European smallarms drill, one might find an individually more effective position in which to hold a muzzleloader before priming the pan and loading a ball, but there was no magic or racially exclusive 'Western way'. The mechanical requirements and ergonomic realities of weapons' systems dictated a certain degree of commonality in the behaviour of the weapons' operator. Those physical realities, when combined with natural patterns in human movement and motor ability, also suggest that parallel independent behavioural developments were indeed possible and that this entire 'who was first, which was best?' debate is a culturally motivated military canard. Western military historians have not fully explored cross-cultural models that would undermine their pet theories of military superiority as related to discipline, drill and military technology. Where is their analysis of crossbow drill training as applied to fourteenth-century European companies of crossbowmen? Why have they escaped comparison to the crossbow drill and discipline exercised by Han Dynasty Chinese crossbowmen in the first century BCE? Are you comfortable with 'who was first' and 'what that means about the inevitability of political ascendance' or would you prefer to think that the physical demands of a crossbow ensure a certain amount of commonality with regard to how the weapon is carried, loaded and its projectile released? For that matter why not consider the degree to which the Han Chinese attained a 'revolution in military affairs' by mass producing interchangeable parts for their crossbows in the form of cast-bronze trigger assemblies? Asian military achievement in the pre-European period has not been judged fairly by Western military historians.

Once the Portuguese established direct sea-borne contact with South Asia, a level of operational commonality was virtually ensured as South Asian and European mercenaries cross-fertilized the military knowledge base. The 'superior training, discipline and drill' explanation does not consider the degree to which the Portuguese played a direct bridging role between South Asia and Europe that pre-dates northern European colonial expansion in the eighteenth century. As cited earlier, David Harding's exhaustive research of EIC records showed that by the mid-seventeenth century Topasses were serving with the British as well as the Portuguese and learning their respective drill procedures. The Topasses in turn were regular recruits in the military forces of various South Asian powers including the Marathas. Economic considerations in the successful and profitable prosecution of South Asian warfare drove the search for cost-effective competitive advantage. And in that regard the market dynamics of the South Asian military economy worked towards commonality and the sharing of military resources such as weapons, personnel and the military intangibles we might label as techniques and tactics. In simplistic terms, if your opponent had something that seemed to offer a military advantage, one of the easiest ways to remain competitive – or deny him that advantage – was to incorporate that same knowledge or item. This worked in both directions across cultural lines, as we can see in Arthur Wellesley's employment of *pindaries* numbering in the thousands.

In 1803 both Sindia's 'regular corps' sepoys and those of the EIC were interchangeable with regard to discipline and drill, as demonstrated by their immediate ability to serve either force. Sindia's sepoys had, in effect, the same training as the EIC's sepoys; many had been in service to the

EIC before the demobilization of 1802. This reflected the ebb and flow of human resources within the South Asian military economy. The men who made up the bulk of sepoys in the Bengal and Bombay Presidencies in 1803 were from pre-existing and overlapping military labour markets commonly shared by the Marathas and the British. As Charles Stuart testified in his diary, Sindia and Perron were happy to employ the EIC's sepoys when they were discharged. They could easily be slotted in to fill casualty-driven vacancies in Maratha battalions. That the commonality of Maratha sepoy drill was sufficiently equal to that of the British is reflected in General Lake's retention and immediate deployment of Maratha sepoys without sending them to rear areas for retraining. The 1803 Anglo-Maratha Campaigns demonstrate a commonality extending right down to the sharing of specific individual personnel. One could also make a case that the Maratha sepoys of Sindia's older battalions had much more combat experience than some of the younger EIC battalions owing to the constant use of the 'regular corps' in Sindia's battles against Jeswant Rao Holkar and the countless refractory *zamindars*.

In addition to Sindia hiring discharged EIC sepoys, we also must consider the commonality of training, drill and discipline that resulted from the Marathas employing British mercenaries who had been officers in His Majesty's army as well as the military forces of the EIC. In the eyes of British military men of 1803, discipline and drill were reckoned to be the greatest guarantors of unfaltering tactical precision on the battlefields of that era. Comments that individual sepoy units needed discipline and drill, expressed in the language of the day, were a subjective analytical judgement on military preparedness and/or disposition. Britons such as Hugh Sutherland, who rose to be the second most powerful mercenary leader in Sindia's chain of command in 1803, had an advantageous position when it came to improving training and drill. As a brigade commander for Daulat Rao Sindia he had the power to modify and adapt various European and South Asian models of drill to incorporate local knowledge and that special tactical combat experience known to many soldiers as fieldcraft. Sutherland could take Maratha sepoys and put them through the same training process as that experienced by British recruits. But he was not locked into a specific or proscribed doctrine of drill like his EIC counterparts, who had to conform to Bengal, Bombay or Madras regulations. Since the time of Mahadji Sindia the commanders of Maratha sepoy battalions had been given a greater degree of leeway in setting their own battalion drill standards. By integrating his expertise and combat experience in India, a man like Sutherland could fine-tune the knowledge so that regardless of whether a sepoy had started his military career with the EIC or with Daulat Rao Sindia, the result was an infantry soldier equal to any in India.

Major Ferdinand L. Smith – who was rated by Colonel Collins as one of Sindia's finest combat leaders – observed that one Maratha Telinga battalion, composed of sepoys from Awadh, was drilled in 'the old English exercise of 1780'.[5] These slight variations in drill were not of major importance as evidenced by the fact that each of the EIC's armies had its own drill, separate from that of His Majesty's troops. The drill exercises and theories of Colonel Dundas* proved very popular and by 1792 a revised version of his 'Principles of Military Movements' became the British army's regulation drill book.[6] The 'Dundas reforms' were greatly influenced by the belief that the British army's service in America during the Seven Years' War and the American War of Independence had led to an overall decline in British heavy infantry. Much to the chagrin of Britons who advocated greater use of the rifle, Dundas criticized what he saw as the light infantry fad that had proliferated in the American wars. Modern advocates of a 'Western victory through superior drill theory' often forget that flintlock smooth-bore muskets and their theory of deployment were 'old news' by 1800. As a smallarms system the flintlock had been in general circulation for a century. The massing of men in formation to achieve a ballistic effect, analogous to a gigantic shotgun blast, was greatly expedited by military drill. But which system one used to achieve the effect was of secondary importance to those on the receiving end. True, there were degrees of effectiveness and some variations may have offered particular advantages for particular scenarios. As David Harding has pointed out, knowing which system of 'manoeuvres' and 'firings' to employ at a given time came down to the individual judgement exercised by officers. However, in comparing British and Maratha infantry standards in 1803 there was insufficient overall difference in my opinion to make the claim that superior drill was a serious reason for British victory in 1803. The military skills of sepoys in the Marathas' 'regular corps' were essentially identical to those of corresponding EIC battalions. That the Maratha experience produced sepoys of the first order is indicated by their interchangeability and battlefield recruitment at places like Agra.

International military trends move through armies – as they always have – but they do not always move from west to east. During the first year of America's Civil War (1861–5) both the Union and Confederate armies possessed elite Zouave rifle units. These American Zouaves

---

* Later, General Sir David Dundas.

modelled themselves on the original Zouaves of the French colonial armies. The latter were Algerian light infantry troops famous for their drill and characteristic, gaudy uniform featuring bright colours, baggy trousers, gaiters, short and open jacket, and a turban or fez. The original Zouaves were also famous for their ability to fire and reload the musket from the prone position.

US militia units followed them 'down to the detail of shaving their heads like the North Africans'.[7] Were the American Zouaves imitations or copies, as if the American commanders and troops were simple children with naive beliefs that red pants make a rifle shoot more accurately? Of course not, and what foolhardy soul would have had the audacity to suggest to historically contemporary American military leaders that such a ridiculous observation was valid because the latest trend originated outside their own cultural environment?

If we observe military history across culturally specific divides we can see that trends or patterns were often deemed worthwhile because they were believed to have military merit. In the case of the Zouaves it was *esprit de corps* and *élan*, as well as a series of techniques that included the prone firing and reloading of muzzle-loading weapons, which represented a potential battlefield advantage. For Sindia, the battalions of the 'regular corps' in 1803 represented a means of competing with the British on the basis of organizational and tactical equality – symmetrical warfare. By relying on force symmetry the Marathas minimized any advantage in organizational structure that the British might have exploited to their benefit. The training of the Maratha regulars allowed them to compete against their opposite numbers and that is reflected in the surprise shown at Assaye when the Marathas 'changed front' not once but twice in the face of Arthur Wellesley's deployment. They also managed to do this at Laswari and the depth as well as quality of their training was indicated by the fact that the commands were executed despite the absence of the units' usual officers. Since the Proclamation process had already stripped the battalions of their usual officers, such as Pohlman and Smith, this suggests the Maratha sepoys were not dependent on their European officers for 'discipline'. What the Maratha sepoys needed their officers for, on a vital performance basis, was offensive combat leadership under fire. We must not confuse absence of command and control with lack of discipline and drill.

Superior British artillery

The Marathas could readily attain superior firepower by concentrating on production of advanced artillery systems in an era when comparatively modern weapons technology still reverted to knowledge of age-old alloys and basic metalworking techniques. In terms of their artillery's modernity, the Marathas benefited from the fact that eighteenth-century European artillery science was relatively primitive and static. Muzzle-loading smooth-bore (MLSB) artillery was based on weapons principles that had remained virtually unchanged for almost 300 years and were known across Europe and Asia. The Marathas faced no technological hurdle of competition, in the form of high technology or new technology, to compete with or surpass the quality of British artillery. Several basic factors in MLSB artillery remained virtually unchanged until the decline of that weapon in the latter half of the nineteenth century.[8] These 'factors' might be said to have included:

- the metallurgy of construction materials
- the fabrication techniques
- the physical laws governing mobility
- the chemical nature of propellants
- the design and configuration of projectiles
- the projectiles' kinetic energy transfer potential and its ballistic ramifications

Claims of Western technical superiority in a major weapons production sector that suffered from comparative stagnation, for three centuries, remain vehement but hollow.

The capture of small casting centres with one or two furnaces and two to four casting pits, during the period from 1800 to 1803, deceived some foreign observers into thinking that South Asian production capacity was low.[9] They did not consider that the variety and number of casting centres was far greater than in European countries. The accumulative production level of South Asian foundries became apparent when the five months of campaigning against the Marathas in 1803 resulted in the capture of an estimated 1,000 pieces of artillery.[10] In writing to John Malcolm, Arthur Wellesley let it be known that a great number of the Maratha guns were indeed superior to anything the British or EIC armies possessed. 'Colonel Close will have informed you of our victory on the 23rd [Assaye]. Our loss has been very severe; but we have got more than 90 guns, 70 of which are the finest brass ordnance I have ever seen.'[11]

Arthur Wellesley had seen the superiority of Maratha artillery at Assaye and almost a year after the battle he wrote a letter indicating he had revised his thinking on the ability of Maratha infantry and their artillery in particular. His letter noted that 'all' the Marathas seemed to possess artillery so powerful that it rendered the holding of British positions in the field as untenable. A transformation had taken place. No longer Arthur Wellesley the innocent as authority, this was the voice of experience, having sustained more than 33 per cent casualties in the attack on Sindia's camp at Assaye.[12] He wrote the following paragraph to Colonel Murray almost a year after the near-run battle of 23 September 1803:

You must by all means avoid allowing him to attack you with his infantry. There is no position in which you could maintain your camp against such powerful artillery as all the Marhattas have. If you should not hear of their approach until they are close to you and coming to attack you, it would be better to secure your baggage in any manner, and move out to attack them. Do not allow them to attack you in your camp, on any account.[13]

Ironically the experience at Assaye and all this attention to artillery may have left Arthur Wellesley better equipped for one of the jobs he later held in England – Master-General of Ordnance.*

Following the Battle of Laswari, C-in-C Gerard Lake, who had been on the receiving end of artillery in several wars, sent a secret dispatch to Governor-General Richard Wellesley. In forty years of soldiering Lake had faced death many times on foreign battlefields. Not even the pounding he took at Yorktown could measure up to what he endured at the hands of Sindia's battalions. His encounter with massed batteries at Delhi and Laswari had made him think twice about his own mortality. South Asian gunners in Maratha service had reduced him to prayers for his future safety.

These battalions are most uncommonly well appointed, have a most numerous artillery, as well served as they possibly can be, the gunners standing to their guns until killed by the bayonet. All the sepoys of the enemy behaved exceedingly well . . . I never was in so severe a business in my life, and pray God I never may be in such a situation again. Their army is better appointed than ours; no expense is spared whatever, and they have three times the number of men to a gun we have. Their bullocks, of which they have many more than we have, are of a very superior sort. All their men's knapsacks and baggage are carried upon camels, by which means they can march double the distance . . . These fellows fought like devils, or rather heroes, and had we not made a disposition for attack in a style that we should have done against the most formidable army we could have been opposed to, I verily believe, from the position they had taken, we might have failed.[14]

In 1803, quick-firing breech-loading guns were less than three generations away.[15] The Marathas were ahead of the learning curve in putting greater tactical emphasis on artillery firing huge numbers of anti-personnel projectiles. Western historians are uncomfortable with the idea of South Asian leadership in military science but that is what Maratha artillery doctrine represented. Using a greater volume of artillery fire to

* Appointed on 26 December 1818.

dominate the battlefield and increase lethality through projectile selection meant being a step closer to the industrialization of killing. The thought of this trend as having been present in a society that Westerners would prefer to label as pre-industrial is difficult for Western military historians in particular to concede. But then cross-cultural confusion over the issue of Maratha artillery superiority existed in 1803 as well. Historically contemporary observers such as Thomas Munro were at a loss to understand the meaning of Maratha doctrine. Munro thought that the fact that more British troops were killed by artillery fire than any other cause at Assaye was a poor reflection on the Marathas and he viewed cavalry as Sindia's only chance for victory.[16] Like many Britons, Munro believed in the moral and military superiority of 'cold steel'. To his British military culture and its way of thinking, Assaye underscored the Maratha military experience as one that had failed to learn the fundamental lesson of European-style infantry warfare as closing with the bayonet. However, the Marathas were not crystal-ball gazing when they put their faith in more firepower. They were pragmatists who knew that if you filled the air with projectiles there was less chance that your opponent could advance against your position successfully. With artillery, it took fewer men to achieve a volume of fire equal to a battalion of regulars. But there was another factor that underscored the Marathas' need to use artillery in 1803. That was the Proclamation and the subsequent loss of infantry leadership through defections.

When the bulk of the Maratha infantry officers defected in 1803 they took with them their skill and experience in leading their own units under fire. As studies by twentieth-century authorities such as S. L. A. Marshall have demonstrated, infantry performance, and in particular forward movement under fire, is critically linked to the personal and direct example of infantry officer leadership.[17] Those few individuals who tried to fill the vacuum left by the defectors could issue orders to the infantry but they could not obtain peak motivational performance – particularly in an offensive scenario. Men like Perron, Sutherland and Pohlman had years of experience leading their men successfully in battle. Loyalty to them took the form of a willingness to go on the offensive and engage the enemy. With no immediate substitute for the years of accumulated trust and belief in those father-figure officers who had departed, offensive capability suffered.[18] One remedy for this loss was to try and use the tremendous firepower of the artillery as a substitute for diminished infantry capability. Battles like Delhi, Laswari and Assaye were characterized by an abundance of Maratha artillery firepower and limited infantry advances. The lack of offensive Maratha infantry movement in 1803 is further contrasted by specific examples at Delhi and Laswari where the

Marathas sought to use their artillery in ambushes to neutralize British advantage.

The Maratha Campaigns of 1803 made it apparent to British artillery-men that they were understaffed and 'out-gunned'. Artillery duty was hard, heavy work and the chances of injury from accidents were great. This dangerous duty was made more so in 1803 owing to the Maratha predilection for counter-battery fire as a means of ensuring domination of the battlefield. Once they had the location and range of an enemy gun they went out of their way to render it and the crew 'unserviceable'. The demands of the 1803 Maratha Campaigns had stripped the EIC's Presidencies of their artillerymen. As casualties among artillery personnel mounted, the shortfalls were felt from the junior to the senior ranks. Good gunners were difficult to find and to train. Varanasi* and Calcutta were so short of gunners that invalids had to be mobilized until a new artillery company could be raised.[19] In September 1803 the Governor-General authorized a bonus of Rs. 100 for each recruit that volunteered for the artillery.[20] Lieutenant-Colonel John Horsford – General Lake's senior artillery officer in the field – knew his craft well and kept himself out of unnecessary interservice infighting. So it was something of a noteworthy occasion when Horsford took time out from the war, while encamped at Agra, to sit down and address a memorial to Lake as C-in-C. He pleaded the case for an immediate increase in the number of field officers in the artillery corps.[21] Artillery officers, like their comrades the engineers, were rightly considered men of science who needed greater training time than infantry officers. Lake passed the document on to Richard Wellesley and, as if by way of omen at Agra, acquired the Maratha artillerymen of the 'Madaghur' as well as its thirty-two guns – thanks to the negotiations of 'Afzul the Tindal'.

The ultimate case against technological determination as an explana-tion for British victory should be strengthened with the observation that if wars had their outcome guaranteed by technological superiority, then the Marathas should have won the war in 1803 owing to their superior ar-tillery firepower. And it is somewhat discouraging to think that Western military culture remains so arrogant as to believe that the outcome of wars is determined by the technical quality of weapons. As we have seen in Arthur Wellesley's comments on the quality of Maratha guns, there was a glaring inconsistency between the historic record and any mod-ern pronouncement that the British had superior artillery. The theory of 'technological determination', or in this case the specific belief that the British had superior artillery, remains a popular explanation for British

* 25°+, 83°+/−. Aka Banaras, Benares.

victory. That an Asian enemy could defy the stereotyping of Western military culture remains a point of bitterness for the ethnocentric. Given this 'credibility impasse' it seems appropriate to argue the point of Maratha technological superiority from a different position.

Major George Constable* was one of Britain's leading experts on general-purpose field artillery in India during the 1803 campaigns. In previous years he had seen extensive service in Bengal and Awadh. Judging from his favourable interaction with South Asian gun crews, particularly during his years of Bengal field trials and gunnery instruction, he had developed a respect for the capabilities of South Asian gunners. George Constable saw the overwhelming performance of the Maratha guns in 1803 and that inspired him to take particularly detailed notes when he inventoried captured Maratha guns. He realized that those Maratha guns with laminated iron and brass barrels were far superior to anything Britain possessed. Often described as 'bar guns' owing to their interior hexagonal bore cylinder being constructed from longitudinally welded iron or steel bars, these specialized cannon barrels represented advanced weapons technology possessing a key advantage over British ordnance.

Generally speaking, an iron barrel was subject to much less internal damage or wear and tear during firing owing to the hardness of iron compared to brass. When loaded, shot resting in the bore was in direct contact with the barrel's interior. This metal-to-metal contact was the source of significant barrel erosion during firing. The relative lifespan of an iron barrel was also superior for firing 'fixed' or 'semi-fixed' rounds in which sheet metal bands fixed the projectile to a wooden sabot. The wooden sabot, attached to the projectile as a gas-seal, was manufactured to fit the bore of the gun fairly tightly to derive maximum advantage from the gas pressure generated by the rapidly burning black powder charge.[22] The sheet-metal bands often scored the interior of brass bores and repeated firing shortened the brass barrel's life and accuracy considerably. Many gunners preferred iron barrels for repeated firing. But iron barrels were a problem for two reasons. Cast-iron barrels could become brittle and hairline fractures could worsen and ultimately lead to barrel separations and explosions under repeated high-pressure loads. A greater problem with iron barrels, for daily field service, was their weight. Iron gun tubes outweighed brass barrels and weight was a logistical consideration of major proportion. Weight affected the number of guns you carried into the field, the size and type of gun carriage, the number of draught animals, the volume of fodder required to feed the transport animals and the number

_____

* Later, Lieutenant-Colonel George Constable.

of men required to manhandle the piece under battlefield conditions where survival might depend on movement.

The laminated barrel, consisting of an interior iron-bore cylinder with a brass sleeve cast over it, was a major design improvement. The inner liner was iron and that allowed extended barrel life, but the supporting or exterior barrel sleeve was brass, which reduced weight. The brass sleeve also proved sufficiently 'elastic' to minimize the potential threat posed by barrel brittleness, hairline casting flaws and crystalline-like fractures associated with manufacturing iron cannon barrels. George Constable knew that the Marathas' 'bar guns' were of a superior design and that Britain might do well to copy them. But he was also representative of those dynamic EIC officers who realized that technical discourse concerning design principles might not be sufficient to motivate action back in England. This was especially true of anything that might change procurement or production. He effectively argued that he should be seconded from the Bengal Artillery to England's Board of Ordnance to conduct trials on this construction technique and assess its viability for the purposes of British manufacturing. In other words, he convinced the EIC and the ordnance establishment that this system was so superior that the British should try to copy it to see if it might better meet the needs of their military forces. The 'East India Military Calendar' described it thus:

In 1806 Lieut.-Col. Constable obtained permission of the court of directors to put himself under the control of the Board of Ordnance, (the Marquis of Hastings then Master-Gen.) to forge and cast ordnance on the same principle as manufactured in Asia, and taken at Allygurh, Delhi, Agra and Laswarree. From having been a member of a committee for the survey of all the captured guns and stores, etc. he obtained a thorough knowledge of the enemy's brass ordnance with iron cylinders, their nature, properties, and formation, possessing advantage and superiority over the guns of Europe. In this object he succeeded, after experiencing innumerable difficulties, but which he was enabled to surmount through the assistance of the Board of Ordnance, and under the auspices of General Sir Thomas Blomfield, who furnished him with materials from Woolwich. The guns were cast by Lieut. Col. Constable in London, put to proof, and surveyed by a committee of artillery field-officers at Woolwich, and the thanks of the Honourable Board of Ordnance were conveyed to him for his trouble.[23]

The significance of the Maratha laminated guns was summarized in a 'Description of Brass Guns with iron cylinders as manufactured in Asia and cast in England in 1806 under the direction of Lieut-Col. George Constable of the Regiment of Bengal Artillery, by the order of the Honourable Board of Ordnance, the Marquis of Hastings then Master General.'

The gun metal is a composition of brass and iron: the cylinder smooth as glass, and formed of metal of a distinct quality: vent of solid iron, and gun made after the English model.

The advantages of the Asiatic ordnance are strength and lightness. In strength equal to iron ordnance. In lightness less than brass. In proof of the latter position a three pounder of the above consistency, proved at Woolwich, weighed 2 cwt.,* 3 qrs.[†] and 1 lb.: an English 3 pounder weighs 3 cwts. being a difference in metal of 27 lbs. The advantages in respect to weight are of the greatest importance: viz. facility of movement, light and easy exercise in the field and in garrison, and having a consequent saving both in men and horses. On shipboard a reduction of one-fourth to one-fifth in weight of metal must be of incalculable service.

It is notorious to officers who have seen much service that brass guns are, owing to their fusibility, often rendered in the field and in batteries totally unserviceable. From the running and melting of the guns, increase of windage, etc. the shot is fired without a certainty of direction or distance and hence it is evident that a 'brass train' of artillery can never be relied on.

Footnote: The necessity of resorting to brass guns with iron cylinders for all services by sea and land as recommended by Lieut.-Col. Constable appears deserving the serious attention of the British Government. J. P.[24]

However, there would be an entire series of problems if the British set about any large-scale replacement of their existing artillery pieces. Cost, production facilities, delivery dates and ammunition supply requirements were all very real worries if modernization were attempted on a grand scale. It would be an enormous logistical problem and who would pay the bill? Some members of the Board of Ordnance preferred not to think about such a massive undertaking. Perhaps it was best to consider alternative ways to maximize the effectiveness of artillery pieces already in service. The answer to increasing effectiveness – as opposed to a solution for problems of mobility, longevity and durability – was to try changing the nature of ammunition. There follows an extract of a letter from the EIC's Court of Directors, dated 29 April 1808, to Lord Minto, then Governor-General in Bengal. It refers to George Constable returning to India as an authority on the deadly new artillery shells championed by Henry Shrapnel (1761–1842) and destined to change the nomenclature of war.

In our letter of the 2nd of May, 1806, you were informed that we had shipped for your presidency a quantity of shells of a new and improved construction, for the use of field artillery; and as we conceive that the effect of these shells would be better appreciated were there an officer on the spot experimentally acquainted with their construction, to be occasionally advised with and consulted, and who would also be able to instruct others in their use, we recommend to your notice

* cwt = hundred weight.   † qr = quarter (28 lb).

Major Constable, belonging to the corps of artillery on your establishment, who, during his stay in England, has been employed under the orders of his Majesty's board of ordnance, and has therefore acquired such a knowledge of the principles of the construction of these shells, and of the improvements lately introduced into the Royal Arsenal at Woolwich, as to render him well qualified for the purpose above-mentioned. Major Constable has also evinced a laudable zeal in having successfully introduced into the Royal Arsenal the Asiatic mode of casting brass ordnance, and we direct that you communicate to him our approbation of his conduct in this respect.[25]

We do know that Constable returned to Bengal in the *Hugh Inglis*, having received orders to 'proceed to Allahabad where he practised and exercised the artillery corps in Colonel Shrapnel's experimental shot and shells the principles of which he had acquired of that judicious and scientific officer whilst in England'.[26] The British had seen the undeniable merits of South Asian laminated artillery systems but opted for Shrapnel's projectiles as a cost-effective means of upgrading the lethality of their existing artillery.

The Marathas had embraced artillery as the result of their military evolution and not from any perception of a military revolution. Muzzle-loading smooth-bore artillery was old but effective design technology. In order to face large enemy armies, lay siege to fortresses and hold isolated outposts, the Marathas saw a need to incorporate more cannon as force-multipliers. On a cultural basis they did not advocate artillery predominance for reasons of technological determination. They did not affix chauvinistic theories of scientific or cultural superiority to their military evolution. Maratha artillery proficiency evolved from the domestic arms races that characterized nucleated warlord environments and the Marathas' competitive position in the South Asian military economy. If we are looking for reasons as to how or why the British triumphed militarily over the Marathas in 1803, we have to look to something other than artillery superiority as an explanation. There is no credible evidence that the British possessed artillery parity, let alone superiority, in the 1803 Anglo-Maratha Campaigns.

Superior British credit

As in any other economic market place, there were some amazing opportunities within the South Asian military economy, which only presented themselves to players who could afford to indulge in economies of scale. To deliver a victory British commanders were authorized to buy out military forces or hire them on credit underwritten by the EIC. The promises of pay were enough to lure away entire battalions who had seen Maratha leaders stripped of their capital and credibility. For their part many of the

South Asian troops had little problem with a promissory agreement since they had often served with their pay in arrears.[27] The campaigns of 1803 had exposed the EIC to the system on a level not previously known. In the cross-cultural context of competing military cultures and economic systems, the British began to rely more extensively on the South Asian entrepreneurial approach to warfare. Fortresses, sepoy battalions, cavalry units, mercenary officers, intelligence – it all had a price if you knew how to manipulate the greater markets. However, the economies-of-scale approach as found in the EIC's hiring of thousands of *pindaries*, sepoy battalions and logistical military support systems in 1803 was not cost-effective on a short-term basis; it took time for the British to discern the most cost-beneficial control mechanisms.

In learning how to use their 'superior credit' the British also learned how to manipulate the rise and fall of the South Asian military economy. This was an economic environment of tremendous dimension and liquidity. The military economy generated as well as consumed vast revenues. To adequately explore the ramifications of historically contemporary defence economics one must think in terms of salaries, outfitting costs, logistical expenditures and budgetary allocations for armies numbering hundreds of thousands of people on a subcontinental basis. There were clever strategies to be exploited in this market as well. It was common to charge Subsidiary Treaty allies, like the *Peshwa* and the *Nizam*, inflated rates for the upkeep of EIC troops 'provided' to them by the British under a treaty agreement. This helped to defray EIC fixed military costs, while treaty provisions usually stipulated that the troops could be seconded as needed for EIC military operations as part of a less than forthright reciprocal defence clause embodied in many treaties. The supply of artillery to allies could also be coupled with the acceptance of accompanying EIC approved artillery personnel, as was the case with the *Peshwa* in 1803. They, like the artillery, were added to the bill and settled by way of specified payments made in cash, promissory notes based on projected land revenues, direct revenue (tax) collection, or – in the event of default – the assignment of the land itself. This also served as another mechanism to keep South Asian artillery under British control.

It is hard to conceive of an economic argument that would dare place the Marathas on an equal footing with the British as an economic superpower.[28] The EIC was, after all, one of the first publicly traded companies to systematically develop global markets as an integral means of increasing shareholder value in the process of exponentially increasing corporate wealth. Selling Bihar's opium in China or Bengal's saltpetre to Portugal was profitable, but selling a chance to share in the profits did more to enhance the EIC's leverage in international money markets. Its governmental partner in the imperial joint venture of India was the British Crown, no lightweight when it came to international financial clout in underwriting wars from Lisbon to Moscow during the Napoleonic period.

An economic security model would be an interesting variation on the superior credit model in examining reasons for British victory. It might reveal the imbalance in the degree to which the EIC and the Marathas could interfere with or destabilize each other's economic resources. In several respects the EIC's armies had the advantage of proximity to the Maratha enemy while the EIC's ultimate headquarters in London enjoyed the safety of distance. Maratha financial service centres in Pune, Ujjain and Mathura were all within Richard Wellesley's striking distance – even those Parsi connections to the Persian Gulf were vulnerable to British pressure. The Marathas had no international means of menacing the EIC. The British could play upon clan rivalries in an economic context and threaten competing revenue districts in a manner that the Marathas could not hope to match. Daulat Rao Sindia's holdings in Ujjain and Gwalior could be directly invaded or reduced to ungovernable turmoil by backing the rival claims of other members of the *durbar* such as Ambaji Inglia. And while the Sindias, Bhonsles and Holkars could menace the hinterlands of the Bombay, Bengal and Madras Presidencies, they could not inflict the same economic depth of penetration if the British sought reciprocal action in disrupting the Marathas' economic livelihood. The Marathas could not threaten the ultimate economic security of the British because they did not have access to London half a world away, whereas the economic basis of Sindia's administration in North India remained vulnerable. The European mercenaries had resisted relinquishing control of their economic holdings and *jaidads* until the last possible minute. They feared unbridled asset stripping by greedy *sardars* would leave the system defenceless, as sepoys would desert a leader that did not reward them with regular pay. The *jaidad* as an economic building block for military funding within the South Asian military economy was too archaic to compete against an internationally based crown corporation with access to a global line of credit. Maratha revenue collection within a war zone remained precarious, while the British could use less physically vulnerable international capital markets to dominate the South Asian military economy, purchasing victory and then imposing a form of peace that would feed the process by restoring revenue collections to be ploughed back into the EIC's coffers.

The British, without designating it as a formal objective, attained control of the South Asian military economy in the form of weapons production centres, military labour markets, logistical networks and intelligence systems. These acquisitions all pointed to organizational gains leveraged through superior financial networking. The EIC possessed unparalleled access to international markets. Additional opium could be shifted to Canton and bullion could be diverted from the China trade en route to London. The dominant player in the South Asian military economy was now the EIC as the local representative of a joint-stock multinational functioning in England, the Persian Gulf, South Asia (including Sri Lanka), China and Malaysia. The sepoys who filled the military labour market were being funnelled into EIC armies for service from Egypt to the Philippines. Perhaps the EIC as a corporate entity was better suited than the British government to seeing the value of human and material resources. This was not so much the Europeanization or foreign control of the South Asian military economy as it was the capitulation of that market to the embryonic forces of globalization. The globalization of EIC business demanded additional military growth to provide physical security and the vast dimensions of the EIC's Asian military commitment generated additional economic growth in British logistical and financial services. This was an expanding British network that diversified its risk by trading in many countries and economies around the world. There was no way for the Marathas to sabotage the money pump that fed Richard Wellesley the cash he spent creatively in his war effort.

The East India Company had superior access to credit and specie from the following sources:
- taxes and tariffs collected in India by the EIC
- trade revenues generated within India by the EIC
- the liquidation of goods shipped by the EIC from London
- EIC bullion shipments diverted from China (derived from opium sales)
- direct Loans in the form of bullion shipments from London to India
- specie and booty captured from the Marathas
- loans, 'donations' and subsidies, from native allies such as the *Nawab* of Awadh and the *Nizam* of Hyderabad
- loans and 'subscriptions' from private citizens and soldiers

The short-term deficit spending that made control possible, when placed on 'annual return' ledger sheets, led to demands for the sacking of Governor-General Richard Wellesley. With George Barlow's eventual appointment* as Wellesley's successor, it became official policy to rein-in the spending. Payrolls were cut and budgets were slashed with the result that many of the Marathas' military assets were put back on the block for the highest bidder. And, irony of ironies, several of those individual assets

---

* Cornwallis technically followed Richard Wellesley as Governor-General but he died before being able to fully assume the duties required of him in India.

wound up back in the hands of Maratha *sardars*. However, there were provisions and safeguards to prevent the Marathas from regaining majority controlling positions in the South Asian military economy. Regional alliances proved particularly difficult for the Marathas to rebuild, as the EIC was a much more appealing ally from the standpoint of its overall financial stability. The EIC was not likely to demand *chauth* or tribute from its allies the way certain Maratha powerbrokers had done.

British domination of the military economy, following 1803, meant no single indigenous power would be able to claim the lion's share of South Asia's military resources. Regional military resistance would be possible among the Sikhs, Afghans and even the disgruntled *Peshwa* would offer a last gasp of defiance to his subordinate status, but there would be nothing on a countrywide scale that one might portray as analogous to a modern bid for national liberation. From this point forward the EIC could, if it chose to do so, interfere in attempts to raise standing conventional armies of the sort needed to do lasting damage to British power on a subcontinental basis. By the time of the Anglo-Maratha and *Pindari* War (1817–19), the remaining Maratha threat had been downgraded in threat-level status with many *pindari* bands being classified as criminal insurgents. Portuguese mercenaries and a number of Euro-Asian officers attempted to synthesize a conventional response but they could not command large-scale access to the regular infantry resources as Sindia had done in 1803. Without superior credit systems or access to the greater South Asian military economy, that the British had come to dominate, the Maratha military response in 1817 became largely ineffective. One could argue that the Anglo-Maratha and *Pindari* War of 1817–19 indicated the EIC did not 'get it right' in 1803 and failed to halt the need for further battles. But evidence suggests that the Maratha military challenge of 1817 was not a threat to EIC expansion – that was largely accomplished. Rather, the final Maratha war was a mopping-up operation designed to complete the EIC's unfinished expansion of 1803–5, which had been cut short by London's budgetary worries about balancing Richard Wellesley's deficits against funding needed for the Napoleonic Wars. Central India fell to the EIC as an overdue military inheritance as opposed to a coveted economic objective. As for the later Sikh challenge to British authority, it was conventional and reminiscent of 1803 in terms of artillery power and infantry capability.[29] But British positioning in the North Indian labour market and an easily defined strategy of containment meant this was geographically condemned to being a limited regional challenge. An independent Punjab was conceivably possible, but by 1845–9 no indigenous South Asian military power was going to drive the British back to what had been the original coastal enclaves of the EIC's three Presidencies.

Prohibitions against employing unapproved European mercenary officers in British India and criminal statutes against *pindari* behaviour were all part of the effort to permanently alter the traditional South Asian military economy so that it better served the EIC's cause. It is doubtful that the British military writers of the nineteenth century would have readily seen the assimilation of South Asian military culture in terms of an irreversible British victory. For them the words 'British victory' were defined in decisive and absolute battlefield terms – such was the fiction of imperial military history. Nevertheless, in using economic incentives as a means of conflict resolution and to control military behaviour – buying out opposition, hiring enemy units and employing enemy officers – the British were shortening the process of military conquest. They were making it less sanguinary by employing non-destructive South Asian means of integration and incorporation. With all due respect to Colonel Malleson and his study of 'decisive battles' for control of colonial India, real control ultimately had more to do with the British triumph over the South Asian military economy than the efforts of a few white men wearing red coats.[30]

Having established that the issue of 'superior credit' is without dispute, there remains the question of how it factored into the mechanics of British victory in 1803. As we have seen, Maratha horsemen as well as regular and irregular infantry were all subject to purchase via contractual offer for continuation of their employment in EIC or allied service (i.e. Gokhale and his cavalry, the sepoys at Agra, Colonel Sheppherd's brigade, etc.). This was in accordance with the prevailing military culture and the historic functioning of the South Asian military economy as a free-market system in which military goods and services were valued or priced through an informal competitive bid process analogous to other commodities markets. However, this study has focused on the purchased defection of mercenary officers in Maratha service. Purchase here is taken in the broadest possible context to include outright cash offers for service, commutation of pension schemes and offers based on transference of rank as well as continuation of command with specific units brought over as part of the defection process. The purchased defection of Sindia's mercenary officer corps had a three-fold impact that directly contributed to Maratha defeat on the battlefield in 1803.

First, it completely destroyed Sindia's command and control capability at the strategic theatre level. Perron's defection set off a chain of events well calculated to collapse the existing system by depriving it of the men it had evolved to depend upon. The process of trying to rebuild command and control in the field during August 1803, as Lake and Wellesley advanced, meant men were tossed together in temporary positions. There was no time to build trust or familiarization with those who now assumed leadership titles. The loyalist *sardars* and European renegades, thrown into brigade command positions left vacant by the defections, had no time to fall back and consolidate their forces under new junior officers and NCOs. Communication was also adversely affected as the Marathas scrambled to find a viable means of coordinating their defensive reaction to British advances in a multi-front war. These new men who were supposed to lead had no time to establish a system of regular communication with their counterparts in other battalions. This process of collapse had the effect of decapitating what passed as the Maratha regular army by depriving it of overall command and control.

Second, the defections robbed Sindia's 'regular corps' of its offensive potential by depriving it of effective combat leadership under fire. As we have seen, the British troops in 1803, both European and sepoy, directly benefited from inspired combat leadership. The degree to which British officers gave a part of themselves is reflected in their casualty rates. We know from extensive combat studies in twentieth-century military staff colleges that the role of the individual infantry leader is crucial to the battlefield performance of subordinate troops. With the departure of their usual officers, Maratha sepoy morale suffered a devastating blow. The defections also meant fewer officers per battalion, therefore less effective leadership ratios as calculated by the number of enlisted men per officer. The presence of unproven *ersatz* officers and/or light horse leaders seconded as infantry officers meant the posting of individuals of unproven ability – to say nothing of undetermined bravery – in the context of combat infantry leadership under fire. The latter point compounded the fall in morale with a liability factor that compromised the dependability of normally steadfast battalions. Without dependable infantry Sindia's tactical options became much more limited.

That partially accounts for the degree to which artillery was the mainstay of Maratha resistance at Assaye, Delhi and Laswari. Maratha artillerymen found strength in their massive firepower and at Laswari the additional comfort of the temporary shelter offered by an anchored line of guns. Their cannon were a reassuring presence as they blew holes in the British formations that opposed them. A gun crew, particularly one that was stationary, was a team that functioned with a lesser degree of personal leadership than infantry of the line. Gunners were often cross-trained and could assume different duties within the team as required. Given the order to load and fire at will, the artillerymen proceeded with their duty, often continuing until bayoneted, as testified to by C-in-C General Lake. Maratha artillery doctrine, with its volume of fire dogmatism, could not help but prove devastating to British infantry given the tactics of the day. And it can be argued that the prevailing infantry tactics were greatly responsible for the high British officer casualty rates as the Maratha gun crews only had to wait for their targets to be thrown at them.

The defensive spirit of Sindia's men would have been bolstered by their proximity to the artillery when they were defending their artillery lines. But that would not be adequate tactical compensation for their loss of officer leadership when they needed to go on the offensive. The infantryman of that era was not trained as an independent thinker with discretionary judgement. He was specifically trained as an automaton in order that his actions might be delivered in unison as part of the orchestra of fire that was conducted and led by his officers. Without their accustomed officers these infantry units could not achieve peak proficiency on the attack, particularly if replacement officers did not situate themselves directly at the head of a body of men and lead them forward. The absence of Maratha infantry offensives in this war, or for that matter effectively controlled counter-offensives at Assaye and Delhi, underscored the absence of decisive combat leadership. Alighar and Agra were different in that they represented the defence of a fortified position. But Assaye, Delhi and Laswari represented a waste of fine Maratha infantry in that their tactical potential was denied. Without effective combat leadership, Sindia's battalions were relegated to an increasingly defensive role and he ultimately forfeited the strategic option of an offensive war.

Third, the defections represented an insurmountable intelligence loss. The amount as well as the value of intelligence sold out to the British was beyond all calculation. How do you compensate for the loss of information concerning the location of your troops, the quantity of food rations you have on hand, the distribution of your ammunition? The intelligence loss occurred on the tactical level with men like Lucan at Alighar and on the strategic level with Perron. Many armies would not be able to sustain the loss of their C-in-C on the eve of their first major battle of a war. Perron and his ADC Beckett rode away from Coel carrying political, military and economic intelligence of the highest magnitude. Aside from knowing who was Sindia's greatest Maratha rival and how many guns the Marathas possessed, they knew which battalions were likely to follow their defecting officers. Perron and Beckett also carried the all-important revenue data that Richard Wellesley needed to defend himself against London's charges of waging an unnecessary and unaffordable war. But Perron's departure posed a second problem beyond the turning over of brigade ledgers and revenue accounts. There was the secondary damage to the Marathas of not knowing the extent of disclosure and collaboration. Who had helped Perron and the other foreign officers? Were all the *sardars* who worked with foreigners compromised? It was the 'not knowing' that triggered a secondary flood of activity to minimize potential damage. Over-reaction itself became an enemy as purges were conducted and mistrust spread within Sindia's *durbar*. Was Ghatke to be trusted with any new information if Rotton as his brigade commander had 'gone over'? Was Ambaji stepping into the breach to save Sindia by purchasing the *Soubaship* of Hindustan or preparing to carve out another puppet state as *Peshwa* Baji Rao II had apparently done? Had Ambaji already paved the way for his own defection through the efforts of Colonel Sheppherd? These were real questions for what remained of Sindia's *durbar*. As the British reaped their triple bounty of military, political and economic intelligence, they were able to construct a blueprint to better understand the functioning of Sindia's North Indian administration and its army. By the time the 1803 campaigns drew to a close, they were in an ideal position to pick and choose which components they wished to salvage.

The British had a head start on how to use their intelligence system for maximum political gain and not all their good fortune was owed to the Proclamation's seemingly serendipitous windfall. In engineering an intelligence system that would assist them in achieving supremacy, the British used a number of their mercenary contacts to actively steer the course of political events. One of the most fascinating examples of that power was to be found in the case of Colonel J. F. Mieselbach* who was responsible for handling the EIC's buy-out of a Maratha ally for a guaranteed *jaidad* yielding annual revenues of Rs. 20 *lakhs*. Like many of the mercenaries placed in the role of spy, Mieselbach later had reason to complain about his own pension settlement when the budget-conscious administration of Richard Wellesley's successor, Governor-General George Barlow, tried to reduce pension entitlements as a cost-cutting move. The Colonel was one of the mercenaries listed in EIC ledgers as a 'Class 1' pension claimant, meaning he had not just defected under recognized terms but rather had done something so special he was to be granted maximum financial benefits.[31] Mieselbach's successful appeal, to be kept on a pension at Rs. 1,000 per month, went all the way to the EIC's Court of Directors.[32] In reviewing their extensive official files, which documented and verified the Colonel's claim, the EIC acknowledged that they had in effect purchased Bundlekund with their financial line of credit.

Governor-General Barlow was forced to admit that the Colonel had attached himself to the EIC's interests, 'not only before the Proclamation was promulgated but before the War with the Marhattas commenced'.[33] Mieselbach argued that those mercenaries, who 'came in' after the war

---

* Aka Mieselback, Meiselbeck.

started, did so because they really had no choice. It was not safe for them to stay with the Marathas past 29 August 1803 as anti-foreign senti-ment was rampant and operatives like Pohlman, who refused to fight in the Deccan, were incarcerated. Mieselbach, although not unique among those mercenaries who were actively spying before 1803, had been a well-documented source of excellent intelligence since 1801. The Colonel was a battalion commander in Bundlekund, which was con-sidered as strategic by the British for the purposes of containing the Marathas' northern access routes. And although the details of British military operations in Bundlekund have been omitted from this book, it is safe to say that military requirements to annex that region were directly reduced as the result of Mieselbach's actions. The Colonel was technically one of Ambaji Inglia's men but he had been seconded by the Marathas and assigned with his battalion to work with the regional ruler Himmat* Bahadur Gosavi.[34] This was a logical trade-off since the Marathas needed significant numbers of Himmat Bahadur's Ghosseins in Bundlekund as auxiliary and proxy troops to supplement a number of positions there, which the Marathas deemed to be vulnerable.

Colonel Mieselbach soon endeared himself to Himmat Bahadur and became his closest political advisor as well as head of his personal honour guard. It was quite fashionable to have foreign mercenary troops in that role – after all, Daulat Rao had Perron. Within his trusted advisory po-sition, Mieselbach worked closely under the Bengal government's direc-tion. As an intelligence operative he was much more proactive than those mercenaries at battalion level who were passive intelligence gatherers, *ears to the ground* if you will. Mieselbach's immediate handler, on behalf of the EIC, was the District Collector Mr Richard Ahmuty. But this operation remained extremely close to Richard Wellesley since Ahmuty reported directly to Richard's brother Henry Wellesley – Lieutenant-Governor of the Ceded Districts.

During 1802 Himmat Bahadur was preoccupied with the chaos in Pune as the Holkar–Sindia war could have gone either way and Himmat did not want to be caught out in the shifting alliances. Ahmuty applied pressure to Mieselbach to use the distraction to get Himmat Bahadur to agree to a Subsidiary Treaty Alliance with the British. Although the bid was not successful in 1802 the arguments put forward by Mieselbach gave Himmat Bahadur cause to listen to him as a voice of reason or counter-poise to those *sardars* who advocated war. The Colonel's calming influence helped him gain increasing access to the affairs of Bundlekund. On 1 June 1803 he alerted the British to a proposed coalition of Holkar,

---

* Aka Himmant.

Sindia and Ambaji Inglia. Mieselbach's instructions were to remain in position and monitor the reactions of the local leaders.[35] Ahmuty downplayed the chances of a coalition but urged Mieselbach to stay on top of events – the British were still hoping to annex the region and convert it into a subservient buffer state. In July 1803, as the Colonel ascertained the feelings of the local leaders in Bundlekund to British control, Ahmuty held out the offer of a direct monetary reward for Mieselbach's continuing efforts.[36] He was worth far more to the British if he remained in place rather than 'coming in' the following month under the Proclamation.

The British gamble, to leave Mieselbach in place as a covert operative, proved a masterstroke. As Sindia and Bhonsle committed to war against the British, Himmat Bahadur granted full treaty-negotiating powers to Colonel Mieselbach and Haji Huddin Khan.* On 2 September 1803, the newly sworn envoys travelled to Shahpur on the banks of the Yamuna River to meet Richard Ahmuty and his associate Mr Mercer. Later the Colonel admitted that he had received a counter-offer from Himmat's rival Shamshir† Bahadur. While camped at Cullinjur en route to the Yamuna River meeting, Shamshir had sent a messenger to Mieselbach. He offered Rs. 1 *lakh* and an increased *jaidad* if the Colonel would defect to the rebel cause with his troops.[37] Mieselbach's loyalty to the British did not waver. In short order, using his pre-approved signing authority from Himmat, he arranged a provisional treaty inviting the British to enter Bundlekund. In return for the agreement, Himmat Bahadur was to receive a British-sanctioned *jaidad* in Bundlekund with an annual revenue return of Rs. 20 *lakhs*.

The *jaidad* negotiations gave Colonel Mieselbach cause to consider his personal duty to his men and their fate in view of the treaty settlement. Without their own *jaidad* provisions they would have no income or employment. Mieselbach felt a particular obligation to his battalion officers who ensured his safety during the preceding months of secret negotiations. The Colonel bargained hard for their fair treatment under the Proclamation even though it had yet to find its way to Bundlekund. Like other mercenaries, he had known for quite some time that a deal was in the works. As a valuable intelligence agent he also knew he would be looked after – his silence would have to be purchased. However, he made Mercer and Ahmuty aware that by arranging the subjugation of Bundlekund, including his battalion's *perghannas*, he was acting against his battalion's best interests. The Colonel stood fast and in his subsequent affidavit, filed with the EIC's Court of Directors, he made known his relief

* Aka Wajehooddeen Khan.   † Aka Shumsheer, Shumshir, Shamsheer, Shamsher.

at seeing that his discussions with Mercer and Ahmuty had not fallen on deaf ears.

Two or three days after I had signed all the articles of agreement Mr. Mercer received the Proclamation of the Governor General in Council, of 29th August 1803 and on delivering me some copies to be Sent into the Marhatta Country, he mentioned, that as my provision had been already settled, the Proclamation then received would enable the Officers of my Battalion to be provided for.[38]

Foreseeing the need for a military presence to mop up opposition generated by Shamshir Bahadur, Ahmuty and Mercer urged Mieselbach to remain in service to Himmat. Once again Mieselbach was more valuable right where he was. If the Colonel had opted to 'come in' – claiming his pension and immediately retiring as permitted under the Proclamation – the British would have lost their continuing voice of influence in local affairs as well as a battalion that could be used to facilitate the strategy of containment; at that time Delhi had yet to be captured. On 16 September 1803 a British detachment under Lieutenant-Colonel Powell linked up with Mieselbach's men and pressed the attack against the rebel forces of Shamshir Bahadur who was driven from the field after what was termed 'feeble resistance'. Mieselbach had initiated, managed and delivered a political victory, capping his achievement by inflicting a military defeat on the rebel faction.

Conclusion

This chapter has been dedicated to examining the four explanations usually offered for British victory in the 1803 Anglo-Maratha Campaigns. They were assessed with the following findings:

(1) There is no evidence that British success in 1803 resulted from 'the momentum of British victory'; indeed the entire supposition of momentum is questionable given an overview of Britain's colonial wars 1792–1815. Future historians attempting to validate the theory in the context of the greater Napoleonic struggle will have to address the issue of how such momentum, if it ever existed in a pre-1803 context, was regained following the 1802 Peace of Amiens.

(2) As for 'superior British training, discipline and drill', this seems to have evolved as a nineteenth-century military variation on Social Darwinism.[39] Ethnocentric twentieth-century Western military analysts kept the myth alive rather than doing their intellectual duty to expose it as a racist diatribe. Among those who profited most from sustaining it were the advocates of the so-called European 'Military Revolution'. They made an unapologetically biased attempt to float technological determination

as a viable explanation for the macro-cycle of European economic ascendance from 1500–1800. Rather than associate economic domination with trade-based European economic expansion in global markets and the displacement of indigenous economies, the group felt compelled to label the era as the 'rise of the West' and place it in the context of a military rationalization.

(3) When we come to the theory of 'superior British artillery' in 1803, it can be categorically said the British lagged behind the Marathas in both the quality and quantity of artillery that they brought into the field. Beyond that, this book has also argued that the British ran a poor second to the Marathas in the sophistication of their artillery doctrine. The British military of that era had a very limited view of the role of artillery on the battlefield. Fire mobility with the cavalry's regimental gallopers was still very new, although the formation of designated horse artillery regiments was fast approaching. The British military establishment of 1803 was more keen on the basics of the bayonet for advancing under fire rather than revising its thinking to see artillery as a dynamic component for tactical initiative or strategic advantage. The correspondence of the Board of Ordnance with the EIC's Court of Directors concerning George Constable's work reveals the British military establishment was under no delusion with regard to the status of its weapons' design principles. The modern persistence of 'technological determination' as an explanation for British victory is inexcusable given the documentation that shows British military authorities of the day were quite willing to acknowledge the superiority of specific types of Maratha artillery as well as its overall quality. The inability of the British to fully read the evolutionary simplicity of the Maratha artillery or fully extrapolate on its theatre-based lessons is reflected in the later rise of the powerful Sikh artillery and its role in Anglo-Sikh conflict.

(4) The essential point concerning the superiority of British credit is without contention given the global commercial network that the British enjoyed in the Maratha Campaigns of 1803. But it remains a qualified explanation for British victory in 1803 owing to the creative ways in which economic and financial leverage were used. And the linkage of superior credit with the attainment of Richard Wellesley's military objectives should not be reduced to an equation of *more money = purchasing victory*. The buying-out of light horse and *pindari* units struck both Bhonsle and Sindia, while the purchased defection of the mercenary officer corps and several regular battalions temporarily crippled Sindia to the point of making his an unwinnable war. On an immediate basis, it reduced the duration and severity of combat with the Marathas in 1803 by diminishing the threat level of the 'regular corps'. Taking entire battalions out of

enemy service meant a shorter and less sanguinary war. Wages, pensions and offers of employment could only be tendered by the EIC under their Proclamation owing to their possession of a credit system strong enough to embrace the temporary deficits needed to underwrite the plan. To a large degree, the 1803 campaigns represented a short-term credit trap for the EIC. Richard Wellesley was chastised for deficit spending and EIC surpluses were eventually restored. In the long term, it gave Britain majority control over the South Asian military economy and its vital component parts.

The EIC actively sought to discourage military competitors after 1803. Britain irrevocably changed the South Asian military economy by seeking to control it. Denying widespread indigenous patronization dismantled aspects of it that ran counter to British political objectives. While in the case of domestic weapons' production, much of the sector dried up after being prohibited or bypassed by the EIC's rigidly controlled centralized sourcing system. Logistical transport systems were absorbed and refined by way of British military codification. But the most readily recognizable surviving component of the South Asian military economy was the military labour market. Soldiers from that source formed the armies needed to garrison the countryside and represented British military and foreign policy abroad. The North Indian military labour market was destined to grow tremendously after the martial races debate of post-1857 India and, numerically speaking, overall South Asian recruiting did not reach its pinnacle until well into the twentieth century. The real profitability in controlling the South Asian military economy could only be judged in relationship to the long-term economic and regional security benefits the British derived from ruling India up until 1947. British dominance of the traditional South Asian military economy, as reflected in the North Indian military labour market, produced its own control-oriented changes over time – but the system staggered on in one form or another. In fact the cultural legacy of the South Asian mercenary system and military labour markets, as components of the historic (as opposed to modern) South Asian military economy, can still be found today at the dawn of the twenty-first century. The contemporary armies of the United Kingdom and the Republic of India both retain Nepalese Gurkhas, while American military operations in Afghanistan featured the purchase of regional commanders as local bodies of troops were bought out of service for an agreeable price.[40]

Appendix I:
Chronology of Anglo-South Asian wars

The Anglo-Bengal Wars
First   1686
Second   1756–7
Third   1763–5

The Anglo-Karnatak Wars
First   1744–8
Second   1749–54

The Anglo-Mysore Wars
First   1766–9
Second   1780–4
Third   1789–92
Fourth   1799

The Anglo-Maratha Wars
First   1775–6, 1779–82
Second   1803–5
Third   1817–19

The Anglo-Gurkha War   1814–16

The Anglo-Burmese Wars
First   1823–6
Second   1852–3
Third   1885

The Anglo-Afghan Wars
First   1839–42
Second   1878–80
Third   1919

The Anglo-Sikh Wars
First   1845–6
Second   1848–9

Note This list is far from being definitive. Numerous wars, campaigns and
other types of military actions have been omitted such as the Poligar War
and the Kandian War. Nor was any attempt made to categorize or include
the events of 1857–8. For a discussion of the debate surrounding the
numbering of the Anglo-Maratha Wars see endnote 1 of the Introduction
to this book.

Appendix II:
British troop strengths and casualties for the Hindustan and Deccan Campaigns 1803

The following statistics do not include figures for the 1803 operations in Bundlekund, Cuttack, Gujarat or for Major-General Campbell's Hyderabad Reserve Force.

General Lake's Grand Army in 1803[1]

Cavalry
3 Regt.s European Cavalry
5 Regt.s Native Cavalry

Infantry
H.M.'s 76th Regt.
11 Battalions of sepoys
Artillery

Total	10,500

Major General Arthur Wellesley's Force in 1803[2]

	Europeans	Natives	Total
Cavalry			
H.M.'s 19th Dragoons	384	1,347	1,731
4th, 5th & 7th Regt.s of Native Cavalry			
Infantry			
H.M.'s 74th & 78th Regt.s	1,368	5,631	6,999
6 Battalions of Sepoys			
Artillery			172
		Total	8,902

With 357 Artillery Lascars and 653 Madras Pioneers

Colonel Stevenson's Hyderabad Subsidiary Force[3]

	Europeans	Natives	Total
Cavalry		900	900
Infantry	778	6,113	6,891
Artillery	120		120

Total 7,911

With 276 Artillery Lascars and 202 Pioneers

British battle casualties as compiled from the *Calcutta Gazette*
Extraordinary
Reprinted in
*House of Commons Account, Bengal, Fort St. George and Bombay Papers
Presented to the House of Commons, Pursuant to their orders of the 7th of
May last, from the East
India Company, relative to the Mahratta War in 1803,*
Printed by Order of the House of Commons 5th and 22nd June 1804[4]

The Deccan Campaign

Ahmadnagar[5] Combining the Action against the Pettah with that against
the Fortress

Total Killed (European & Native)	30
Total Wounded (European & Native)	111

Assaye[6]

Total Killed (European & Native)	428
Total Wounded (European & Native)	1138
Total Missing (European & Native)	18
Horses Killed	325
Horses Wounded	111
Horses Missing	2

Asirghar[7]

Total Killed	2
Total Wounded	6

Argaum[8]

Under Wellesley

Total European Killed	13
Total European Wounded	101
Total Natives Killed	21
Total Natives Wounded	93
Total Natives Missing	4

Horses Killed	6
Horses Wounded	3
Horses Missing	9

Under Stevenson

Total European Killed	2
Total European Wounded	44
Total European Missing	2
Total Natives Killed	10
Total Natives Wounded	55
Total Natives Missing	1
Horses Killed	18
Horses Wounded	6
Horses Missing	2
Total Men Killed, Wounded and Missing	346
Total Horses Killed, Wounded and Missing	44

Gawilgarh	data missing

The Hindustan Campaign

Coel[9]

Total Killed (European & Native)	1
Total Wounded (European & Native)	4
Horses Killed	3
Horses Wounded	8
Horses Missing	10

Alighar[10]

Total Killed (European & Native)	59
Total Wounded (European & Native)	212

Delhi[11]

Total European Killed, Wounded and Missing	197
Total Native Killed, Wounded and Missing	288
Total Horses Killed, Wounded and Missing	173

Agra
Action of 10 Oct. 1803[12]

Total European and Native Killed, Wounded and Missing	229

Siege of 17 Oct. 1803[13]

Total European and Native Killed, Wounded and Missing	6

Laswari[14]

Total Killed (European & Native)	172

Total Wounded (European & Native)	652
Horses Killed	277
Horses Wounded	154
Horses Missing	122

NOTES

1. Thorn, *War in India*, p. 75. Note Thorn's failure to identify specific sepoy infantry units. The total figure does not include the 3,500-men Bundlekund Reserve Force stationed at Allahabad.
2. Ibid., p. 70.
3. Ibid., p. 71.
4. House of Commons, Misc. Papers, Session 1803–4, vol. XII.
5. Ibid., vol. XII, p. 267.
6. Ibid., vol. XII, p. 281.
7. Ibid., vol. XII, p. 291.
8. Ibid., vol. XII, p. 295.
9. Ibid., vol. XII, p. 268.
10. Ibid., vol. XII, p. 269.
11. Ibid., vol. XII, p. 271.
12. Ibid., vol. XII, p. 278.
13. Ibid., vol. XII, p. 279.
14. Ibid., vol. XII, p. 289.

Appendix III:
Governor-General Wellesley's 'Maratha' Proclamation of 1803

Some British government agents in North India received copies of the Proclamation as early as 1 August 1803 and that helps to explain how a significant number of soldiers in Maratha service came to know about the East India Company's offer before it was publicly posted.[1] It was impossible for the British to achieve uniform distribution of the Proclamation and the erratic release of information was later cited as a reason for extending the terms of the Proclamation to those men who did not leave the Marathas until after September 1803.[2]

On the 1st August, I have the honour of acknowledging the proclamations which His Excellency the Most Noble the Governor General in Council was pleased to direct me to issue, in the European and Native languages as soon as hostilities should have commenced I also received your subsequent letter, of the 3rd August containing an additional sentence to be inserted in one of the proclamations in question. I now have the honour to acquaint you that, on the 31st of August these proclamations were issued at this place, and published throughout the zellas of Allahabad, Cawnpoor, Etawak, Furruckabad and Morabad, in the manner we best adopted to give them speedy and extensive circulation. That parts of the notification which relates to foreign Europeans in the service of the hostile Marhatta Chieftans, was translated into the French and German languages as I was given to understand, there were many Germans in that Predicament.[3]

The following notices appeared in the *Calcutta Gazette* on Thursday 8 September 1803.[4]

PROCLAMATION,

BY HIS EXCELLENCY THE MOST NOBLE

THE GOVERNOR GENERAL IN COUNCIL,

WHEREAS THE GOVERNOR GENERAL IN COUNCIL *has deemed it to be necessary to provide effectual security for the defence of the British Possessions against the designs of Dowlut Rao Scindiah, and of the Rajah of Berar.*

His Excellency in Council hereby requires all British Subjects holding employment in the Military Service of Dowlut Rao Scindiah, or of the Rajah of

Berar, or of any Marhatta Chief, or of other Power, or State, confederated with Dowlut Rao Scindiah, or with the Rajah of Berar, forthwith to relinquish the Service of such Chief, Power, or State, respectively, and to repair to such Places as shall have been appointed by the Officers commanding the British Forces in Hindostan and the Dekkan, for the purpose of receiving all such British Subjects, as shall retire from the Service of the said Chiefs, Powers, or States, in obedience to such Proclamations or Orders, as may have been issued by the said Commanding Officers, in conformity to the Instructions of the Governor General in Council, or in obedience to this Proclamation. And the Governor General in Council is hereby further pleased to declare, that all British subjects who have retired, or who may retire from the Service of the said Chiefs, Powers, or States, in obedience to any Proclamation issued by the said Commanding Officers, or in obedience to this Proclamation, who shall have been, or shall be admitted by the said Commanding Officers to the Protection of the British Government, shall receive from the Honourable Company, a provision, equal to the amount of the fixed Pay and Allowances, which such British Subjects have received in the Service of the said Chiefs, Powers, or States respectively. The said provision to continue during the continuance of Hostilities between any of the said Chiefs, Powers, or States, and the British Government, and so long as such British Subjects shall be employed in the Service of the Honorable Company; and all such British Subjects after having quitted the Service of the Honorable Company, shall receive a reasonable remuneration, and every indulgence, which their respective situations may appear to require, and which maybe consistent with the principles and regulations of the British Government. And the Governor General in Council further declares, that all British Subjects, who shall remain in the Service of Dowlut Rao Scindiah, or of the Rajah of Berar, or of any Marhatta Chief, or other Power or State, confederated with Dowlut Rao Scindiah, or with the Rajah of Berar; and all British Subjects, who shall bear arms against the British Government, shall be considered to have forfeited all Right and Claim to the Protection of the British Government, and shall be treated accordingly.

The Subjects of France, or of any other foreign European, or American State, holding employments in the Military Service of Dowlut Rao Scindiah, or of the Rajah of Berar, or of any Marhatta Chief, or of any Power or State confederated with Dowlut Rao Scindiah, or with the Rajah of Berar, who may be disposed to relinquish the Service of the said Chiefs, Powers, or States, and to repair to such Places, as shall have been appointed by the Officers commanding the British Forces in Hindostan and the Dekkan, for the purpose of receiving such British Subjects as shall retire from the Service of the said Chiefs, Powers, or States, shall be admitted to the benefits extended by this Proclamation to all British Subjects.

By Command of
His Excellency the Most Noble
The Governor General in Council,
  J. Lumsden,
  Chief Sec. to the Govt
Fort William, August 29, 1803.

~

PROCLAMATION,

BY HIS EXCELLENCY THE MOST NOBLE

THE GOVERNOR GENERAL IN COUNCIL,

It is hereby signified to all Non-Comissioned Officers and Sepoys, formerly in the Service of the Honorable Company, or of His Excellency the Nawaub Vizier, and to all other Persons natives [sic] of the British Territories in India, or of the Territories of His Excellency the Nawaub Vizier, holding employment in the Military Service of Dowlut Rao Scindiah, or of the Rajah of Berar, or of any Marhatta Chiefs, or other Powers or States confederated with Dowlut Rao Scindiah, or with the Rajah of Berar, that they are required to quit the service of those Chiefs, Powers, or States, and that on repairing to such places as shall have been appointed by the Officers Commanding the British Forces in Hindostan and the Dekkan, for the express purpose of receiving all Persons of the above description, who shall retire from the Service of the said Chiefs, Powers, or States, they will be received into the Service of the Honorable Company, or otherwise will obtain a provision equal to the amount of their Pay and Allowances in the Service of the said Chiefs, Powers, or States, and will be entitled to every indulgence, consistent with the Principles and Regulations of the British Government. Such Persons will be required to produce to the Officer stationed at the Places appointed for the express purpose of receiving them, authentic proofs of their having quitted the service of the said Chiefs, Powers, or States, in consequence of this Proclamation, previously to their being considered to be entitled to the benefits tendered to their acceptance by the Terms of this Proclamation.

By Command of
His Excellency the Most Noble
The Governor General in Council,
  J. Lumsden,
  Chief Sec. to the Govt
Fort William, August 29, 1803.

NOTES

1. NAI, Foreign Dept Files, Secret Series, Consultations 22 Sept. 1803 nos. 23–8, p. 25 (page number in pencil), From Archibald Seton, Agent at Barelli to the Gov. Gen., to Neel Benjamin Edmonstone, Sec. to Gov't. in the Secret and Political Dept at Fort William Bengal.
2. In several cases the cover letters, telling of the benefits to be derived from the government's offer, were unexplainably detached and delivered alone without the Proclamation. This was the case with numerous copies sent to local North Indian magistrates 'by Dawk Baughies' which supposedly accounted for their 'non-arrival'. See NAI, Secret Index to the Foreign Dept 1803 (aka: Foreign Dept – Secret Index 1803), p. 119 Judges and Magistrates 12 Nov. 1803 nos. 138, 143–8.
3. NAI, Foreign Dept Files, Secret Series, Consultations 22 Sept. 1803 nos. 23–8, p. 25 (page number in pencil), From Archibald Seton, Agent at Barelli to the Gov. Gen., to Neel Benjamin Edmonstone, Sec. to Gov't in the Secret and Political Dept at Fort William Bengal.
4. *Calcutta Gazette* vol. XL, Thurs. 8 Sept. 1803, no. 1019. Technically speaking, the paper carried four such documents that day. The first proclamation was printed in English and intended for officers; while the second, third and fourth editions (printed in English, Persian and Devanagari script respectively) were intended for sepoys and non-commissioned officers. The fourth proclamation, printed in Devanagari script, differed slightly in that N.B. Edmonstone signed it as 'Sec. To Govt. Sec. Pol. & For. Depts.'.

Appendix IV:
Mercenary pension records

This chart was taken from BL: OIOC 0/6/6 Personal Records. It should not be construed as a complete list of mercenaries in Maratha service during 1803.[1] Rather, it should be considered an incomplete record of those who had a pension or settlement claim that was adjusted in 1805–7 and reviewed circa 1817. As a 'list of mercenaries in Maratha service' it fails to include a great number of individuals, such as those who died before the review, those who were 'captured' as renegades or POWs, and those who had their service commuted to allied forces such as the *Nizam* or the *Peshwa*. As for being an indicator of nationality and/or race, the list does not provide notation of race or country of origin. Comments about the impossibility of accurately judging the mercenaries' nationality according to their surnames are included elsewhere in this book. In accordance with the quotation practice established in this book, all original spellings were retained as copied and researchers following leads in this document are urged to consult the main body of this text for notes on anomalies in spelling (i.e. Beckett vs. Becket, etc.).

Name & rank	Amount of pensions first granted in Rs. per month	Amount of pensions permanently fixed (Bengal) in Rs. per month	When fixed	Place of residence & where the pensions were ordered to be paid
Almeida, F. D., Lieut.	100	48	20 Oct. 1806	Behar
Armstrong, R., Maj.	—	240	28 Aug. 1806	Calcutta: He has since been permitted to come to England & draw his pension by his agents at Calcutta.
Adkins, Capt.	—	180	10 Sept. 1806	Calcutta
Bruce, Robert, Capt.	500	180	26 Feb. 1807	Agra: From 18th April 1808 allowed to reside & draw his pension at Calcutta

(cont.)

(cont.)

Name & rank	Amount of pensions first granted in Rs. per month	Amount of pensions permanently fixed (Bengal) in Rs. per month	When fixed	Place of residence & where the pensions were ordered to be paid
Birch, J., Ensign	—	60	2 Oct. 1806	Delhi: From 18th May 1807 allowed to reside & draw his pension at Calcutta
Birch, S., Lieut.	—	180	2 Oct. 1806	Furruckabad
Butterfield, Capt.	—	180	24 Dec. 1806	Agra
Becketts, J., Capatin	—	120	30 Jan. 1809	Goruckpore: By Minutes 10 Oct. 1809 & 15 Nov. 1811 His pension payable at the Presidency
Burwell, Lieut.	—	60	5 Sept. 1808	Bundelcund
Camera, D., Capt.	200	*[blank]*	*[blank]*	*[blank]*
Choblet, Lieut.	200	60	2 Oct. 1806	Behar
Collins, Capt.	—	180	24 Dec. 1806	Cawnpore
Cleops, John, Quarter Master	—	40	26 Feb. 1807	Cawnpore
Cleops, P., Sergeant Major	—	30	26 Feb. 1807	Cawnpore
Deassiz, Silva	100	60	2 Oct. 1806	Behar
Derema, Manuel Capt.	400	400	2 Dec. 1806	Delhi
Deremao, Domingo	—	60	2 Dec. 1806	Delhi
Derridon, L., Maj.	800	180	2 Oct. 1806	Agra
Darson, Major	200	180	29 Jan. 1807	Calcutta
Dudrenec, M.	—	300	15 Jan. 1807	Calcutta
Diremao, Anthony	—	40	2 Dec. 1806	Delhi
Draper, George	—	120	19 Feb. 1807	Hyderabad
De Camera, Capt.	—	120	29 Jan. 1807	Behar
Dugremon, Capt.	—	120	2 April 1807	Behar
De Camera, J.	—	48	4 Jan. 1808	Behar
Edwards, J.	—	180	4 Dec. 1806	Calcutta
Fisson, J. B.	—	200	18 Feb. 1807	Calcutta
Fanthome, B., Capt.	411(Lieut)	120	2 Oct. 1806	Behar
Fergusson, Lieut	—	120	2 Oct. 1806	Benares
Francois, Ensign	—	60	24 Dec. 1806	Cawnpore
Geriad, Joseph	200	60	16 Oct. 1806	Furruckabad
Garman, John, Cadet	100	48	26 Feb. 1807	Agra
Gardiner, Lieut. Col.	900	300	2 Oct. 1806	Behar
Grant, R. Lieut.	400	120	16 Oct. 1806	Agra
Harriott, Maj.	2,600	*[blank]*	*[blank]*	*[blank]*
Hessing, G. W. Col.	—	300	26 Feb. 1807	Calcutta
Hitchcock, Lieut.	—	120	28 Aug. 1806	Calcutta

(cont.)

Name & rank	Amount of pensions first granted in Rs. per month	Amount of pensions permanently fixed (Bengal) in Rs. per month	When fixed	Place of residence & where the pensions were ordered to be paid
Henry, A., Capt.	—	120	9 Oct. 1806	Bundlecund: By Minute of 28th July 1807 to reside & draw his pension at Calcutta
Hearsy, H., Capt.	—	180	2 Oct. 1806	Boglepore
Henry, Thomas, Sergeant Major	—	30	26 Feb. 1807	Cawnpore
Jennings	—	60	22 Feb. 1808	Boglepore
Koine, F., Ensign	80	48	16 Oct. 1806	Allyghur
La Foucarde, John De, Lieut.	200	*[blank]*	*[blank]*	*[blank]*
L'Epinette, L.	300	120	2 Oct. 1806	Behar
Loyed, Thomas Capt.	—	120	28 Aug. 1806	Calcutta
La Fleur, Maj.	—	240	24 Dec. 1806	Cawnpore
Long, Lieut.	—	120	2 Oct. 1806	Allyghur
MacCallangh, J., Ensign	225	60	16 Oct. 1806	Furruckabad
Martin, Commissary	—	100	18 Dec. 1806	Delhi
Martin, F., Lieut.	—	120	24 Dec. 1806	Agra
Meiselbeck, John, Col.	—	1,000	24 April 1807	Calcutta
Martin, J., Lieut.	—	120	1807	Cawnpore
Netelk, Angelo	200	60	2 Oct. 1806	Agra
Nuremberg, J.	—	60	4 Jan. 1808	Behar
Pedron, A., Lieut.	150	60	16 Oct. 1806	Allyghur
Pedron, Col.	300	*[blank]*	*[blank]*	*[blank]*
Pedron, M., Ensign	80	48	16 Oct. 1806	Allyghur
Port, J. W.	—	120	19 March 1807	Calcutta
Phillips, Lieut.	—	120	24 Dec. 1806	Cawnpore
Pohlman, Col.	—	300	16 Oct. 1806	Agra
Piaggio, G., Ensign	—	48	4 Dec. 1806	Allygur: The 28th Dec. 1807 allowed to reside & draw his pension at Calcutta
Rennick, A., Capt.	400	180	16 Oct. 1806	*[blank]*
Rennick, Capt.	400	180	16 Oct. 1806	Calcutta
Rigault, A., Lieut.	300	60	5 March 1807	Calcutta
Royal Ally Louis, Mons.	100	60	16 Oct. 1806	Agra
Rotton, R. W.	—	180	2 Dec. 1806	Lucknow
Reckett, John, Capt.	—	120	2 Dec. 1806	Boglepore
Rogers, J., Lieut.	—	60	16 Oct. 1806	Furruckbad: The 3rd Sept. 1807 allowed to Reside & Draw his pension at Calcutta
Scott, G., Capt.-Lieut.	400	120	2 Oct. 1806	Behar

(*cont.*)

(cont.)

Name & rank	Amount of pensions first granted in Rs. per month	Amount of pensions permanently fixed (Bengal) in Rs. per month	When fixed	Place of residence & where the pensions were ordered to be paid
Silva, F. D. A., Capt.	100	*[blank]*	*[blank]*	*[blank]*
Smith, L. F., Maj.	1,200	240	28 Aug. 1806	Calcutta This Gentlemen Has since been allowed to come to England & Draw His Pension Through Agents in Calcutta
Sutherland, Henry, Maj.	800	240	28 Aug. 1806	Calcutta This Gentlemen Has since been allowed to come to England & Draw His Pension Through Agents in Calcutta
Scott, J. W., Lieut.	—	120	10 Sept. 1806	Calcutta
Skinner, J., Capt.	—	300	2 Oct. 1806	Delhi
Skinner, A., Lieut.	—	120	2 Oct. 1806	Delhi
Stewart, K. B. Capt.	—	180	2 Oct. 1806	Behar
Sheppard, J. R. Lieut.-Col.	—	300	24 Dec. 1806	Cawnpore
Stewart, D., Capt.	—	180	6 May 1809	Calcutta
Turnbull, P., Capt.	—	180	7 Dec. 1806	Delhi

NOTE

1. In Smith, *A Sketch of the Regular Corps*, there is a list of subscribers ranging from EIC officials to British regular army officers. However, readers who are interested in tracing individual mercenaries should be aware that a considerable number of the names located in the subscribers' list are those of mercenaries who 'came over'.

Appendix V:
The Marathas' employment of mercenaries in historic perspective

As this book is being completed during the bicentennial anniversary of the 1803 Anglo-Maratha Campaigns, it seems appropriate to say something about the mercenary issue that continues to haunt Maratha military history.

Currently Gurkhas from the Hindu Kingdom of Nepal can be found in the armed forces of Great Britain and the Republic of India with significant numbers stationed in strategic positions that include the petroleum-rich Muslim Sultanate of Brunei. In that respect it can be said that South Asian mercenaries play a role in contemporary international security, albeit a comparatively minor role in the overall global military balance. This dual employment of Gurkha troops by Western and Asian powers suggests the acceptability of mercenaries continues on a cross-cultural basis of outstanding military professionalism and proven ability.

The use of mercenaries has never been limited to one military culture or one period of time. Mercenaries have been used in a culturally parallel fashion by Western and Asian armies since ancient times. One only needs to read the *Campaigns of Alexander* by Lucius Flavius Arrianus to see that Western invasions of South Asia have been dependent upon mercenary employment since the time of Alexander the Great.[1] Over centuries the use of mercenaries became customary in South Asia as it did in virtually every other great civilization. Unfortunately it was in the twentieth century that the word mercenary took on a decidedly negative connotation and it was often thought to be synonymous with the term 'professional killer'. However, that emotive phrase is somewhat absurd in that virtually all professional soldiers who go to war and kill their country's enemies can be termed professional killers. In this book the term mercenary is used, in a much less judgemental historic context, as one who accepts payment or reward in return for military service outside his immediate family or under a banner that is not generally accepted as his own. This may sound somewhat ambiguous but other traditional definitions (i.e. fighting for a country other than one's own) pose problems in an era when several key nation states had yet to be defined. The word mercenary remains a perfectly adequate generic term to use when referring to the historic military employment of millions of men who have plied their trade in times of peace and war. Within Maratha history we can broadly apply the term mercenary to any group of independent military entrepreneurs with no particular connotation as far as clan, caste, colour, race, religion or ethnicity.

Historically speaking, the South Asian military economy thrived on the perception of physical security as well as its antithesis – physical insecurity. Armies with militarily specific human resource requirements did not always have the option of conscription and many South Asian armies resorted to hiring from existing labour pools as found in military labour markets. The contractual underpinning of mercenary employment was quite simple. Military service for reward, or remuneration, was predicated on the common understanding that military labour contracts, be they written, verbal or implied, were like civilian labour contracts. One party offered a service (work/labour/soldiering) and the other party utilized it, or retained the option on it, in exchange for something of value. Perhaps the straightforward nature of contractual service reduced sectarian tendencies within South Asian military history. Aurangzeb, often described as the most ardent Muslim emperor of the Mughal dynasty, showed a consistent willingness to employ Hindu Rajput generals at the highest operational level of theatre command. Aurangzeb even selected them to lead campaigns against their fellow Hindus – enemies like Shivaji.

The extensive use of mercenaries by the Marathas from 1600 to 1800 coincided with their political ascendancy but it also paralleled the extensive employment of mercenaries in Western nations. Field Marshal Lord Carver cautioned his readers that the national heritage of the British Army was much more short-lived than they might think. As recently as 1625 it had used impressment to obtain soldiers, but the nucleus of Britain's army was mercenary. Veterans of French, Dutch, Swedish and Spanish service were common in the seventeenth century and the preparations for war in 1756 included a fierce debate on whether to entrust the fate of England to mercenaries, particularly those from Hanover and Hesse.[2] Mercenary units were employed by the British around the world, from India to America, but their service in the American Revolutionary War (circa 1776) was particularly noteworthy. We tend to forget the degree to which Germany's eighteenth-century military labour market, as characterized by Hessian troops for hire, had enabled the British to fight wars of a global dimension in that era.* But not all German

---

* The British were said to have paid £3,191,000 to Frederick II for the use of 22,000 Hessians in America.

mercenaries were contractually bound to British employment during the Revolution. The Prussian Baron von Steuben* drilled and disciplined American troops and wrote what became known as the Continental Army's official drill book. Von Steuben was with the Americans at Yorktown, a battle where Gerard Lake served in the opposing British trenches and which was made famous by Cornwallis' surrender.

The Marathas, being major consumers as well as 'drivers' of the South Asian military economy from 1600 to 1800, were keen to exploit the military and economic advantages offered by mercenaries. Chapter 1 made the case that Maratha military leaders were pragmatic people who saw the sense in hiring well-qualified soldiers to complete the military tasks at hand. From the Marathas' own standpoint, soldiers were always needed to fill various military roles and the hiring of mercenaries (South Asian or European) provided potential solutions that could be tailored to budgetary limits. In modern terminology one could try and match military requirements with military occupational specialties when looking for infantry to garrison a fort, officers to lead infantry, or cavalry to launch interdiction raids. Mercenary employment was another means of outsourcing military labour requirements and some contractual arrangements probably represented cost-effective compromises. When volleys of smallarms fire were needed at the Battle of Dabhoi in 1731 (see chapter 1), it made little difference if Kolis produced those discharges with matchlocks or Marathas with flintlocks. The military effect – that of a cloud of smallarms projectiles released on command – was the same.

The Marathas' military needs did not remain static. Certain types of mercenaries enjoyed popularity cycles in Maratha service at different times in history and hiring trends could also be linked to major events. Take for example the combat leadership vacuum created by the Marathas' loss to the Afghans at Panipat in 1761. The horrendous number of deaths among Maratha forces, in excess of 25,000 individuals, created something of a crisis in military leadership at the junior to middle-ranking officer level. Many post-Panipat Maratha clan armies lacked an adequate number of effective and responsible officers. And the period from 1761–1800 also corresponds to the rise in the number of Northern European[†] and Euro-Asian[‡] officers in service to the Sindia and Holkar clans as well as non-Marathas such as the *Nizam* of Hyderabad.

---

[*] Aka Frederick William Augustus Henry Ferdinand Steuben, born Magdeburg, Prussia, 1730.

[†] Northern European is specified here because many of these Europeans were from Britain, Germany, Holland and France.

[‡] Military service was a socially and culturally acceptable profession for these soldiers of cross-cultural heritage – men like James Skinner.

The Marathas' increasing retention of Northern European mercenaries during the second half of the eighteenth century coincides with an increasing supply cycle. In turn, that supply cycle was tied to the South Asian extension of Europe-based conflict as outlined in the Introduction to this book. The end of the Seven Years' War and its affiliated South Asian conflicts meant there were veteran officers to be had – including those with experience leading South Asian troops under fire. Some were French officers of the old order who saw no particular reason to return to their homeland, which was becoming increasingly divided by revolutionary sentiment. There were other Frenchmen who had served as soldiers, tradesman-technicians, or merchants in service to various 'country powers' located in Mysore, Lucknow and Hyderabad. There was also a wave of opportunists who fit the general description of 'European adventurers'. These included ship-jumping former servicemen of the working class who thought they would do better for themselves in India. They became familiar with the stories of others who had 'made it', men from Armenia, Greece and Switzerland to name but a few.

As for British mercenaries in Maratha service, they had similar experiences. Some were veterans of transplanted European wars; others were former employees of the East India Company and the armies of the Bengal, Bombay and Madras Presidencies. Many Britons retained by the Marathas were trading on their military experience for lack of any other equivalent economic option. Mahadji Sindia was known to pay well and the frequency of his campaigns meant lucrative chances for plunder and prize funds that surpassed those of the EIC's armies, which made the meagre returns of King George's troops look like petty change in comparison. But the number of British adventurers also owed something to the peculiar recruiting policy of the EIC itself. As outlined in the letters of George Carnegie (chapter 5), many Britons had travelled to India thinking they could land lucrative positions with the EIC. To their dismay most found that one could not apply for such positions from within India and that you had to be accepted for EIC employment in Britain and subsequently sent out to India in possession of a posting. Having travelled to India on speculation of employment, many found themselves nearly destitute and willing to take service in a so-called 'native army' in order that they might at least earn their passage back home if not make a fortune in service.

In speculating on reasons why the Marathas failed to achieve complete control of India, some historians have focused on the issue of foreign mercenary employment. This would seem to be an outgrowth of questions concerning the depth of Maratha dependence on foreigners and the degree to which they could be trusted. Much has been made of the fact that Europeans served Mahadji Sindia and Daulat Rao Sindia as commanders (i.e. de Boigne and Perron) but that was far from a historic precedent. As we saw in chapter 1, the great Shivaji had used the accomplished Portuguese officer Rui Leitao Viegas as commander of the Maratha navy and Portuguese sources indicate over 300 Portuguese were in service to the Marathas at that time. But Shivaji's power base may have been more secure than that of the Sindia clan in the period from 1790–1803. The examples set by de Boigne and Perron, as guardians of the clan leader, suggest there were times when a European non-Maratha mercenary identity was advantageous. Generals de Boigne and Perron were more readily trusted since there was little incentive for them to launch a coup as usurpers. Their European identity denied them political stature sufficient to carry the *durbar*, which was dominated by clan politics driven by Maratha identity. The main difference in comparing Shivaji's use of European mercenaries to that of Daulat Rao Sindia seems to lie in the extent to which the latter gave his mercenaries custody of strategic information sources and placed them in positions of dual economic and military importance. Shivaji's European mercenaries did not have control of his key fortress chain and/or his most prized revenue producing lands; they tended to be long on military skill and short on exposure to Shivaji's most valuable strategic assets. By way of contrast, when General Perron left service, he rode into a warm British welcome carrying vital military and economic intelligence that included defensive plans as well as revenue return figures for the Gunga-Yamuna Doab. With intelligence like that Richard Wellesley could assess target priority and calibrate the economic as well as political push needed to swing remaining enemy allegiance to the British side.

What was it that motivated so many of the European mercenaries to leave Maratha service at such a militarily crucial time in 1803?

(1) National pride and racial kinship. Many British mercenaries did not want to fight their fellow countrymen. Many joined the service of Marathas like Sindia under the contractual stipulation of not having to fight against the EIC or His Majesty's troops (see chapter 5). The linkage to nationality also had the potential to bridge racial barriers. Some mixed-race or Euro-Asian soldiers reflected upon their father's British identity. Having been raised as the 'soldier sons' of soldiers, an identity that carried the blessing of both South Asian and British warrior societies, it made a great deal of sense to honour the military culture and tradition of one's father.

(2) The chance for continuing monetary or material gains. If the European mercenaries stayed in 'native service' and the Marathas lost the war, their Maratha pensions might be worthless and there

would be no chance to collect the British pension offered to them under terms of Richard Wellesley's Proclamation. However, if they resigned Maratha service they had several options. They could retire from soldiering altogether on their new British pension. But, if they became proactive and 'brought over' other troops, or were willing to fight their former Maratha employers, they could volunteer for combat and take an EIC army posting and enjoy additional rewards as well as the financial security of continuing employment. Granted there is some irony in speaking of financial security and volunteering for combat in the same breath – but that was the way mercenaries viewed their employment options in that era.

(3) Fear of legal action. British officers such as John Pester believed that the fear of being labelled traitors influenced some of their mercenary opponents to defect before the fighting started (see the entry dated 19 Aug. 1803 as noted in chapter 4). But the degree to which a fear of hanging persuaded men is unknown. The EIC's finest legal minds were tasked with trying to ascertain whether or not any recalcitrant mercenaries may have broken the law by remaining in either Sindia's or Bhonsle's service. Robert Smith, the EIC's Advocate General in Bengal, went so far as to draft a treason warning.[3] However, the legal option of treason trials was never used; lucrative Proclamation terms, with minor exceptions, remained virtually open for all ranks and races throughout the 1803 campaigns.* The risk of trying to prosecute men who had arguably defended the sovereign borders of 'native powers' was probably unwise. It had not been that long since the British government had demonstrated its willingness to turn on Governor-General Warren Hastings for his role in what many opponents termed an 'illegal war of aggrandizement'.

(4) Fear of detection as spies. As noted elsewhere in this book, several mercenaries were actively serving as British intelligence sources. If they stayed in service and were identified as spies betraying their terms of service, they stood a very good chance of being executed by the Marathas. Sargi Rao Ghatke was among those Maratha leaders in Sindia's *durbar* that considered the option of a blood purge.

(5) Fear of future incrimination. Some of the undecided mercenaries left after seeing their comrades go. They felt that if they stayed and things did not go well, they might be suspected as turncoats and that might

---

* Only a few mercenaries such as Bourquin (aka Louis Bernard) were taken as renegade prisoners.

result in their incarceration. This fear was quite real in the Deccan where Pohlman was seized and sent to prison in Ujjain before he had the opportunity to formerly resign and leave Sindia's service.

Did the Marathas sow the seeds of their own destruction in 1803 by using European mercenaries? Well it is true that the Western European mercenaries were particularly prone to changing sides during the 1803 Anglo-Maratha Campaigns. Their willingness to 'go over' to the British side was largely the result of a combination of national, racial, economic and legal reasons outlined above. These factors were consciously embodied in the Proclamation as a dedicated British stratagem to detach the mercenaries as well as the sepoy battalions from Maratha service (see the Proclamation in appendix III). Perhaps the question would be better phrased as, 'Did the fluid nature of the military labour market, within the South Asian military economy, leave the Maratha armies susceptible to subversion?'

What was the impact of removing the foreign mercenaries from Maratha service? Did it constitute the reason the Marathas lost the battles of 1803? And on a larger scale, does it explain why the Marathas lost what amounted to any hope of controlling India? This book has presented evidence that the loss of strategic intelligence and tactical leadership under fire was enough to have deprived the Marathas of any realistic expectation of winning against Lake's Grand Army in Hindustan and Arthur Wellesley's Army of the Deccan. That argument can be made effectively without the additional factors that included:

- Colonel Mieselbach's secret pact and the delivery of Bundlekund;
- the degree to which the EIC continued to recruit during remaining battles, such as Agra and Laswari, thereby creating military labour spot markets that further destabilized the remaining Maratha efforts to challenge the British militarily.

However, I do not believe that the eventual domination of the South Asian military economy by the British can be turned around into an argument that the Maratha military system was tragically flawed by mercenary employment and that the situation was aggravated by the employment of foreigners to such an extent that it prevented the Marathas from becoming the dominant power in India. Speculation as to why the Marathas failed to attain lasting control of the subcontinent falls into a larger 'what if?' school of history. That line of questioning would have to address underlying problems with Maratha clan unity and historic shifts in the power structure of the Maratha polity. As suggested in chapter 1, the equitable division of civil–military power within the Maratha polity had

been a problem since the rise of Baji Rao I when the delicate balance between the *Chhatrapati*, *Senapati* and *Peshwa* was permanently thrown out
of kilter. Ironically, the over-concentration of civil–military power under
Peshwa Baji Rao I (a period free from powerful European mercenaries)
may have proven the subcontinental potential of Maratha military power
while dooming the Maratha polity to the cyclical infighting we associate
with the later *Peshwa* period (circa 1761–1818).

NOTES

1. *Arrian: the Campaigns of Alexander*, trans. De Selincourt.
2. Carver, *The Seven Ages of the British Army*, pp. 1–3, 67.
3. NAI, Foreign Dept Files, Secret Series for 1803, Consultations 12 Nov. 1803,
   Nos.139–41.

Glossary

Glossaries are often less than perfect tools because the meanings of words change over time. The glossary that follows was constructed to apply to this specific analysis of the Anglo-Maratha Campaigns of 1803. These explanatory notes should not be taken as comprehensive definitions and readers are advised to consult reference works for the broader historic and geographic variations in the use of these terms.

arrack: a generic term for indigenous liquor often added to water by the British for the purposes of purification or killing bacteria.

bana: rockets, the body of which was formed by rolling sheet iron to form a tube. The rocket tubes were filled with a gunpowder-based propellant and usually lashed to a length of bamboo that served as a stabilizing tail. The Indian rocket was studied and modified by Congreve at Woolwich, where he produced the first widely used British military rockets.

banjaras: an itinerant socio-ethnic group who specialized in the transportation of grain and salt. *Banjaras*, most often acting as independent entrepreneurs in the South Asian military economy, played a vital role in the logistical survival of armies in India as they moved food supplies over great distances.

bargi-giri: a form of light horse doctrine associated with warfare in the Deccan and the fortress of Ahmadnagar in particular. The tactical basis of this doctrine emphasized the avoidance of direct frontal assault or stationary battle in which the advantages of speed, manoeuvrability and surprise would be sacrificed. Perhaps the greatest use of *bargi-giri* tactics was in logistical interdiction and the physical isolation of enemy units.

bargir: a horseman associated with the doctrine outlined above. Chapter 1 suggests a rather more detailed explanation for the evolution of the term *bargir*, which is often listed as a Persian 'loan word' defined as meaning 'a soldier who enlisted without a horse'.

batta: this refers to a vast assortment of special payments and field allowances found in armies such as those of the EIC. Complicating our understanding of the word *batta* are regional differences in how armies were paid and compensated, particularly for their time in the field. For example, some troops that experienced interruptions in their grain supply while on campaign might have adjustments made in their *batta* or so-called 'forage money' to allow them to purchase supplementary grain from local bazaars.

beasties: water carriers or water bearers.

burkundauze: Colonel Henry Yule and A. C. Burnell, in their glossary of Indian words known as *Hobson-Jobson*, speak of *burkundauze* as armed retainers or policemen in service to a civil authority. In British dispatches from North India in 1803 the term most often refers to local men armed with matchlocks. They were termed 'irregulars' in comparison to uniformly dressed and armed sepoy regulars. Many *burkundauze* were hired to serve in *sebundi* units (see *sebundi* below).

campoo: a term used to describe a Maratha infantry brigade of the type associated with Mahadji Sindia and Daulat Rao Sindia. In comparison to similar British forces in 1803, the Maratha *campoos* were characterized by a smaller officer-to-enlisted-man ratio and a greater volume of organic firepower in the form of the Marathas' more numerous artillery.

chauth: often termed Maratha 'tribute money' or tribute payments made to Maratha overlords. Failure to pay *chauth* could be interpreted as a failure to acknowledge the Maratha claim to control a given area. This type of resistance to authority invited retaliatory raiding by the Marathas as they sought to forcibly extract the tribute payment 'owed' to them. As a result of that rather extortionate alternative means of revenue collection, *chauth* has also appeared in many modern glossaries as 'protection money' but that is a rather pejorative term given the bureaucratic institutionalization of *chauth* in Maratha history.

chhata: coats of mail used as general-purpose body armour.

chhatra: an umbrella; however, it was used symbolically to denote or imply kingship. The *chhatra* could be found in stylized form in art and architecture as well.

Chhatrapati: historically the Marathas reserved this term for their highest-ranking leader, their Lord of the Umbrella – King of Kings. It first gained widespread notoriety as the title associated with Shivaji.

coss (aka kos): a unit of linear distance measurement with great regional variation. Many British sources opt for the standard of the Bengal Presidency and assign a rough approximation of 2 miles (or 3.2 kilometres) to the *coss*. However, any serious attempt to discern precise distances from historic documents must be based on a common understanding of which regional *coss* was being applied at the time of writing.

cossminars: distance (*coss*) markers erected during the Mughal era.

dawk: regularly referred to as the 'country' postal system – it was perhaps more accurately described in 1803 as a relay system for the forwarding of goods, information, mail and messages.

durbar: usually translated as court – as in the 'court of a ruler'.

feringhi: foreigner and used to describe Europeans who were also known in some older accounts as 'hat-wearing people'.

gajnal: this was a so-called 'camel gun' – a small-bore cannon mounted on the back of a beast of burden such as a camel (aka *jejala*).

garbhandi: small cannons or more precisely mortars from which pebbles (*gar*) were fired for close-range anti-personnel application. Pebbles and stones could be gathered in the field and that saved the trouble as well as expense of transporting special-purpose ammunition.

golundauze: approximately a gunner or cannoneer. The word was reported to have been derived from a term meaning 'ball thrower'. Artillerymen of this description should not to be confused with other lower-ranking members of the artillery corps such as *lascars* (see separate entry).

harkarrahs: although often translated as messenger, the term *harkarrah* has changed over time and several late twentieth-century texts implied that *harkarrahs* were spies. The confusion comes from their service in the field. Moving ahead of advancing armies relaying information and intelligence, they were vital information links but they were not spies as we know them in the context of covert operations. Their garb and/or badges of rank made many *harkarrahs* quite distinguishable while in service to regional armies or the British. Arthur Wellesley's *harkarrahs* from Mysore readily stood out in the interior of Maharashtra during 1803 and consequently, as Wellesley's trusted messengers, they were fed misinformation by the Marathas to hamper the British advance.

howdah: the compartment or seating arrangement carried on the back of an elephant. Some regal ceremonial *howdahs* were bejewelled and screened, other military *howdahs* were armoured and defended with rockets or light artillery. In early sixteenth-century European accounts *howdahs* were sometimes described as 'castles' on the backs of elephants and rendered as the 'elephant and castle' on some military emblems.

hundi: a written indigenous financial instrument regularly used for the transference of money. While some translate this simply as a cheque, there were differences in negotiable and non-negotiable *hundi* depending on the issuer and associated financial terms.

jaghir: assigned revenue-producing lands that often became hereditary. The *jaghirdar* or holder of the *jaghir* derived social, political and military recognition from this reward but it also carried subsequent obligations. One could draw a point of parallel evolution between Europe's feudal system and the Mughal's *mansabdari* system. The revenue generated by the *jaghir* – particularly tax revenue in the 1803 period – was to be used for the maintenance and payment of a body of armed horsemen. A *jaghirdar*'s horsemen could then be seconded to the aid of the central authority. Many of the Marathas' so-called 'southern *Jaghirdars*' became auxiliary cavalry under the overall command of Arthur Wellesley in 1803. Students of the 1803 period should avoid confusion of the term *jaghir* with the term *jaidad* as found in numerous British primary sources – see below.

jaidad: in previous examinations of the 1803 Anglo-Maratha Campaigns there has been a temptation to say that *jaidad* had the same meaning as *jaghir*. However, research for this work revealed multiple cases where the term *jaidad* – although similar as a unit of revenue-producing land assigned for the up-keep and supply of troops – specifically meant an allocation of land for the revenue-based maintenance of infantry as opposed to mounted troops. EIC records reveal appeals for horsemen from southern Maratha *jaghirdars* and arrangements for the transfer of Maratha infantry officers' *jaidads* and their accompanying sepoy brigades in Hindustan.

jambur: a type of Maratha field gun. In artillery-specific eighteenth-century Maratha dispatches, *jamburs* were associated with the use of grape shot.

jangi topha: taken in Maratha documents to mean intermediate-range/size field artillery.

Jemadar: in 1803 a *Jemadar* in one of the three EIC armies (Bengal, Madras, or Bombay) would have been a middle-ranking indigenous South Asian infantry officer, with the rank equivalency of lieutenant in a sepoy unit.

jezail: one of many terms for a South Asian matchlock – this word was often associated with the very accurate long-barrelled matchlocks of western India and Afghanistan.

jowari: the grain commonly known in the West as millet.

Khalsa: used in this text to refer to the 'Sikh warrior brotherhood of the pure'.

khilat: a ceremonial robe of honour with significance as an indicator of social and political recognition.

killadar: the commandant of a fort.

lakh: one hundred thousand – most often seen in reference to valuations expressed in hundreds of thousands of *rupees* or simply *lakhs* of *rupees*.

lascars: in the armies of 1803, *lascars* were artillerymen tasked with man-handling the artillery into position. Many were needed to man the heavy drag ropes used to pull the guns forward.

matross: an artilleryman who served as part of the gun crew subordinate to the gunner or cannoneer. Their duties included ammunition handling, loading and ramming as well as swabbing the barrel after firing to prevent premature ignition when loading subsequent rounds.

nullah: a watercourse or land feature (i.e. ditch or depression) associated with water. Not all *nullahs* were water-filled and dry *nullahs* often lent themselves to being used as field-expedient trenches during battles.

palkis: a means of conveyance or carriage sometimes likened to a palanquin coach or shuttered litter.

palpati: a tax paid by *pindaries* for the right to plunder.

perghannas: administrative districts often referred to in local calculations of revenue value.

Peshwa: contemporary discussions of the late eighteenth- and early nineteenth-century Maratha political system often refer to the term *Peshwa* as meaning 'the Maratha Prime Minister'. However, this was not always the case and those unfamiliar with the evolution and institutionalization of the office are urged to use the term Prime Minister with caution when discussing the period prior to *Peshwa* Baji Rao I.

pettah: the town affixed to, or adjacent to, a fort. In some cases the *pettah* had separate defensive outer works.

pindaries: irregular light horse troops who traditionally paid a fee to plunder, or provided their retainers with a percentage of their booty, for the right to plunder. Maratha powerbrokers found *pindaries* to be cost-effective since they did not require regular monthly wages. *Pindaries* were used in a variety of roles including that of reconnaissance, logistical interdiction, screening the movement of infantry, or as sack men against civilians in areas withholding *chauth* payments. Although *pindaries* found their way into the Maratha order of battle, the *pindaries* themselves often came from various ethnic and religious backgrounds. The *pindari* leader Amir Khan was a Pathan soldier of fortune.

pounder: it was customary in the era of muzzle-loading artillery to list cannon bore diameters in a manner that expressed calibre as a theoretical projectile weight in pounds. As a result of this practice, the standard 3.6-inch bore diameter British battalion support gun (in 1803) became known as the 6 pounder. The 3 pounder was a light gun, while 18 and 24 pounders were favoured by the British as siege guns to smash fortress walls.

pucka: although commonly taken to mean 'proper', many military documents from 1803 use *pucka* to mean brick and mortar buildings or fortifications such as those used to reinforce the western coast of India. In the latter case, *pucka* fortifications – proper brick and mortar defensive walls – were superior to those made of wood, sandbags or earth-filled wicker baskets.

puckallies: those individuals who filled the large water butts or water skins that held a military unit's water supply on the march.

puja: Hindu worship or performance of Hindu religious ceremony as in 'to make *puja*'.

rakhale: cart-mounted horse-drawn guns in the 1–3 pounder size range. The *rakhale*'s use in combat can be traced as far back as Mughal Emperor Babur's First Battle of Panipat in 1526.

rissalah: most often taken to mean a large body of horse as in a *rissalah* of cavalry.

rupee: the rupee remains India's standard unit of currency. Rupees are abbreviated as Rs. Transactions in rupees during 1803 were often listed in EIC documents as three-figure citations, i.e. Rs. 3 : 2 : 1, which would correspond to the currency units 3 rupees, 2 annas and 1 pice. However, it is not safe to assume that all three-figure expressions were equivalent in 1803. The value of rupees is difficult to establish unless the specific type is stated, i.e. Bombay, Arcot or Sonaut rupees. The complexity was exacerbated by the imposition of a standard known as the Sicca Rupee and the book keeping of the Bengal Presidency relied on 'imaginary coins', called Current Rupees, Current Annas and Current Pice. These Bengal 'Current' designations were the book-keeping standard.

sardar: readily translated as officer. When the term is used generically to discuss Maratha military powerbrokers in this text, it could mean military rank or political rank. Most senior ethnic Maratha *sardars* in Sindia's court (*durbar*) would have had large entourages of armed retainers – often kinsmen – suggesting some original form of military underpinning for their political positions.

sebundi: so-called 'revenue militia'. Within British-controlled areas they were usually under the command of local magistrates to aid in tax collection. They ranged in quality and ability from peasant rabble armed with matchlocks to highly proficient militia regulars trained and drilled according to prevailing sepoy standards. During the mobilization of 1803 several North Indian *sebundi* units went through systematic upgrading from *sebundi* to irregular infantry to regular infantry in reserve.

Senapati: from the time of Shivaji to that of his grandson Shahu the formal Maratha title for Commander-in-Chief was *Senapati*.

shroffs: most glossaries indicate that a *shroff* was a 'money-changer' or 'banker'. However, that explanation is less than satisfactory in that many *shroffs* made their living largely by charging commissions for facilitating various financial and economic transactions. They were not necessarily direct employees of a 'bank' or members of so-called banking clans and so the term 'banker' is misleading. *Shroffs* serving on campaigns with armies were financial specialists who offered a variety of economic products and services. Their network and reputation were often directly related to their effectiveness ranking. A *shroff* could arrange currency conversion as in the case of Bengal rupees being converted to a more readily useable form of local currency. And *shroffs* might also arrange monetary transfers by way of credit notes or *hundi* that may or may not have been drawn upon banking centres in Varanasi or Surat. In the 1803 Anglo-Maratha Campaigns *shroffs* also advanced credit to soldiers who could not wait for their arrears in pay to catch up with them. Several British diary references indicate *shroffs* loaned money against jewellery taken as plunder. Such sources suggest the *shroffs* helped individual soldiers in a manner we might liken to that of a pawnbroker. *Shroffs* were integral to the South Asian military economy in terms of facilitating liquidity during periods of conflict.

silladar: a term used to describe horsemen (often 'irregular horse') who reported for duty/employment with their own horses and weapons. Employing *silladar* horse was an option for those powerbrokers who sought to avoid the economic and logistical problems associated with the central supply of horses and weapons.

Soubahdar: in 1803 a *soubahdar* in one of the three EIC armies (Bengal, Madras, or Bombay) would have been an indigenous senior South Asian infantry officer, with the rank equivalency of captain in a sepoy unit.

Soubha of Hindustan: Viceroy of North India.

tindal: a leader of *lascars* in the artillery. A *tindal* could be likened to a *lascar* non-commissioned officer in being in charge of work details in which *lascars* were employed (i.e. preparation of defensive artillery positions).

Topass: the term first gained wide-scale use in reference to those offspring who resulted from relationships between Portuguese men and South Asian women. At one time these Euro-Asians represented a sizeable portion of the population in Goa and other Portuguese colonies. Many assumed their father's religion and profession as soldiers with the result that they became known as 'black Christians', valued in infantry and artillery units.

topha thorli: great guns as described in Maratha documents – taken to mean large-calibre long-range field guns as well as siege guns.

tulwars: swords.

vakils: it was once popular to list this word as meaning ambassador. But in the battlefield negotiations of 1803 the term emissary is perhaps more appropriate, as in the case of the Maharajah of Berar sending a *vakil* to Arthur Wellesley to ask about the British General's military intentions.

zamindars: in the simplest form – landlords. In the economic competition for North India, *zamindars* readily assumed the role of middlemen. They collected taxes from local peasant farmers and were, in theory, supposed to meet the tax obligations imposed by higher authorities such as the Mughals, Marathas or British. Those *zamindars* who withheld the money owed to central authorities, such as the EIC's Bengal Presidency, became infamous as 'refractory *zamindars*'.

Notes

INTRODUCTION

1. The number of Anglo-Maratha Wars remains open for debate. Everyone seems to agree that the First Anglo-Maratha War started in 1775. Historically there were those who called the second round of fighting in the First Anglo-Maratha War, meaning the battles of 1779–82, the Second Maratha War. Others see 1779–82 as the conclusion of the First Anglo-Maratha War, a second campaign (as opposed to war) brought about by a British unwillingness to accept the outcome of conflict in 1775. This book refers collectively to the Anglo-Maratha battles of 1803 as the northern and southern campaigns and avoids labelling them the Second Anglo-Maratha War. The problem with labelling the 1803 campaigns as the Second Anglo-Maratha War is that many history books note the second conflict as ranging from 1803 to 1805 and yet there is not historic consensus of opinion about how that definition evolved. A number of EIC sources from 1804 refer to the 'late Maratta War in 1803'. Major William Thorn, stated in his memoirs that the 'Second Maratha War' concluded at the end of 1803. See Thorn, *Memoir of the War in India Conducted by General Lord Lake, Commander-in-Chief, and Major-General Sir Arthur Wellesley, Duke of Wellington; from its Commencement in 1803, to its Termination in 1806, on the Banks of the Hyphasis* (T. Egerton, London, 1818) (hereafter referred to as Thorn, *War in India*). Thorn offered no explanation as to how or why the subsequent 1804–5 Holkar Campaign came to be included in his work. But his title suggests that the combined actions from 1803 to 1806 constitute a single 'late' war in India. It seems that in the period from 1805 to 1817, the collective actions of 1803 to 1805 had become known as the 'Second Maratha War'. John Blakiston, who fought beside southern campaign commander Arthur Wellesley, wrote about his experience long afterward. Blakiston's work, published in 1829, referred to the 'War in 1803', with no specific indication that it was linked to the 1804–5 campaign against Jeswunt Rao Holkar as part of the same war. See Major John Blakiston, *Twelve Years' Military Adventure in Three Quarters of the Globe: Or Memoirs of an Officer who served in the Armies of His Majesty and of the East India Company, between the years 1802 and 1814, in which are contained the Campaigns of the Duke of Wellington in India, and his last in Spain and the South of France*, 2 vol. set (London, 1829) (hereafter referred to as Blakiston, *Twelve Years*'). Some modern historians began to label the campaign against Holkar in 1804–5 as the Third Anglo-Maratha War. This compounded the

confusion as most British imperial historians recognize the 'Anglo-Maratha and *Pindari* War of 1817–1819' as the Third Anglo-Maratha War. Therefore it should come as no surprise that I have tried to specify the campaign and year, wherever possible, rather than referring to the First, Second, Third, or even Fourth, Anglo-Maratha Wars. See appendix I for what passes as a standard chronology of British wars in South Asia.

2. This view contrasts those of histories which assert that the Battle of Panipat in 1761 represents the Marathas' 'high-water' mark.

3. Although not covered in this work, British victory in 1803 also meant the seizure of Maratha ports and the furtherance of British plans for the naval containment of the subcontinent.

4. William H. McNeill, *The Age of Gunpowder Empires 1450–1800* (American Historical Association, Washington DC, 1989) and *The Pursuit of Power: Technology, Armed Force, and Society since A.D. 1000* (University of Chicago Press, 1982); Paul Kennedy, *The Rise and Fall of Great Powers: Economic Change and Military Conflict from 1500 to 2000* (Random House, New York, 1987); Geoffrey Parker, *The Military Revolution: Military Innovation and the Rise of the West, 1500–1800* (Cambridge University Press, 2nd edition 1996).

5. In 1955 Michael Roberts delivered his inaugural lecture as Professor of Modern History at the Queen's University, Belfast. In that lecture he argued that changes in tactics, strategy, the scale of warfare and its impact upon society, which had their origin in the United Provinces and culminated in the Swedish army of Gustavus Adolphus, were deserving of the label 'revolutionary'. Roberts' idea that a military revolution occurred in the early modern period from 1560–1660 remained largely intact until Geoffrey Parker revised and relaunched the theory in 1984 (the Lees Knowles Lectures given at Trinity College, Cambridge). That Parker's rendition propagated a more culturally chauvinistic dogma can be detected from the title of his book, *The Military Revolution: Military Innovation and the Rise of the West, 1500–1800*, first published in 1988.

6. Parker spoke of 'the absolute or relative superiority of Western weaponry and Western military organization over most others'. And that, 'in large measure "the rise of the West" depended upon the exercise of force, upon the fact that the military balance between the Europeans and their adversaries overseas was steadily tilting in favour of the former; and it is the argument of this book that the key to the Westerners' success in creating the first truly global empires between 1500 and 1750 depended upon precisely those improvements in the ability to wage war which have been termed "the military revolution" '. Parker, *The Military Revolution*, pp. 115 and 4 respectively.

7. 'Imitating the western way of war involved adaptation at many levels. Simply copying weapons picked up on the battlefield could never suffice; it also required the "replication" of the whole social and economic structure that underpinned the capacity to innovate and respond swiftly.' See Geoffrey Parker, 'The Western Way of War', in Geoffrey Parker (ed.), *The Cambridge Illustrated History of Warfare: the Triumph of the West* (Cambridge University Press, 1995), p. 7. John Lynn added: 'Britain's decisive superiority lay in a cultural and

institutional approach to war that its native foes could not imitate with the same ease with which they could fire – or manufacture – fusils. European attitudes toward combat ran wholly counter to native Mongol notions of horse-archer warfare, which stressed individual prowess and minimal casualties. The battle culture of forbearance was counter-instinctual, a product of long experience on the gunpowder battlefield. Once acquired, it had to be taught and replicated through relentless drill. Traditions of this kind were wholly alien to the states of South and East Asia.' John A. Lynn, 'The Seventeenth-Century Military Change: "The Western Way of War", and South Asia', in MacGregor Knox and Williamson Murray (eds.), *The Dynamics of Military Revolution 1300–2050* (Cambridge University Press, 2001), p. 54.

8. As with the Military Revolution, there is disagreement over defining the RMA. Many believe that RMA theory encompasses the idea that information-based technology will transform the future course of war. And within that majority, most believe the West not only leads the RMA in weaponry and research but sets the guidelines for its further development. Perfecting the RMA, according to a minority of so-called 'utopian' RMA advocates, will reduce the likelihood of war because potential enemies will supposedly be unwilling to start wars they cannot hope to win. For a more weapons-oriented consideration of the RMA as encompassing no fewer than five Western-based military revolutions, see Murray and Knox's essay 'Thinking about revolutions in warfare', in Knox and Murray (eds.), *The Dynamics of Military Revolution 1300–2050*, pp. 1–14. Knox and Murray caution against technological determination and their analysis 'suggests that two very different phenomena have been at work over the past centuries: "military revolutions", which are driven by vast social and political changes, and "revolutions in military affairs", which military institutions have directed, although usually with greater difficulty and ambiguous results'.

9. For an account of how RMA thinking came to be 'sold' as part of the *Western way of war* at the close of the twentieth century, see Michael Ignatieff, *Virtual War: Kosovo and Beyond* (Metropolitan Books, London, 2000).

10. John Plowright, 'Revolution or Evolution? A Review Article', *British Army Review*, 90 (Dec. 1988), pp. 41–3. Jeremy Black, *A Military Revolution? Military Change and European Society 1550–1800* (Macmillan, London, 1991). Bert S. Hall and Kelly DeVries, 'The Military Revolution Revisited', *Technology and Culture*, 31 (1990), pp. 500–7. C. J. Rogers (ed.), *The Military Revolution Debate: Readings on the Military Transformation of Early Modern Europe* (Boulder, 1995). In releasing the second edition of *The Military Revolution* (1996), Parker saw fit to defend his assertions by way of a new chapter 'Afterword: in defence of *The military revolution*', pp. 155–75. Nevertheless, European-oriented criticism continued as found in Bert S. Hall, *Weapons & Warfare in Renaissance Europe: Gunpowder, Technology, and Tactics* (Johns Hopkins University Press, Baltimore and London, 1997).

11. For an overview of the military dichotomy in the historic period of transition (USSR-to-Russia) see Jacob W. Kipp, 'The Nature of War: Russian Military Forecasting and the Revolution in Military Affairs: a Case of the Oracle of

Delphi or Cassandra?', *The Journal of Slavic Military Studies*, 9, no. 1 (March 1996), pp. 1–45.

12. The second edition of Parker's book noted: 'Only military resilience and technological innovation – especially the capital ship, infantry firepower and the artillery fortress: the three vital components of the military revolution of the sixteenth century – allowed the West to make the most of its smaller resources in order to resist and, eventually, to expand global dominance.' *The Military Revolution*, p. 175.

13. V. G. Dighe, *Peshwa Bajirao I and Maratha Expansion* (Karnatak Publishing House, Bombay, 1944).

14. Some historians have rejected the term 'Confederacy' as pejorative. But in the nineteenth century it was quite common for the British to refer to the Maratha polity as the Maratha Confederacy. We should remember that a number of British officers, who had lived or served in North America before they went on to soldier in India, would have been familiar with the Iroquois Confederacy: 'a political system granting the autonomy of each nation over local interests whilst deciding general confederacy matters, such as foreign and military policy, through a "grand council." This form of federalism was strengthened by the clan system which threaded through the whole confederacy and linked the various nations to each other.' See Robert S. Allen, *His Majesty's Indian Allies, British Indian Policy in the Defence of Canada, 1774–1815* (Dundurn Press, Toronto and Oxford, 1992), pp. 13–14. For references on the Confederacy's basis in law, see footnote 4, p. 228.

15. For an account examining the war's political as well as military complexity, see chapters 2, 3 and 4 of G. S. Sardesai, *The New History of the Marathas*, vol. III, *Sunset Over Maharashtra 1772–1848* (Phoenix Publications, Bombay, 1968), pp. 37–132. If nothing else, the war should have demonstrated to British observers that the Marathas posed a credible conventional military threat. M. R. Kantak, *The First Anglo-Maratha War 1774–1783: a Military Study of Major Battles* (Bombay, Popular Prakashan, 1993), upholds the traditional belief that the Marathas' chief claim to fame was mounted warfare. But in this book he also did an admirable job of showing us that the First Anglo-Maratha War embodied a very sizeable amount of siege warfare – the domain of the infantry.

16. Raghuji Bhonsle was descended from a different family line than the famous Maratha leader Shivaji Bhonsle and he should not be confused as heir to that family. For notes on the two distinct Bhonsle clans see the 'Genealogy of the Maratha chiefs' section and in particular pp. 663–4 in Sir George William Forrest (ed.), *Selections from the Letters, Despatches, and Other State Papers Preserved in the Bombay Secretariat, Maratha Series*, vol. I, parts I–III (Bombay, 1885) (hereafter referred to as Forrest, *Maratha Series*).

17. Later, after the defeat of Daulat Rao Sindia and Raghuji Bhonsle in 1803, Holkar would be involved in a separate and distinct campaign against the British in 1804–5.

18. For letters concerning the affairs in Gujarat see, in particular, W. H. Gense and D. R. Banaji (eds.), *The Gaikwads of Baroda, English Documents*, vol. VI (Bombay, n.d.). For an English-language background to information about Orissa and Cuttack contained in the Marathi letters of the Menavli Dafter,

see T. S. Shejwalkar (ed.), *Nagpur Affairs: Selection of Marathi Letters from the Menavli Daftar*, Deccan College Monograph Series 9 (Poona, 1954), [vol. I], pp. xxiii–xxv. The area had relied for some time on the local Udiya militia for defence and was said to have succumbed 'to the British after nominal fighting'. *Nagpur Affairs*, Deccan College Monograph Series 14 (Poona, 1959), vol. II, p. xvi.

19. Thorn, *War in India*, ch. 8, pp. 253–65.
20. For notes regarding Portuguese observations on Shivaji's 'Maratha navy', see P. S. Pissurlencar, *Portuguese-Mahratta Relations*, translated by T. V. Parvate (Bombay, 1983), pp. 35–40, 63.
21. See *A Review of the Origin, Progress and Result of the Late Decisive War in Mysore, in a Letter from an Officer in India: with Notes and an Appendix, comprising the Whole of the Secret State Papers found in the Cabinet of Tippoo Sultaun, at Seringapatam; taken from the Originals*, by M. Wood, Esq. M. P. Colonel and Late Chief Engineer, Bengal (printed by Luke Hansard, for T. Cadell, London, 1800). See also *Select Letters of Tippoo Sultan to Various Public Functionaries: including his Principal Military Commanders; Governors of Forts and Provinces; Diplomatic and Commercial Agents*, arranged and translated by Colonel William Kirkpatrick (Black, Parry and Kingbury, London, 1811).
22. In addition to their French army commissions issued under General Caen's authority, they carried proclamations inciting resistance as well as letters to the *Peshwa*, Sindia and Holkar in BL: Manuscript Collections, Wellesley Papers, Add. MS 13,876, ff.11–20. In particular note ff.13–18, 'Memorandum Respecting the Examination of the French Prisoners Courson, Durhone, and Dauble'. Lieutenant Dauble was particularly well suited for a role as a fifth columnist as he spoke the 'country language' (ff.11–12), and as a young man had been Claude Martin's deputy serving as a superintendent of the arsenal belonging to the Nawab of Lucknow.
23. In August 1803, the month the war with the Marathas commenced, the French garrison at Pondicherry became prisoners of war and were taken into captivity by the EIC. BL: OIOC, Board's Collections, F/4/200 no. 4530, ff.21–23, Instructions From Bengal Relative to Prisoners of War, F. Thompson Town Major Fort St George to the Chief Sec. of Govt., 6 June 1805.
24. Thorn, *War in India*.
25. Ibid., p. vii.
26. Ibid., p. 7.
27. Ibid., p. viii.
28. The Duke of Wellington died in 1852.
29. See 'Works by G.W. Forrest, C.I.E., Opinions of the Press', an addendum to G. W. Forrest, *Sepoy Generals: Wellington to Roberts* (London, 1901), p. 6.

1 MARATHA MILITARY CULTURE

1. As found in the dedicated curriculum provisions of institutions such as the Royal Military College of Science (Shrivenham) and tenured professorial appointments in that discipline by civilian centres of higher education, i.e. Cranfield University, United Kingdom.

2. See Gunther D. Sontheimer, 'Hero and Sati-stones of Maharashtra', with photographs in the appendix labelled as 'Fig. 7. Attack on a fort, Bavde' and 'Fig. 8. Conquest of a fort, Akluj' in S. Settar and Gunther D. Sontheimer, *Memorial Stones: a Study of Their Origin, Significance and Variety* (published jointly by the Institute of Indian Art History, Karnatak University, Dharwad, and South Asian Institute, University of Heidelberg, Germany, Mainpal Power Press, 1982), pp. 261–81.

3. The Maharashtrian stones come from not one but a series of time periods and cover offensive and defensive land warfare as well as naval battles. Some of the most intricate are those of the seventh-century Deccan-based Chalukya Empire.

4. The Mesopotamian example was exploited in particular by William H. McNeill, *Keeping Together in Time: Dance and Drill in Human History* (Harvard University Press, Cambridge, MA, 1995), p. 106.

5. I have used broker/dealer here as meaning individuals that brokered military employment contracts and dealt in military goods and services. These individuals may, or may not, have been warrior participants in the subsequent battles. Later, in the 1803 period, I cite specific cases of broker/dealers who were also officers of specific units. This definition stands in contrast to other authors who have used the term 'jobber-commanders', which specifically implies a commander/officer who brokers employment (jobs) for himself and his soldiers.

6. George Michell and Mark Zebrowski, *The New Cambridge History of India*, I. 7: *Architecture and Art of the Deccan Sultanates* (Cambridge University Press, 1999), p. 9. 'Their territories more or less coincided with the Marathi, Kannada and Telugu countries.'

7. See James Cuninghame Grant Duff, *A History of the Mahrattas*, vol. I, p. 73 (Longman, Rees, Orme, Brown, and Green, London, 1826).

8. Appendix V contains a working definition for the term mercenary as used in this book.

9. *Maharashtra State Gazetteers, Maharashtra Ahmadnagar District* (revised edition, Bombay, 1976), p. 64.

10. Ibid., p. 62.

11. The greater Islamic military world incorporated military knowledge drawn from a tremendous cross-section of the 'known' world. In the fifteenth and sixteenth centuries that meant contact ranging from the Iberian Peninsula to Africa, Southwest Asia to Central Asia, as well as South and Southeast Asia. By the time of official Mughal court codification in the *Namas* (i.e. Babur's military doctrine as expressed in the *Baburnama*), we can detect the regional origin or association of specific tactics as exemplified by such descriptive terms as the *Anatolian method* or in the *Rumé style*. For a refreshing look at the *Baburnama* as seen through a translation of the original language text, as opposed to Persian, see Wheeler M. Thackston's rendering: *The Baburnama, Memoirs of Babur, Prince and Emperor* (Oxford University Press, 1996).

12. Albuquerque, then Portugal's pre-eminent soldier-adventurer in India, believed the Indian weapons were better and deemed the issue to be classified military intelligence. He sent samples of Indian-made weapons to King Dom

Manoel of Portugal in 1510 to underscore his assertion of South Asian technical superiority. See his letter of 22 Dec. 1510 in Frederick Charles Danvers, *The Portuguese in India Being a History of the Rise and Decline of their Eastern Empire* (1894, reprinted Asian Educational Services, New Delhi, 1988), vol. I, pp. 211–12. Albuquerque's first-hand account directly contradicts Cipolla's assertion that the Portuguese artillery of the period was superior to that of India. Carlo M. Cipolla, *Guns and Sails in the Early Phase of European Expansion 1400–1700* (Collins, London, 1965), p. 107. In 1537 the Portuguese soldier de Cunha found indigenous bronze as well as iron cannon that indicated to him that Indian casting techniques were well in advance of the Europeans'. The iron gun tubes were not simple wrought iron barrels with hoop and stave construction as often seen in Europe. The Indians had perfected cast-iron cannon barrels that far exceeded their Western counterparts. See Danvers, *The Portuguese in India*, vol. I, p. 420. The deficiencies of sixteenth-century European iron ordnance are covered in Cipolla, *Guns and Sails*, pp. 41–3. Most of the issues surrounding superior military technology in the weapons trade of that day were based on design differences, basic metallurgy or fabrication techniques that emphasized a weapon's longevity.

13. By that time India had been producing steel and exporting pig iron to the Middle East's Damascus steel market for over 1,500 years and analysis of historic South Asian slagheaps reveals that iron production was not localized but rather carried on across the subcontinent. K. N. P. Rao, J. K. Mukerjee and A. K. Lahiri, 'Some Observations on the structure of ancient steel from South India and its mode of production', *Bulletin of the Historical Metallurgy Group, UK*, 3, no. 2 (1969), p. 12. For a chemical and qualitative spectrograhic analysis of ancient steel from Mysore see p. 16.

14. By way of contrast, in 1574 Queen Elizabeth issued an order restricting the number of guns to be cast in England to only those for the 'use of the Realm'. Cipolla, *Guns and Sails*, p. 45.

15. Contrary to the belief of many Western historians as expressed by McNeill, *The Age of Gunpowder*, p. 1.

16. The urgent need to finish casting before the arrival of the monsoon can be traced in Maratha *Chhatrapati* Shahu's later letters of April 1735. See G. S. Sardesai (ed.), *Selections from Peshwa Daftar*, vol. XVII (Bombay, 1931), pp. 26–7. My thanks to N. K. Wagle for translating these letters.

17. My thanks to Dr J. A. Charles (St John's College Cambridge) for his diagrams relating the molecular contribution of monsoon moisture to hydrogen bubble formation in casting. I also owe a special debt of thanks to Yaduendra Sahai (City Palace Museum, Jaipur) for showing me a cleverly disguised observatory, relating Indian astronomy and weather forecasting to cannon casting. The addition of a *chhatra* dome to the observatory is architecturally misleading but its remains can be seen a short distance outside the Jaigarh Fort, which has one of the world's best-preserved historic cannon-casting facilities.

18. *Maharashtra State Gazetteers*, p. 76.

19. Western historians, particularly those studying the Renaissance period or those advocating a 'military revolution', usually compare the cost of training

and equipping foot soldiers with firearms, to the time and effort required historically to train proficient archers. This has led them to consistently arrive at the conclusion that mass smallarms training was a more efficient utilization of resources than the tendered alternative of infantry armed with the bow; a supposedly balanced comparison in the analysis of the West's evolution of missile weapons. I am suggesting the analytical model may not be as relevant in South Asia where major field artillery firepower was available much earlier in a manner that drastically altered the force-to-space ratio.

20. Hall, *Weapons and Warfare in Renaissance Europe*, p. 158.

21. *Maharashtra State Gazetteers*, p. 79. In Daulatabad fortress I viewed surviving iron cannon cast in sections that matched this general description.

22. The same holds true when we speak of 'British troops' in South Asian history. A sepoy in the service of the Madras Presidency in 1803 may well have been a South Indian Muslim, but his identity would have been conveniently merged if he fought under Wellesley in 1803 since he would have been referred to generically as 'one of the British troops'.

23. Within this work, I admit to taking extensive liberties with the term 'military labour market' as originally introduced and applied to North India by Dirk H. A. Kolff in his tremendously valuable pioneering work *Naukar, Rajput and Sepoy: The Ethnohistory of the Military Labour Market in Hindustan, 1450–1850* (Cambridge University Press, 1990). Not wishing to limit the concept of military labour markets to one geographic region such as Hindustan, I have applied the essence of Kolff's theory to the larger South Asian military environment. This is reflected in the institutionalization of military employment in the Deccan as well as South India and it can be detected in the huge South Indian 'trade' armies that served overseas in ancient times, as noted in Mines Mattison, *The Warrior Merchants: Textiles, Trade and Territory in South India* (Cambridge University Press, 1984). Military labour markets were but one facet or component of the Asian military economy.

24. The EIC is thought to have had troops at Surat from 1612 onwards. By 1641–2 the British were using sepoys in their battles against the Portuguese. The free-agent approach to military employment via 'jobber-commanders' was also evident in the Madras Presidency. As of 1664/5–76 large numbers of 'native troops' using British drill appear in official EIC documentation. And by 1664 the Surat garrison was deemed large enough to defend British interests against Maratha incursions led by Shivaji. In 1673 we find the British forces in the Madras Presidency had supplemented their sepoy forces by adding Portuguese *topasses* to their numbers. By the 1680s the British troops at Surat included two companies of 'mainland Rajputs', 'Moores', 'Canoreens', and 'Topasses' from the Portuguese colonies. See D. F. Harding, *Smallarms of the East India Company 1600–1856*, vol. IV: *The Users and Their Smallarms* (Foresight Books, 1999), pp. 150–1, 237–8.

25. This latter point remains a source of historic irony. Some fervent Hindu nationalists in Maharashtra advocate a form of Hindu chauvinism that stands as the antithesis to the historic practice of peripheral incorporation found in traditional 'Maratha' armies. This makes the modern Hindu nationalist association with the Maratha warrior king Shivaji all the more interesting. For a somewhat different view of Shivaji and the process of peripheral

incorporation, as seen from the vantage point of post-Ambedkar Maharashtra, see K. L. Mahaley, 'Shivaji and the Downcasts', in *Shivaji: the Pragmatist* (Vishwa Bharati Prakashan, Nagpur, 1969), pp. 36–51.

26. For official court-sanctioned accounts of Shahuji (aka Shahji) Bhonsle's mercenary relationship with the Mughal authorities, see W. E. Begley and Z. A. Desai (eds.), *The Sha Jahan Nama of 'Inayat Khan: an Abridged History of the Mughal Emperor Shah Jahan, Compiled by His Royal Librarian; the Nineteenth-Century Manuscript Translation of A. R. Fuller (British Library, Add. 30,777)* (Oxford University Press, Delhi, 1990). (Hereafter this volume is referred to as *'Inayat Khan's Shah Jahan Nama*.)

27. See the numerous references including pictures by Gunther D. Sontheimer in Settar and Sontheimer, *Memorial Stones*.

28. *Julius Caesar: The Battle for Gaul*, a new translation by Anne and Peter Wiseman (Chatto and Windus, London, 1980), pp. 152–4, book VII 'The Attack on Gregovia', p. 161; at the siege of Alesia, Vercingetorix ordered his cavalry force of 15,000 to assemble. 'Since he was strong in cavalry, it would be very easy he said, to stop the Romans getting supplies of grain and forage.'

29. *'Inayat Khan's Shah Jahan Nama*, pp. 68–9.

30. Ibid., p. 98.

31. The *banjaras* can still be found in India, many still using their distinctively decorated carts. Colonel Hugh Pearse, writing in the *Memoir of the Life and Military Services of Viscount Lake, Baron Lake of Delhi and Laswaree 1744–1808* (London, 1908), p. 158, noted: 'The *Brinjaris* (to use the Modern spelling) still exist, and carry on their business of transporting grain from place to place, though the spread of railways has lessened their sphere of utility.' For an older but nonetheless interesting historic overview, see Captain John Briggs, 'Account of the Origin, History, and Manners of the Race of Men Called Bunjarras' (original reading 25 May 1812), *Transactions of the Literary Society of Bombay*, 1 (1819, reprinted 1877), pp. 170–97.

32. Stewart Gordon, *The New Cambridge History of India*, II. 4: *The Marathas, 1600–1818* (Cambridge University Press, 1993), pp. 44–5. While some authors define *bargi-giri* as guerrilla warfare others seem to prefer the term 'predatory warfare' but retain guerrilla as a specific description of Shivaji's method of warfare. See R. C. Majumdar, H. C. Raychaudhuri and Kalikinkar Datta, *An Advanced History of India* (Madras, 4th edition 1978, reprinted 1986), p. 460.

33. Gordon, *The Marathas*, p. 45. In fairness, however, one might speculate that it could have reflected the state's wish not to risk direct equipment loss in pitched battle.

34. William Irvine, *The Army of the Indian Moghuls: Its Organization and Administration* (London, 1903, reprinted Eurasia Publishing House, New Delhi, 1962), p. 37.

35. Colonel Henry Yule and A. C. Burnell, *Hobson-Jobson: a Glossary of Colloquial Anglo-Indian Words*, 2nd edition, ed. William Crooke (1903); new edition with Foreword by Anthony Burgess (Routledge and Kegan Paul, London, 1985), p. 69, see definition of 'BARGEER'. Jadunath Sarkar followed orthodoxy in defining *bargi* as 'a corruption of *Bargir* (a Persian loanword in Marathi), meaning a horseman supplied with his mount and arms by

Government'. But then Sarkar went out of his way to paint Maratha *bargis*, in the mid-eighteenth-century Bengal campaign, as using gang rape as a standard tactic. See Sarkar, *Fall of the Mughal Empire* (hereafter referred to as Sarkar, *Mughal Empire*) (4th edition, Orient Longman, New Delhi, 1991), vol. I, pp. 44–5. This pejorative example presented by Sarkar is typical of the contrast found in Maratha military histories. The Marathas were often portrayed as geniuses in terms of doctrine but evil in their conduct towards non-Maratha peoples.

36. There was the option of using *rakhale*, cart-mounted horse-drawn guns in the 1–3 pounder size range. The *rakhale* had been known since Emperor Babur's time (*c*. 1526), but rockets (*bana*) offered light horsemen greater mobility, as cartwheels were prone to sticking in the mud, breaking on rocks and in general slowing horsemen down when they needed replacing.

37. See *Inayat Khan's Shah Jahan Nama*, pp. 100–1. *Gajnals* also known as *jejala*. For pictures of artillery in Daulatabad and a surviving example of the 'camel gun' complete with iron 'saddle mount' as photographed by the author, see Randolf G. S. Cooper and N. K. Wagle, 'Maratha Artillery: From Dabhoi to Assaye', *Journal of the Ordnance Society*, 7 (1995).

38. That the process of linking Maratha history to guerrilla activity and wars of liberation was largely completed by the 1940s is indicated by the 1944 preface to V. G. Dighe's excellent study of Maratha expansion under *Peshwa* Baji Rao I. He wrote of the Marathas as throwing off a Deccani 'inferiority complex' and striking 'boldly for the liberation of their homeland'. He also commented on the greater trend that had emerged. 'Their early struggle for swaraj under Shivaji's leadership is now familiar to students of history in the works of Sir Jadunath Sarkar, Prof. Rawlinson, Mr. Chintaman Waidya, the two Kincaids and the essay collection published by the Shivaji Karyalaya.' Dighe, *Peshwa Bajirav I*, pp. viii–ix.

39. Napoleon is alleged to have referred to his festering Peninsular War as 'the Spanish ulcer'. The French spent a great deal of their time hunting down the insurgent bands (*partidas*) that had been organized to harass them by the British. A high percentage of those original guerrillas were members of criminal bands recruited under pardon to harass the French. These *guerra* or small wars within the larger war effort lead to the insurgents being called *guerrillas*. Charles J. Esdaile's exhaustive research in the Spanish archives has done much to enhance our understanding of how these guerrillas fitted into the history of that war; see his *The Spanish Army in the Peninsular War* (Manchester University Press, 1988).

40. Mao Ze Dong, thought by many to be the twentieth century's greatest authority on guerrilla warfare, wrote about it in terms of the evolution of warfare with guerrilla activity representing an initial phase of military development. In his own People's Liberation Army he traced the three stages of evolution from *guerrilla* to *irregular* to *conventional*. Only gradually did they assume the trappings of regular armies, acquiring more captured weapons, donning uniforms and conducting irregular warfare until they had built from strength to strength and could engage the Nationalists (KMT) in set-piece battles where Mao's forces ultimately challenged fixed fortifications in conventional positional warfare.

41. Ian Raeside in *The Decade of Panipat (1751–61)*, Marathi Historical Papers and Chronicles (Popular Prakashan, Bombay, 1984), p. 44.

42. Ibid., pp. 140–1, footnote 14.

43. Sam C. Sarkesian (ed.), *Revolutionary Guerrilla Warfare* (Precenent Publishing Inc., Chicago, 1975), p. 7.

44. Shivaji's army was visually identifiable in the conventional context of uniforms. His enlisted men wore uniforms, emphasizing standardized turbans and tight-fitting trousers, while his officers were issued with helmets, quilted armour and chain mail. The inventory of Shivaji's possessions at his time of death was contained in *Dattaji-Malkare Bakhar*. His personal armoury, the contents of which ranged from grapeshot to bugles, included such standardized uniform equipment listings as Cuirass: 4,000 pieces, Coats of mail (*chhata*): 1,000 pieces, Helmets: 4,000 pieces. French, English, Dutch and Portuguese observers among others noted these trappings. See Jadunath Sarkar, *House of Shivaji Studies and Documents on Maratha History: Royal Period* (reprinted Orient Longman, Delhi, 1978), pp. 163–9. Shivaji may have had occasion to use disguise or deception and try and pass himself off as something other than a soldier. However, such isolated cases of subterfuge cannot be construed as proof of a 'guerrilla identity' when they were merely mission-oriented ploys.

45. R. C. Majumdar (ed.), *The History and Culture of the Indian People* (Bombay, 1974), vol. VII, pp. 254–5.

46. Ibid., p. 257.

47. BL: Warren Hastings Papers, Add. MS 29,209, see 'Histories of the Two Maratta Wars & of the Rise & Declension of the House of Sewajee', ff.309–310.

48. The importance of maintaining a monsoon fighting capability was of utmost importance to the survival of the Marathas. This was reflected in the fact that portions of the Maratha state edict known as the *Ajnapatra*, issued under *Chhatrapati* Shahu, specifically noted defence preparedness in relationship to the artillery (i.e. the battle-ready weatherproofing of guns and the maintenance of gunpowder stores during the rains). The *Ajnapatra* or 'Handbook of Maratha Statecraft' appeared in the first quarter of the eighteenth century (1715–16). A portion of the manual was published in the *Journal of Indian History*, 8 part I, no. 22 (Madras, April 1929), pp. 207–33. Unfortunately this was only a partial translation lacking technical military terms and items such as artillery preparation for the monsoon season. P. N. Ghoshi published a highly detailed Marathi version (Pune, 1960), made available to the author by way of readings given by Professor N. K. Wagle.

49. Dr V. T. Gune (ed.), *Gazetteer of the Union Territory Goa, Daman, and Diu*, District Gazetteers Part 1 (Goa, Panaji, 1979), p. 166.

50. Ibid., p. 168.

51. BL: Add. MS 29,209 'Warren Hastings Papers', 'Histories of the Two Maratta Wars & of the Rise & Declension of the House of Sewajee', ff.309–310.

52. See Francois Bernier, *Travels in the Mughal Empire AD 1656–1668* (1891, reprinted Delhi, 1989), pp. 187–90.

53. Jagadish Narayan Sarkar, *The Military Despatches of a Seventeenth Century Indian General* (Calcutta, 1969), p. 55. Anees Jahan Syed, *Aurangzeb in Muntakhab-Al Lubab* (Bombay, 1977), p. 220. Maratha forts surrendered by Shivaji to Jai Singh/Aurangzeb in the 1665 Treaty of Purandhar:

1	Rudramala or Vajragarh	13	Tulsikhul
2	Purandhar	14	Nardurg
3	Kondana	15	Khaigarh or Ankola
4	Rohira	16	Marggarh or Atra
5	Lohgarh	17	Kohaj
6	Isagarh	18	Basant
7	Tanki	19	Nang
8	Tikona in Konkan	20	Karnala
9	Mahuli	21	Sangarh
10	Muranjam	22	Magarh
11	Khirdurg	23	Khandkala at Kondana
12	Bhandardurg		

A. J. Syed used a different form of transliteration than Stewart Gordon, but many of the forts listed above can be found on Gordon's map (see map 2 in chapter 1).

54. One could make a strong case that as Shivaji competed with the Mughals directly, his army began to look more and more like that of the Mughals. And those who have argued that Shivaji's army was a *guerrilla force* composed of *fleet-footed horsemen* usually avoid mentioning the eyewitness description of Shivaji's military entourage as it travelled to meet Aurangzeb. It displayed all the earmarks of classic Mughal armies including Turkish-style headgear for his footmen, officers in *palkis*, *banjaras*, baggage-laden camels and an infantry force marching behind a standard-bearing elephant. See Sarkar, *House of Shivaji*, pp. 144–5.

55. Pissurlencar, *Portuguese–Mahratta Relations*, p. 36.

56. Radhabai Balkrishna, *Shivaji the Great, Part IV Shivaji, the Man and His Work* (Arya Bhanu Press, Kohlapur, 1940), p. 103. On p. 90 the author asserts that European experts were sought for the casting of the largest cannon.

57. Pissurlencar, *Portuguese–Mahratta Relations*, p. 36.

58. Some *pindaries* were ethnic Marathas and some ethnic Marathas led *pindari* bands. However, it would be wrong to assume that all *pindaries* were Marathas, or vice versa, as several of the leading nineteenth century *pindari* leaders and their bands proved to be of Pathan origin (e.g. Amir Khan). Many *pindari* bands took refuge in Maratha territories when there were chances for legitimate employment as auxiliaries.

59. 'Kartoji Gujar was made by Shivaji his *Senapati* or Commander-in-Chief with the titles of Pratap Rao and *sar naubat* of horse.' *Senapati* Kartoji Gujar's 'meteoric career' was 'cut short in a rash charge on the Bijapuri army' in 1674. Perhaps it pre-empted any crisis in civil–military affairs that might have occurred with regard to the management of the military under so powerful a *Chhatrapati* as Shivaji. See Sarkar, *House of Shivaji*, pp. 162–3.

60. I was extremely fortunate to have worked with Professor N. K. Wagle on the military significance of an eyewitness account concerning the great battle between *Peshwa* Bajirav I and *Senapati* Trimbuk Rao Dabhade fought

at Dabhoi. I also made extensive use of Professor Wagle's translation of artillery-relevant entries from the *Ajnapatra, Peshwa Daftar* and the *Shahu Daftar* in the supporting footnotes of the article listed previously as Cooper and Wagle, 'Maratha Artillery'. Those primary sources revealed the extent to which Maratha state policy had evolved towards artillery. Official documentation, much of it in *modi* script painstakingly translated by Professor Wagle to retain its peculiarly Maratha *militarese* context, covered issues concerning cannon casting (metallurgy and mould construction), weatherproofing for monsoons, ammunition procurement (i.e. intercommunal subcontracting) as well as funding and officially regulated ballistic testing.

61. *Organic firepower* refers to the amount of firepower contained within a specified unit. In essence, the firepower the unit normally carried with it into battle as opposed to secondary firepower assigned or seconded to it. For example, British battalion guns – the two field pieces assigned to an infantry battalion in the eighteenth century – along with the massed ranks of infantry, would represent the *organic firepower* for that battalion.

62. 'One of the earliest kinds of scatter projectiles was case shot, or canister, used at Constantinople in 1453. The name comes from its case, or can, usually metal, which was filled with scrap, musket balls, or slugs. Somewhat similar, but with larger iron balls and no metal case, was grape shot, so-called from the grape-like appearance of the clustered balls . . .' Albert C. Manucy, *Artillery Through the Ages* (Washington, 1962), pp. 68–9.

63. Reprint by P. N. Ghoshi (Pune, 1960).

64. Later 'fixed ammunition' became quite popular with European nations. It had the projectile, sabot and bagged powder charge all attached or *fixed* in one unit, a concept akin to the cartridge. This was usually accomplished by attaching the projectile and bagged powder charge on either side of a wooden sabot.

65. See *Selections from the Peshwa Daftar*, vol. 41, letter no. 87, as cited in Cooper and Wagle, 'Maratha Artillery'.

66. I submit that overall European trends in adopting a volume of fire approach to artillery doctrine did not develop in earnest until the advent of breechloading artillery and could not be said to have reached their full potential until World War One (1914–18).

67. Dighe, *Peshwa Bajirao I and Maratha Expansion*, p. 24.

68. A number of British irregular units kept their matchlocks until well into the 1830s. See Harding, *Smallarms*, vol. IV, pp. 460–3. The retention of matchlocks by the EIC's irregular troops did not represent the pawning-off of inferior weapons on secondary forces. For those irregulars responsible for their own weapons' replacement costs, it was a case of economics since they could perform simple repairs and maintain their arms.

69. Harding, *Smallarms*, vol. III, p. 204.

70. Among the best surviving examples that I have seen are the spiral bound-cotton match-chords of Jaipur. Match-chord material, regardless of composition, was usually soaked in a potassium nitrate solution to regulate the burning. Harding also pointed out other types of tree bark were used as well. See Harding, *Smallarms*, vol. II: *Catalogue of Patterns*, p. 494.

71. Matchlock men were often organized, trained and drilled in the same fashion as flintlock-carrying infantry. The addition of a bayonet lug to their matchlock barrels enabled matchlock-armed infantry to use European-style socket bayonets. Thorn, *War in India*, p. 78.

72. A battlefield analogy could be drawn with the American war in Vietnam circa 1970. Most US infantry units by then carried the Colt M-16 (US M16A1), a gas-operated assault rifle, but army snipers depended on a bolt-action rifle (US M40A1) based on Remington's adaptation (Model 700) of the classic Mauser action which was patented in the nineteenth century.

73. Hall, *Weapons and Warfare in Renaissance Europe*, p. 3.

74. Innes Munro, *Narrative of the Military Operations on the Coromandel Coast Against the Combined Forces of the French, Dutch and Hyder Ally Cawn from 1780 to the Peace in 1784* (London, 1789), p. 131. And for a nineteenth-century example of matchlock snipers in action against the British, see Alfred Clarke (aka 'Carnaticus'), *Summary of the Mahratta and Pindarree Campaign, During 1817, 1818 and 1819 under the Direction of the Marquis of Hastings: Chiefly embracing the Operations of the Army of the Deckan under the command of His Excellency Lieut.-Gen. Sir T. Hislop* (London, 1820), p.146.

75. See Harding, *Smallarms*, vol. III: *Ammunition and Performance*, p. 377.

76. Ibid., p. 378.

77. Dr Peter Krenn, the Director of the Styrian Provincial Armoury at Graz, Austria, compiled some of the most dramatic European matchlock test data. In addition to firing selected sixteenth–eighteenth-century black powder weapons for accuracy and chronographed velocity, Krenn went several steps further to test wound channel ballistics. An examination of his ballistic tables reveals a startling comparison between a seventeenth-century European matchlock musket and the state-of-the-art Austrian AUG (StG 77) assault rifle firing the NATO standard 5.56 mm cartridge. The matchlock firing a 17.38 gram projectile at a muzzle velocity of 449 metres per second had an impact energy at the muzzle of 1752 joules and 1242 joules at a distance of 30 metres from the muzzle. The 5.56 mm AUG assault rifle fired a 3.6 gram projectile at a muzzle velocity of 990 metres per second to produce a muzzle energy of 1764 joules and an energy of 1642 joules at a distance of 30 metres from the muzzle. In comparing the two we find that the tested seventeenth-century matchlock had 75.6 per cent of the impact or 'knock-down' energy at 30 metres ($1242 \div 1642 \times 100 = 75.63946\%$). That percentage constitutes more than enough for a margin of lethality given an impact in a vital area such as the torso. At a distance of 100 metres the matchlock's 17.38 gram projectile still had sufficient energy to penetrate 93 mm in a dry spruce wood block. The most complete range of data was contained in the book *Von Alten Handfeuerwaffen, Entwicklung Technik Leistung* (Graz, Austria, 1989), a compilation of five essays by four authors in 148 pages. English-language readers may access a portion of this data in Krenn, 'Test-Firing Selected 16th-18th Century Weapons', translated by Erwin Schmidl, *Military Illustrated Past & Present*, 33 (February 1991), 34–8.

78. Two months before the outbreak of hostilities in 1803, Arthur Wellesley wrote: 'The greater experience I gain of Marhatta affairs, the more convinced

I am that we have been mistaken entirely regarding the constitution of the Marhatta empire. In fact, the Peshwah never has had exclusive power in the state.' See Arthur Wellesley to Major Malcolm, Camp, 20 June 1803 in *Selections From the Dispatches and General Orders of Field Marshal the Duke Of Wellington*, ed. Lt-Col. Gurwood (London, 1843), p. 58.

79. The battle was popularized in North Indian songs and oral tradition. A modest English-language contribution to the bardic tradition was made by Rudyard Kipling with his 1890 poem 'With Sindia to Delhi', which was told from the perspective of a Maratha horseman. See the *Collected Verse of Rudyard Kipling* (London, 1912), pp. 244–9.

80. Two of the more useful accounts are Tryambak Shankar Shejwalkar, *Panipat: 1761* (Deccan College, Poona, 1946) and Sarkar, *Fall of the Mughal Empire*, vol. II: *1754–1771*.

81. The gun Rudyard Kipling referred to as *Zamzamah* was said to have been used at Panipat. It was attributed to an Armenian craftsman in Mesrovb Jacob Seth's, *Armenians in India* (reprinted Delhi, 1983), pp. 115–19. The Armenians brought a number of casting secrets with them to Persia as well as Afghanistan and these Armenian casting experts formed an identifiable community of some influence. Their specialized artillery knowledge was thought to have been gained in the Russo-Turkish Wars as well as various Austrian Wars. Many of the Armenian technicians were given 'Persian names' which confuses their identification on 'signed' cannon barrels and on memorial stones in the grave yards of Lahore as well as Kabul.

82. Raeside, *The Decade of Panipat*, p. 95, commented on the Marathas preferring their 'old system'.

83. Western military historians who doubt the ability of clan-based Asian warrior societies to have moved voluntarily towards systematic officer training should consider the complex model posed by Japan. Clans in Satsuma and Choshu were aware of differences in English, French and Prussian military theory. They had already begun modernizing military professionalism and training prior to the introduction of wide-scale institutionalized officer leadership training under Yamagata Aritomo during the Meiji period.

84. In September 1787 the British Resident with Sindia reported that the 'Princes of Hindostan' had a 'passion for artillery and large bodies of infantry (formed somewhat on the model of ours) with which all or most of the Hindustan Chieftans have of late years been inspired'. Sir Jadunath Sarkar, *English Records of Maratha History: Poona Residency Correspondence*, vol. I, *Mahadji Sindhia and North Indian Affairs 1785–1794* (Government Central Press, Bombay, 1936) (hereafter referred to as PRC, *North India Affairs*, vol. I), p. 254, Resident W. Kirkpatrick to Governor General Earl Cornwallis, 'Fathgarh', 14 Sept. 1787.

85. P. M. Joshi's Foreword in *Persian Records of Maratha History*, general editor: P. M. Joshi, vol. II: *Sindia as Regent of Delhi (1787 & 1789–91)*, translated by Jadunath Sarkar, published by the Director of Archives, Government of Bombay, 1954.

86. Harding, *Smallarms*, vol. IV, p. 238. In the interest of accuracy, it should be mentioned that this early British practice of drilling South Asian troops was

not limited to Bombay. See also Harding's remarks on p. 155 regarding the 1687 orders of Madras Governor Elihu Yale on the discipline and drill of 'native troops'.

87. PRC, *North India Affairs*, vol. I, pp. 203–5, W. Kirkpatrick, British Resident with Sindia, to Gov. Gen. Earl Cornwallis, 'Fathgarh', 19 July 1787.

88. PRC, *North India Affairs*, vol. I, pp. 137–8, Resident W. Kirkpatrick to Gov. Gen. Earl Cornwallis, 'Safdar Jang's tomb, near Delhi', 10 March 1787.

89. PRC, *North India Affairs*, vol. I, pp. 38–40, Resident James Anderson to Gov. Gen. John Macpherson, 'Sindhia's Camp, Shergarh', 2 Jan. 1786. Sardesai, *History of the Marathas*, vol. III, p. 147 noted: 'By the close of 1784 he had run into a debt of 80 lacs. His own force of 30 thousand with his artillery cost him 7 lacs a month and the imperial contingents . . . added about three lacs to his monthly cost.'

90. PRC, *North India Affairs*, vol. I, p. 76, de Boigne to J. A. Pow, 'Dated Camp near Kalinjar', 3 July 1786.

91. PRC, *North India Affairs*, vol. I, pp. 214–17, 'Paper of Intelligence from Sindia's Army, Relative to the Action of the 12th Shawal or 28th July (received at Poona)', 25 Aug. 1787. Only one European mercenary was reported to have defected while Lestineau and Vasseult remained loyal at that time. Monsieur Lestineau was later said to have defected but the data was incomplete and reports were conflicting. See pp. 261–2, Resident W. Kirkpatrick to Gov. Gen. Earl Cornwallis, 'Havilganj', 21 Sept. 1787 and compare to pp. 210–14.

92. PRC, *North India Affairs*, vol. I, p. 76, de Boigne to J. A. Pow, 'Dated Camp near Kalinjar', 3 July 1786.

93. PRC, *North India Affairs*, vol. I, pp. 210–14, 'Intelligence from the Camp of Sindia, Written 12th Shawal 1201 Higiree (28th July 1787), Near the Hill Jowana (Sent by Colonel Harper to G.G. Received 15th August 1787)'. It should be noted that later, near Agra in 1788, de Boigne was still reported as operating with Lestineau in an action against Ishmail Beg Khan; see pp. 306–7, 'Translation of a Letter received by Bhagwant Row and Delivered to the Calcutta Council', 16 July 1788.

94. PRC, *North India Affairs*, vol. I, pp. 258–9, W. Kirkpatrick to C.W. Malet, 'Fathgarh', 15 Sept. 1787.

95. PRC, *North India Affairs*, vol. I, pp. 269–70, 'Translation of Proposals Delivered in Writing By Bhow Bakhshy to Major William Palmer', 'Cawnpore', 14 Nov. 1787.

96. PRC, *North India Affairs*, vol. I, pp. 360–1, Resident W. Palmer to Gov. Gen. Earl Cornwallis, 'Sindia's Camp', 28 Jan. 1790.

97. PRC, *North India Affairs*, vol. I, pp. 390–1, Resident W. Palmer to Gov. Gen. Earl Cornwallis, 'Fathgarh', 21 July 1793.

98. PRC, *North India Affairs*, vol. I, pp. 390–1, Resident W. Palmer to Gov. Gen. Earl Cornwallis, 'Fathgarh', 21 July 1793, says 27 *lakhs* while a figure of 35 *lakhs* is given in C. W. Malet to the Court of Directors, 'Poona', 5 Feb. 1794, p. 391.

99. Grant Duff, *A History of the Mahrattas*, vol. III, p. 33.

100. Major Lewis Ferdinand Smith, a British veteran of Sindia's service, observed that Sindia's Telinga battalion was armed with flintlock muskets made at Agra. *A Sketch of the Rise, Progress and termination, of the Regular Corps, Formed & Commanded by Europeans, In the Service of the Native Princes of India* (Calcutta, 1805), pp. 50–1. Mercenary officer William Henry Tone, 'Illustrations of Some Institutions of the Mahratta People', *The Asiatic Annual Register, for the Year 1799* (London, 1800), p. 143, observed: 'The late regulations of the company, respecting the return to Europe of all unserviceable arms, may for a time prevent the increase of native infantry corps, but then it will drive them to the expedient of making their own firelocks, as Scindeah has done, and his are very excellent ones, far superior to the ordinary Europe arms to be met with in the bazars.'

101. The policy for return of unserviceable arms noted by Tone followed a large arms-dealing scandal in Bombay in the late 1790s. Chests of EIC muskets had been found among the possessions left by the mercenary officer Capt. Peter Gossan who died in Maratha service in 1797. Charges were laid against EIC employees for illegal arms trafficking. For details see BL: OIOC, Proceedings of the Bombay Military Council, P/354/1, ff.1518–1612, 1676–1679, 1974–1978.

102. PRC, *North India Affairs*, vol. I, pp. 392–7, entitled 'Mahadji Sindhia's General Col. De Boigne's Troops, 1793'. It was originally an enclosure for C. W. Malet to the Court of Directors, 'Poona', 5 Feb. 1794. By way of clarification, these 'stock makers' were probably specialist wood workers who restocked weapons broken in service or those decommissioned by the EIC as unfit for service. During field research I visited gun makers in Udaipur who still fashion museum-grade flintlocks for sale to the West as authentic reproductions. I watched a local stock maker turn out a mango-wood stock for a flintlock using nothing more sophisticated than an adze and three chisels. The process from plank to 'rough in-letted stock' took less than twenty minutes.

103. All spellings used in this troop *breakdown* were taken directly from source. See PRC, *North India Affairs*, vol. I, p. 391, C. W. Malet to the Court of Directors, 'Poona', 5 Feb. 1794.

104. Many *sardars* as well as Europeans in Sindia's service held Mughal ranks reflecting their role in the Mughal infrastructure used in Sindia's control of the Doab.

105. For an example of a *hundi* being used to cover Rs. 25,000 for two months' wages in arrears see BL : OIOC MSS. EUR. D. 547, General B. de Boigne to Major Gardner, Coel, 30 June 1795, ff.27–28.

106. BL: OIOC MSS. EUR. D. 547, General B. de Boigne to Major Gardner, Coel, 30 June 1795, ff.27–28.

107. BL: OIOC MSS. EUR. D. 547, General B. de Boigne's 'Statement of the Articles Sent by Dolpore Batt'n', Coel, 30 June 1795, f.24. Despite being out of sequence in this volume of correspondence, this was originally an enclosure for Coel, 30 June 1795, ff.27–28 which refers to the Light Infantry being issued specially commissioned light muskets and bayonets. Any left over light muskets were to be assigned to Sindia's grenadier officers.

108. Sindia's battalions continued to enjoy a 5:2 ratio in organic artillery compared to the British battalions who had two battalion guns each. The Wellington Papers at the University of Southampton indicate that after having withdrawn the diminutive 3 pounders from service as battalion guns, the EIC tried to reissue them on different carriages as 'gallopers' or horse artillery (1799–1802). But they proved too light for that purpose and a number of them were then pawned-off on South Asian allies like the *Nizam* of Hyderabad who were billed accordingly under terms of the Subsidiary Alliance Treaty.

109. BL: OIOC MSS. EUR. D. 547, General B. de Boigne to Captain Robert Sutherland, Camp, 19 Nov. 1795, ff.22–23.

110. BL: OIOC MSS. EUR. D. 547, General B. de Boigne to Major Gardner, Coel, 30 June 1795, f.28.

111. BL: OIOC MSS. EUR. D. 547, General B. de Boigne to Captain Robert Sutherland, Camp, 19 Nov. 1795, ff.25–26.

112. BL: OIOC MSS. EUR. D. 547, General B. de Boigne to Captain Robert Sutherland, Camp at Coel, 25 Dec. 1795, f.29.

113. BL: OIOC MSS. EUR. D. 547, ff.3–4. J. W. Hessing to Major Sutherland, Fort at Agra, 22nd *Jemad ul Awul* 1215 (16 Jan. 1801).

114. Although much is made here of recruiting from the lower strata of the caste system, the Marathas also drew military men from various Brahmin communities. British observers also noted the potential for Brahmin control or domination within the *Peshwa*'s army. Prior to the 1803 war, Captain Hemming in Hyderabad reported that the *Peshwa* intended to raise several battalions of sepoys to match Sindia. But unlike Daulat Rao Sindia, the *Peshwa* planned to use Brahmins rather than Europeans for the leading appointments of his officer corps. See T. E. Colebrooke, *Life of the Hon. Mountstuart Elphinstone* (John Murray, London, 1884), vol. I, p. 34.

115. For an analysis, which suggests the tactical use of the infantry column was becoming passé among the Marathas by 1725, see the footnotes in Cooper and Wagle, 'Maratha Artillery'.

116. The charge of inferior discipline remained a catchall indictment levelled against military rivals and it would periodically re-emerge in history. See the use of the term in an 1864 case of British interservice rivalry as cited by Peter Stanley, *White Mutiny, British Military Culture in India, 1825–1875* (Hurst and Co., London, 1998), p. 272.

2 BRITISH PERCEPTIONS AND THE ROAD TO WAR IN 1803

1. John Henry Grose, *A Voyage to the East Indies*, vol. II (London, 1766), pp. 91, 140, 222.

2. Ibid., vol. II, p. 140.

3. Ibid. Commodore James also had a chance to observe similar Maratha field craft at Fort Goa as directed by a commander identified as Ramaji Punt, p. 218.

4. *Reports from the Committee of Secrecy Appointed to Enquire into the Causes of the War in the Carnatic and of the Condition of the British Possessions in those Parts*, vol. I, 1781.

5. Munro, *Narrative of the Military Operations*, p. 103.

6. Tone, 'Institutions of the Mahratta People', p. 130. William Henry Tone, brother of Irish rebel Wolfe Tone, wrote his observations on the Marathas in a letter to a friend in the Madras army. While it is usually dated in accordance with the publication date, contextual as well as circumstantial evidence indicate parts of it were written earlier (some items parallel, or originated in, Major Alexander Dirom, *A Narrative of the Campaign in India which Terminated the War with Tippoo Sultan in 1792* (London, 1793; reprinted Asian Educational Services, Delhi, 1985)). Maratha mercenary officer Major Lewis Ferdinand Smith reported W. H. Tone was shot in the head and killed while serving Holkar in the 1802 battles between Sindia and Holkar. See Smith, *A Sketch of the Regular Corps*, p. 14.

7. Major James Rennel, Surveyor-General in Bengal, *Memoir of a Map of Hindoostan; of the Mogul Empire* (London, 1792), p. lxx.

8. Noted here as the previously cited Dirom, *A Narrative of the Campaign*. For rank and duty specific references see pp. 144, 185.

9. Ibid., pp. 7–8.

10. Ibid., pp. 1–2. Some of the British may have also harboured negative opinions towards the Marathas as the result of the uneasy Anglo-Maratha alliance against Haider Ali in 1766–7. The British believed the Marathas had cut short their military efforts in that earlier war after having been 'bought out' by Haider.

11. Ibid., p. 2.

12. A sketch of this scene was included in the early editions of Dirom's book.

13. Ibid., p. 9.

14. Western cannon were also given names (i.e. 'Mons Meg', the fifteenth-century cannon made in Flanders, or 'Big Bertha', the massive gun associated with World War One). The practice of painting individual names on pieces of military technology was common in the West during the twentieth century with the aircraft *Enola Gay* gaining notoriety owing to the association of its name with the destruction generated by the dropping of the atomic bomb known as 'Little Boy' on Hiroshima.

15. Dirom, *A Narrative of the Campaign*, p. 11. Once again there are parallel, or perhaps *repeated*, observations on the same subject ten years later by Tone in 'Institutions of the Mahratta People', pp. 143–4. However, the wheel makers of India varied on an individual as well as regional basis and there was no one single pattern used then on a subcontinental basis.

16. Following campaign experience in the Maratha country in 1803 the Bombay army developed a marked preference for the wheels made by the Maratha craftsmen in Pune. See *Supplementary Despatches and Memoranda of Field Marshal Arthur Duke of Wellington, K.G. edited by his son the Duke of Wellington* (London, 1858) (hereafter referred to as *Supp. Desp.*). This reference is found in vol. XIII, p. 232, [no. 490], Deputy Adjutant General to Lt Col. Colman [Aka Coleman], 21 March 1804, as well as [no. 492], 23 March 1804.

17. Dirom, *A Narrative of the Campaign*, p. 12.

18. Prior to the capture of Tipu's bullock-breeding facilities at Srirangapatnam in May 1799, Arthur Wellesley had noted: 'The want of speed in the artillery of this country has been the cause that many advantages have been missed, many opportunities of bringing the enemy to action have not been taken because the artillery could not be brought up in time; and, for some unaccountable reason the Native armies, having had better draught-bullocks and larger establishments, have been able to draw off their artillery when that of the British Army could not be moved.' Arthur Wellesley to Major General St Leger, Fort William, 11 April 1799, *Supp. Desp.*, p. 1.

19. Dirom, *A Narrative of the Campaign*, p. 11.

20. Tone had identified the *Peshwa*'s forces as being made up of Hindus and Muslims. He noted the Hindus as northerners primarily, 'Raaj Poote and Purvia cast', meaning Rajput and Purbiya. Tone, 'Institutions of the Mahratta People', p. 139.

21. Dirom, *A Narrative of the Campaign*, p. 11.

22. Ibid., pp. 103–4.

23. Captain John Little had noted a total lack of ammunition among 300 Christian Maratha infantry from Goa. See his report to the Governor-General as commander of 'the Bombay Detachment in the Maratha Army', 31 Dec. 1791, pp. 534–5, in Forrest, *Maratha Series*.

24. Sardesai noted Dhundia was descended from an old Pawar family that had soldiered for the Adil Shahi rulers. Dhundia was an accomplished Maratha light horse mercenary who served a variety of masters from the Raja of Kolhapur to Haidar Ali and Tipu Sultan. And it is believed that it was Tipu that converted Dhundia to Islam. Sardesai, *History of the Marathas*, vol. III, pp. 360–2.

25. BL: Add. MS 13,644, 'Malcolm & Lambton's Journals in 1799', ff.87–89.

26. *Supp. Desp.*, vol. II, p. 28, [no. 454] Arthur Wellesley to Lieutenant-Colonel Dalrymple, Camp at Hurryhur, 21 June 1800.

27. Ramchandra Appa and other Patwardhans had joined Arthur in the summer of 1800, but the Gokhale clan made a particular sacrifice in this campaign. Dhondopant Gokhale was killed in the action near Kittur and his nephew Bapu Gokhale (the then future general of Baji Rao II) was wounded. Sardesai, *History of the Marathas*, vol. III, p. 362.

28. Lieutenant-Colonel John Gurwood, *The Dispatches of Field Marshal the Duke of Wellington K. G., during his various campaigns in India, Denmark, Portugal, Spain, and the Low Countries and France, from 1799 to 1818*, vol. I (1st edition, London, 1834) (hereafter referred to as *Wellington Dispatches*), p. 120. Maj. Gen. Arthur Wellesley to Lt Gen. Stuart, Camp 12 miles north of the Gulpurba, 29 March 1803.

29. *Wellington Dispatches*, vol. I, p. 122, Engagement given by Maj.-Gen. Arthur Wellesley to the Vakeel of Appah Sahib. It was an enclosure that accompanied Wellesley's letter to Lt-Gen. Stuart, 29 March 1803.

30. 'As before long we may look to war with the Marhattas, it is proper to consider of the means of carrying it on. The experience which has been acquired in the late contest with Dhoondiah Waugh, of the seasons, the nature of the country, its roads, its produce, and its means of defence, will be of

use in pointing them out. I shall detail my observations upon each of these points, for the benefit of those in whose hands may be placed the conduct of the operations of the army in case of such a war, as I have above supposed we may expect.' 'Memorandum upon Operations in the Marhatta Territory, 1801', *Wellington Dispatches*, vol. I (2nd edition, London, 1837), p. 357.

31. *Wellington Dispatches*, vol. I, p. 129, April 1803. For a more detailed list of the *saradars* and the forces that joined Arthur Wellesley, see BL: Add. MS 13,748 'Correspondence of Sir John Malcom 1803–1810', ff.111–131.

32. Arthur Wellesley to Major-General Brathwaite, Camp at Jellahall, 30 Aug. 1800, *Supp. Desp*, vol. I, [no. 561], pp. 132–3. The manufacturing facilities Wellesley indicated were those associated with wheel production and located in Tipu's fortress of Srirangapatnam. By 1803 the factory produced a number of wheeled vehicles ranging from gun carriages to fire engines as noted in BL: OIOC, Board's Collections, F/4/200 no. 4515, ff.1–42, 'Report on Gun Carriage Manufacturing at Seringapatam'. Pontoon trains were a common sight in many of the European armies by 1800. The technical details of pontoon carriage construction were made widely available to European armies in T. Jeffrey's etchings for Guillaume Leblond, *A Treatise of Artillery: or, Of the Arms and Machines Used in War Since the Invention of Gunpowder* (London, 1746, reprinted Museum Restoration Service in Ottawa, 1970).

33. BL : Add. MS 13,644 'Malcolm & Lambton's Journals in 1799', f.87.

34. Ibid., f.83.

35. Tarasankar Banerjee conducted a short analytical study of sources dealing with the Maratha invasion of Bengal 1742–4. In particular, he compared the historically contemporary Bengali account written by 'Gangaram' in *Maharashtra Puran* with Persian, Marathi and English sources. Gangaram noted that in the rainy season of 1742, Maratha troops demonstrated advanced skills in building pontoon bridges to facilitate their operations in eastern Bengal. Boats were lashed together and bamboo laid across them. These were covered with mats that supported a layer of earth. When completed, the bridge was sufficiently strong to allow Maratha mounted units as well as infantry to cross. See Tarasankar Banerjee, 'Maratha Invasions of Bengal', 1742–1744', in A. G. Pawar (ed.), *Maratha History Seminar Papers* (Shivaji University Press, Kolhapur, 1971), p. 186.

36. See Arthur Wesllley to Major-General Brathwaite, Camp at Hoobly, 22 Oct. 1800, *Supp. Desp.* vol. I, [no. 643], p. 229.

37. Arthur had written a memorandum on the local manufacture of basket boats, but he stopped short of recommending their use in lieu of pontoons as Maratha and Mysore armies had done on numerous occasions. See *Supp. Desp.*, vol. IV, pp. 55–6, 'Memorandum Respecting Basket Boats', Camp, 27 March 1803. He seems to have been less than convinced about the strength and capacity of basket boats, as reflected in his letter to Bombay's Governor Jonathan Duncan, p. 106 [no. 1585], Camp at Bardoly, north of the Beemah, 5 June 1803. Ironically, five years later while writing to the Right Honourable Henry Dundas about the threat to British India from invasion across the rivers of the Punjab, Arthur would claim that the best

pontoons were those made using basket boats. See Arthur Wellesley to Dundas, *Supp. Desp.*, vol. IV, [no. 1897], pp. 592–7, Dublin Castle, 20 April 1808.

38. Wellesley's letters acknowledge the extensive theoretical testing and field experiments that his ideas had forced upon the Bombay army. See *Supp. Desp.*, vol. IV, [no. 1572], p. 95. Arthur Wellesley to Major-General Nicholls, Camp at Poonah, 25 May 1803. The paper trail can be further traced in *Supp. Desp.*, vol. IV, [no. 1590], pp. 109–10, Arthur Wellesley to Jonathan Duncan, Camp, 10 June 1803 and Maharashtra State Archives, Mumbai, Bombay Military Dept., Military Board Diary, 1804, no. 73, ff.889–898, 12 July 1804. 'Report compiled on the status of pontoon bullocks including inventory and disposition'.

39. Arthur Wellesley to Major-General Nicholls, Camp at Poonah, 14 May 1803, *Supp. Desp.*, vol. IV, [no. 1559], pp. 80–1.

40. BL: Add. MS 13,722 'Political Papers on Sir G. H. Barlow', see 'Hasty Notes on the Account of the revenues of the land & dependencies Held by General Perron'. Also note ff.15–26, the 'Political Memorandum' by George Barlow to Richard Wellesley, Burhanpore, 15 Oct. 1803 with notations in the margin by Wellesley. This is the draft covering revenue-generating lands and the disposition of 'Shah Alum's pension' or – as it came to be known – 'the Mughal settlement'.

41. In 1802 Colonel Close, the British Resident at Pune, had feared that if the Emperor died Richard Wellesley might use that event as an excuse to intervene in force – but that scenario could spark a succession dispute. Sindia's hold on the Doab was firm while Richard Wellesley's ambition was suspect. Dedicated members of the Wellesley clique confided to each other that there was no foreseeable end to this process and that was worrisome as it could only mean war at a time when the Bengal army was rife with insubordination in the aftermath of a mutiny. See Mounstuart Elphinstone's notes on his conversation with Colonel Close in Pune, 1802, including his (Elphinstone's) apparent admission of having privately ridiculed the Governor-General for making his brother Henry Wellesley the Lieutenant-Governor of the land ceded by the Nawab of Awadh, in Colebrooke, *Mountstuart Elphinstone*, vol. I, p. 42.

42. Thackston, *The Baburnama*, pp. 352–3.

43. Patience was a virtue and there was no rush to change the *window dressing* of Mughal power. It was not until 20 Nov. 1837, by legislative order, that Persian ceased to be a court language in British India and the Mughal Empire could be said to have 'lasted' until 1857.

44. Thorn, *War in India*, pp. 75–6.

45. For an overview of this period see ch. 4, 'Confrontation (June, July 1803)', in Anthony S. Bennell, *The Making of Arthur Wellesley* (Sangam Books, London, in conjunction with Orient Longman, Hyderabad, 1997), pp. 47–67. Anthony Bennell also did an excellent job of laying out Arthur Wellesley's essential correspondence from this period in *The Maratha War Papers of Arthur Wellesley, January to December 1803* (Sutton Publishing, for the Army Records Society, Gloucestershire, 1998), ch. 2, pp. 55–197.

46. Montgomery Martin (ed.), *The Despatches, Minutes and Correspondence of the Marquess Wellesley, K.G. During His Administration in India*, 5 vols. (Wm. H. Allen, London, 1837) (hereafter referred to as Martin, *Desp. of Marquess Wellesley*), vol. III, [no. XXXV], Marquess Wellesley to Maj.-Gen. Wellesley, Fort William, 27 June 1803, paragraph 18, pp. 153–8.

47. Ibid., p. 154.

48. Ibid., paragraphs 11 and 12, p. 156.

49. The bridging of civil–military affairs by British generals is quite evident in colonial warfare from the America Revolutionary War to the 'Malayan Emergency'. However, for a more relevant model with an interesting set of examples which fall back on a later military comparison with 'Wellington in Europe', see Wesley B. Turner, *British Generals in the War of 1812, High Command in the Canadas* (McGill-Queen's University Press, Montreal, 1999). For the definitive EIC collection of documentation concerning Arthur's powers see BL: OIOC, Board's Collections, F/4/166, no. 2874(a), ff.1–212, 'Major General Arthur Wellesley's Separate Powers for the Direction and Control of all Military and Political Affairs in Hindostan & the Deccan for Making Peace or War With the Mahrattas'.

50. C-in-C Lake's authorization, as well as Arthur Wellesley's, was juxtaposed with the precedent for this action, which dated back to Goddard's mandate in the First Anglo-Maratha War. See BL: OIOC, Board's Collections, F/4/166, no. 2874(b), ff.1–248, 'Delegation of Powers by the Supreme Council of Bengal'.

51. The 'extraordinary powers' were questioned in the context of granting *de facto* 'governing' powers to an individual who was not a member of the EIC. See BL: OIOC, Home Misc. Series, H/481 ff.823–831, 'Mr. Adam's Opinion on the Powers granted to the Major General Wellesley'.

52. Martin, *Desp. of Marquess Wellesley*, vol. III, [no. XXXV], Marquess Wellesley to Maj.-Gen. Wellesley, Fort William, 27 June 1803, paragraph 18, p. 158.

53. *Selections from the Wellington Dispatches*, [no. 67], p. 66, Arthur Wellesley to General Lake, Camp at Sangwee, 29 July 1803.

54. Ibid. [no. 107], paragraph 3, p. 100, Arthur Wellesley to Major Kirkpatrick, Camp at Adjuntee, 25 Oct. 1803.

55. Martin, *Desp. of Marquess Wellesley*, vol. III, p. 167.

56. Despite Richard Wellesley's eventual success at negotiating a buffer-state treaty network, his successor Governor-General Barlow issued orders to dissolve the alliances with Jaipur, Bharatpur and Matcheary. BL: OIOC, Board's Collections, F/4/195 no. 4431, f.7, Extract of Political Letter from Bengal to the Secret Committee, 24 Dec. 1805.

57. Martin, *Desp. of Marquess Wellesley*, vol. III, p. 168.

58. Blakiston, *Twelve Years'*, p. 145.

3 THE DECCAN CAMPAIGN OF 1803

1. James Welsh, *Military Reminiscences Extracted from a Journal of Nearly Forty Years Active Service in the East Indies*, 2 vols., 3rd edition (Smith, Elder and

Co., 1830), vol. I, p. 147. Welsh published as Colonel Welsh but he was a captain in 1803, eventually rising to the rank of major-general.

2. Ibid., pp. 155–6.

3. Gokale or 'Goclah' was reported killed in operations against General Smith in 1816, Blakiston, *Twelve Years'*, vol. I, p. 90.

4. BL: Add. MS 13,644 'Malcolm & Lambton's Journals in 1799', ff.87–89.

5. One of Gokale's men (a *carcoon*) later told the British that although his master extracted two months' pay for his services, he wanted a third month's wages in advance or he was threatening to withdraw. BL: Add. MS 13,599. See in particular the summaries of letters by Close and Stevenson which formed the basis of Wellesley's report: Arthur Wellesley to Military Secretary M. Shawe, Camp, 23 Aug. 1803, ff.15–19.

6. Even Wellesley's executive intelligence officer could not tell Gokale's men from enemy *pindaries*. See Mounstuart Elphinstone to Edward Strachey, 'Camp at Midgaon (the old place)', 16 Sept. 1803, Colebrooke, *Mountstuart Elphinstone*, vol. I, pp. 61–2. Differentiating between the *Peshwa*'s troops and those of Bhonsle and Sindia would remain a problem during the campaign. Blakiston, *Twelve Years'*, vol. I, pp. 214–15. Later in the 1804–5 campaign against Holkar several 'friendly fire' incidents were reported when British troops shot members of the allied contingent sponsored by the 'Nawab of Bareitch'. Eventually they were issued pennants to display in the hope that their British allies would not shoot them so readily.

7. Blakiston, *Twelve Years'*, vol. I, p. 85.

8. For notes on Bombay's climatic disaster in 1803, see Major Jasper Nicolls, 'Remarks Upon the Temperature of the Island of Bombay During the Years 1803 and 1804', *Transactions of the Literary Society of Bombay*, vol. I (London, 1819), pp. 4–9. Nicolls, in reference to the famine conditions known as 'crop failure', makes the distinction that an associated 'monsoon failure' means a reduced amount of rain and not the total absence of rain.

9. *Supp. Desp.*, p. 63, [no. 1541], vol. IV, G.O. Camp at Panowullah, 27 April 1803.

10. Blakiston, *Twelve Years'*, vol. I, p. 93. For references on the economic short-comings of bullock transport in the Deccan Campaign of 1803, see R. G. S. Cooper, 'Beyond Beasts and Bullion: Economic Considerations in Bombay's Military Logistics', *Modern Asian Studies*, 33, part 1 (1999), pp. 160–3.

11. Blakiston, *Twelve Years'*, vol. I, p. 97.

12. Ibid., vol. I, p. 109.

13. Welsh, *Military Reminiscences*, vol. I, p. 152.

14. See the 'Copy of a Memorial addressed to Maharage Dowlut Rao Scindiah, by Colonel Collins', Martin, *Desp. of Marquess Wellesley*, vol. III, pp. 172–3.

15. Grant Duff, *History of the Mahrattas*, vol. III, p. 235.

16. Sardesai, *New History*, vol. III, p. 409.

17. See BL: OIOC, Political & Secret Dept., L/PS/5/91 (10a), Gov. Duncan to Col. Clarke, Bombay, 16 Oct. 1801. Col. Arthur Wellesley to Col. William Clarke, Seringapatam, 13 Nov. 1801. Secret Dept., Lord Clive, J. Stuart, Wm. Petrie, and M. Duke to V.P. in Council G. Barlow, 14 Dec. 1801. For a background of the events in 1801, see Randolf Cooper, 'Amphibious

Options in Colonial India: Anglo-Portuguese Intrigue in Goa 1799', in William B. Cogar (ed.), *New Interpretations in Naval History* (United States Naval Institute Press, Annapolis, Maryland, 1997), pp. 95–113.

18. University of Southampton Wellington Papers, WP/3/3/78, f.37, Maj. Gen. Arthur Wellesley to Col. Murray, Camp, 2 July 1803.

19. Published in *Supp. Desp.*, p. 101, [no. 1579]; a hand-written version dated 23 May 1803 can be found in University of Southampton Wellington Papers, WP/1/150.

20. The old Grand Trunk Road was a road system Sher Shah linked, as opposed to having completely 'built', from Bengal to the Indus River. It was said to have extended 1,500 *coss* (aka *kos*) by the time of his death in 1545.

21. Fath Singh Mane, one of Holkar's *sardars*, had successfully attacked the *pettah* in July 1802 during Holkar's invasion of Sindia's territories. Jadunath Sarkar, *Fall of the Mughal Empire*, vol. IV, *1789–1803*, p. 163.

22. University of Southampton Wellington Papers, WP/3/3/53, ff.167–168, Col. J. Collins Res. to the Court of D. R. Scindiah to Gov. Gen. Richard Wellesley, Camp near Aurungabad, 11 Aug. 1803.

23. Blakiston, *Twelve Years'*, vol. I, pp. 129–30, 140. Blakiston noted that both the British and Marathas used Arabs. The EIC was left so short of men during the 1803 Maratha Campaigns that it considered large-scale recruiting of Arab mercenaries for duty in Ceylon during the 'Kandian War' as indicated by Governor North's inquiry cited in University of Southampton Wellington Papers, WP/3/3/18, f.286, Gov. Jonathan Duncan to Maj. Gen. Arthur Wellesley, Bombay Castle, 8 Oct. 1803.

24. Major R. G. Burton, *Wellington's Campaigns in India*, Division of the Chief of the Staff Intelligence Branch, 'For Official use only' (Superintendent Government Printing, Calcutta, 1908), p. 53, footnote †. Both the *Peshwa* and Bhonsle later used Arab mercenaries against the British; see Clarke, *Summary of the Mahratta and Pindarree Campaign, During 1817, 1818 and 1819*, pp. 264–5.

25. These matchlocks fit the general description of those identified by Major Innes Munro in the Second Anglo-Mysore War as 'carlise' matchlocks. Munro, *Narrative of the Military Operations*, p. 131. They were generally of a small-bore diameter for muzzle-loaders (some described as a pistol-sized bore) and about 6 feet in length, but reported by historically contemporary observers to be ballistically much superior to the standard British musket in accuracy and range. Arab sharpshooters were known to hold six to eight lead balls in their mouths and spit one into the muzzle when loading for each shot. Some believe this practice, combined with the wearing of a powder flask on a lanyard, reduced their loading time and fumbling with paper cartridges in the heat of battle.

26. Thorn, *War in India*, pp. 266–7. Although he published as a major in 1818, Thorn was a captain at the time of the 1803 campaign.

27. Arthur Wellesley to 'The Officer Commanding the Troops in the Territories of Anand Rao Gaikwad, Baroda', Camp, 6 Aug. 1803 and the 'ultimatum delivered to Sindhia', Raghubir Singh (ed.), *English Records of Maratha History Poona Residency Correspondence*, vol. X, *The Treaty of Bassein and the*

Anglo-Maratha War in the Deccan 1802–1804 (Sri Gouranga Press, Bombay, 1951), pp. 118, 120–1.

28. Thorn, *War in India*, p. 267.
29. Welsh, *Military Reminiscences*, vol. I, p. 158. Later references by various authors suggest this to mean the 3,000 light horse were those noted as having camped under shelter of the walls.
30. Thorn, *War in India*, p. 267. A search of statements by defecting European mercenary officers reveals no mention of a deployment here by the 'regular corps'. However, it is possible there were other Maratha infantry present. As noted in chapter 1 above, the generic description of 'infantry', in a Maratha context, covered a variety of troops. Thorn served in the northern theatre with Lake and so the observation of a battalion from the 'regular corps' should be viewed as hearsay evidence.
31. Welsh, *Military Reminiscences*, vol. I, p. 164.
32. Thorn, *War in India*, p. 267.
33. Welsh, *Military Reminiscences*, vol. I, p. 157.
34. Ibid., vol. I, p. 157.
35. Thorn, *War in India*, p. 268.
36. Grant Duff, *History of the Mahrattas*, vol. III, p. 237. Many of the differing statistics come down to whether we are talking about the capture of the *pettah*, the fort, or both. Some of the smaller British figures occurred when authors listed only European casualties and did not include South Asian EIC troops.
37. Thorn, *War in India*, p. 268. This agrees with official figures as listed in appendix II.
38. Blakiston, *Twelve Years'*, vol. I, p. 134.
39. Ibid., vol. I, p. 136.
40. Ibid., vol. I, p. 135. This was apparently a reference to John Müller, *A Treatise on Artillery* (London, 1757; re-printed by the Museum Restoration Service, Ottawa, 1965).
41. Blakiston, *Twelve Years'*, vol. I, p. 139.
42. Welsh, *Military Reminiscences*, vol. I, p. 162.
43. Maj. Gen. Arthur Wellesley to Gov. Gen. Richard Wellesley, Camp at Ahmadnagar 12 Aug. 1803, Forrest, *Maratha Series*, pp. 605–7.
44. Welsh, *Military Reminiscences*, vol. I, p. 163.
45. *Supp. Desp.*, vol. IV, [no. 1579], p. 100.
46. Welsh, *Military Reminiscences*, vol. I, pp. 180–2.
47. Sardesai, *New History*, vol. III, p. 410. British sources are less than revealing on this point. However, the June 1803 'Memorandum', *Supp. Desp.*, vol. IV, [no. 1579], p. 100, noted the Arabs in the *pettah* 'commanded by three French officers, a little dark-coloured, and who wear blue clothes'. Sarkar, *Mughal Empire*, vol. IV, the footnote on p. 265, lists '3 French mestizo officers-outside the fort'. But none of the sources have ever offered evidence in support of statements about their nationality or ethnicity.
48. These passages are noted as from V. V. Khare and Y. V. Khare (eds.), *Aitihasik Lekh Sangraha*, vol. XIV, refs. 6683 and 6685. See the footnote on p. 265, of Sarkar, *Mughal Empire*, vol. IV.
49. Welsh, *Military Reminiscences*, vol. I, p. 165.

50. Ibid., vol. I, p. 164.
51. *Supp. Desp.*, vol. IV, [no. 1623], p. 151, G.O. Camp at Ahmednuggur, 12 Aug. 1803.
52. Ibid., p. 152.
53. Blakiston, *Twelve Years'*, vol. I, p. 141.
54. Welsh, *Military Reminiscences*, vol. I, p. 164.
55. Burton, *Wellington's Campaigns*, p. 56.
56. Welsh, *Military Reminiscences*, vol. I, p. 165, identified Graham as having been the 'Pay-master' of Wellesley's army. Sarkar stated that Wellesley 'hanged the revenue collector' of Sindia's ancestral village of Jamgaon. See Jadunath Sarkar, *Mughal Empire*, vol. IV, p. 266. The status of Sindia's ancestral holdings would ultimately become an issue in peace negotiations and a part of the subsequent treaty settlement. See *Supp. Desp.*, vol. IV, [no. 1690], pp. 248–9 (11 Dec. 1803), and pp. 254–5 (26 Dec. 1803) 'Memorandum of the Conferences with Jeswunt Rao Goorparah and Naroo Punt Nana, Vakeels on the Part of Dowlut Rao Scindiah', and 'Treaty of Peace and Friendship with Dowlut Rao Scindiah', art. III re. 'Ahmednuggur' with specified 'enaum' lands in art. VII, vol. IV, pp. 264–5.
57. Thorn, *War in India*, p. 270.
58. *Supp. Desp.*, vol. IV, [no. 1623], p. 152, G.O. Camp at Ahmednuggur, 12 Aug. 1803.
59. Thorn, *War in India*, p. 270. The second of the two logistical convoys arrived under Major Hill on 18 Sept. 1803.
60. Sardesai, *New History*, vol. III, p. 410. Thorn, *War in India*, p. 271. Stevenson came to rely quite heavily on the 'Rohilla leader Salabat Khan' and his irregular horsemen who had great success in ambushing the Marathas. That was particularly evident at Jalna on 10 Sept. 1803. See the reports forwarded to Lt. Col. Close in ff.110–112, BL: Add. MS 13,599.
61. Elphinstone ultimately expressed great admiration for Arthur Wellesley but he wrote to his friend Strachey, 'The river, I am sorry to say, is fordable in several places, which is unusual . . . Sindia has certainly sent for some of his infantry, I believe Begum Sumroo's.' 3 Sept. 1803, 'Camp on the Godavery, forty miles S.S.E. of Aurungabad'. Four days and twelve miles later, he wrote to Strachey, 'I do not know why you think the General deficient in intelligence. There is always speedy information of every movement of the enemy, who, you know, are a mighty army of sixteen thousand horse, very ill-mounted, and without one gun.' Camp, 7 Sept. 1803. See Colebrooke, *Mountstuart Elphinstone*, vol. I, pp. 55 and 59.
62. Grant Duff, *History of the Mahrattas*, vol. III, p. 238.
63. Thorn, *War in India*, p. 271.
64. Blakiston, *Twelve Years'*, vol. I, p. 92.
65. Thorn, *War in India*, p. 271.
66. University of Southampton Wellington Papers, WP/3/3/70, see f.7, in 'Memorandum' by Major John Malcolm, Poonah, 18 June 1803, ff.5–9.
67. *Maharashtra State Gazetteers, Maharashtra Ahmadnagar District* (revised edition, Bombay, 1976), pp. 136, 140, 397.
68. Thorn, *War in India*, p. 271.

69. See B. Close to J. Collins, Poona, 15 July 1803, in Forrest, *Maratha Series*, pp. 601–2.

70. See Mounstuart Elphinstone to Edward Strachey, 'Camp near Peepulgaon', 11 Sept. 1803, Colebrooke, *Mountstuart Elphinstone*, vol. I, pp. 59–60. Further *pindari* indications of effectiveness and Wellesley's unwillingness to trust his own intelligence staff with information are found in Mounstuart Elphinstone to Edward Strachey, 'Camp at Peepulgaon', 12 Sept. 1803, Colebrooke, *Mountstuart Elphinstone*, vol. I, p. 61.

71. Not all members of the Bombay Presidency's government were inclined to give Wellesley's requests priority. R. G. S. Cooper, 'Indian Army Logistics 1757–1857: Arthur Wellesley's Role Reconsidered', in Alan J. Guy and Peter B. Boyden (eds.), *Soldiers of the Raj: The Indian Army 1600–1947* (National Army Museum, London, 1997).

72. Maharashtra State Archives, Mumbai: Goa Envoy's Records Diary, no. 3/605 of 1801–06, pt. III, pp. 518–519, Courts Martial.

73. Ironically, the Battle of Assaye proved only a temporary reprieve for the condemned man. Sheik Daud, also identified as Sheik David, was shot in front of his unit on 16 Oct. 1803, as was Mohamed Reeza of the [2/12th?]. NAM, A copy of 'Arthur Wellesley's Order Book, 7 Feb. 1803 – 21 June 1804', Accession no. 6308–11, p. 24, G.O., 15 Oct. 1803. Notations on this order book indicate it was the personal property of Jasper Nicolls, later Lt-Gen. Sir Jasper Nicolls, KCB, C-in-C India.

74. Thorn, *War in India*, p. 272.

75. For the wider implications of multiple *harkarrah* failures that day see C. A. Bayly, *Empire and Information: Intelligence Gathering and Social Communication in India, 1780–1870* (Cambridge University Press, 1996), p. 68.

76. *Supp. Desp.*, vol. IV, [no. 1681], p. 210, Arthur Wellesley to Lt-Col. Munro, Camp at Cheesekair. In contrast, the 'regular corps' in Maratha service made provisions for assigned battalion *harkarrahs*. See the listing for two such individuals in the 'Workmen/Writers' column of the ledger, 'Return for the Month of June of the Radge Battalion of Maharage Dowlut R. Sindia Commanded by Lieut. Col. Pohlman, 30th June 1803', signed by the British mercenary Ensign Marrs, University of Southampton Wellington Papers, WP/3/3/47.

77. Before the Lord's Committee on Indian Affairs in 1830, Elphinstone modestly described his duties 'as a sort of political assistant or secretary'. See Colebrooke, *Mountstuart Elphinstone*, vol. I, p. 49. He boldly admitted his failings as a Marathi interpreter and lack of control over the Intelligence Department to his dear friend Edward Strachey; see Elphinstone to Strachey, 'Camp at Paloor, 13 miles south of Adjuntee', 9 Oct. 1803, pp. 75–8.

78. The shortcomings were evident in August 1803 as Elphinstone wrote 'We are ill off for intelligence.' See Elphinstone's letter of 30 Aug. 1803, Camp, twelve miles east of Aurungabad, Colebrooke, *Mountstuart Elphinstone*, vol. I, pp. 53–4. Unfortunately the information Wellesley received often conflicted so badly that he became suspicious of those around him. He was also cautious of writing about issues that might be construed as failures. This tended to jeopardize communication with other officers as his information disclosure became more guarded. That is evident in Elphinstone's comment to

Strachey: 'Is it possible that General W. should write to Colonel Close [British Resident at Pune] every day and never mention the enemy's having invaded the country?' See Mounstuart Elphinstone to Edward Strachey, 'Camp at Midgaon (the old place)', 16 Sept. 1803, Colebrooke, *Mountstuart Elphinstone*, vol. I, pp. 61–2.

79. 'I think, if anyone in this line were to apply, he might improve the intelligence; but I had some people given me, and a way shown me, and so fell into the habit of jogtrottery, the great foe of improvement.' Mounstuart Elphinstone to Edward Strachey, 'Camp at Paloor, 13 miles south of Adjuntee', 9 Oct. 1803, Colebrooke, *Mountstuart Elphinstone*, vol. I, pp. 75–8.

80. Several of Elphinstone's original *harkarrahs* deserted and at least one was captured, giving him cause to ask for 'intelligent fellows'. Mounstuart Elphinstone to Edward Strachey, 'Camp at —, twelve miles from Midgaon', 22 Sept. 1803, Colebrooke, *Mountstuart Elphinstone*, vol. I, pp. 62–3.

81. Mounstuart Elphinstone to Edward Strachey, 'Camp at Midgaon (the old place)', 16 Sept. 1803, Colebrooke, *Mountstuart Elphinstone*, vol. I, pp. 61–2.

82. Mounstuart Elphinstone to Edward Strachey, 'Camp at Paloor, 13 miles south of Adjuntee', 9 Oct. 1803, Colebrooke, *Mountstuart Elphinstone*, vol. I, pp. 75–8.

83. *Supp. Desp.*, vol. IV, [no. 1687], p. 219, Arthur Wellesley to Sir William Clarke, Camp, 7 Nov. 1803.

84. Blakiston, *Twelve Years*', vol. I, p. 107.

85. BL: OIOC, Board's Collections, F/4/200, no. 4512, 'The Establishment of Native Intelligencers formed by General Stuart Cont'd for the Present' ff.8–9, see the 'Minute' by Lt-Gen. Sir John Craddock, C-in-C Madras, Fort St George, 27 Nov. 1805.

86. Welsh, *Military Reminiscences*, vol. I, p. 171. In the months immediately following the battle, stories persisted that Arthur had only learned of the Marathas' location when he encountered two defecting European mercenaries who had just left Sindia's forces. After Arthur's rise to fame, these accounts all seem to have disappeared from the popular 'battle piece' of Assaye.

87. Sarkar, *Mughal Empire*, vol. IV, footnote † on p. 270.

88. For comments on the incident at 'Sultanpettah Tope', see Blakiston, *Twelve Years*', vol. I, pp. 82–3. Field Marshal Lord Carver, *The Seven Ages of the British Army* (London, 1984), p. 84, commented on the potential meaning of the incident to Arthur's later career.

89. An interesting though somewhat tenuous case could be made that Arthur's career, as exemplified by Assaye in 1803 and Waterloo in 1815, holds examples of two separate infantry warfare doctrines. The lesson of Assaye, when combined with Arthur's memorandum on Monson's retreat before Holkar, became the cornerstone of what later soldiers (like Napier of Sind) would interpret as Wellington's Indian doctrine – never retire before a native opponent. That doctrine, if it can be said to exist as such, may reveal deep-seated racial assumptions, which Wellesley held about the need to always go on the offensive against a 'native' enemy. But it may also be that it was not until after his departure from India that Arthur Wellesley reflected on the human cost of offensive infantry warfare and rethought what was to become his European infantry doctrine.

90. It is the Kailna River in Grant Duff, *History of the Mahrattas*, vol. III, p. 239. Burton, *Wellington's Campaigns*, p. 60, noted: 'This is incorrectly printed Kaitna in the Wellington Despatches – and the error is repeated in all previous accounts of the battle.'

91. Lt-Col. Wilson was one of the very few secondary source authors to note that the action at Assaye was directed only against the army of Sindia and not, as is so often said, 'the combined armies of Sindia and Bhonsle'. See Lt-Col. W. J. Wilson, *History of the Madras Army*, vol. III (Madras, printed by E. Keys, at the Govt. Press, 1883), p. 402.

92. Thorn, *War in India*, p. 274.

93. Ibid., p. 271.

94. Sarkar, *Mughal Empire*, vol. IV, p. 132. Blakiston, who was present at Assaye, noted Pohlman as a German mercenary commanding the Marathas at Assaye. Blakiston, *Twelve Years'*, vol. I, p. 176.

95. Welsh, *Military Reminiscences*, vol. I, p. 174.

96. Thorn, *War in India*, p. 274.

97. In describing the hordes of the 'Maratha–Pindari War' (aka Third Anglo-Maratha War 1817–19), Lt-Col. Fitzclarence said: 'it is supposed, that out of every five men, two were fighting men, two others mere plunderers, mounted on inferior horses, and the remaining one on a pony'. See Lt-Col. Fitzclarence, *Journal of A Route Across India . . .* (London, 1819), p. 11.

98. *Wellington Dispatches*, vol. II, p. 324, Arthur Wellesley to the Governor General, Camp at Assye, 24 Sept. 1803.

99. Sarkar noted, 'There was a ford over the Khelna at Lingwari a mile west of Wellesley's first point of observation.' Sarkar, *Mughal Empire*, vol. IV, p. 272.

100. Jac Weller, *Wellington in India* (Longman, London, 1972), p. 178.

101. Mountstuart Elphinstone, writing the day before battle, noted, 'I hope confidently that we shall have an engagement on the day after to-morrow. Even *they* [Sindia] talk of fighting on Saturday, and this is Thursday.' Mountstuart Elphinstone to Edward Strachey, 'Camp at —, twelve miles from Midgaon', 22 Sept. 1803, Colebrooke, *Mountstuart Elphinstone*, vol. I, pp. 62–3.

102. V. V. Khare and Y. V. Khare (eds.), *Aitihasik Lekh Sangraha*, vol. XIV, 3 Nov. 1803, p. 7908, with thanks to Professor N. K. Wagle for translating.

103. Sarkar, *Mughal Empire*, vol. IV, pp. 270–1. Note in particular the footnote reference to Khare's use of the term *dharna*.

104. Mountstuart Elphinstone, 'Camp near Assye, ten miles from Jafferabad', 27 Sept. 1803, Colebrooke, *Mountstuart Elphinstone*, vol. I, pp. 64–9. See also Colonel John Biddulph, *The Nineteenth and Their Times* (John Murray, London, 1899), p. 139.

105. Burton, *Wellington's Campaigns*, p. 61.

106. Brevet Lt-Col. W. D. Bird, 'The Assaye Campaign', *Journal of the United Service Institution of India*, 41, no. 187 (1912), pp. 112–16.

107. Mountstuart Elphinstone to Edward Strachey, 'Camp at Cadool, or Cadouly', 24 Sept. 1803, Colebrooke, *Mountstuart Elphinstone*, vol. I, pp. 63–4.

108. Biddulph, *The Nineteenth and Their Times*, p. 140.

109. Blakiston, *Twelve Years'*, vol. I, pp. 161–3.

110. Ibid., vol. I, pp. 164–5.

111. Ibid., vol. I, pp. 174–5. While I was conducting interviews with British Indian Army veterans, at the Royal Hospital in Chelsea during 1991, two NCOs stationed in Bombay mentioned that until their December 1941 embarkation they regularly had Indian bearers carry their kit. With typical sarcasm and a wink, one turned and said, 'The cheek of them making me walk up the gang plank carrying my own kit.'

112. *Supp. Desp.*, vol. IV, [no. 1870], p. 495. Arthur Wellesley to the Adjutant-General of the Army, Fort St George, 27 Sept. 1803. Confirmation of the claim and the granting of a pension on full pay were corroborated in Maharashtra State Archives, Mumbai, Goa Envoy's Records Diary, no. 3A/606 of 1803–09, Pt. II, printed page no. 1494 (pencilled-in page no. 361), a G.O. by Gov't., Signed by George Buchanan, Chief Secretary to Gov't., Fort St George, 22 March 1805.

113. NAM, Accession no. 7807–90, *Early Days of the 74th Highland Regiment*, compiled by Colonel John Grahame (DSO) (printed regimentally, 1951), p. 182. Although I saw and photographed the Wellesley gun noted by Grahame (in Warur), there is no historically contemporary account that confirms Grahame's assumption that this piece had remained there as the result of being stuck in the mud. It may have merely been positioned there to commemorate the village as the crossing point. For photo see Cooper and Wagle, 'Maratha Artillery', p. 58.

114. British ammunition was 'fixed', meaning the charge, sabot and ball of the 6 pounder were complete as a cartridge. The Maratha 'semi-fixed' system meant that the powder was pre-bagged but not attached to ball, chain, canister or grape. They could single or double load as needed. For notes on some of the other differences in EIC vs. Maratha artillery components, see 'Report of the Ordnance . . . captured opposite Delhi on the 11th of September, 1803' by Lt-Col. John Horsford, Commanding the Artillery in the Field, Appendix P, Martin, *Desp. of Marquess Wellesley*, vol. III, p. 668.

115. Mounstuart Elphinstone, 'Camp near Assye, ten miles from Jafferabad', 27 Sept. 1803, Colebrooke, *Mountstuart Elphinstone*, vol. I, pp. 64–9.

116. Blakiston, *Twelve Years'*, vol. I, pp. 173–4.

117. Thorn, *War in India*, p. 275.

118. 'Narrative drawn up by Captain (afterward Sir Colin) Campbell relative to the Battle of Assaye – at the time of the transaction', University of Southampton Wellington Papers, WP/1/150, 7 folio sheets signed by Captain Campbell, Brigade Major to Gen. Wellesley. Campbell's words also found their way into print, see *Supp. Desp.*, vol. IV, pp. 184–7. (Hereafter referred to as Campbell, 'Narrative'.)

119. Mounstuart Elphinstone, 'Camp near Assye, ten miles from Jafferabad', 27 Sept. 1803, Colebrooke, *Mountstuart Elphinstone*, vol. I, pp. 64–9.

120. Grant Duff, *History of the Mahrattas*, vol. III, p. 241.

121. 'The line moved forward rapidly (I may say without firing two rounds) & took possession of the first line of Guns', as found in Campbell, 'Narrative'.

122. James Skinner, founder of Skinner's Horse, spent hours practising with the double-edged broad-bladed Maratha spear when he first joined Sindia's army. The weapon remained a favourite of Skinner's for close-quarters combat as it was just as lethal for cutting as stabbing which gave it an advantage over the bayonet. Also in its favour was the fact that it could be wielded equally well at arm's length as a pike or by placing the shaft under the arm and 'choking-up' on the broad bladed point so it could be swung in a hacking motion or turned to parry a sword cut.

123. Thorn, *War in India*, pp. 275–6.

124. Bird, 'The Assaye Campaign', p. 116.

125. Blakiston, *Twelve Years'*, vol. I, p. 176.

126. Thorn, *War in India*, p. 277.

127. Campbell, 'Narrative'. The thorny milk hedge or milk bush is the toxic *Euphorbia Tirucalli*; see Yule and Burnell, *Hobson-Jobson*, p. 568. Although noted in several texts for being suitable for military application as a barrier, it was often planted to delineate fields and discourage the unwanted entry of livestock or other animals. When I visited Assaye it was still to be found adjacent to an unattended grove.

128. Blakiston, *Twelve Years'*, vol. I, p. 165.

129. Grant Duff, *History of the Mahrattas*, vol. III, p. 241. Technically speaking all of H.M.'s Dragoons in India at this time were 'Light Dragoons'. However, it has become customary to refer to this unit simply as the '19th Dragoons'.

130. NAM, 'Acct. From NCO Swarbruck 19th Dragoons', Accession no. 8207–64.

131. *Supp. Desp.*, vol. IV, [no. 1660], p. 189, Arthur Wellesley to Lt-Col. Collins, 3 Oct. 1803.

132. BL: OIOC, *Madras Courier*, XIX, no. 942, 26 Oct. 1803.

133. NAM, 'Acct. From NCO Swarbruck 19th Dragoons', Accession no. 8207–64.

134. Grant Duff, *History of the Mahrattas*, vol. III, p. 242.

135. 'Captain Sale and others, afterwards saw him when in hospital, blow out a candle from his lungs: – the reader will be pleased to learn that the gallant serjeant recovered.' Ibid., vol. III, p. 242.

136. Thorn, *War in India*, p. 276.

137. Ibid., pp. 276–7.

138. Wellesley had two horses killed under him at Assaye, and of the cavalry's 1,200 mounts Welsh reported 113 killed and 325 wounded. Welsh, *Military Reminiscences*, vol. I, p. 177. Wellesley's other horse had a leg blown-off by cannon fire. Blakiston, *Twelve Years'*, vol. I, p. 97.

139. Blakiston, *Twelve Years'*, vol. I, pp. 170–1.

140. Following the battle, word went out from Bombay that the Deccan army was in urgent need of cavalry mounts. Maharashtra State Archives, Mumbai: Military Dept., Military Board Diary, 1803, no. 64, 7 Oct. 1803, ff.3680–3681, General Order by Government.

141. Blakiston, *Twelve Years'*, vol. I, p. 177.

142. See appendix II.

143. Blakiston, *Twelve Years'*, vol. I, pp. 173–4. 'In the eighteenth century, grape shot was a general term embracing all forms of artillery ammunition made

up of small shot.' The weapon's effect was like that of a gigantic shotgun blast with a single round of 12 pounder case shot broadcasting up to 84 balls weighing 2 ounces each. Adrian B. Caruana, 'Tin Case-Shot in the 18th Century', *Canadian Journal of Arms Collecting*, 28, no. 1 (Feb. 1990), pp. 11, 13.

144. Blakiston, *Twelve Years'*, vol. I, p. 179.
145. Mounstuart Elphinstone, 'Camp near Assye, ten miles from Jafferabad', 27 Sept. 1803, Colebrooke, *Mountstuart Elphinstone*, vol. I, pp. 64–9.
146. 'The following alterations will take place tomorrow [28 Sept.] in the details for the infantry piquets, and continue until further orders. The 74th regiment is to be struck off the roster.' *Supp. Desp.*, vol. IV, [no. 1656], p. 182, Camp at Assaye, 27 Sept. 1803.
147. Blakiston, *Twelve Years'*, vol. I, p. 165.
148. Mounstuart Elphinstone to Edward Strachey, 'Camp at Paloor, 13 miles south of Adjuntee', 9 Oct. 1803, Colebrooke, *Mountstuart Elphinstone*, vol. I, pp. 75–8.
149. NAM, A copy of 'Arthur Wellesley's Order Book, 7 Feb. 1803–21 June 1804', Accession no. 6308–11, p. 20, G.O., Assaye, 27 Sept. 1803.
150. Welsh, *Military Reminiscences*, vol. I, p. 183.
151. NAM 'Acct. From NCO Swarbruck 19th Dragoons', Accession no. 8207–64.
152. Mounstuart Elphinstone, 'Camp near Assye, ten miles from Jafferabad', 27 Sept. 1803, Colebrooke, *Mountstuart Elphinstone*, vol. I, pp. 64–9.
153. Mounstuart Elphinstone to Edward Strachey, 'Camp at Assye', 3 Oct. 1803, Colebrooke, *Mountstuart Elphinstone*, vol. I, pp. 74–5.
154. Elphinstone provided some additional detail in writing: 'Our sick and wounded are in the tolerably good fort of Nizamabad, or Adjuntee, with three companies. They have excellent accommodation.' Mounstuart Elphinstone to Edward Strachey, 'Camp at Paloor, Eighteen miles from Aurungabad, near Poolmurry', 14 Oct. 1803, Colebrooke, *Mountstuart Elphinstone*, vol. I, pp. 78–80. See also NAM, A copy of 'Arthur Wellesley's Order Book, 7 Feb. 1803–21 June 1804', Accession no. 6308–11, p. 20, G.O., Assaye, 29 Sept. 1803. Wellesley had originally ordered his regimental surgeons to remain at the newly created hospital at Pune, p. 3, G.O., 1 June 1803. Earlier in the campaign, with the capture of Ahmadnagar, which was also intended to accept sick and wounded, Wellesley had considered closing his hospital at 'Erroor' and relocating the staff to Pune. See *Supp. Desp.*, vol. IV, [no. 1629], p. 159, Arthur Wellesley to Lt-Col. Colman, Camp North of the Nimderrah Ghaut, 18 Aug. 1803. In addition to the hospitals that Wellesley established there were also those of Colonel Stevenson's force noted in Lt-Col. John Collins to Maj.-Gen. Arthur Wellesley, Camp Rakasban, 12 Sept. 1803. Sinh (ed.), *Poona Residency Correspondence*, vol. X, pp. 146–7.
155. NAM, A copy of 'Arthur Wellesley's Order Book, 7 Feb. 1803–21 June 1804', Accession no. 6308–11, p. 21, G.O., Assaye, 30 Sept. 1803.
156. *Supp. Desp.*, vol. IV, [no. 1658], pp. 182–3, G.O., 30 Sept. 1803, as also found with slight variations in NAM, A copy of 'Arthur Wellesley's Order Book, 7 Feb. 1803–21 June 1804', Accession no. 6308–11, p. 20.

157. Maharashtra State Archives, Mumbai: Military Dept., Military Board Diary, 1803, no. 64, 7 Oct. 1803, f.3683, Commanding Officer of the Forces Maj. Gen. Oliver Nicolls to the Bombay Military Board, Bombay.

158. University of Southampton Wellington Papers, WP/3/3/84, ff.30–1, 'Memorandum from Alex Kennedy Staff Surgeon to the Subsidiary Force', Camp, 30 April 1803. To his credit, Arthur Wellesley had ordered the establishment of a field hospital at Pune during June of 1803. See his General Order of 1 June 1803 in Gurwood (ed.), *Selections from Wellington Dispatches*, p. 56.

159. Maharashtra State Archives, Mumbai: Bombay Pres., Military Dept., Military Board Diary, 1803, no. 64, 19 Oct. 1803, f.3932, Maj. Gen. Arthur Wellesley to James Augustus Grant, Sec. to Govt., Camp, 13 Oct. 1803. With regard to the Bombay Govt.'s existing arrangements (4 Oct. 1803), see University of Southampton Wellington Papers, WP/3/3/18, f.282.

160. Maharashtra State Archives, Mumbai, Bombay Pres., Military Dept, Military Board Diary, 1803, no. 64, 19 Oct. 1803, f.3935, Proceedings of the Bombay Military Board. For an examination of the civil–military implications of this action and its logistical ramifications, see Cooper, 'Beyond Beasts and Bullion', pp. 164–7.

161. Maharashtra State Archives, Mumbai, Bombay Pres., Military Dept, Military Board Diary, 1803, no. 64, 28 Oct. 1803, p. 4061, Maj. Gen. Arthur Wellesley to James Augustus Grant, Sec. to Govt, Camp, 17 Oct. 1803.

162. For a somewhat biased summation on the part of Bombay officials see Maharashtra State Archives, Mumbai: Bombay Pres., Military Dept, Military Board Diary, 1803, no. 64, 28 Oct. 1803, ff.4063–4065, Lt. Col. Henry Woodington to the Bombay Military Board, Bombay, 25 Oct. 1803. Transportation of the wounded remained a problem for Wellesley throughout the war; see his General Army Order of 1 Dec. 1803 in Gurwood (ed.), *Selections from Wellington Dispatches*, p. 124.

163. BL: OIOC, Board's Collections, F/4/200, no. 4516, f.145, George Strachey Sec. to Madras Govt, to the President and Members of the Madras Military Board, Fort St George, 27 Nov. 1805.

164. NAM, A copy of 'Arthur Wellesley's Order Book, 7 Feb. 1803–21 June 1804', Accession no. 6308–11, p. 21.

165. Representatives of the revisionist school, favouring smaller hospitals, included Pringle, Brocklesby, Guthrie, Jackson and McGrigor. Although these men were not present in India, Guthrie and McGrigor lived long enough to risk disciplinary action by opposing Wellesley in a later European war. See Louis C. Duncan, *Medical Men in the American Revolution 1775–1783* (first published 1931, reprinted New York, 1970), pp. 20–1.

166. For notes on Panvel's role in British logistical plans for the 1803 campaign, see Cooper, 'Beyond Beasts and Bullion', p. 163.

167. NAM, A copy of 'Arthur Wellesley's Order Book, 7 Feb. 1803–21 June 1804', Accession no. 6308–11, p. 52, G.O., Camp at Panowly, 24 May

1804. Maharashtra State Archives, Mumbai, Military Dept, Military Board Diary, 1804, no. 71, 8 June 1804, f.1510, Arthur Wellesley to Bombay Military Authorities, Camp at Cheerpoor, 4 June 1804.

168. Prize money was not limited to just EIC soldiers. Arthur Wellesley was able to live very well on his share as one of His Majesty's soldiers, although granted his 'take' was proportionately larger than most. He later admitted: 'When I came from India I had 42 or 43,000 Pounds which I made as follows. I got 5,000 Prize money at Seringapatam; 25,000 £ Prize money in the Mahratta War; the Court of Directors gave me 4,000 £ for having been a Comr. in Mysore; & the Govt. paid me about 2,000£ in one Sum the arrears of an Allowance as Comg. Officer at Seringapatam; & the remainder was Interest upon these Sums Savings & c during the time I was in India.' Badajoz, 13 Sept. 1809. Sir Charles Webster (ed.), 'Some Letters of the Duke of Wellington to his Brother William Wellesley-Pole', *Camden Miscellany*, 43, Camden third series vol. LXXIX, (London, Royal Historical Society, 1948), sect. II, p. 24.

169. See Maharashtra State Archives, Mumbai: Military Dept, Military Board Diary, 1803, no. 64, 7 Oct. 1803, ff.3707–3715, Report of Mr J. M. Thriepland to James Augustus Grant, Sec. to Govt, 6 Oct. 1803.

170. NAM, A copy of 'Arthur Wellesley's Order Book, 7 Feb. 1803–21 June 1804', Accession no. 6308–11, p. 28, G.A.O., 7 Nov. 1803.

171. Ibid., p. 22, G.O., Camp at Assaye, 1 Oct. 1803.

172. University of Southampton Wellington Papers, WP/3/3/53, f.35, Col. Collins Res. to the Court of D. R. Scindiah, to Maj. Gen. Arthur Wellesley, Camp near Chickly, 4 June 1803.

173. University of Southampton Wellington Papers, WP/3/3/84, f.423, 'Intelligence by Colonel Stevenson's Hircarrahs'.

174. University of Southampton Wellington Papers, WP/3/3/84, f.427, 'Intelligence from the Enemy's Camp by Colonel Stevenson's Hircarrahs', S. Johnstone, Persian Interpreter, Subsidiary Force, Camp, 1 Oct. 1803.

175. Word went out from Bombay that the Deccan army was in urgent need of replenishment following the battle. Horses from Gujarat were sought in particular as remounts. Gujarati horses were believed to have particularly good blood lines owing to regular infusions of breeding stock from the Persian Gulf (Arab stallions). Maharashtra State Archives, Mumbai: Bombay Military Dept, Military Board Diary, 1803, no. 64, 7 Oct. 1803, ff.3680–3682.

176. In addition to University of Southampton Wellington Papers, WP/3/3/84, f.427, see f.434, 'Hircarrah Report', S. Johnstone, Persian Interpreter, Subsidiary Force, 4 Oct. 1803.

177. Ibid., ff.433 and 436 (sequential numbering askew), Colonel Stevenson to Maj. Gen. Wellesley, Camp at Jamnair, 4 Oct. 1803.

178. Ibid., f.454, 'Intelligence from Scindiah's Camp', S. Johnstone, Persian Interpreter, Subsidiary Force, Camp, 9 Oct. 1803.

179. 'Memorandum by Lieut.-Gen. Lake with the Marquess Wellesley's observations thereon, 18th July, 1803', in particular item number 14, Martin, *Desp. of Marquess Wellesley*, vol. III, p. 192.

180. *Supp. Desp.*, vol. IV, [no.1679], p. 208, Arthur Wellesley to the Adjutant-General, Camp at Ferdapoor, 24 Oct. 1803.

181. Thorn, *War in India*, p. 295.

182. University of Southampton Wellington Papers, WP/3/3/84, f.479, 'Roll of Europeans in the Service of Dowlut Rao Scindiah who have surrendered themselves to Colonel James Stevenson', Camp at Berhanpoor, 16 Oct. 1803, signed J. Colebrooke Deputy Adjutant General, Subsidiary Force. This list includes name, rank and country of origin.

183. In Colebrooke's above-noted 'Roll of Europeans . . .' dated 16 Oct. 1803, one NCO was of undetermined national origin but a comparison with Colebrooke's revised 'Roll of Europeans . . .' of 13 Nov. 1803, as found in WP/3/3/84, overleaf of f.476, reveals that this individual was Sergeant Joseph Vareicemonto of Portugal. Over the years some scholars have attempted to guess mercenary nationality based on the 'appearance' of names found in surviving 'Pension Records'. The resulting estimates proved to be invariably dominated by the British and French with little attention paid to the extensive, and historic, contribution of Portuguese mercenaries who had served the Marathas collectively since the days of Shivaji. Many of the Portuguese mercenaries from 1803 found their way back to Goa or to Hyderabad and enlisted in the armies of other 'princes' without benefit of a pension. Ensign John Berdard and Sergeants Anthony Dalmaid and Joseph Roman were among the Portuguese listed in several of the Deccan Rolls.

184. University of Southampton Wellington Papers, WP/3/3/84, overleaf of f.476, 'Roll of Europeans late in the Service of Dowlut Rao Scindiah who have surrendered themselves as Prisoners of War to Colonel J. Stevenson, Commanding the Sub. Force', signed J. Colebrooke Deputy Adjt. Genl. Sub. Force, Camp 2 Miles N.E. of Mukapoor, 13 Nov. 1803. This list includes their date of surrender and the monthly sum they were receiving from the Marathas at the time they left service. John Roach and Scotsman George Blake, who was misidentified on this list as Englishman George Black (someone in copying the list must have dropped the 'e' from Blake), left the service of Begum Sumru's *campoo* near Aurungabad on 31 Aug. 1803. They had not deserted but 'left service'; under the standing provision that they not be employed in combat against their countrymen. A 'Deposition' or synopsis of their debriefing is found in University of Southampton Wellington Papers, WP/3/3/53, f.195, Camp at Aurungabad, 2 Sept. 1803.

185. University of Southampton Wellington Papers, WP/3/3/84, f.478, 'Extract of D.O. By Colonel James Stevenson Comg. the Subsidiary Force . . . A true extract', J. Colebrooke, Deputy Adjt. Genl. Sub. Force, Camp at Burhanpoor, 16 Oct. 1803.

186. Bird, 'The Assaye Campaign', p. 118. Thorn, *War in India*, p. 296.

187. It was known as the 'Key of the Deccan'. James Mill, *The History of British India*, 5th edition with notes by H. H. Wilson (London, 1858: reprinted by Chelsea House, New York, 1968), vol. VI, p. 369.

188. University of Southampton Wellington Papers, WP/3/3/84, f.474, Col. J. Stevenson to Maj. Gen. Wellesley (signed by J. Colebrooke), Camp

2 Miles N.E. of Mukapoor, 13 Nov. 1803. The inscription, if authentic, would indicate this ring had at one time belonged to the Mughal Emperor. The British were keen to inventory imperial items in Maratha possession as it helped to bolster their accusation that Sindia and his men had plundered the Mughal Emperor while they were supposed to be looking after his interests. Needless to say, this particular piece of evidence was circumstantial and there was no immediate way of knowing how or why the ring came into Maratha possession.

189. University of Southampton Wellington Papers, WP/3/3/84, f.475, Col. J. Stevenson to Maj. Gen. Wellesley, Camp at Iddelabad, 9 Nov. 1803.

190. Welsh, *Military Reminiscences*, vol. I, p. 187.

191. For a comprehensive treatment of Arthur Wellesley's diplomatic efforts during this period see Bennell, *The Making of Arthur Wellesley*, in particular ch. 6, 'Negotiations in the General's Tent', pp. 94–114. Many of the supporting documents for Bennell's astute summation can be found in his *The Maratha War Papers of Arthur Wellesley*. See in particular 'An Interim Settlement – October to December, 1803', pp. 291–425.

192. Welsh, *Military Reminiscences*, vol. I, p. 188.

193. Thorn, *War in India*, p. 300; Bird, 'The Assaye Campaign', p. 120; contrasted by Weller, *Wellington in India*, p. 201.

194. Some reports (i.e. Elphinstone) say elements of Sindia's infantry were camped behind them or on the periphery but they did not take part in the action. Just as Bhonsle's infantry sat out Assaye, so apparently Sindia's infantry sat out Argaum.

195. Welsh, *Military Reminiscences*, vol. I, p. 190; Blakiston, *Twelve Years'*, vol. I, p. 197.

196. Blakiston, *Twelve Years'*, vol. I, pp. 197–8.

197. Ibid., p. 197.

198. Ibid., pp. 198–9.

199. Welsh, *Military Reminiscences*, vol. I, p. 190.

200. The young man was found with a wound in the thigh and was fortunate not to have been bayoneted. As it turned out he was a former employee of Mr Cherry, the British Resident at Lucknow. On recovering his health the man was granted a permanent position by Elphinstone. Journal entry for 30 Nov. 1803, Colebrooke, *Mountstuart Elphinstone*, vol. I, p. 90. Elphinstone had heard Beni Singh's infantry numbered as many as 7,000.

201. Elphinstone noted that none of Sindia's troops were engaged in the battle except Gopal Rao's horsemen. See journal entry for 30 Nov. 1803, ibid., vol. I, p. 91.

202. *Wellington Dispatches*, vol. II, p. 561, Arthur Wellesley to Major Shawe, Camp at Akote, 2 Dec. 1803.

203. Blakiston, *Twelve Years'*, vol. I, pp. 200–2. William Henry Springer proclaimed this as the first time the tactic was used by Wellesley although later, as the Duke of Wellington, he became famous for it. See W. H. Springer, *The Military Apprenticeship of Arthur Wellesley in India, 1797–1805*, PhD Dissertation, Yale University, 1965 (University Microfilms International, Ann Arbor, Michigan, 1987), p. 152.

204. Reported to have been used on 10 Sept. 1780, following the orders of Colonel William Baillie. See Brig.-Gen. Sir F. B. Norman, 'Medals and Honourary Distinctions Granted Under the Orders of the Government of India', *Journal of the United Services Institution of India*, Calcutta, 15, no. 69 (Feb., 1887), p. 349.

205. Welsh, *Military Reminiscences*, vol. I, p. 190.

206. Ibid.

207. Grant Duff, *History of the Mahrattas*, vol. III, p. 262 and Welsh, *Military Reminiscences*, vol. I, pp. 190–1. Blakiston, *Twelve Years'*, vol. I, p. 204, identified these men as the Farsi Rizulla ('Pharsee Risaulah') or Persian Battalion, adding that they dropped their matchlocks and advanced with swords and shields. The origin of the name in an infantry context is not clear. Farsi, relating to Persian, could have mistakenly been linked to Arabs of the Persian Gulf but 'Risaulah' would seem to be a corruption of 'razzia' from the Arabic *ghaziya*. The term *ressala* or *risalah* is often reserved in North India for irregular horse units but in *Hobson-Jobson* (see RESSALDAR, p. 762) there is evidence of its rank-related use in South India as applied to infantry. The day after the battle Elphinstone rode over the ground and described the dead as 'all Mussulmans, dressed in blue. They have long beards and fine countenances. There are many old men among them. Three of the group are almost as fair as the fairest Europeans, except in the parts exposed to the sun. They say this party was called the Farsi Risaleh. Others say they were Arabs. There are three or four hundred of these fellows lying close to one another.' See Journal Entry for 30 Nov. 1803, Colebrooke, *Mountstuart Elphinstone*, vol. I, p. 90.

208. Welsh, *Military Reminiscences*, vol. I, p. 191. This early description of the Ghosseins in uniform is of particular interest because they are listed in some nineteenth-century accounts of other South Asian battles as entering battle naked, their bodies smeared in ashes.

209. Blakiston, *Twelve Years'*, vol. I, pp. 202–3.

210. Ibid., pp. 207–8. Elphinstone saw the bodies of the dead enemy cavalrymen and commented, 'On our way back to where the charge began we saw another European. The whole four were dressed exactly like natives, and two, probably all, were circumcised.' Journal entry for 30 Nov. 1803, Colebrooke, *Mountstuart Elphinstone*, vol. I, p. 90.

211. Mill, *The History of British India*, vol. VI, p. 372. Wilson, *History of the Madras Army*, vol. III, p. 405 notes Wellesley's human casualties as 243 (that is killed and wounded in action combined including South Asians as well as Europeans) and Stevenson's as 118 (that is killed and wounded in action combined including South Asians as well as Europeans) for a total of 361 human casualties and 39 horse casualties. Note these figures do not match those listed in Parliamentary Papers as noted in appendix II.

212. Almost a week before Argaum Wellesley wrote giving his blessing to the plan to divide Sindia and Bhonsle by means of a separate peace treaty process. Arthur wrote, 'My motives for agreeing to this suspension of hostilities are, First; that I have no power of injuring Scindiah any further. I have Taken

all he had in the Deccan . . .' Arthur Wellesley to Lt-Gen. Stuart, Camp at Rajoora, 23 Nov. 1803, dispatch entry 126 in Gurwood (ed.), *Selections from Wellington Dispatches*, p. 117.

213. Maharashtra State Archives, Mumbai: Bombay Military Dept, Military Board Diary, 1804, no. 68, 3 Jan. 1804, f.21, Maj. Gen. Arthur Wellesley to James Augustus Grant Sec. to Govt, Camp, 13 Dec. 1803.

214. The British had already begun a campaign of systematic hill fort demolition in South India as a counterinsurgency measure. Maj. Norris and Lt Hilary Harcourt Torriano of the Corps of Engineers were sent from Madras and they received assistance from Capt. Bently and Capt. Dardel of the Bombay army. Lt Torriano was particularly noted for his technique, which was emphasized in reports as economical. He made it a habit to use stores of black powder which had been deemed unfit for the manufacture of paper musket cartridges. However, to offset the cost of these operations the EIC conducted sales of building materials derived from the demolition. See BL: OIOC, Board's Collections, F/4/193 no. 4397, ff.1–20, 'Fort Demolition in Southern Territories Under Maj. Norris and Lt Torriano'. These reports are highly detailed with reference to specific forts and their physical dimensions.

215. Some military historians attributed the Gwalior victory to the willingness of British troops to engage in perilous vertical assaults. Perhaps a blatant retrofitting of a 'Who Dares Wins' credo, but the accounts also dovetail with the earlier mid-eighteenth-century accounts of Wolfe's troops ascending the cliffs at Quebec above the St Lawrence River at the Plains of Abraham.

216. Colebrooke, *Mountstuart Elphinstone*, vol. I, pp. 91–5.

217. Blakiston, *Twelve Years'*, vol. I, p. 223.

218. Ibid., p. 221.

219. Journal entry for 14 Dec. 1803, Colebrooke, *Mountstuart Elphinstone*, vol. I, pp. 99–101.

220. Ibid.

221. Welsh, *Military Reminiscences*, vol. I, p. 195.

222. Ibid.

223. Journal entry for 10 Dec. 1803, Colebrooke, *Mountstuart Elphinstone*, vol. I, pp. 94–5.

224. Elphinstone saw them covered with tree boughs.

225. Welsh, *Military Reminiscences*, vol. I, p. 195.

226. Journal entry for 13 Dec. 1803, Colebrooke, *Mountstuart Elphinstone*, vol. I, pp. 98–9.

227. Ibid.

228. Ibid., pp. 101–4.

229. Mounstuart Elphinstone to Edward Strachey, 'Camp at Ellichpoor', 18 Dec. 1803, Colebrooke, *Mountstuart Elphinstone*, vol. I, pp. 104–7.

230. Ibid.

231. Blakiston, *Twelve Years'*, vol. I, p. 229.

232. Mounstuart Elphinstone to Edward Strachey, 'Camp at Ellichpoor', 18 Dec. 1803, Colebrooke, *Mountstuart Elphinstone*, vol. I, pp. 104–7.

233. Thorn, *War in India*, p. 307.

234. Weller, *Wellington in India*, p. 225. Jac Weller had some rather fanciful theories about the defenders *sliding down their turbans* [sic] to escape.
235. Blakiston, *Twelve Years'*, vol. I, pp. 230–1.
236. The recipients of those instructions included men like Major-General Deane (Chunar), Colonel Fenwick (Midnapur) and Broughton with the Ramghar Battalion. BL: Add. MS 13,472. 'Secret Despatches to Various Officers July–August 1803'. This slim-bound collection of correspondence pertains specifically to the irregular units posted as rapid reaction forces. It contains their 'secret' interception orders from Lumsden, Chief Secretary to the Supreme Government in Bengal. Their instructions were to 'pull in' all regulars on loan to the civil authority and prepare to use them in a blocking action intended to trap the Marathas. The corresponding copies for Broughton, including notes on up-grading local troops to provincial troops and irregulars to regular force status, are found in BL: Add. MS 13,723.

4 THE HINDUSTAN CAMPAIGN OF 1803

1. For an example of a discharge order for sepoy privates, including settlement of their clothing allowance and arrears in pay, see BL: Add. MS 13,726 f.140, 'General Orders' by the C-in-C Head-Quarters (HQ), Cawnpore, 8 April 1802, signed by Adjutant General Lt Col. Gerrard.
2. NAM, Acc. no. 9204–121, Major Charles Stuart's Diary (hereafter referred to as NAM, Acc. no. 9204–121, Stuart's Diary). This quotation was taken from Stuart's Introduction. Stuart's diary was carefully rewritten and is very legible but the pages are without numbers and the entries are not consistently chronological on a day-by-day basis. While the majority of entries have some form of date attached there is often reason to believe that it was the date of entry as opposed to the date of occurrence. In the first pages, or what serves as the Introduction, there is a limited political overview. The diary is on paper watermarked 1841 and references in the work suggest that Stuart rewrote it from his personal notes, taken in the field in 1803–5. Stuart died in Hillingdon Grove, Middlesex, on 29 Aug. 1854. Charles Stuart (1776/7–1854) was a natural son of Baron Blantyre. Although listed as Major Stuart in some documents, he held the rank of lieutenant during the 1803 campaign and that is how he is referred to here. In later service Major Stuart was a valued member of the Bengal army and he was made head of the EIC's Cadet College in Calcutta in 1809. He retired from the military 16 July 1823 (biographical data courtesy of NAM's Hodson Card Index). For a published account of Stuart's service in the Bengal Army, see *The East India Military Calendar Containing the Services of General and Field Officers of the Indian Army*, vol. 1 (hereafter referred to as *East India Military Calendar*) by the Editor of the Royal Military Calendar (Printed for Kingsbury, Parbury & Allen, Leadenhall Street, London, 1823), pp. 292–6.
3. BL: Add. MS 13,416. 1 Jan. to 22 April 1803, 'Report To the Secret Committee of the Court of Directors', ff.5–8, paragraphs 18–25.

4. Lieutenant F. G. Cardew, *A Sketch of the Services of the Bengal Native Army to the Year 1895* (Calcutta, 1903; reprinted Today & Tomorrow's Publishers, New Delhi, 1971), p. 76.
5. The Nawab-Vizir of Awadh had entered into a new subsidiary treaty with Richard Wellesley's government in Nov. 1801.
6. The process of ceding territory to the EIC was quite common and during the course of British history in India we find other 'Ceded Territories', i.e. in the Karnatak.
7. There has been much confusion over St Leger's identity for years and I am indebted to David F. Harding for his help in researching this matter and clarifying the issue. The officer at Kannauj was William St Leger and should not be confused with Barry St Leger who served much earlier in America, nor should he be confused with George St Leger of the Madras establishment who led cavalry in pursuit of the Maratha horse following Wellesley's victory at Argaum.
8. Thorn, *War in India*, p. 80.
9. BL: Add. MS 9905, ff.58–60, 'Memorial on the Subject of Flying or Horse Artillery'.
10. For *rakhale* references see chapter 1.
11. The gallopers in service included the 3 pounder, 6 pounder and limited numbers of 4.5-inch howitzers. See BL: Add. MS 9905 'Army Stores and Artillery Practice in India 1778–1796', f.8. Maratha units who had previously 'upgunned' were able to tell a 3 pounder by its diminutive signature (sound and muzzle-flash on firing). Maratha horsemen became adept at staying beyond 1,000 yards where the ballistics of a 3 pounder round shot proved somewhat marginal – slowing to the point of projectile visibility. The 6 pounder could fire round shot at a greater range but the preferred load (for closer engagement) against enemy horse was grape shot since it could take down animals and men with equal proficiency and increased hit probability. An enemy horseman, when dismounted, was relatively easy prey for British or EIC cavalry. As indicated in the endnotes of chapter 1 of this book, canister or case shot carried more individual projectiles at the sacrifice of size and range. See BL: Add. MS 9905, ff.48–56, 'Proceedings of a Committee Assembled at Dum Dum 18 Dec. 1792'. Extensive trials were conducted on 12 pounder gallopers but the gun proved too heavy for practical use.
12. Mud War, 'by which term the operations attending the reduction of certain mud forts in the Jumna Doab were long known.' Cardew, *A Sketch of the Services*, p. 76.
13. If you look carefully in the countryside between Dieg and Bharatpur you can still find remnants of these forts and their earthen perimeter walls. I had the chance to see two in 1990 and the earthen berms had stood the weathering of time remarkably well. The existence of packed-earth outer works readily clarifies the remarks of Bengal engineers like Galloway who said the forts had an amazing ability to simply 'suck up' or absorb iron round-shot. It certainly makes evident the manner in which 'Refractory Zamindars' simply dug spent British shot out of the berms and, without undue ceremony, fired them back.

14. In writing the foreword for Aniruddha Ray, *The Rebel Nawab of Oudh: Revolt of Vizir Ali Khan 1799* (Calcutta, 1990), p. xi, Barun De pointed out that the nucleated nature of defiance, on the part of these rebel 'chieftains', prevented them from organizing any greater form of resistance.

15. E. John Pester, *War and Sport in India 1802–1806: an Officer's Diary*, ed. J. A. Devenish (Heath, Cranton and Onseley, London, 1912), pp. 27–68. Sasni and Bijighur are close to Alighar. Kachaura lies about 6 miles east of Sikandra Rao and is situated slightly north of the Grand Trunk Road that ran from Kannauj to Delhi via Alighar; an ideal pre-existing military road which the British put to good use in 1803.

16. NAM, Acc. no. 9204–121, Stuart's Diary, Introduction.

17. 'If opposition is ultimately to be expected from M. Perron's force, a detachment of an adequate strength, formed at Sarsney, or Bidjeeghur, might either attack M. Perron, at Coel, or by an easy change of position, might intercept his communication with Agra', taken from 'Note by the Marquess Wellesley', Fort William, 28 June 1803, in Martin, *Desp. of Marquess Wellesley*, vol. III, p. 166.

18. Pester, *War and Sport in India*, p. 62.

19. Ibid., pp. 63, 65, 67.

20. Ibid., p. 67.

21. NAM, Acc. no. 9204–121, Stuart's Diary, Introduction.

22. Lake's efforts to balance troop distribution had begun months earlier with men being shifted up-country from Bengal. He had been given carte blanche for redistribution between the upper and lower provinces, as can be seen in, BL: OIOC Board's Collections, F/4/174, no. 3078 'Measures adopted on the Army Assembly in the Upper Provinces'.

23. Pester, *War and Sport in India*, pp. 135–7.

24. Colonel Hugh Pearse, *Memoir of the Life and Military Services of Viscount Lake, Baron Lake of Delhi and Laswaree 1744–1808* (London, 1908), pp. 154–5.

25. Thorn, *War in India*, p. 78. If you follow the Kali Nadi up from the Gunga at Kannauj you pass Etah, Alighar and Kurja.

26. Ibid., p. 80.

27. Pester, *War and Sport in India*, p. 142. The files of the EIC's Secret and Political Department clearly reveal that some judges and magistrates in North India had received their copies of Governor-General Richard Wellesley's Proclamation by 1 Aug. 1803. As we shall see in chapter 5, this helps to account for how a significant number of mercenaries knew about the offer before it was officially posted. Some confusion had been precipitated by instructions that the notices were to be posted 'when hostilities commence'; see National Archives of India, New Delhi (hereafter referred to as NAI), Secret Index to the Foreign Department, 1803 (Aka: Foreign Dept. – Secret Index 1803), p. 9, Agent in the Ceded Provinces, 29 Dec. Archibald Seton wrote, 'On the 1st August, I have the honour of acknowledging the proclamations which His Excellency the Most Noble the Governor General in Council was pleased to direct me to issue, in the European and Native languages as soon as hostilities should have commenced I also received your subsequent letter, of the 3rd August containing an additional sentence to be inserted in

one of the proclamations in question. I now have the honour to acquaint you that, on the 31st of August these proclamations were issued at this place, and published throughout the zellas of Allahabad, Cawnpoor, Etawnak, Furruck-abad and Morabad, in the manner we best adopted to give them speedy and extensive circulation. That parts of the notification which relates to foreign Europeans in the service of the hostile Marhatta Chieftans, was translated into the French and German languages as I was given to understand, there were many Germans in that Predicament.' See NAI, Foreign Dept. Files, Secret Series, Consultation 22 Sept. 1803 No. 23–28, p. 25 (page numbers in pencil), from Archibald Seton, Agent (at Barelli) to the Gov. Gen., to Neel Benjamin Edmonstone – Sec. to Gov't. in the Secret and Political Dept. at Fort William Bengal.

28. See appendix III: Governor-General Wellesley's 'Maratha' Proclamation of 1803.
29. Thorn, *War in India*, p. 83.
30. Pester, *War and Sport in India*, p. 137.
31. Ibid., p. 138. Lieutenant Pester had served with Cunynghame and Wemyss in the 'Mud War' where Cunynghame possessed full civil powers from Governor-General Richard Wellesley in Council 'to treat and settle letters with all the Rajahs in this part of the Ceded Provinces', p. 25. John Pester shared a bungalow with brother officers Middleton and Peyron during their leave in April 1803. Bareilly was an ideal place to strengthen social contacts as it was the regional seat of government selected for Henry Wellesley, Lieutenant-Governor of the Ceded Provinces, p. 92. Henry Wellesley was Richard and Arthur Wellesley's younger brother.
32. BL: OIOC Board's Collections, F/4/174, no. 3079 'Corps of Sebundies Raised and Augmented', ff.1–10.
33. Pester, *War and Sport in India*, p. 138.
34. Ibid., p. 139.
35. Ibid., p. 144. Blair's link to the First Anglo-Maratha War via Gwalior later proved somewhat ironic for Pester. In January 1804 Pester was sent to Gwalior to aid in its capture but Blair begrudgingly resisted his transfer there, pp. 240–72.
36. Ibid., p. 143. Blair had earned a reputation for saving his unit under fire when his commanding officer 'froze' in action. Blair had taken the virtually unheard of step of arresting his senior and then directing his unit to safety.
37. One example of the modification of Bombay army drill was the insertion of a fourth command, that being 'Load', during musket firing drill to ensure that troops 'pause for precision', i.e. remain on aim after firing until the command 'Load'. Maharashtra State Archives, Mumbai: Goa Envoy's Records Diary, no. 3A/606 of 1803–09, Pt. III, (pencilled-in p. 441).
38. This included members of Lt Charles Stuart's 3rd Bengal NC.
39. Pester, *War and Sport in India*, p. 147.
40. Ibid., p. 148.
41. Wemyss's brother was General Wemyss with whom Lake was quite close back in England. Although Pester's friend Wemyss was officially an EIC civil servant based in Mainpuri (Aka 'Mynpoorie'), he was taken on campaign

'with General Lake, living in his Excellency's family'. It was an appropriate description since Lake was accompanied by his son as ADC, but Lake's staff system was very much like that of an extended family with both EIC and King's officer components. Pester, *War and Sport in India*, p. 142.

42. Ibid., p. 149.
43. NAM, Acc. no. 9204–121, Stuart's Diary.
44. The selected galloper gun crews had to be cavalrymen first and foremost. But by designating them as galloper personnel they were taken out of 'normal' cavalry use. For example in a charge, where every blade counted, they were absent as 'gun personnel'. It did not mean they were unemployed. Many times they were busy laying covering fire or providing a distraction for their comrades. The issue was that reconfiguration of personnel to provide galloper crews lessened the actual numbers of 'sabres in hand' at critical moments. Later, there was further resentment as cavalrymen were seconded to become mounted gunners outside their regiments; this in part explains the mixed reception of some horse artillery units that were separate from the cavalry regiments whose numbers they had depleted. Interservice rivalry and unit rivalry were endemic.
45. Cardew, *A Sketch of the Services*, p. 76, G.G.O. 21 April 1802 and G.O.C.C. 1 May 1802.
46. NAM, Acc. no. 9204–121, Stuart's Diary.
47. NAM, Acc. no. 6807–211, Notes on Various Campaigns in India, Papers compiled by Sir John Fortescue, RUSI, *Mahratta War 1803–1805*, p. 7.
48. Presumably 'bunga' was a variation on *bhungy* or sweeper caste. In this case the individual appears to have been earning money as a coolie by carrying supplies into camp.
49. The intelligence was not just limited to North India. Elphinstone, in a letter of 22 Aug. 1803 to Edward Strachey, noted that they were under the impression General Perron had already 'come over [the Yamuna] with his cavalry'. Elphinstone was then serving in the Deccan as Arthur Wellesley's secretary and executive intelligence officer in Malcolm's absence. He must have been cross-examined, by Wellesley, about spreading such an irresponsible story and the dependability of his sources for he mentions being 'interrogated'. This made Elphinstone very uncomfortable as Wellesley's executive intelligence officer because he could not confirm the news. He wrote about his *faux pas* humorously, adding, 'People in the secret are all sanguine. I am with the vulgar, not sanguine.' Colebrooke, *Mountstuart Elphinstone*, vol. I, pp. 52–3.
50. NAM, Acc. no. 9204–121, Stuart's Diary.
51. Pester, *War and Sport in India*, p. 152.
52. Ibid., p. 150.
53. NAM, Acc. no. 9204–121, Stuart's Diary.
54. Thorn, *War in India*, p. 92.
55. NAM, Acc. no. 9204–121, Stuart's Diary, 29 Aug. 1803.
56. Pester, *War and Sport in India*, p. 150.
57. Thorn, *War in India*, p. 92.
58. NAM, Acc. no. 9204–121, Stuart's Diary.

59. Pester, *War and Sport in India*, p. 150.
60. Thorn, *War in India*, p. 92.
61. William Hodges had noted the popularity of the 'half-pike' during a time period that coincided with the closing years of the First Anglo-Maratha War. See William Hodges, *Travels in India During the Years 1780, 1781, 1782, and 1783* (London, 1793), p. 30. The so-called Rajput pikes are fairly common in arms collections and usually measure approx. 32–36 inches in length with a cleaver-sized cutting blade shaped and affixed in the curved style of conventional pikes. Variations include the 'bird's eye', 'clipped point' and a ceremonial 'shaped blade' not unlike the central design of a paisley print. Collectors should be forewarned that I visited an antique arms 'manufacturer' in Udaipur who still makes 'genuine antique' (sic) pikes in the traditional manner, including very highly priced museum-grade Damascus steel pikes with gold inlaid handles. As for Maratha spears, they are much more difficult to find today. During one of my visits to Pune I visited a shrine near the central post office where there was a perfectly preserved pair. This was the style of weapon praised by James Skinner for its versatility in infantry as well as mounted combat, but the drill for its use was said to be quite complex. For those unfamiliar with the point and cutting edges, the closest analogy I can make is that the broad spearhead is not unlike a full-bodied *fleur-de-lis*.
62. Thorn, *War in India*, p. 92.
63. Pester, *War and Sport in India*, p. 150.
64. BL: Add. MS 13,742. See the reference to *harkarrah* reports in Lt Col. Ochterlony to General Lake, Bidjaghur, 23 July 1803, f.100.
65. NAM, Acc. no. 6807–211, Notes on Various Campaigns in India, Papers compiled by Sir John Fortescue, RUSI, *Mahratta War 1803–1805*, p. 7.
66. NAM, Acc. no. 9204–121, Stuart's Diary, 29 Aug. 1803.
67. Pester, *War and Sport in India*, p. 151.
68. Thorn, *War in India*, p. 94.
69. NAM, Acc. no. 9204–121, Stuart's Diary, Camp before Alighar, 29 Aug. 1803. Pester, *War and Sport in India*, p. 151. Camp before Alighar, 30 Aug. 1803.
70. NAM, Acc. no. 6807–150, Journal of Captain George Isaac Call (hereafter referred to as NAM, Acc. no. 6807–150, Call's Journal), vol. I, p. 9. Call's diaries were first given widespread attention in T. H. McGuffie, 'Lake's Mahratta War Campaigns, Report on the Call Journals, 1803 to 1805, Now in the Royal United Services Institution', *Journal of Army Historical Research*, 29 (1951), pp. 55–62. Call probably rewrote them or at the very least annotated them six years later, as there is a reference, in vol. I, p. 108, to a building 'now falling to ruins, 1809'. At the time the 1803 campaign began, George Isaac Call held the rank of cornet, which obviously frustrated him from time to time as evidenced by his remarks indicating that the veteran sepoys did not always give him the serious respect he thought he deserved. In some secondary sources Call is listed as an ensign in 1803, ensign being the infantry equivalent of cornet.
71. 31 Aug 1803, Pester, *War and Sport in India*, pp. 151–2.
72. Ibid., p. 152.

73. There is a problem in the ethnic vs. racial descriptions concerning the stand-off among the defenders of Alighar. These officers were South Asians who served Daulat Rao Sindia but that did not necessarily mean these individuals were all ethnically Marathas. Some accounts labelled them 'Rajput' but that may have been a simplified historically contemporary generalization used interchangeably with the expression 'native officer' with no pejorative context implied. The confusion often stemmed from Western observers who associated one South Asian warrior group with a given military occupational speciality or rank. If one were to select a comparative European analogy, for the German officer corps of World War I (1914–18), the generalization – right or wrong – might be 'Prussian'.

74. NAM, Acc. no. 9204–121, Stuart's Diary, 1 Sept. 1803.

75. The stories of some, such as the chief of police for Delhi, can be traced through their family's appeal for later pension adjustments as found in the BL: OIOC, F/4 Series of Board's Collections.

76. Pester, *War and Sport in India*, p. 153.

77. Thorn, *War in India*, p. 93.

78. Pester, *War and Sport in India*, p. 153.

79. Pearse, *Memoir of the Life and Military Services of Viscount Lake*, p. 170.

80. NAM, Acc. no. 9204–121, Stuart's Diary, 29 Aug. 1803.

81. Pester, *War and Sport in India*, p. 153.

82. Pearse, *Memoir of the Life and Military Services of Viscount Lake*, p. 168. Pearse identified the second in command as a 'Rajput officer' named Baji Rao [sic]. Thorn, while failing to name the individual, reported that he was a 'Maratha chief' who was later killed in this action. Thorn, *War in India*, p. 99.

83. Thorn, *War in India*, p. 95.

84. BL: Add. MS 13,597. See ff.266–268, Military Secretary Merrick Shawe to Lt Col. Close, Calcutta, 16 Sept. 1803. The engineers considered themselves extremely lucky in having gained the fort before the drawbridge was complete, as they had no contingency plan for dealing with the moat. It was of such great proportion that they remarked 'a seventy four could sail around the fort', meaning it was big enough to hold a battleship of that era.

85. Thorn, *War in India*, p. 96.

86. The difficulty in using the petard can be judged by the former popularity of the expression 'hoist with one's own petard'. The only surviving example of this weapon I could find during research was that located at the Fort St George Museum in Chennai (viewed in Dec. 1986). Diagrams of the weapon and method of its use survive in Parlby, *The British Indian Military Repository*, 5 vols. (Samuel Smith, Calcutta, 1822–7) and 'blowing the gates with a gun' is clearly diagrammed in the *Madras Artillery Proceedings* (1826).

87. Thorn, *War in India*, p. 96.

88. NAM, Acc. no. 6807–150, Call's Journal, vol. I, pp. 7–10. Pester, *War and Sport in India*, pp. 154–5. Pester's published account differs from the others in that it does not mention the early failure of a 6 pounder at Alighar.

89. NAM, Acc. no. 6807–150, Call's Journal, vol. I, p. 6.

90. Pester, *War and Sport in India*, p. 155.
91. Thorn, *War in India*, p. 97.
92. Pester, *War and Sport in India*, p. 155.
93. The wound proved so severe that Colonel Monson did not rejoin Lake's army on the march until 2 Oct. 1803 at Mathura. Pester, *War and Sport in India*, p. 188.
94. NAM, Acc. no. 6807–150, Call's Journal, vol. I, p. 11.
95. Thorn, *War in India*, p. 98.
96. NAM, Acc. no. 9204–121, Stuart's Diary. It is not clear if this individual was the same man who opposed Pedron's surrender plan and was identified in Pearse, *Memoir of the Life and Military Services of Viscount Lake*, p. 168, as Baji Rao. Thorn, while failing to name the same individual who quarrelled with Pedron, referred to him as a 'Maratha chief' who was later killed in this action. Thorn, *War in India*, p. 99.
97. Pester, *War and Sport in India*, p. 157.
98. Ibid.
99. Pester, *War and Sport in India*, p. 156.
100. NAM, Acc. no. 6807–150, Call's Journal, vol. I, p. 12.
101. Pester, *War and Sport in India*, p. 156.
102. Ibid.
103. NAM, Acc. no. 9204–121, Stuart's Diary.
104. Gov.-Gen. Richard Wellesley's 'General Orders' of 15 Sept. 1803 Fort William, cited in Thorn, *War in India*, p. 103.
105. NAM, Acc. no. 9204–121, Stuart's Diary.
106. Thorn, *War in India*, p. 100 and Grant Duff, *History of the Mahrattas*, vol. III, p. 249, respectively.
107. NAM, Acc. no. 6807–150, Call's Journal, vol. I, p. 12.
108. Pester, *War and Sport in India*, p. 157.
109. NAM, Acc. no. 6807–150, Call's Journal, vol. I, p. 12, regarding cash in Alighar. Some soldiers carried off 400 to 500 dollars each. One officer was said to have 'made off' with a tumbrel holding 7,000 dollars but subsequent prize fund notations suggest it was recovered.
110. Sir Penderel Moon indicated that this victory might have had later repercussions in terms of Lake's estimation of North Indian fortifications. The implication is that Lake became conditioned to success against fortresses based upon this outcome and that it clouded his judgement and thinking, leading to failure at Bharatpur sixteen months later. See Sir Penderel Moon, *The British Conquest and Dominion of India* (Duckworth, London, 1989), p. 324.
111. Some of Sindia's sepoys wore scarlet uniforms that were virtually indistinguishable from British forces on the battlefield and any generalizations about the uniforms of his troops must be confined to a unit-by-unit analysis.
112. Thorn, *War in India*, p. 101.
113. Pester, *War and Sport in India*, p. 160.
114. NAM, Acc. no. 6807–150, Call's Journal, vol. I, p. 20. In Call's diary Ram Narrain is identified as 'Ramnerain'. For the purposes of this chapter's narration one might also take *shroff* to mean banker, but that is equally

unsatisfying if we look at the overall role of *shroffs* in South Asian history. In many cases they managed exchange transactions and *hundi* (promissory notes akin to bank drafts).

115. NAM, Acc. no. 9204–121, Stuart's Diary, 2 Sept. 1803.
116. Pester, *War and Sport in India*, p. 157. It is interesting to observe that George Isaac Call later footnoted the word *pindari* as meaning 'incendiary'. While it was not correct in terms of the word's meaning or origin it did often fit the *modus operandi*. See NAM, Acc. no. 6807–150, Call's Journal, vol. II, Pt. 1, p. 25.
117. 'Livesey and myself having a pipe of Maderia at Futty Ghur, agreed that it would be better for it to remain there, as the risk of losing it would be less than at Shikoabad (close upon the borders of the Mahratta territory).' Pester, *War and Sport in India*, pp. 134–5.
118. Thorn, *War in India*, p. 106.
119. BL: OIOC Board's Collections, F/4/168, no. 2952, 'Disturbances in the Provinces of Etawah and Cawnpore excited by the incursion of a body of the Enemy's Horse', ff.1–39. This report contains reference to the insurgency that Fleury's raid triggered. The 'Rajah of Chuttra Saul' had hidden his prohibited artillery in the *zenana* (women's quarters) of his 'mud fort', to elude the detection of British arms inspectors. But encouraged by Fleury's daring raid, the Rajah brought out his cannon and began to make preparations for further resistance. Some District Collectors urged the creation of a separate army for counterinsurgency operations but this idea was vetoed as impractical. With the Marathas' subsequent losses at Delhi and Agra the *zamindar* revolt quickly lost momentum.
120. NAM, Acc. no. 6807–150, Call's Journal, vol. I, p. 18.
121. Thorn, *War in India*, p. 107.
122. Ibid.
123. Pester, *War and Sport in India*, p. 158.
124. NAM, Acc. no. 6807–150, Call's Journal, vol. I, p. 21. Call noted the monsoon season in the Doab that year was very hot and he claimed it did not rain until 10 Sept. and then only lightly. It was 26–27 Sept. 1803 before heavy rains arrived.
125. Thorn, *War in India*, p. 85.
126. Pester, *War and Sport in India*, p. 160.
127. NAM, Acc. no. 9204–121, Stuart's Diary, 7 Sept. 1803.
128. Thorn, *War in India*, p. 108.
129. NAM, Acc. no. 9204–121, Stuart's Diary, 8 Sept. 1803.
130. Stuart also made a cryptic reference to a distant cannonade which was supposedly an artillery duel between two competing Maratha mercenaries, Monsieur Bourquin and another presumed to be Dussein (?).
131. Pester, *War and Sport in India*, p. 163.
132. NAM, Acc. no. 6807–150, Call's Journal, vol. I, p. 22.
133. Ibid., p. 24.
134. Ibid., p. 27.
135. Thorn, *War in India*, p. 110.
136. Ibid., p. 111.

137. NAM, Acc. no. 6807–150, Call's Journal, vol. I, pp. 24–5.
138. Ibid., p. 28.
139. NAM, Acc. no. 9204–121, Stuart's Diary.
140. Thorn, *War in India*, p. 111.
141. Pester, *War and Sport in India*, p. 165.
142. NAM, Acc. no. 6807–150, Call's Journal, vol. I, p. 25. Call also recorded Ware's narrow escape from a spent projectile: 'Major General Ware was struck by a matchlock ball in the middle of the stomach, but from his corpulency yielding to the ball, passed off, without doing any other injury than that of a severe bruise' (p. 29).
143. Pester, *War and Sport in India*, p. 166.
144. Ibid., p. 167.
145. Ibid., p. 167.
146. NAM, Acc. no. 6807–150, Call's Journal, vol. I, p. 25.
147. Thorn, *War in India*, p. 113.
148. NAM, Acc. no. 6807–150, Call's Journal, vol. I, pp. 25–6.
149. Pester, *War and Sport in India*, p. 167.
150. NAM, Acc. no. 9204–121, Stuart's Diary.
151. Ibid. By Stuart's calculation it was 'spread over 8 or 9 *coss* of country'. See *Hobson-Jobson*, p. 261, Bengal vs. Doab *coss*. Even at the most conservative of estimates it was 10 miles but may have been as much as 20 miles.
152. Grant Duff, *History of the Mahrattas*, vol. III, p. 251. A casualty chart is available in Thorn, *War in India*, p. 115.
153. Thorn, *War in India*, p. 114.
154. NAM, Acc. no. 6807–150, Call's Journal, vol. I, p. 26.
155. Ibid.
156. Ibid.
157. NAM, Acc. no. 9204–121, Stuart's Diary, 13 Sept. 1803.
158. Ibid.
159. Pester, *War and Sport in India*, p. 169.
160. Ibid., p. 174.
161. NAM, Acc. no. 9204–121, Stuart's Diary.
162. NAM, Acc. no. 9204–121, Stuart's Diary, 13 Sept. 1803.
163. Pester, *War and Sport in India*, p. 170.
164. Thorn, *War in India*, pp. 116–17.
165. See 'Report of the Ordnance, & c. captured opposite Delhi on the 11th of September, 1803' by Lt. Col. John Horsford, Commanding the Artillery in the Field. Martin, *Desp. of Marquess Wellesley*, vol. III, p. 668, appendix P.
166. NAM, Acc. no. 9204–121, Stuart's Diary.
167. During the training period at Kannauj, Stuart's 3rd Bengal NC was off supporting infantry operations in the 'Mud War'.
168. NAM, Acc. no. 9204–121, Stuart's Diary, 13 Sept. 1803.
169. NAM, Acc. no. 6807–150, Call's Journal, vol. I, p. 27. NAM, Acc. no. 9204–121, Stuart's Diary, 12 Sept. 1803 noted some members of the 6th NC had escorted in the 10,000 *banjara* bullocks laden with grain for Lake's army.
170. NAM, Acc. no. 9204–121, Stuart's Diary, 14 Sept. 1803.

171. Thorn, *War in India*, p. 118. Lewis Ferdinand Smith, who had left Maratha service earlier, wrote that Bourquin's real name was Louis Bernard. Smith, *A Sketch of the Regular Corps*, p. 22.
172. NAM, Acc. no. 6807–150, Call's Journal, vol. I, p. 30.
173. Ibid., p. 29.
174. NAM, Acc. no. 9204–121, Stuart's Diary, 15 and 16 Sept. 1803.
175. Pester, *War and Sport in India*, p. 171.
176. NAM, Acc. no. 9204–121, Stuart's Diary, 23 Sept. 1803.
177. Thorn, *War in India*, p. 175.
178. NAM, Acc. no. 6807–150, Call's Journal, vol. I, p. 46. Thorn, *War in India*, p. 175.
179. NAM, Acc. no. 9204–121, Stuart's Diary, 25 Sept. 1803.
180. Tughluqabad was a fortress-city founded by Ghiyas-ud-din near Delhi.
181. Thackston, *The Baburnama*, p. 275.
182. Pester, *War and Sport in India*, pp. 181–2.
183. Ibid., p. 183.
184. NAM, Acc. no. 9204–121, Stuart's Diary, 24 Sept. 1803.
185. NAM, Acc. no. 6807–150, Call's Journal, vol. I, p. 46, 26 Sept. 1803, 'offering the whole up to our service: terms not agreeable'.
186. NAM, Acc. no. 9204–121, Stuart's Diary.
187. NAM, Acc. no. 6807–150, Call's Journal, vol. I, pp. 47–8, 2 Oct. 1803.
188. Thorn, *War in India*, p. 176.
189. Ibid., p. 177.
190. NAM, Acc. no. 9204–121, Stuart's Diary, 2 Oct. 1803.
191. Later, in the campaign against Holkar in 1804, Lake instituted a ban on cow slaughter. In the 1803 campaigns, it appears Arthur Wellesley's Madras army sepoys were primarily Muslims while Lake's Bengal army sepoys were predominantly Hindus.
192. NAM, Acc. no. 9204–121, Stuart's Diary, 2 Oct. 1803.
193. NAM, Acc. no. 6807–150, Call's Journal, vol. I, p. 48.
194. NAM, Acc. no. 9204–121, Stuart's Diary, 5 Oct. 1803. Pester noted the camp as two and a half miles from the southeast face of the fort and that the Maratha guns were throwing round shot above their heads which indicates their range capability. Pester, *War and Sport in India*, p. 195.
195. Thorn, *War in India*, p. 181.
196. NAM, Acc. no. 9204–121, Stuart's Diary, 7 Oct. 1803.
197. Re. *Masjid-i-Jahan Nama*, see Majumdar and Raychaudhuri, with Datta and Kalikinkar, *An Advanced History of India*, pp. 586–7. For an additional note on construction see *'Inayat Khan's Shah Jahan Nama*, pp. 205–6 as well as plate 37. The mosque was said to have been the pet project of Shah Jahan's daughter Princess (Begum) Jahanara between 1643 and 1648. According to H. R. Nevill, *Agra: a Gazetteer Being Vol. VIII of the District Gazetteers of the United Provinces of Agra and Oudh* (Government Press, Allahabad, 1905), pp. 199–200, it took five years to build at a cost of Rs. 5 *lakhs*. The mosque proved a highly defensible position with two large octagonal towers. It had great structural integrity owing to its twenty-three domes. Charles Stuart apparently misidentified the mosque as the 'Jumna Masjid'.

198. On 11 July 1803 Lieutenant Pester had received a letter from his friend Peyron of the 3rd Bengal NC (the same unit that Charles Stuart served in). The letter said that Major Middleton, the 3rd NC's commanding officer, 'desired my company at dinner in the Tauge at Agra on the 5th of October next!'. The prophetic letter is made all the more interesting by the fact that the proposed date was less than a week out of synchronization with events that unfolded. See Pester, *War and Sport in India*, pp. 129–30.

199. Pre-war intelligence from Agra warned that the walls of the mosque were so strong that they were virtually cannon proof. BL: Add. MS 13,742 Lt. Col. D. Ochterlony to General Lake, Bidjaghur, 23 July 1803, ff.81–82.

200. Pester, *War and Sport in India*, pp. 198–9.

201. NAM, Acc. no. 9204–121, Stuart's Diary, 10 Oct. 1803.

202. Thorn, *War in India*, p. 183. Negotiations were briefly mentioned in Pester, *War and Sport in India*, p. 200.

203. Pester, *War and Sport in India*, p. 203. As of 26 Oct. 1803 (Pester, p. 215), the enemy dead around the ravines were still not buried. By then they were badly decomposed in addition to having been ripped apart by the dogs and birds of prey.

204. This was Colonel H. Sutherland and he should not be confused with R. Sutherland as was apparently done in Herbert Compton's character sketch as found in *A Particular Account of the European Military Adventurers of Hindustan from 1784 to 1803* (T. Fisher Unwin, London, 1892), appendix, pp. 410–16. William Dalrymple implied that Robert was a cousin, a 'ne'er-do-well member of the clan'; see the footnote on pp. 382–3 of his *White Mughals: Love and Betrayal in Eighteenth-Century India* (Harper-Collins, London, 2002). However, I have opted to believe that Robert and Hugh Sutherland were brothers based on the letters to them – including one consoling Hugh on Robert's death – found in BL: OIOC MSS. EUR. D. 547. There remains the possibility that I am mistaken and 'brother' was an assumptive term by a third party or a term of endearment for a brother-in-arms, as many of H. Sutherland's letters are congenial exchanges with other mercenary officers. Hugh (aka Henry) was older and was said by some to have arrived in India as a King's officer but I was unable to positively confirm that in the army lists. Both Sutherlands, Robert and Hugh, served the Sindia clan under de Boigne and then Perron. As we shall see in the next chapter, the Sutherlands' relationship to this story is extremely complex and important.

205. Thorn, *War in India*, p. 184.

206. Ibid., p. 187. This probably reflected the inability of British supporting artillery to safely limit their fire at night and since work on the grand battery continued around the clock there was no safe time for the British to call supporting artillery fire on this position.

207. A 12 pounder charged with grape shot was of tremendous use in 'sweeping' an area or keeping the enemy pinned down. Although we equate the bore diameter with the weight of a round shot – meaning a 12 pounder had a bore diameter judged to be the calibre of an iron ball weighing 12 pounds – the charge of grape shot was heavier. The standard British EIC grape shot load

for the 12 pounder was composed of 40 balls weighing 9 ounces each, for a total of 22.5 pounds of metal projectiles. By the time the sabot and 'stem' were added, to hold the cluster of grape together, the projectile weighed 24 pounds and was propelled by 3 pounds of black powder. Field trials with this load were held in 1796. The first discharge was at an angle of 15° 45' and the second was at 17° 51'. The minimum range was 1,434 yards and the maximum was 1,676 yards with the pieces of grape grazing the ground at approximately 1,000 yards. The lateral dispersion, or width of the blast from the 40 balls each weighing more than half a pound, was a remarkable 365 yards at the mid-point between the minimum and the maximum range. See BL: Add. MS 9905 'Army Stores and Artillery Practice in India 1778–1796', f.46.

208. Pester, *War and Sport in India*, p. 206.
209. Ibid., pp. 190 and 207.
210. Ibid., p. 209. In Western terms Rs. 2,440,000. As an EIC lieutenant, Pester expected 'six thousand (equal to £800) to be my share'; an additional reference is on p. 220.
211. Major, later Lieutenant-Colonel, George Constable inventoried the piece after the fall of Agra. 'The fruits of this glorious conquest were from twenty-two to twenty-four lacs of rupees, seventy-six brass guns, and eighty-six iron guns of different calibres, mortars, howitzers, carronades and gallopers: total 164, with their tumbrils. One of their guns, surveyed by Lieut.-Col. Constable, was of a most extraordinary nature, viz. brass of one cylinder, calibre twenty-three inches, metal at the muzzle eleven and a half inches, diameter of the trunions eleven inches, length fourteen feet two inches, length of the bore eight feet eight inches, of the chamber four feet four inches, diameter of the ditto ten inches, length of the cascabel one foot two inches, weight of the gun 1207 1/2 maunds, equal to 96,600 pounds. The ball, made of cast iron, weighed 1500 pounds and 108 pounds of gunpowder were requisite to fill the chamber. The enemy shot balls of stone. The gun was valued at 100,000 rupees.' *East India Military Calendar*, vol. I, p. 60.
212. Thorn, *War in India*, p. 188.
213. Pester, *War and Sport in India*, pp. 213–14.
214. Ibid., pp. 212 and 216.
215. Grant Duff, *History of the Mahrattas*, vol. III, p. 253.
216. Thorn, *War in India*, pp. 210–11.
217. Grant Duff, *History of the Mahrattas*, vol. III, p. 253.
218. Pester, *War and Sport in India*, pp. 212–13.
219. Thorn, *War in India*, pp. 213–14.
220. The French traveller Francois Bernier had a chance to observe the Mughal armies in action during the wars of succession waged by Shah Jehan's sons Aurangzeb and Dara Sukoh. In the following description of Dara's order of battle, we find a linear artillery deployment with the large guns fixed as stationary. 'He placed the whole of his cannon in front, linked together by chains of iron, in order that no space might be left for the entrance of the enemy's cavalry.' See Francois Bernier, *Travels in the Mogul Empire AD 1656–1668*, pp. 47–8. Thackston, *The Baburnama*, asserts the tactic

was known to Babur as being from Anatolia; see footnote 18 of part three: 'Hindustan', which discusses subtle differences between this tactic and that used at the First Battle of Panipat.

221. Thorn, *War in India*, p. 214.
222. Ibid., pp. 214–15.
223. NAM, Acc. no. 6807–150, Call's Journal, vol. I, p. 74, 'Laswaree – November 1st 1803'.
224. Grant Duff, *History of the Mahrattas*, vol. III, p. 253.
225. Thorn, *War in India*, p. 218.
226. Ibid., p. 219.
227. Ibid.
228. Ibid., p. 226.
229. Ibid., p. 220. George Lake eventually recovered but was killed while storming the heights of Roleia in Portugal in Aug. 1808, during the Iberian phase of the Napoleonic Wars (p. 226).
230. Ibid., p. 220.
231. NAM, Acc. no. 6807–150, Call's Journal, vol. I, p. 76.
232. Thorn, *War in India*, p. 221.
233. Pester, *War and Sport in India*, pp. 219–20.
234. Thorn, *War in India*, pp. 225–6.
235. Ibid., p. 232.
236. Ibid., p. 233. 'the mortars and howitzers being furnished with elevating screws, made by a simple and ingenious adjustment, to give either of them the double capacity of mortar and howitzer.'
237. Ibid., p. 233.
238. As usual, the sources conflict on casualty figures for this battle. NAM, Acc. no. 6807–150, Call's Journal, vol. I, p. 77. At Laswari enemy killed and wounded were estimated in excess of 6,600 (16 battalions × 550 = 8,800 – 2,000 survivors = 6,800 dead and/or wounded). The British forces captured 72 cannon and 36 stands of colours were taken. There was an inconclusive reference in Call's diary that the Marathas' 'Chief Sardar' was killed in this action. But Call's words are confusing in that Ambaji Inglia survived the battle. Call's diary also indicates that three or four Europeans were taken prisoner. Presumably these were renegade mercenaries who, for one reason or another, had not taken advantage of the Proclamation to 'come in'. British losses were stated as 800 casualties and 36 officers – 13 of the latter listed as dead. Grant Duff said the British lost 824 combined casualties.
239. Thorn, *War in India*, p. 227.
240. Ibid., p. 222.
241. Ibid. The Maratha forces were well supplied with smallarms. In addition to the 5,000 smallarms, presumably removed from among the casualties on the field, 57 carts or hackeries containing muskets and matchlocks were also captured, p. 233.
242. Ibid., p. 223.
243. Grant Duff, *History of the Mahrattas*, vol. III, pp. 256–7. Given the Marathas' military participation ratio in World War II (1939–45) and the number of decorations won by distinguished infantry units (i.e. the Mahratta Light

Infantry), one can only dismiss Duff's comment as historically short-sighted.

244. Thorn, *War in India*, p. 233.
245. Ibid., p. 234.
246. Ibid., p. 236.

5 'COMING IN'

1. Alexander Allan Cormack, *The Mahratta Wars 1797–1805: Letters from the Front by Three Brothers Nicholas, George and Thomas Carnegie of Charleton, Montrose* (published by the author, printed by Banffshire Journal Ltd, Banff, 1971) (hereafter referred to as *Letters from the Front*), George Carnegie to Susan Carnegie, 17 Oct. 1799, Calcutta, pp. 25–7.
2. Ibid., Susan Carnegie to Alexander Abberdein, circa Oct. 1800, Charleton, Montrose, pp. 28–9.
3. Ibid., Thomas Carnegie to Susan Carnegie, 19 Feb. 1801, London, pp. 93–5.
4. Ibid., Thomas Carnegie to Susan Carnegie, 8 Aug. 1801, Bombay, pp. 100–2.
5. Ibid., Thomas Carnegie to Susan Carnegie, 15 Aug. 1801, Bombay, pp. 102–3.
6. Collins' appointment as military secretary to Gov.-Gen. Shore was dated 28 Oct. 1793. See *List of the Military Secretaries to the Governors-General and Viceroys from 1774–1908* (Calcutta, Superintendent Government Printing, 1908), pp. 2–3, 43–4.
7. NAI, *Foreign Dept. Secret Index, for the year 1799*, p. 208, 30 July 1799, no. 13.
8. Nevill, *Agra: a Gazetteer*, p. 165.
9. Compton, *European Military Adventurers*, p. 364.
10. A small portrait of Hessing senior survives in the BL: OIOC Collection and is cross-referenced with the Sutherland correspondence by way of the prints and drawings database.
11. Compton, *European Military Adventurers*, p. 364.
12. Some nineteenth-century British authors criticized George Hessing for the loss on 18 July 1801. Herbert Compton accused him of behaving 'in a most cowardly manner' for not having remained on the field to become a casualty as did eleven of his fellow mercenaries in that action. Compton, *European Military Adventurers*, p. 363. Jadunath Sarkar, using the accounts of Holkar and Amir Khan, painted a rather more sympathetic picture of a valiant effort against all odds. Monsoon torrents forced many of Hessing's sepoys to rely on the bayonet but the vast numbers of Holkar's horse overwhelmed the Dutchman's bedraggled brigade. See Sarkar, *Fall of the Mughal Empire*, vol. IV, pp. 152–5.
13. Compton, *European Military Adventurers*, p. 364. There is some indication that Metcalfe did not fit as well into his position as he might have; see Colebrooke, *Mountstuart Elphinstone*, vol. I, p. 50.

14. Readers interested in the fascinating aspects of cross-cultural relations during this period are advised to consult William Dalrymple's monumental contribution to the field, *White Mughals*. Many of the characters in this book, such as Colonel Collins, were romantically involved with South Asian women.

15. Compton, *European Military Adventurers*, p. 363.

16. Ibid., p. 365.

17. The biographical sketch of 'Colonel Robert Sutherland' can be found in Compton, *European Military Adventurers*, pp. 410–16. Correspondence addressed to Robert and Hugh Sutherland can be found in BL: OIOC MSS Eur. D. 547. Much of Compton's account is actually the story of Hugh Sutherland's life and not Robert's. As the senior officer, Hugh (aka Henry) was more well known. Robert died before his pension could be reviewed, which explains why only Hugh is noted on the Maratha mercenary pensions review list found in BL: OIOC 0/6/6 Personal Records.

18. Compton, *European Military Adventurers*, p. 412. Skinner is the attributed source for the linkage of Filoze and Bourquin to Perron.

19. BL: OIOC MSS Eur. D. 547, Major L. Derridon to Major H. Sutherland, Muttra, 20 April 1806, ff.12–13.

20. BL: OIOC MSS Eur. D. 547, L. Derridon to Major H. Sutherland, Cole, 21 May 1806, ff.14–15. Contextual references taken from ff.12–15 collectively indicate Hessing, Pohlman, Sutherland, Derridon and Pedron had all survived the war with their mutual friendship intact.

21. NAI, Secret Proceedings [S. No. 169], 16 Aug. 1802, Letter no. 169, p. 7402, Col. J. Collins Resident with Sindia to Gov. Gen. Richard Wellesley, Futtyghur, 19 Sept. 1801.

22. Ibid.

23. It has been hard for many Western writers to avoid negative descriptions of Ghatke. Several Maratha historians have singled him out for abuse in what amounted to his duplicitous conduct and its negative impact on the interests of the Sindia clan. Govind Sakharam Sardesai had no qualms in labelling Ghatke an 'evil genius' in the context of his later alliance with Ambaji. Sardesai, *History of the Marathas*, vol. III, p. 430.

24. NAI, Secret Proceedings [S. No. 169], 16 Aug. 1802, Letter no. 174, pp. 7411–13, Col. J. Collins Resident with Sindia to Gov. Gen. Richard Wellesley, Futtyghur, 23 Oct. 1801.

25. Ibid., Letter no. 175, pp. 7424–5. Col. J. Collins Resident with Sindia to Gov. Gen. Richard Wellesley, Futtyghur, 27–31 Oct. 1801.

26. Ibid., 16 Aug. 1802, Letter no. 169, pp. 7404–5, Col. J. Collins Resident with Sindia to Gov. Gen. Richard Wellesley, Futtyghur, 19 Sept. 1801.

27. BL: Add. MS 13,601 See in particular ff.12–13 (old pagination pp. 25–6), N. B. Edmonstone Sec. to Gov't to Col. Collins Resident with Scindiah, Cawnpoor, 15 Jan. 1802.

28. Ibid., paragraphs 20 and 21, ff.31–32.

29. Ibid., see in particular the 4th point, paragraph 7 on f.12, and the reiteration of the '4th condition to be the most important of any', in paragraph 14 on f.13.

30. Ibid., paragraph 16, ff.28–29.

31. Ibid., paragraph 16, f.28.
32. Ibid., paragraph 17, ff.29–30.
33. Ibid., paragraph 15, f.13. There is a confusion in transliteration and Perron appears as 'Piron' in this manuscript.
34. Ibid., paragraph 22, f.32.
35. NAM Acc. no. 5205–19, Skinner Diary.
36. NAI, Secret Proceedings [S. No. 169], pp. 7571–8, Letter no. 192, Col. Collins Resident with Sindia to Gov. Gen. Richard Wellesley, Camp near 'Oujein', 30 March 1802.
37. Perron and Sutherland subsequently reached an accommodation which defused the situation and there was no lasting animosity detectable in subsequent correspondence.
38. NAI, Secret Proceedings [S. No. 169], pp. 7561–70, Letter no. 190, Col. Collins Resident with Sindia to Gov. Gen. Richard Wellesley, Camp near 'Oujein', 28 March 1802.
39. Robert Sutherland was Perron's nephew by marriage to the General's niece, according to Compton, *European Military Adventurers*, Appendix, p. 412.
40. *Letters from the Front*, George Carnegie to Susan Carnegie, 12 Dec. 1801, Ougein, pp. 38–40. See also Susan Carnegie to George Carnegie, 23–25 Dec. 1804, Benholm Castle, pp. 61–4.
41. Ibid., George Carnegie to Susan Carnegie, 25–26 July 1802, Camp in the Mewat Country 50 miles from Delhi, pp. 40–6.
42. Ibid., p. 46.
43. NAI, Secret Proceedings [S. No. 169], pp. 7561–70, Letter no. 190, Col. Collins Resident with Sindia to Gov. Gen. Richard Wellesley, Camp near 'Oujein', undated but the context indicates not later than 18 April 1802.
44. Ibid., p. 7583, Letter no. 190.
45. Ibid.
46. Ibid., 16 Aug. 1802, Letter no. 184, p. 7490, Col. J. Collins Resident with Sindia to Gov. Gen. Richard Wellesley. Martin, *Desp. of Marquess Wellesley*, vol. III, pp. 327–8. Richard Wellesley to Lord Hobart, 25 Sept 1803, Fort William. Hobart had written on 17 May 1803 to tell Richard about the end of the Peace (Amiens) and resumption of war with France. Richard Wellesley received word on 11 Sept 1803, well after he had launched his Maratha war.
47. NAI, Secret Proceedings [S. No. 169], 16 Aug. 1802, Letter no. 184, p. 7490, Col. J. Collins Resident with Sindia to Gov. Gen. Richard Wellesley.
48. BL: Add. MS 13,644, 'Malcolm & Lambton's Journals in 1799', f.4. Malcolm's portion of this work is found on ff.1–60.
49. In the Third Anglo-Mysore War the British had solicited the defection of Tipu Sultan's European mercenaries. The first week of February 1792 saw a total of 57 European mercenaries leave Tipu's service to take up a British offer. Dirom, *A Narrative of the Campaign*, p. 183.
50. Edward Ingram (ed.), *Two Views of British India. The Private Correspondence of Mr. Dundas and Lord Wellesley: 1798–1801* (Bath, 1970), p. 22, the Earl of Mornington to the Rt. Hon. Henry Dundas, 23 Feb. 1798.
51. Martin, *Desp. of Marquess Wellesley*, vol. III, p. 167, Marquess Wellesley to Lieut.-General Lake, Fort William, 28 June 1803.

52. Ibid., p. 168.

53. It was rumoured de Boigne's asking price was Rs. 3.5 *lakhs* for the Mughal 'rissalah'. William Palmer described the force as elite cavalry and it was hoped they might serve as a particularly vigilant Governor-General's horse guard. 'They are all armed with light small matchlocks purposely made for them, in the use of which they are dexterous with a pair of Trooper's Pistol's, and Broad Swords. They are dressed uniformly from head to foot of green broad cloth but in the Mughal dress with European boots and belts. They ride in their own way on country saddles, called Basrarees with saddle cloths which, and saddles are all alike for the regiment. All the arms, accoutrements and furniture for the horses and men are almost or quite new.' BL: OIOC, Board's Collections, F/4/9 no. 712, Maj. William Palmer Resident at the Court of D. R. Sindia to Gov. Gen. John Shore.

54. Martin, *Desp. of Marquess Wellesley*, vol. III, pp. 62–3. The Marquess Wellesley to his Excellency Lieut-General Lake, the Commander-in-Chief, &c., (Private), Barrackpoor, 27 March 1803.

55. Ibid., 'proposition' number 7.

56. Ibid., see the final paragraph.

57. University of Southampton Wellington Papers, WP/3/3/53, 'No. 246', To Richard Marquis Wellesley Gov. Gen. from Col. Collins, 30 April 1803, Camp near Burhanpour.

58. BL: Add. MS 13,742 ff.20–21, General Perron to Commander-in-Chief General Lake, Coel, 9 April 1803.

59. *Letters from the Front*, George Carnegie to Susan Carnegie, 25 Nov. 1803, Camp at Delhi, pp. 48–52. This reference taken from p. 48.

60. University of Southampton Wellington Papers, WP/3/3/53, 'No. 247', To Richard Marquis Wellesley Gov. Gen. from Col. Collins, 2 May 1803, Camp near Burhanpour.

61. Ibid.

62. NAM, Acc. no. 9204–121, Stuart's Diary, Introduction.

63. BL: Add. MS 13,743, Capt. L. Hook, Sec. to Gov't., Military Dept. to Lt. Col. Gerard (aka Gerrard) Adjutant General, Fort William, 13 July 1803, ff.4–6.

64. Ibid., Maj. Frith to Richard Wellesley, 21 June 1803 as noted in N. B. Edmonstone Sec. to Govt to Capt. Lake Sec. to C-in-C, Secret Dept, Ft. William, 13 July 1803, ff.7–10.

65. BL: Add. MS 13,597. See ff.69–70, N. B. Edmonstone Sec. to Gov't., to Col. B. Close Resident at Poona, 15 July 1803.

66. According to contextual references in letters of introduction issued by Hugh Sutherland and found in BL: OIOC MSS Eur. D. 547, Mr Boyd was one of the few American-born mercenaries serving with the Marathas.

67. BL: Add. MS 13,597. See ff.69–70, N. B. Edmonstone Sec. to Gov't., to Col. B. Close Resident at Poona, 15 July 1803.

68. University of Southampton Wellington Papers, WP/3/3/47, 'Report taken down from Mr. Stuart by Lieut. Colonel Close on the 15th October 1803'.

69. Ibid.

70. John Pemble, 'Resources and Techniques in the Second Maratha War', *Historical Journal*, 19, no. 2 (1976), p. 394.
71. Ibid. Compare to Martin, *Desp. of Marquess Wellesley*, vol. III, p. 285. Pemble apparently had not seen the correspondence of Carnegie or Ochterlony or the original Skinner manuscript.
72. BL: Add. MS 13,742, Lt. Col. D. Ochterlony to General Lake, Bidjaghur, 23 July 1803, f.80.
73. Rotton was one of the mercenaries sensitive to this as he had two children in a Calcutta boarding school. BL: OIOC, F/4/267, no. 5880 'Transmitting a Memorial of Mr. Rotton', ff.1–29, Auditor's Office 1808.
74. BL: Add. MS 13,742, ff.92–95, Lt. Col. Ochterlony to General Lake, Bidjaghur, 21 July 1803.
75. Ibid., f.94.
76. Ibid.
77. The *Calcutta Gazette* was more than a local newspaper; it was the official platform for publishing the military and political notices of the Bengal government. Eventually it carried the Proclamation signalling the end of this uneasy period of clandestine negotiation. (See appendix III.)
78. BL: Add. MS 13,742. See the reference to *harkarrah* reports in Lt. Col. Ochterlony to General Lake, Bidjaghur, 23 July 1803, f.100.
79. Ibid., ff.79–80.
80. Martin, *Desp. of Marquess Wellesley*, vol. III, p. 193, 'Memorandum by Lieut.-General Lake with the Marquess Wellesley's observations thereon', 18 July 1803; see item number 19.
81. BL: Add. MS 13,742, George Carnegie to Lt. Col. D. Ochterlony, Camp at Secundra, 18 Aug. 1803, ff.149–50. This letter was forwarded to Lake and, in turn, copied and sent to Richard Wellesley as an enclosure on 22 Aug. 1803. The eight battalions each had the standard Maratha complement of 5 guns for a total of 40 cannon. See Lt. Col. D. Ochterlony to General Lake, Bidjaghur, 23 July 1803, f.80.
82. BL: Add. MS 13,742, George Carnegie to Lt. Col. D. Ochterlony, Camp at Secundra, 18 Aug. 1803, ff.149–150.
83. Ibid.
84. Smith, *A Sketch of the Regular Corps*, p. 31.
85. Ibid., p. 32.
86. Ibid., pp. 34–6.
87. Ibid., p. 35.
88. Pester, *War and Sport in India*, p. 173.
89. BL OIOC H/485, ff.263–6, HQ Camp Near Secundra, C-in-C Gerard Lake to Gov. Gen. Richard Wellesley, 25 Aug. 1803.
90. Ibid., ff.270–4, Coel, Gen. Perron to C-in-C Gerard Lake, 23 Aug. 1803.
91. Ibid., ff.274–5, HQ of the British Army, Gen. Lake, to Gen. Perron, 27 Aug. 1803. For the timing of the receipt of Perron's letter vs. this reply see context of, ff.268–9, 27 Aug. 1803, HQ Camp Near Bidgighur, Gen. Lake to Gov. Gen. Richard Wellesley.
92. Ibid., f.281, Coel, Gen. Perron to C-in-C Gerard Lake, 28 Aug. 1803.
93. Ibid., f.258.

94. Exact numbers are very difficult to establish for a number of reasons including rank and accompanying followers. However, George Carnegie indicates he 'came in' with ten fellow officers. See *Letters from the Front*, George Carnegie to Susan Carnegie, 25 Nov. 1803, Camp at Delhi, pp. 48–52.

95. On K. B. Stuart's identity, see Dennis Holman, *Sikander Sahib: the Life of Colonel James Skinner 1778–1841* (Kingswood, Surrey, 1961), p. 73. On George Carnegie's familial connection, see *Letters from the Front*, p. 114.

96. See Cormack's editorial notes, *Letters from the Front*, pp. 8, 51.

97. Martin, *Desp. of Marquess Wellesley*, vol. III, p. 322, Lieut-Gen Lake to Richard Wellesley, Delhi, 23 Sept. 1803.

98. *Letters from the Front*, George Carnegie to Susan Carnegie, 25 Nov. 1803, Camp at Delhi, pp. 48–52, this reference p. 48. Some accounts such as H. G. Keene, *Hindustan under Free Lances 1770–1820* (London, 1907), p. 128, insist that once Stuart and Carnegie had applied to leave Maratha service it caused Perron to order all the rest of the British officers to leave camp. That is simply not true as is revealed in subsequent evidence presented in this chapter. The story seems to have originated in J. Baillie Fraser, *Military Memoir of Lt. Col. James Skinner*, 2 vols (London, 1851).

99. *Letters from the Front*, George Carnegie to Susan Carnegie, 25 Nov. 1803, Camp at Delhi, p. 49. Captain Stuart differed from Carnegie in claiming he did not physically see the Proclamation until he read that which was published in the *Calcutta Gazette* of 8 Sept. 1803. NAI Foreign Dept Secret Series for 1803, Consultations 29 Dec. 1803, No. 9–11. Kenneth Bruce Stuart to Chief Sec. of the Gov't Lumsden, 21 Dec. 1803, Calcutta. Captain Stuart's request caused secretary Neel Edmonstone of the Secret and Political Department to write to Major G. A. F. Lake, military secretary to his father, C-in-C Lord Lake. Edmonstone advised Major Lake that Stuart had been forwarded Rs. 1,600 but he also asked how much Stuart was entitled to. See NAI, Foreign Dept Secret Branch, Consultation 29 Dec. 1803, No. 9–11. Edmonstone, Sec. to the Gov't (Bengal) to Major G. A. F. Lake, Military Secretary to his father Commander in Chief Gen. Lake, 22 Dec. 1803.

100. NAI, Foreign Dept Secret Branch, Consultation 29 Dec. 1803, No. 9–11. Kenneth Bruce Stuart to Chief Sec. of the Gov't Lumsden, 2 Dec. 1803, Calcutta.

101. Examples of pension grading can be viewed in BL: OIOC 0/6/6 Personal Records.

102. NAM, Acc. no. 9204–121, Stuart's Diary, 29 Aug. 1803.

103. *Letters from the Front*, George Carnegie to Susan Carnegie, 25 Nov. 1803, Camp at Delhi, p. 49.

104. Baillie Fraser, *Skinner*.

105. Skinner's use of Persian script in formal correspondence is indicated in his surviving letters in the British Library. My use of the term 'regional form of Urdu' is derived from consulting a Cambridge Persian scholar who examined a 'Persian' cavalry manual attributed to Skinner circa 1820s. It is in Persian script but liberal use is made of Turkish military 'loan' words. The coexistence of Turkish and Persian *militarese* should come as no surprise

given that (a) many of Skinner's later horsemen (users of the drill) came from the South Asian cultural frontier and (b) the mounted tactics of Central Asia as well as those of the Mughal era shared much that was non-Persian (not unlike the *Baburnama*'s mention of 'Anatolian' tactics).

106. NAM, Acc. no. 5205–19, Skinner Diary; see unpaginated appendix.
107. BL OIOC, H/485, ff.360–361, Camp, 5 Sept. 1803, Gen. Perron to Gen. Lake.
108. Ibid., ff.363–365, HQ British Army, 6 Sept. 1803, Gen. Lake to Gen. Perron.
109. Keene, *Hindustan under Free Lances*, p. 130.
110. BL: Add. MS 13,743, 'Letters to Lt. Col. Lake 1803–1805'. This collection refers to General Lake's son George who served as his father's most confidential military secretary as well as one of his aides. See f.25, N. B. Edmonstone Sec. to Gov't to Capt. Lake, Ft William, 1803. Note that many men like the younger Lake were promoted repeatedly between 1802 and 1805 owing to war service. The result was some confusing but not necessarily contradictory listing of different ranks for them. So in the letters for the period we see three ranks listed for young Lake: his rank initially on service, his rank after staff promotion and the rank he held at the time his correspondence was catalogued in Richard Wellesley's papers.
111. NAM, Acc. no. 6807–150, Call's Journal, vol. I, p. 17.
112. *Letters from the Front*, George Carnegie to Susan Carnegie, 1 March 1805, 100 miles N-W of Delhi, p. 65.
113. NAM, Acc. no. 9204–121, Stuart's Diary.
114. NAM, Acc. no. 6807–150, Call's Journal, vol. I, p. 16. Thorn, who also accompanied Lake, cites Gov.-Gen. Richard Wellesley's 'General Orders' of 15 Sept. 1803 Fort William; see Thorn, *War in India*, p. 103.
115. NAM, Acc. no. 6807–150, Call's Journal, vol. I, p. 17. The order was listed as 'Camp at Coel September 5th, 1803'.
116. Ibid., p. 1.
117. Compton, *European Military Adventurers*, appendix, p. 370.
118. Pester, *War and Sport in India*, p. 159.
119. NAM, Acc. no. 6807–150, Call's Journal, vol. I, p. 24.
120. NAM, Acc. no. 9204–121, Stuart's Diary.
121. NAM, Acc. no. 6807–150, Call's Journal, vol. I, p. 24.
122. Lieutenant John Pester was a quartermaster of brigade and had been out scouting with the quartermaster-general. His story differs slightly in that he says they captured two enemy horsemen who after questioning revealed Sindia's troops were drawn up for battle less than five miles away. Pester noted, 'The Quarter-Master-General did not credit this account, and it appeared improbable to us that they should quit the banks of the Jumna.' Pester stated that the men, whom he called 'spies', were placed under guard and a report on the incident sent back to Lake. Pester, *War and Sport in India*, p. 164.
123. NAM, Acc. no. 6807–150, Call's Journal, vol. I, p. 24.
124. Ibid., pp. 27–8.
125. *Letters from the Front*, Susan Carnegie to George Carnegie, 23–25 Dec. 1804, Benholm Castle, p. 61.

126. Ibid., George Carnegie to Susan Carnegie, 1 March 1805, 100 miles N–W of Delhi, p. 65.

127. Ibid., George Carnegie to Susan Carnegie, 25 Nov. 1803, Camp at Delhi, p. 50.

128. NAM, Acc. no. 6807–150, Call's Journal, vol. I, p. 30. Presumably the other officers were Gessin, Guerinnier, Del Perron, and Jean Pierre, as identified by Thorn who was serving with Lake. See Thorn, *War in India*, p. 118.

129. NAM, Acc. no. 6807–150, Call's Journal, vol. I, p. 30. This would explain the presence of the two cash tumbrels that Charles Stuart (see chapter 4) mentioned as being found by the cavalry after the battle.

130. Thorn, *War in India*, p. 181.

131. Compton, *European Military Adventurers*, appendix, p. 343.

132. NAM, Acc. no. 6807–150, Call's Journal, vol. I, p. 51.

133. Smith, *A Sketch of the Regular Corps*, p. 38.

134. Ibid.

135. Thorn, *War in India*, pp. 191–2. The value of the money when converted to pounds sterling for 1803 was said by Thorn to equal £280,000. The funds were eventually distributed as prize money to the army although some wondered why it was not used to offset the war debt.

136. NAM, Acc. no. 6807–150, Call's Journal, vol. I, p. 49.

137. Ibid., pp. 49–51. *Hobson-Jobson*, pp. 923–4, indicates the term Tindal could be applied to a Corporal of Lascars or leader of an artillery working party. Given the context of Call's report it would seem a linkage to the lascars or pioneers is probable. As the result of this successful negotiation he was rewarded by promotion to *Serang* which *Hobson-Jobson*, pp. 812–13, lists as the chief of a lascar crew.

138. NAI Foreign Dept. Secret Series for 1803, Consultations 12 Nov. No.139, Fort William, 15 Sept. 1803. To Robert Smith the Advocate General from John Lumsden the Chief Secretary to Government.

139. Pester, *War and Sport in India*, p. 202.

140. *Letters from the Front*, George Carnegie to Susan Carnegie, 25 Nov. 1803, Camp at Delhi, p. 51.

141. Ibid., George Carnegie to Susan Carnegie, 17 Nov. 1804, Delhi, p. 57.

142. Ibid., George Carnegie to Susan Carnegie, 1 March 1805, 100 miles N–W of Delhi, p. 66.

143. BL: OIOC, F/4/267, no. 5881 'Colonel Sheppherd's Claim for a Superior Allowance of Pay', ff.1–71.

144. Thorn, *War in India*, pp. 245–6.

145. BL: OIOC, F/4/267, no. 5881. For the verification of dates see 'The Memorial of Colonel J. R. Sheppherd', ff.22–31. With regard to gallopers, see ff.58–59.

146. Ibid., 'Return of Guns, Musquets Ammunition Accoutrements Etc. attached to the Late brigade of Col. Sheppherd', ff.56–57.

147. Ibid., no. 5880 'Transmitting a Memorial of Mr. Rotton', ff.1–29, Auditor's Office 1808. See ff.17–18, R. W. Rotton to Marquis Wellesley Governor General, Lucknow, 9 Nov. 1803.

148. Ibid. See ff.18–19, Merrick Shawe Secretary to R. W. Rotton, Barrackpoor, 20 Nov. 1803.

149. University of Southampton Wellington Papers, WP/3/3/53, Col. J. Collins to Maj. Gen. Arthur Wellesley, Camp at Aurangabad, 22 Aug. 1803.

150. Ibid. As we now know, by the time Collins sent his details on Sutherland's resignation, the Major had been taken captive by the Marathas and imprisoned at Agra.

151. *Supp. Desp.*, vol. II, pp. 141–2, Arthur Wellesley to Captain Brownrigg, Camp at Kanagherry, 7 Sept. 1800 and p. 143, Arthur Wellesley to Colonel Sutherland, Camp at Kanagherry, 7 Sept. 1800.

152. University of Southampton Wellington Papers, WP/3/3/53, Col. J. Collins to Maj. Gen. Arthur Wellesley, Camp at Aurangabad, 22 Aug. 1803.

153. Ibid.

154. Ibid., Col. J. Collins to Maj. Gen. Arthur Wellesley, Camp at Aurangabad, 2 Sept. 1803.

155. Given that the Begum was leading negotiations for her battalions' defection, it is not surprising that Roach and Blake were kept in the dark with regard to the Proclamation. The Begum could not guarantee to deliver over her battalions if the men had already wandered off in large numbers. This suggests the Begum may have exercised some caution in restricting information on negotiations.

156. BL: Add. MS 13,775, 'Enclosures in Secret Letters to Maj. Gen. Wellesley 1803–1804', ff.49–51, G. Lake to Richard Wellesley Governor General, Camp at Pukessur, 22 Nov. 1803.

157. BL: Add. MS 13,775, 'Enclosures in Secret Letters to Maj. Gen. Wellesley 1803–1804', f.52, Begum Sombre to R. Saleur, 13 Nov. 1803. This letter (in French) was an enclosure to ff.49–51 noted above.

158. Thorn, *War in India*, p. 234.

159. Lieutenant Shipp observed that a number of Maratha female *pindaries* were exceptional riders and many of them masters of the matchlock and sword. See Lieutenant John Shipp, *Memoirs of the Extraordinary Military Career of J. Shipp, Late Lieutenant in His Majesty's 87th Regt.* (London, 1843, 2nd edition, 1894), p. 85. While Mounstuart Elphinstone was surprised to see that the *Nizam* used women, dressed in the style of Madras sepoys, as his palace guard as late as 1801. See Colebrooke, *Mountstuart Elphinstone*, vol. I, p. 36. Other European observers had reported seeing the *Nizam's female battalion* in the 1790s when it was reported that their precision marching and marksmanship were outstanding.

160. *Supp. Desp.*, vol. IV, pp. 184–90, Maj. Gen. Arthur Wellesley to Lt-Col. Collins, Camp, 3 Oct. 1803. Also noted in Mounstuart Elphinstone to Edward Strachey, 'Camp at Assye', 1 Oct. 1803, Colebrooke, *Mountstuart Elphinstone*, vol. I, pp. 72–3.

161. Ibid., p. 189, Maj. Gen. Arthur Wellesley, Oct. 1803. Elphinstone wrote of the incident two days prior to Wellesley. He said that Major Malally and some wounded of the 74th heard one of the mercenaries say, 'You *speak* the language better than I do: desire the jemidar of that body of horse to go and cut up those wounded European soldiers.' Elphinstone cast doubt on the story by prefacing it with the line, 'there are reports believed, though I do not think them proved, of one Englishman (or man who spoke the English

language as his own)'. Mounstuart Elphinstone to Edward Strachey, 'Camp at Assye', 1 Oct. 1803, Colebrooke, *Mountstuart Elphinstone*, vol. I, pp. 72–3.

162. Maharashtra State Archives, Mumbai: Goa Envoy's Records Diary, no. 3/605 of 1801–6, Pt. III, p. 517, Camp at Ajunta 8 Oct. 1803, Proclamation by the Honourable Major General Wellesley.

163. University of Southampton Wellington Papers, WP/3/3/47, B. Close to Maj. Gen. Arthur Wellesley, Poona, 26 Oct. 1803.

164. University of Southampton Wellington Papers, WP/3/3/53, Col. J. Collins to Maj. Gen. Arthur Wellesley, Camp near Hyderabad, 12 Oct. 1803.

165. The British Library's handlist of the correspondence known as the 'Grant Family Collection' notes the gentleman in question as 'Charles Grant (1746–1823) Member of the Court Of Directors of the East India Company 1794–1823 and Chairman 1805, 1809, 1815.' See, BL: OIOC, MSS. Eur. E308, Handlist p. 1. For further details on Charles grant's life and service, see Ainslie Thomas Embree, *Charles Grant and British Rule In India* (Columbia University Press, New York, 1962).

166. *Supp. Desp.*, vol. IV, [no. 1678], pp. 206–7, Camp at Ferdapoor, 23 Oct. 1803, to Lt Col. Collins.

167. University of Southampton Wellington Papers, WP/3/3/47, 'Report taken down from Mr Stuart by Lieut Colonel Close on the 15th October 1803'.

168. Smith, *A Sketch of the Regular Corps*, pp. 29–30.

169. University of Southampton Wellington Papers, WP/3/3/47, 'Report taken down from Mr. Stuart by Lieut Colonel Close on the 15th October 1803'.

170. Ibid.

171. Within Smith, *A Sketch of the Regular Corps*, there is a list of subscribers ranging from EIC officials to British regular army officers, but a considerable number of the names are those of mercenaries who 'came over'. BL: OIOC 0/6/6, Personal Records, contains a list of men who had pension adjustments or registered continuations made at the time this list was compiled. It should not be interpreted as anything approaching a definitive list of mercenaries for reasons previously stated.

172. *Letters from the Front*, Thomas Carnegie to Susan Carnegie, 15 Aug. 1801, Bombay, p. 103.

173. Ibid., George Carnegie to Susan Carnegie, 25 July 1803, Camp 50 miles from Delhi, p. 45.

174. Ibid., George Carnegie to Susan Carnegie, 25 Nov. 1803, Camp at Delhi, p. 51. Compare to 'Stuart' as sworn in University of Southampton Wellington Papers, WP/3/3/47, 'Report taken down from Mr. Stuart by Lieut Colonel Close on the 15th October 1803'. The distance – from Carnegie in the Doab to Stewart in the Deccan – would approximate the reference.

175. On George Carnegie's familial connection see *Letters from the Front*, p. 114.

176. Nicholas was destined to become Major-General Officer Commanding the Bengal Artillery. See Cormack's editorial notes, *Letters from the Front*, pp. 8, 51. Daniel Stewart's brother was also an EIC officer and the potential for sharing intelligence on Sindia's army must have been more than tempting.

177. University of Southampton Wellington Papers, WP/3/3/47, 'Report taken down from Mr Stuart by Lieut Colonel Close on the 15th October 1803'.

178. Contrary to popular belief, the body was not that of Dorson as indicated by the fact that he lived long enough after the war to be considered for a pension claim he filed. Compare R. G. Burton's assertion of death, *Wellington's Campaigns in India*, p. 63, with BL: OIOC 0/6/6 Personal Records. Elphinstone noted that the European Maratha officer's body was originally mistaken for that of an officer of the 19th Dragoons dressed with 'much gold lace and ruffles'. See Mounstuart Elphinstone to Edward Strachey, 'Camp at Assye', 3 Oct. 1803, Colebrooke, *Mountstuart Elphinstone*, vol. I, pp. 74–5.

179. Randolf G. S. Cooper, 'Wellington and the Marathas', *International History Review*, 11, no.1 (February 1989).

180. Thorn, *War in India*, p. 271. Lieutenant Daniel Stewart, who served with these men, said 'Dupont' (aka Dupon) was of Dutch extraction and born in Ceylon.

181. University of Southampton Wellington Papers, WP/3/3/47, 'Report taken down from Mr. Stuart by Lieut. Colonel Close on the 15th October 1803'.

182. Ibid.

183. Ibid.

184. Ibid.

185. Ibid., Major Dorson is thought to be the same man listed in appendix IV as Major Darson.

186. BL OIOC, Board's Collections, F/4/439, no. 10616.

187. Ibid., f.5.

188. Ibid., ff.17–20.

189. The EIC eventually recognized three classes of mercenaries for the purposes of pension settlements. Those who were of the greatest use in the context of military and/or intelligence value got the highest settlements which was reflected in the appeals and listings such as those found in BL: OIOC 0/6/6 Personal Records.

190. BL OIOC, Board's Collections, F/4/439, no. 10616, f.15.

191. Ibid., ff.26.

192. John Lynn suggested that the regimental system might deserve more credit than the Military Revolution if one is searching for reasons to explain the so-called 'rise of the West'. He wrote, 'the British did enjoy distinct military advantages in South Asia by the mid-eighteenth century. Their superiority rested at first on battle culture and drill but ultimately depended more upon importing the regiment – another early modern Western invention that owed little to ancient precedent. The regiment provided the foundation for a permanent British/sepoy military establishment in India that defeated the great native state of Mysore, the Maratha warrior confederacy, and ultimately even the tenacious Sikhs.' John A. Lynn's chapter, 'The Seventeenth-Century Military Change, "The Western Way of War," and South Asia', in Knox and Murray (eds.), *The Dynamics of Military Revolution 1300–2050*, p. 55.

193. Mounstuart Elphinstone to Edward Strachey, 'Camp at Firdapoor', 22 Oct. 1803, Colebrooke, *Mountstuart Elphinstone*, vol. I, pp. 82–3.

194. The EIC continued to use 'Rohilla Irregular Horse' from Bundlekund for some time. The distinctive uniform of the 'Second Rohilla Irregular Horse', a long red coat with high blue hat, as well as their eclectic selection of smallarms, was noted by the Earl of Munster who saw them in 1817 while serving as a lieutenant-colonel in the Third Anglo-Maratha and *Pindari* War. See Lieutenant-Colonel Fitzclarence, *Journal of a Route Across India, Through Egypt, to England, in the Latter End of the Year 1817, and the Beginning of 1818* (John Murray, London, 1819), pp. 68–9.

6 THE ANATOMY OF VICTORY

1. Weller, *Wellington in India*, p. 182.
2. In addressing reasons for the ultimate demise of the Maratha polity, Stewart Gordon looked at British military conquest. 'Ultimately, it was superior credit, artillery, and training, and the momentum of victory in the Napoleonic Wars which defeated the Marathas.' See Gordon, The *New Cambridge History of India*, II. 4: *The Marathas, 1600–1818*, p. 194.
3. The explanation pattern for British success (i.e. 'British discipline and British artillery') became standard tools in the writing of colonial history. For the lingering ramifications of this process on the history of Anglo-Maori conflict, see James Belich, *The Victorian Interpretation of Racial Conflict: the Maori, the British, and the New Zealand Wars* (McGill-Queen's University Press, Kingston, Canada, 1989), p. 17. The Marathas and the Maori *were not at all alike* in military profile. The point is that a set of standard explanations evolved in Britain for the defeat of non-white, non-European people.
4. See the references to military organization and South Asian order of battle as found in *Arrian: The Campaigns of Alexander*, translated by Aubrey De Selincourt (Penguin Books, Harmondsworth, revised edition, 1971).
5. Smith, *A Sketch of the Regular Corps*, p. 50.
6. For an examination of Dundas and his work, see J. A. Houlding, *Fit for Service: the Training of the British Army, 1715–1795* (Clarendon Press, Oxford, 1981), pp. 238–45.
7. Mr Grant McCarthy of Toronto was of great help in the development of this example. This quotation is taken from Mark M. Boatner III, *The Civil War Dictionary* (revised edition, Random House, New York, 1991), p. 954.
8. Rifled artillery and modern high-pressure breech-loading artillery would 'come of age' in the second half of the nineteenth century. But the most prolific artillery system of America's Civil War was the MLSB 'Napoleon'. And as late as Britain's Second Anglo-Boer War in 1899, retired veterans writing in journals such as that of the Royal United Services Institute occasionally urged a return to simple MLSB technology as a means of rapidly increasing firepower. This stance was popular among those who had fought in Victoria's colonial wars and become great believers in rugged simplicity, as reflected in their advocacy of 'cold steel and grape shot'.
9. British intelligence estimated that in June of 1803, Sindia's regular infantry alone possessed 444 pieces of artillery. See House of Commons, Misc. Papers, Session 1803–4, vol. XII, p. 296.

10. Thorn, *War in India*, p. 315.
11. *Supp. Desp.*, vol. IV, [no. 1654], p. 180, Camp, 26 Sept. 1803, Arthur Wellesley to Maj. Malcolm, Re. The Battle of Assaye.
12. Arthur's force numbered approximately 4,500 on the day he committed to battle at Assaye. For official casualty figures, see details in appendix II and compare to estimate that casualties were in the neighbourhood of 35.2 per cent.
13. *Wellington Dispatches*, vol. II, p. 393, Arthur Wellesley to Col. Murray, Fort William, 14 Sept. 1804.
14. Martin, *Desp. of Marquess Wellesley*, vol. III, p. 445, C-in-C Gen. Lake's Secret Despatch to Gov. Gen. Richard Wellesley, Re. The Battle of Laswari 1803.
15. The 12-pound Armstrong breech-loader was adopted as a British field gun in 1859.
16. G. R. Gleig, *The Life of Maj.-Gen. Sir Thomas Munro* (2nd edition, London, 1831), vol. I, p. 392: Munro to Col. Read, Punganoor, 6 March 1804.
17. Using post-combat analysis techniques, S. L. A. Marshall demonstrated that peak efficiency could only be derived from infantry when it was personally led into battle. See Marshall, *Men Against Fire* (Wm. Morrow, London, 1947). During the 1960s Marshall's writings, on the necessity for officer leadership under fire, were unsurpassed. His work was in large part responsible for the American army's emphasis on officer leadership under fire during the Vietnam War. Although the veracity of his research techniques fell under question in the 1980s, Marshall's conclusions about the performance of men under fire remain valid.
18. Brian Bond commented on the value of the British officers' 'protective paternalism' as found in the relationship of an officer to his enlisted men during World War I. In speaking about the performance of white Anglo-Saxon troops in that war he noted: 'Without the "creative tension" that existed at unit level between rigorous discipline and paternalism based on common pride in the battalion there would surely have been mutinies in the combat zone.' *The Unquiet Western Front: Britain's Role in Literature and History* (Cambridge University Press, 2002), p. 15.
19. See BL: OIOC Board's Collections, F/4/174, no. 3081, 'Company of Gun Lascars Raised', ff.1–12.
20. Ibid., ff.1–3.
21. BL: OIOC Board's Collections, F/4/192, no. 4291, 'Encrease [sic] of Field Officers to the Artillery Corps recommended', ff.7–14.
22. Black powder does not detonate like nitrogen-based high explosives. Its explosive power is derived from expansive gas pressure generated through rapid burning.
23. *East India Military Calendar*, vol. I, p. 62.
24. Ibid., p. 63. J. P. is presumed to be John Parlby.
25. Ibid., p. 62.
26. Ibid., Constable retired in 1816.
27. Lake was pleased with the 2,500 regulars outside the Fort of Agra who 'came over' and joined the British. Although he told Richard Wellesley that they

accepted terms owing to high casualties, the Maratha sepoys had successfully obtained conditions of service at full pay. See G. Lake to Marquis Wellesley, HQ Camp before Agra, 13 Oct. 1803, House of Commons, Misc. Papers, Session 1803–4, vol. XII, p. 278.

28. For a discussion of Britain as a 'fiscal–military' state, see John Brewer, *The Sinews of Power: War, Money and the English State, 1688–1783* (Unwin Hyman, London, 1989).

29. As early as 1812 there was fear of a Franco-Sikh 'northwest threat' to British India. A. P. Coleman, *A Special Corps: the Beginnings of Gorkha Service with the British* (Pentland Press, Edinburgh, 1999), pp. 42–7. Ranjit Singh, despite having personally observed Maratha armies with powerful artillery, would not mount a military challenge to the British. That would be left for his followers who, like the Marathas, would make use of mercenary officers and high-quality artillery.

30. Colonel G.B. Malleson, *The Decisive Battles of India from 1746 to 1849 inclusive* (W. H. Allen & Co., London, 1883).

31. For the class-rating system explained in terms of the value of intelligence, see BL: OIOC 0/6/6 Personal Records.

32. BL: OIOC, F/4/266, no. 5878, 'Dismissal of the Irregular Corps of Cavalry formerly under Colonel Sheppherd And of Irregulars under Colonel Mieselback, Pension Granted to Col. M. of 300/Rupees a Month, Memorial from Colonel Mieselback there on Pension increased in Consequence to 1,000 Rs. A Month', Auditor's Office 1808, ff.1–93.

33. BL: OIOC, F/4/266, no. 5878, f.15. But it originates in the Governor-General in Council's minute, signed by G. H. Barlow, G. Udny, J. Lumsden, 19 June 1807, ff.55–57.

34. During Mahadji Sindia's days, Himmat Bahadur Gosavi had agreed to try and conquer Bundlekund with Ali Bahadur and hold it in the *Peshwa*'s name. Although nominally brought under the Maratha pennant there were constant outbreaks of insurgency. Ali Bahadur died in 1802 and his son and successor Shamshir had opted to side with Sindia against the British. This state of affairs had led the *Peshwa* to try and trade the area with the British in 1803. The problem was that the most powerful Maratha forces in Bundlekund were those of Colonel Mieselbach, in Ambaji's service. (Passing references to the occasionally stormy relationship between Himmat and the Sindia clan can be found in Colebrooke, *Mountstuart Elphinstone.*)

35. BL: OIOC, F/4/266, no. 5878, ff.63–65. The details surrounding these events are clearly laid out in the memorial which appears on these pages listed as Colonel J. F. Mieselbach to the Honourable Sir George Barlow, Governor General, Banda, 4 June 1807.

36. Ibid., ff.68.

37. Ibid., ff.73–74.

38. Ibid., f.73.

39. The larger ramifications of a 'totally unselfconscious use of an ethnocentric system of measurement' as a 'culture-specific frame of reference' in which 'European styles of military organization and generalship were wrongly assumed to be the only effective forms' was addressed in depth by James Belich,

in ch. 15, section II, 'The Background of Ideas', *The Victorian Interpretation of Racial Conflict*, pp. 321–30.

40. For aspects of continuity in mercenary employment and the larger issue of mercenary employment in Maratha history, see appendix v. For details on the CIA buying-out Afghan warlords and 'sub-commanders with dozens or hundreds of fighters . . . for as little as $50,000 in cash', see Bob Woodward, *Bush at War* (Simon & Schuster, New York, 2002), p. 194.

Bibliography

BRITISH LIBRARY MANUSCRIPT COLLECTIONS

Add. MS 9,905
Add. MS 13,416
Add. MS 13,472
Add. MS 13,597
Add. MS 13,601
Add. MS 13,644
Add. MS 13,722
Add. MS 13,723
Add. MS 13,726
Add. MS 13,742
Add. MS 13,743
Add. MS 13,775
Add. MS 13,876
Add. MS 29,209

BRITISH LIBRARY ORIENTAL AND INDIA OFFICE
COLLECTIONS

Board's Collections, F/4/9
Board's Collections, F/4/166
Board's Collections, F/4/168
Board's Collections, F/4/174
Board's Collections, F/4/192
Board's Collections, F/4/193
Board's Collections, F/4/195
Board's Collections, F/4/200
Board's Collections, F/4/266
Board's Collections, F/4/267
Calcutta Gazette vol. XL, no. 1019
Home Misc. Series, H/481
Home Misc. Series, H/485
Madras Courier, vol. XIX, no. 942
MSS Eur. D. 547
MSS Eur. E. 308
Personal Records 0/6/6

Political & Secret Dept, L/PS/5/91
Proceedings of the Bombay Military Council, P/354/1

MAHARASHTRA STATE ARCHIVES, MUMBAI

Bombay Military Dept, Military Board Diary, 1803
Bombay Military Dept, Military Board Diary, 1804
Goa Envoy's Records Diary, no. 3/605 of 1801–6, Pt III

NATIONAL ARCHIVES OF INDIA, DELHI

Foreign Dept, Secret Index to the Foreign Dept, 1799
Foreign Dept, Secret Index to the Foreign Dept, 1803 (Aka Foreign Dept –
    Secret Index 1803)
Foreign Dept, Secret Series, Consultations for 1803
Proceedings of the Secret Dept, S. No. 169

NATIONAL ARMY MUSEUM, CHELSEA, LONDON

Accession no. 5205–19	Skinner Diary
Accession no. 6308–11	A copy of 'Arthur Wellesley's Order Book, 7 Feb. 1803 – 21 June 1804'
Accession no. 6807–150	Journal of Captain George Isaac Call
Accession no. 6807–211	Notes on Various Campaigns in India, Papers Compiled by Sir John Fortescue, RUSI, *Mahratta War 1803–1805*
Accession no. 7807–90	Early Days of the 74th Highland Regiment
Accession no. 8207–64	Acct. From NCO Swarbruck 19th Dragoons
Accession no. 9204–121	Major Charles Stuart's Diary

WELLINGTON PAPERS, HARTLEY LIBRARY, UNIVERSITY OF SOUTHAMPTON

WP/1/150
WP/3/3/18
WP/3/3/47
WP/3/3/53
WP/3/3/84

GOVERNMENT AND EAST INDIA COMPANY PUBLICATIONS

*House of Commons Account, Bengal, Fort St George and Bombay Papers Presented to
the House of Commons, Pursuant to their orders of the 7th of May last, from the
East India Company, relative to the Mahratta War in 1803*, Printed by Order of
the House of Commons 5th and 22nd June 1804. Listed in Parliamentary
Papers as, House of Commons, Misc. Papers, Session 1803–4, vol. XII.

The Intercepted Correspondence Between Certain Persons in this Country and Their Friends in India, as Published by the French Government, from the Originals Taken on Board the Admiral APLIN Indiaman, translated from French, London, 1804.

List of the Military Secretaries to the Governors-General and Viceroys From 1774–1908, Calcutta, Superintendent Government Printing, 1908.

Maharashtra State Gazetteers, Maharashtra Ahmadnagar District (revised edition), Bombay, 1976.

Notes Relative to the Peace Concluded Between the British Government and the Marhatta Chieftains and to the Various Questions Arising out of the Terms of the Pacification, printed for John Stockdale, Piccadilly, London, 1805.

Reports from the Committee of Secrecy Appointed to Enquire into the Causes of the War in the Carnatic and of the Condition of the British Possessions in those Parts, vol. I, 1781.

ADDITIONAL SOURCES

Ahmad, T., trans., *Ishwardas Nagar's Futuhat-I-Alamgir*, Idarah-I Adabiyat, Delhi, 1978.

Aitchinson, C. U., *A Collection of Treaties, Engagements and Sanads Relating to India and Neighbouring Countries*, vol. VII, Calcutta, 1931.

Aiyangar, S. Krishnaswami (ed.), 'The Ajnapatra or Royal Edict Relating to the Principles of Maratha State Policy', *Journal of Indian History*, 8, part I, no. 22, printed at the Diocesan Press, Madras, April 1929.

Alavi, Seema, 'The Makings of Company Power: James Skinner in the Ceded and Conquered Provinces, 1802–1840', *Indian Economic and Social History Review*, 30, no. 4 (Sage, New Delhi, 1993), pp. 437–66.

Allen, Robert S., *His Majesty's Indian Allies, British Indian Policy in the Defence of Canada, 1774–1815*, Dundurn Press, Toronto and Oxford, 1992.

Arrianus, Lucius Flavius, *Arrian: The Campaigns of Alexander*, translated by Aubrey De Selincourt with an introduction by J. R. Hamilton, Penguin Classics, Penguin Books, revised edition, Harmondsworth, 1971.

Asprey, Robert B., *War in the Shadows: the Guerrilla in History*, William Morrow and Co. Inc., New York, 1994.

Bahura, G. N., and Singh, C., *Catalogue of Historical Documents in Kapad Dwara Jaipur*, Jaigarh Public Charitable Trust, Jaipur, 1988.

Balkrishna, Radhabai, *Shivaji the Great, Part IV Shivaji, the Man and His Work*, Arya Bhanu Press, Kolhapur, 1940.

Banerjee, Tarasankar, 'Maratha Invasions of Bengal, 1742–1744', in A. G. Pawar (convener), *Maratha History Seminar Papers*, Shivaji University Press, Kolhapur, 1971.

Barat, Amiya, *Bengal Native Infantry, its Organization and Discipline*, Firma K. L. Mukhopadhyaya, Calcutta, 1962.

Barnett, Correlli, *Britain and Her Army, 1509–1970: a Military, Political, and Social Survey*, William Morrow, New York, 1970.

Bayly, C. A., *Empire and Information: Intelligence Gathering and Social Communication in India, 1780–1870*, Cambridge University Press, 1996.

The New Cambridge History of India, II. 1: *Indian Society and the Making of the British Empire*, Cambridge University Press, 1988.

Rulers, Townsmen and Bazaars: North Indian Society in the Age of British Expansion, 1770–1870, Cambridge University Press, 1983.

Beaglehole, T. H., *Thomas Munro and the Development of Administrative Policy in Madras 1792–1818*, Cambridge University Press, 1966.

Begbie, Maj. P. J., *Services of the Madras Artillery*, Franck and Co., Madras, 1849.

Begley, W. E., and Desai, Z. A. (eds.), *The Sha Jahan Nama of 'Inayat Khan: an Abridged History of the Mughal Emperor Shah Jahan, Compiled by His Royal Librarian; the Nineteenth-Century Manuscript Translation of A.R. Fuller (British Library, Add. 30,777)*, Oxford University Press, Delhi, 1990.

Belich, James, *The Victorian Interpretation of Racial Conflict: the Maori, the British, and the New Zealand Wars*, McGill-Queen's University Press, Kingston, Canada, 1989.

Bennell, Anthony S., 'The Anglo-Maratha Confrontation of June and July 1803', *Journal of the Royal Asiatic Society* (1962), pp. 107–31.

'Factors in the Marquis Wellesley's failure against Holkar, 1804', *Bulletin of the School of Oriental and African Studies*, 28, part 3 (1965), pp. 553–81.

The Making of Arthur Wellesley, Sangam Books, London, in conjunction with Orient Longman, Hyderabad, 1997.

(ed.), *The Maratha War Papers of Arthur Wellesley, January to December 1803*, Sutton Publishing, for the Army Record Society, vol. 14 in this series, Gloucestershire, 1998.

Bernier, Francois, *Travels in the Mogul Empire AD 1656–1668*, 1891, translation based on Irving Brock's version and annotated by Archibald Constable, 2nd edition revised by Vincent A. Smith, 1934, reprinted complete and unabridged Low Price Publications, Delhi, 1989.

Bhatt, S. K., *Studies in Maratha History, Proceedings of the 4th All India Maratha History Seminar, 11–13/May/1979*, Academy of Indian Numismatics and Sigillography, Indore, 1979.

Bhattacharyya, Pranjal Kumar, *British Residents at Poona, 1786–1818*, Progressive Publishers, Calcutta, 1984.

Biddulph, Col. John, *The Nineteenth and Their Times, being an Account of the Four Cavalry Regiments in the British Army that Have borne the number Nineteen and of the Campaigns in which they Served*, John Murray, London, 1899.

Stringer Lawrence: the Father of the Indian Army, London, John Murray, 1901.

Bilimoria, J. H. (trans.), *Ruka'at-I-Alamgiri or Letters of Aurangzeb*, Idarah-I Adbiyat-I, Delhi, 1972.

Bird, Brevet Lieut Col. W. D., 'The Assaye Campaign', *Journal of the United Service Institution of India*, 41, no. 187 (1912).

Black, Jeremy, *A Military Revolution? Military Change and European Society 1550–1800*, Macmillan, London, 1991.

Blakiston, Maj. John, *Twelve Years' Military Adventure in Three Quarters of the Globe: Or Memoirs of an Officer who served in the Armies of His Majesty and of the East India Company, between the years 1802 and 1814, in which are contained the Campaigns of the Duke of Wellington in India, and his last in Spain and the South of France*, 2 vols., London, 1829.

Boatner III, Mark M., *The Civil War Dictionary*, revised edition, Random House, New York, 1991.

Bond, Brian, *The Unquiet Western Front: Britain's Role in Literature and History*, Cambridge University Press, 2002.

Boxer, C. R., *From Lisbon to Goa 1500–1750: Studies in Portuguese Maritime Enterprise*, London, 1984.

Portuguese India in the Mid-Seventeenth Century, Oxford University Press, Delhi, 1980.

Brand, J., *A Refutation of the Charge Brought Against the Marquis Wellesley on Account of his Conduct to the Nabob of Oude, From Authentic Documents*, P. Stuart, London, 1813.

Brett-James, Anthony (ed.), *Wellington at War 1794–1815: a Selection of His Wartime Letters*, Macmillan and Co., London, 1961.

Brewer, John, *The Sinews of Power: War Money and the English State, 1688–1783*, Unwin Hyman, London, 1989.

Bryant, Arthur, *The Great Duke or the Invincible General*, Collins, London, 1971.

Briggs, Henry George, *The Nizam, His History and Relations with the British Government*, 2 vols., Bernard Quaritch, London, 1861.

Briggs, Capt. John, 'Account of the Origin, History, and Manners of the Race of Men Called Bunjarras' (original reading 25 May 1812), *Transactions of the Literary Society of Bombay*, vol. I, 1819, reprinted 1877.

Broome, Capt. Arthur, *History of the Rise and Progress of the Bengal Army*, vol. I, Smith, Elder and Co., London, 1850.

Buckle, Capt. A., *Memoir of the Services of the Bengal Artillery from the Formation of the Corps to the Present Time with Some Account of its Internal Organization*, ed. J. W. Kaye, William Allen and Co., London, 1852.

Burton, Maj. (eventually Brig.-Gen.) Reginald George, 94th Russell's Infantry, *Wellington's Campaigns in India*, Division of the Chief of the Staff Intelligence Branch, Superintendent Government Printing, Calcutta, 1908.

Burton, Maj., 'Battles of the Deccan', *Journal of the United Services Institution India*, 28 (1899).

'Wellesley's Campaign in the Deccan', *Journal of the United Services Institution India*, 29 (1900).

Butler, Iris, *The Eldest Brother: the Marquess Wellesley, 1760–1842*, Hodder and Stoughton, London, 1973.

Butterfield, Lyman H., 'Psychological Warfare in 1776: the Jefferson-Franklin Plan to Cause Hessian Desertions', in William E. Dougherty and Morris Janowitz (eds.), *A Psychological Warfare Casebook*, published for Operations Research Office, Baltimore, 1958, pp. 62–72.

Cadell, Sir Patrick Robert, *History of the Bombay Army*, London, 1938.

Callahan, Raymond, *The East India Company and Army Reform, 1783–1798*, Harvard University Press, Cambridge, Mass., 1972.

Cambridge Encyclopedia of India, Cambridge University Press, 1989.

Cambridge, R. O., *Account of the War in India between the English and French on the Coast of Coromandel from 1750 to 1760*, London, 1762.

Campbell, Maj. Sir Duncan of Barcaldine, *Records of Clan Campbell in the Military Service of the Honourable East India Company, 1600–1858*, London, 1925.

Cardew, Lt. F. G., *A Sketch of the Services of the Bengal Native Army to the Year 1895*, Calcutta, 1903; reprinted Today & Tomorrow's Publishers, New Delhi, 1971.

Carman, W. Y., *Indian Army Uniforms under the British from the 18th Century to 1947*, Morgan-Grampain, London, 1969.

Caruana, Adrian B., 'Tin Case-Shot in the 18th Century', *Canadian Journal of Arms Collecting*, 28, no.1 (February, 1990), pp. 11–17.

Carver, Field Marshal Lord, *The Seven Ages of the British Army*, London, 1984.

Chandler, David, *The Art of Warfare in the Age of Marlborough*, 2nd edition, Spellmount, Tunbridge Wells, 1990.

Chakravarty, U. N., *Anglo-Maratha Relations and Malcolm 1798–1830*, first published 1937: reprinted Associated Publishing House, New Delhi, 1979.

Choksey, R. D., *A History of British Diplomacy at the Court of the Peshwas, 1786–1818*, Israelite Press, Poona, 1951.

Cipolla, Carlo M., *Guns and Sails in the Early Phase of European Expansion 1400–1700*, Collins, London, 1965.

Clarke, Alfred (Alias 'Carnaticus'), *Summary of the Mahratta and Pindarree Campaign, During 1817, 1818 and 1819 under the Direction of the Marquis of Hastings: Chiefly embracing the Operations of the Army of the Deckan under the command of His Excellency Lieut.-Gen. Sir T. Hislop, (Including 4 letters by 'Carnaticus')*, E. Williams, London, 1820.

Clausewitz, Gen. Carl Von, *On War*, trans. Col. J. J. Graham and notes by Col. F. N Maude, 3 vols., Kegan Paul, Trench, Trubner and Co., London, 1911.

Cockle, Maurice J. D., *A Catalogue of Books Relating to the Military History of India*, Simla, 1901.

Colebrooke, Sir T. E., *Life of the Hon. Mountstuart Elphinstone*, 2 vols., John Murray, London, 1884.

Coleman, A. P., *A Special Corps: the Beginnings of Gorkha Service with the British*, Pentland Press, Edinburgh, 1999.

Compton, Herbert, *A Particular Account of the European Military Adventurers of Hindustan from 1784 to 1803*, T. Fisher Unwin, London, 1892.

Cooper, Leonard, *British Regular Cavalry 1644–1914*, Chapman and Hall, London, 1965.

Cooper, Randolf G. S., 'Amphibious Options in Colonial India: Anglo-Portuguese Intrigue in Goa 1799', in William B. Cogor (ed.), *New Interpretations in Naval History*, United States Naval Institute Press, Annapolis, Maryland, 1997.

'Beyond Beasts and Bullion: Economic Considerations in Bombay's Military Logistics', *Modern Asian Studies*, 33, part 1 (1999).

'Indian Army Logistics 1757–1857: Arthur Wellesley's Role Reconsidered', in Alan J. Guy and Peter B. Boyden (eds.), *Soldiers of the Raj: the Indian Army 1600–1947*, National Army Museum, London, 1997.

'New Light on Arthur Wellesley's Command-Apprenticeship in India: the Dhoondiah Waugh Campaign of 1800 Reconsidered', Alan J. Guy (ed.), *The Road to Waterloo*, National Army Museum, London, 1990.

'Wellington and the Marathas', *International History Review*, 11, no.1 (February 1989).

Cooper, Randolf G. S. and Wagle, N. K., 'Maratha Artillery: from Dabhoi to Assaye', *Journal of the Ordnance Society*, 7 (1995).

Cormack, Alexander Allan (ed.), *The Mahratta Wars 1797–1805: Letters from the Front, by Three Brothers Nicholas, George and Thomas Carnegie of Charleton, Montrose*, published by the author, printed by Banffshire Journal Ltd, Banff, 1971.

Cornwallis, Charles 1st Marquis, *Lord Cornwallis's Plan for Transferring the Indian Army from the Service of the Company to that of His Majesty*, London, 1812.

Dalrymple, William, *White Mughals: Love and Betrayal in Eighteenth-Century India*, HarperCollins, London, 2002.

Daniels, Maj. A. M., *The History of Skinner's Horse*, London, 1925.

Danvers, Frederick Charles, *The Portuguese in India Being a History of the Rise and Decline of their Eastern Empire*, 1894, reprinted Asian Educational Services as 2 vols., New Delhi, 1988.

Das, H. H., *The Norris Embassy to Aurangzib 1699–1702*, Mukhopadhyay, Calcutta, 1959.

Davies, G., *Wellington and His Army*, Oxford, 1954.

Deodhar, Y. N., *Nana Phadnis and the External Affairs of the Maratha Empire*, Popular Book Depot, Bombay, 1962.

Deopujari, M. B., *Shivaji and the Maratha Art of War*, Vidarbha Samshedhan Mandal, Nagpur, 1973.

Deshpande, P. N., 'Maratha Forts: a Study of Forts with Reference to the Royal Period of Maratha History', PhD Dissertation, University of Poona, 1971.

Desika Char, S. V. (ed.), *Readings in the Constitutional History of India 1757–1947*, Oxford University Press, Delhi, 1983.

Deyell, John S., *Living Without Silver: the Monetary History of Early Medieval North India*, Oxford University Press, Delhi, 1990.

Dighe, V. G., *Peshwa Bajirao I and Maratha Expansion*, Karnatak Publishing House, Bombay, 1944.

Dirom, Maj. Alexander, *A Narrative of the Campaign in India which Terminated the War with Tippoo Sultan in 1792*, London, 1793, reprinted Asian Educational Services, Delhi, 1985.

Dodwell, Edward, and Miles, James Samuel, *Officers of the Indian Army*, Longman, Orme, Brown and Co., London, 1838.

Duff, James Cuninghame Grant, *A History of the Mahrattas*, 3 vols., Longman, Rees, Orme, Brown and Green, London, 1826.

Duncan, Louis C., *Medical Men in the American Revolution 1775–1783*, first published 1931, reprinted New York, 1970.

Dundas Col. David, *Principles of Military Movements Chiefly Applied to Infantry*, T. Cadell, London, 1788.

Elers, George, *Memoirs of George Elers, Captain in the 12 Regiment of Foot (1777–1842)*, ed. Lord Monson and George Leveson Gower, William Heinemann, London, 1903.

Elphinstone, the Hon. Mountstuart, *The Rise of the British Power in the East*, ed. Sir Edward Colebrooke, John Murray, London, 1887.

Embree, Ainslie Thomas, *Charles Grant and British Rule in India*, Columbia University Press, New York, 1962.

Esdaile, Charles J., *The Spanish Army in the Peninsular War*, Manchester University Press, 1988.

Fitzclarence, Lt-Col. [Most often listed as the Earl of Munster], *Journal of A Route Across India, Through Egypt, to England, in the Latter End of the Year 1817, and the Beginning of 1818*, John Murray, London, 1819.

Forbes, James, *Oriental Memoirs*, 2 vols., Bentley, London, 1834.

Ras Mala Hindu Annals of Western India, reprinted Heritage, New Delhi, 1973.

Forrest, Sir George William, *Sepoy Generals: Wellington to Roberts*, Edinburgh, 1901.

(ed.), *Selections from the Letters, Despatches, and Other State Papers Preserved in the Bombay Secretariat, Maratha Series*, vol. I, parts I–III, Bombay, 1885.

Fortescue, Sir John W., *The British Army, 1783–1802*, London, 1905.

A History of the British Army, 13 vols., Macmillan, London, 1899–1930.

Wellington, Ernest Benn, London, 1925.

Francklin, Capt. William, *History of the Reign of Shah Aulum*, London, 1794.

Military Memoirs of Mr George Thomas Who by Extraordinary Talents and Enterprise Rose from an Obscure Situation to the Rank of a General in the Service of the Native Powers in the North-West of India, John Stockdale, London, 1805.

Fraser, Hastings, *Memoir and Correspondence of General James Stuart Fraser of the Madras Army*, Whiting and Co., London, 1885.

Fraser, J. Baillie, *Military Memoir of Lt. Col. James Skinner*, 2 vols., London, 1851.

Gat, Azar, *The Origins of Military Thought from the Enlightenment to Clausewitz*, 1989, Oxford (paperback), 1991.

Gates, David, *The British Light Infantry Arm c. 1790–1815*, B.T. Batsford, London, 1987.

The Spanish Ulcer: a History of the Peninsular War, Guild Publishing, London, 1986.

Gense, W. H., and Banaji, D. R. (eds.), *The Gaikwads of Baroda, English Documents*, vol. VI, Bombay, no publication date.

(eds.), *The Third English Embassy To Poona, Comprising Mostyn's Diary September, 1772 – February, 1774 and Mostyn's Letters*, published by D. B. Taraporevala Sons and Co., printed by Karnatik Printing Press, Bombay, 1934.

Ghosh, Biswanath, *British Policy Towards the Pathans and the Pindaris in Central India 1805–1818*, Punthi Pustak, Calcutta, 1966.

Ghoshi, P. N, *Ajnapatra*, Venus, Pune, 1960.

Gleig, Rev. G. R., *The Life of Maj.-Gen. Sir Thomas Munro, Bart. and KGB, Late Governor of Madras with Extracts from His Correspondence and Private Papers*, 2 vols, 2nd edition, London, 1831.

Glover, Michael, *Wellington's Army in the Peninsula*, David and Charles, Newton Abbot, 1977.

Wellington as Military Commander, Batsford, London, 1968.

Glover, Richard, *Britain at Bay: Defence against Bonaparte, 1803–14*, Allen and Unwin, London, 1973.

Peninsular Preparation, the Reform of the British Army, 1795–1809, Cambridge University Press, 1963.

Gommans, Jos J. L. and Dirk H. A. Kolff (eds.), *Warfare and Weaponry in South Asia 1000–1800*, Oxford University Press, 2001.

Gopal, M. H., *Tipu Sultan's Mysore: an Economic Study*, Popular Prakashan, Bombay, 1971.

Gordon, Stewart, *Marathas, Marauders, and State Formation in Eighteenth-Century India*, Oxford University Press, New Delhi, 1994.

The New Cambridge History of India, II. 4: *The Marathas, 1600–1818*, Cambridge University Press, 1993.

'The Slow Conquest: Administrative Integration of Malwa into the Maratha Empire, 1720–1760', *Modern Asian Studies*, 11 (1977), pp. 1–40.

Grey, C., *European Adventurers of Northern India, 1785–1849*, ed. H. L. O. Garrett, London, 1921, reprinted Punjab's Language Dept, Patiala, Offset Master Printers, New Delhi, 1970.

Grose, John Henry, *A Voyage to the East Indies*, 2 vols, London, 1766.

Gune, V. T. (ed.), *Gazetteer of the Union Territory Goa, Daman, and Diu*, District Gazetteers Part 1, Goa, Panaji, 1979.

Gupta, Pratul Chandra, *Baji Rao II and the East India Company, 1796–1818*, Oxford, 1939, this edition Allied Publishers Private Ltd, printed by India Printing Works, Bombay, 1964.

Gurwood, Lt-Col. John (ed.), *The Dispatches of Field Marshal the Duke of Wellington K.G., during his various campaigns in India, Denmark, Portugal, Spain, and the Low Countries and France, from 1799 to 1818*, 13 vols., John Murray, London, 1834–7, vol. I, 2nd edition, 1837.

Selections From the Dispatches and General Orders of Field Marshal the Duke Of Wellington, London, 1843.

Guy, Alan J. (ed.), *The Road to Waterloo*, National Army Museum, London, 1990.

Oeconomy and Discipline Officership and Administration in the British Army 1714–63, Manchester University Press, 1985.

Habib, Irfan, *An Atlas of the Mughal Empire: Political and Economic Maps with Detailed Notes, Bibliography and Index*, Oxford University Press, Delhi, 1982.

Hall, Bert S., *Weapons and Warfare in Renaissance Europe: Gunpowder, Technology, and Tactics*, Johns Hopkins University Press, Baltimore and London, 1997.

Hall, Bert S., and DeVries, Kelley, 'The Military Revolution Revisited', *Technology and Culture*, 31 (1990).

Harding, D. F., *10th Princess Mary's Own Gurkha Rifles: a Short History*, Burgess and Son, Abingdon, Oxfordshire, 1990.

Smallarms of the East India Company 1600–1856: vol. I: *Procurement and Design*: vol. II: *Catalogue of Patterns*: vol. III: *Ammunition and Performance*: vol. IV: *The Users and Their Smallarms*: Foresight Books, London, 1999.

Hitchcock, John T., 'The Idea of the Martial Rajput', in Milton Singer (ed.), *Traditional India: Structure and Change*, Philadelphia, 1959.

Hobson-Jobson, see entry for Yule and Burnell.

Hodges, William, *Travels in India During the Years 1780, 1781, 1782 and 1783*, London, 1793.

Hodson, Maj. V. C. P., *List of Officers of the Bengal Army, 1758–1834*, 4 vols., 1927–47.

Holman, Dennis, *Skinner Sahib: the Life of Colonel James Skinner 1778–1841*, Kingwood, Surrey, 1961.

Holmes, Richard, *Wellington: the Iron Duke*, HarperCollins, London, 2002.

Hook, Theodore, *The Life of General the Right Hon. Sir David Baird, Bart.*, 2 vols, Richard Bentley, London, 1832.

Houlding, J. A., *Fit for Service: the Training of the British Army, 1715–1795*, Clarendon Press, Oxford, 1981.

Hudlestone, F. J., *Warriors in Undress*, John Castle, Boston, 1926.

Hughes, Maj.-Gen. B. P., *Firepower: Weapon Effectiveness on the Battlefield, 1630–1850*, 1974.

Open fire – Artillery Tactics from Marlborough to Wellington, Chichester, 1983.

Hutchinson, Lester, *European Freebooters in Moghul India*, Popular Press, Bombay, 1964.

Ignatieff, Michael, *Virtual War: Kosovo and Beyond*, Metropolitan Books, London, 2000.

Ingram, Edward (ed.), *Two Views of British India. The Private Correspondence of Mr Dundas and Lord Wellesley: 1798–1801*, Adams and Dart, Bath, 1970.

Irvine, William, *The Army of the Indian Moguls: Its Organization and Administration*, London, 1903, reprinted Eurasia Publishing House, New Delhi, 1962.

Joshi, P. M. (ed.), *Selections from the Peshwa Daftar, Revival of Maratha Power (1761–1772)*, Government Central Press, Bombay, 1962.

(ed.) *Persian Records of Maratha History*, vol. II: *Sindia as Regent of Delhi (1787 & 1789–91)*, trans. Jadunath Sarkar, published by the Director of Archives, Government of Bombay, 1954.

Kale, Y. M., *English Records of Maratha History Poona Residency Correspondence*, vol. V: *Nagpur Affairs 1781–1820*, Government Central Press, Bombay, 1938.

Kamble, B. R. (ed.), *Studies in Shivaji and his Times*, Shivaji University, Kolhapur, 1982.

Kangle, R. P., *Kautiliya Arthasastra*, 3 vols., reprinted Motilal Banarsidass, Delhi, 1988.

Kantak, M. R., *The First Anglo-Maratha War 1774–1783: a Military Study of Major Battles*, Popular Prakashan, Bombay, 1993.

Kar, Lt Col. H. C., *Military History of India*, Firma KLM, Calcutta, 1980.

Karandikar, Shivaram Laxman, *The Rise and Fall of Maratha Power*, vol. I, D. D. Gangal, Poona, 1969.

Kaye, Sir John William, *The Life and Correspondence of Major General Sir John Malcolm*, 2 vols., Smith Elder, London, 1861.

Keay, John, *A History of India*, HarperCollins, London, 2000.

Keegan, John, *The Face of Battle*, 1976, reprinted Barrie Jenkins, London, 1988.

Keene, Henry George, *Hindustan under Free Lances 1770–1820*, London, 1907.

Rulers of India, Madhava Rao Sindhia and the Hindu Reconquest of India, Oxford, 1901.

A Sketch of the History of Hindustan From the First Muslim Conquest to the Fall of the Mughal Empire, W. H. Allen and Co., London, 1885.

Kennedy, Paul, *The Rise and Fall of Great Powers: Economic Change and Military Conflict from 1500 to 2000*, Random House, New York, 1987.

Khanna, D. D., *Battle of Assaye 1803*, Dept for Defence Studies, Allahabad, 1963, reprinted University of Allahabad, 1981.

Monson's Retreat in Anglo-Maratha War, 1803–1805, Dept for Defence Studies, University of Allahabad, 1981.

Khanna, D. D., and Tandon, R. K., 'Siege of the Fort of Deeg 9th December to 26 December 1804', *Journal of the Society for Army Historical Research*, 63 (1985).

Kharbas, Datta Shankarrao, *Maharashtra and the Marathas: Their History and Culture – a Bibliographic Guide to Western Language Materials*, G. K. Hall and Co., Boston, 1975.

Khare, G. H., 'The Marathas as "Freebooters"', in A. G. Pawar (convener), *Maratha History Seminar Papers*, Shivaji University Press, Kolhapur, 1971.

Khare, V. V. and Khare Y. V. (eds.), *Aitihasak Lekh Sangatia*, vol. xiv.

Khobrekar, V. G. (ed.), *English Translation of Tarikh-I-Dilkasha*, Department of Archives, Government of Maharashtra, Bombay, 1972.

Kincaid, C. A., and Parasnis, D. B., *A History of the Maratha People*, 3 vols., Bombay, 1925, reprinted Chand, New Delhi, 1968.

Kipling, Rudyard, *Collected Verse of Rudyard Kipling*, Hodder and Stoughton, London, 1912.

Kipp, Jacob W., 'The Nature of War: Russian Military Forecasting and the Revolution in Military Affairs: a Case of the Oracle of Delphi or Cassandra?', *Journal of Slavic Military Studies*, 9, no. 1 (March 1996), pp. 1–45.

Kirkpatrick, Col. William (ed. and trans.), *Select Letters of Tippoo Sultan to Various Public Functionaries: including his Principal Military Commanders; Governors of Forts and Provinces; Diplomatic and Commercial Agents*, Black, Parry and Kingbury, London, 1811.

Knox, MacGregor and Williamson, Murray (eds.), *The Dynamics of Military Revolution 1300–2050*, Cambridge University Press, 2001.

Kolff, Dirk H. A., 'The End of the Ancien Régime: Colonial War in India 1798–1818', in J. A. de Moor and M. L. Wesseling (eds.), *Imperialism and War: Essays on Colonial Wars in Asia and Africa*, Comparative Studies in Overseas History, 8, E. J. Brill, Leiden, 1989.

Naukar, Rajput and Sepoy: The Ethnohistory of the Military Labour Market in Hindustan, 1450–1850, Cambridge University Press, 1990.

Krenn, Peter (ed.), *Von Alten Handfeuerwaffen, Entwicklung Technik Leistung*, Landesmuseum Joanneum Graz, 12 Sonderausstellung im Landeszeughaus Mai–Oktober 1989, Graz, Austria, 1989.

(ed.), 'Test-Firing Selected 16th-18th Century Weapons', trans. Erwin Schmidl, *Military Illustrated Past & Present*, 33 (February, 1991), pp. 34–8.

Krishna, Bal, *Shivaji the Great*, Kolhapur, 1940.

Leblond, Guillaume, *A Treatise of Artillery: or, Of the Arms and Machines Used in War Since the Invention of Gunpowder*, London, 1746, reprinted Museum Restoration Service, Ottawa, 1970.

Longer, V., *Red Coats to Olive Green*, Delhi, 1974.

Low, C. R., *History of the Indian Navy (1613–1863)*, 2 vols., London, 1877.

Lenman, Bruce P., 'The Transition to European Military Ascendancy in India, 1600–1800', in John Lynn (ed.), *Tools of War: Instruments, Ideas, and Institutions of Warfare, 1445–1871*, University of Illinois Press, Urbana, 1990.

Ludovici, Anthony M. (ed. and trans.), *On the Road with Wellington: the Diary of a War Commissary in the Peninsular Campaigns by August Ludolf Friedrich*

Schaumann, Deputy Assistant Commissary-General in the English Army, Alfred A. Knopf, New York, 1925.

Lushington, the Rt Hon. S. R., *The Life and Services of General Lord Harris, Baron of Seringapatam and Mysore, During his Campaigns in America, the West Indies and India*, John W. Parker, London, 1840.

Lynn, John A. (ed.), *Tools of War: Instruments, Ideas, and Institutions of Warfare, 1445–1871*, University of Illinois Press, Urbana, 1990.

Mackenzie, W. C., *Collin Mackenzie, First Surveyor General of India*, Chambers, Edinburgh, 1952.

Macksey, Piers, *The War for America 1775–1783*, Harvard University Press, Cambridge, Mass., 1964.

MacMunn, Lt-Gen. Sir George F., *The Martial Races of India*, Sampson Low, Marston and Co., printed by Purnell and Sons, London, 1933.

MacPherson, C. W., (ed.)., *Soldiering in India, 1764–1787: Extracts from Journals and Letters Left by Lt. Col. Allan MacPherson and Lt. Col. John MacPherson of the East India Company's Service*, London, 1928.

Mahaley, K. L., *Shivaji: the Pragmatist*, Vishwa Bharati Prakashan, Nagpur, 1969.

Majumdar, R. C., (ed.), 'The Mughal Empire', *The History and Culture of the Indian People*, vol. VII, Bharatiya Vidya Bhavan, Bombay, 1974.

Majumdar, R. C., and Raychaudhuri, H. C., and with Datta and Kalikinkar, *An Advanced History of India*, Macmillan, Madras, 4th edition 1978, reprinted 1986.

Malcolm, Maj.-Gen. Sir John, *A Memoir of Central India*, London, 1823.

Report on the Province of Malwa and Adjoining Districts, Calcutta, 1822.

Malleson, Col. G. B., *The Decisive Battles of India from 1746 to 1849 inclusive*, W. H. Allen and Co., London, 1883.

Final French Struggles in India and on the Indian Seas, London, 1878.

A History of the French in India, London, 1868.

Manucci, Niccolao, *Storia do Mogor 1653–1708*, trans. William Irvine, vol. IV, reprinted Calcutta Editions Indian, Calcutta, 1967.

Manucy, Albert C., *Artillery Through the Ages; A short illustrated history of cannon, emphasizing types used in America*, Washington, 1962.

Marshall, S. L. A., *Men Against Fire*, Wm. Morrow, London, 1947.

Martin, Montgomery (ed.), *The Despatches, Minutes and Correspondence of the Marquess Wellesley, K. G. During His Administration in India*, 5 vols, Wm. H. Allen, London, 1837.

Mason, P., *A Matter of Honour: an Account of the Indian Army, Its Officers and Men*, Jonathan Cape, London, 1974.

Mattison, Mines, *The Warrior Merchants: Textiles, Trade and Territory in South India*, Cambridge University Press, 1984.

McGuffie, T. H., 'Lake's Mahratta War Campaigns, Report on the Call Journals, 1803 to 1805, Now in the Royal United Services Institution', *Journal of the Society For Army Historical Research*, 29 (1951), pp. 55–62.

McNeill, William H., *The Age of Gunpowder Empires 1450–1800*, Washington DC, American Historical Association, 1989.

Keeping Together in Time: Dance and Drill in Human History, Harvard University Press, Cambridge, Mass., 1995.

The Pursuit of Power: Technology, Armed Force, and Society since A.D. 1000, University of Chicago Press, 1982.

Merewether, F. H. S., *A Tour Through the Famine Districts of India,* A. D. Innes and Co., London, 1898.

Michell, George and Zebrowski, Mark, *The New Cambridge History of India,* I. 7: *Architecture and Art of the Deccan Sultanates,* Cambridge University Press, 1999.

Mill, James, *The History of British India,* vol. VI, 5th edition with notes by H. H. Wilson, London, 1858, reprinted Chelsea House, New York, 1968.

Misra, B. B., *The Central Administration of the East India Company 1773–1834,* Manchester University Press, 1959.

Montgomery, Field-Marshal Viscount of Alamein, *A History of Warfare,* Collins, London, 1968.

Moon, Sir Penderel, *The British Conquest and Dominion of India,* Duckworth, London, 1989.

Müller, John, *A Treatise on Artillery,* London, 1757, reprinted Museum Restoration Service, Ottawa, 1965.

Munro, Maj. Innes, *Narrative of the Military Operations on the Coromandel Coast Against the Combined Forces of the French, Dutch and Hyder Ally Cawn from the year 1780 to the Peace of 1784, in a Series of Letters,* London, 1789.

Naravane, M. S., *Forts of Maharashtra,* APH Publishing Corp., New Delhi, 1995.

Nevill, H. R., *Agra: a Gazetteer Being Vol. VIII of the District Gazetteers of the United Provinces of Agra and Oudh,* Allahabad, Government Press, 1905.

Nicolls, Maj. Jasper, 'Remarks Upon the Temperature of the Island of Bombay During the Years 1803 and 1804', in *Transactions of the Literary Society of Bombay,* I, London, 1819.

Nightingale, Pamela, *Trade and Empire in Western India, 1784–1806,* Cambridge University Press, 1970.

Norman, C. B., *Battle Honours of the British Army, from Tangier, 1662, to the Commencement of the Reign of King Edward VII,* John Murray, London, 1911.

Norman, Brig.-Gen. Sir F. B., 'Medals and Honourary Distinctions Granted Under the Orders of the Government of India', *Journal of the United Services Institution of India,* Calcutta, 15, no. 69 (Feb. 1887).

Orme, R., *Historical Fragments of the Mogul Empire,* London, 1782; reprinted Associated Publishing, New Delhi, 1974.

History of the Military Transactions, 3 vols, Madras, 1861.

Owen, Sidney J., *A Selection From the Despatches, Treaties and Other Papers of the Marquess Wellesley, During His Government in India,* 2 vols., Mysore Government Press, Bangalore, 1877.

Parihar, G. R., *Marwar and the Marathas 1724–1843,* Hindi Sahitya Mandir, Jodhpur, 1968.

Parker, Geoffrey, *The Military Revolution: Military Innovation and the Rise of the West, 1500–1800,* Cambridge University Press, first published 1988, 2nd edition, 1996.

Parkinson, C. Northcote, *Trade in the Eastern Seas, 1793–1813*, Cambridge, 1937. *War in the Eastern Seas 1793–1815*, George Allen and Unwin, London, 1954.

Parlby, Captain Samuel (ed.), *The British Indian Military Repository*, 5 vols, Samuel Smith, Calcutta, 1822–7.

Pasley, Rodney, *'Send Malcolm!': the Life of Major-General Sir John Malcolm 1769– 1833*, The British Association for Cemeteries in South Asia, printed by H. W. Walden and Co., London, 1982.

Pawar, A. G. (convener), *Maratha History Seminar Papers*, Shivaji University Press, Kolhapur, 1971.

Pearce, Robert Rouiere, *Memoirs and Correspondence of the Most Noble Richard Marquess Wellesley*, 3 vols, Richard Bentley, London, 1846.

Pearse, Col. Hugh, *Memoir of the Life and Military Services of Viscount Lake, Baron Lake of Delhi and Laswaree 1744–1808*, William Blackwood and Sons, London, 1908.

Peers, Douglas M., *Between Mars and Mammon: Colonial Armies and the Garrison State in Early Nineteenth-century India*, I. B. Tauris Publishers, London, 1995.

Pemble, John, 'Resources and Techniques in the Second Maratha War', *Historical Journal*, 19, no. 2 (1976), pp. 375–404.

Perlin, Frank, 'Money-use in Late Pre-colonial India and the International Trade in Currency Media', in J. F. Richards (ed.), *The Imperial Monetary System of Mughal India*, Oxford University Press, Delhi, 1987, pp. 323–73.

Pester, Lt. John, *War and Sport in India 1802–1806: an Officer's Diary*, ed. J. A. Devenish, Heath, Cranton and Ouseley, London, 1912.

Philippart, Sir Joseph, *The East India Military Calendar, Containing the Services of General and Field Officers of the Indian Army*, by the editor of the Royal Military Calendar, 3 vols, Kingsbury, Parbury and Allen, London, 1823–6.

Philips, C. H., *The East India Company, 1784–1834*, Manchester, 1940.

(ed.), *The Correspondence of David Scott Director and Chairman of the East India Company Relating to Indian Affairs, 1787–1805*, 2 vols, London, 1951.

Phul, Raj Kumar, *Armies of the Great Mughals, 1526–1707*, Oriental Publishers and Distributors, New Delhi, 1978.

Pissurlencar, Panduranga Sakharam, *Portuguese-Mahratta Relations*, trans. T. V. Parvate, Maharashtra State Board for Literature and Culture (Sahitya Sanskriti Mandal), Bombay, 1983.

Pitre, Brig. K. G., *Second Anglo/Maratha War 1802–1805: a Study in Military History*, Dastane Ramchandra and Co., Poona, 1990.

Plowright, John, 'Revolution or Evolution? A Review Article', *British Army Review*, 90, Dec. (1988).

Prakash, Om, 'Foreign Merchants and Indian Mints in the Seventeenth and the Early Eighteenth Century', in J. F. Richards (ed.), *The Imperial Monetary System of Mughal India*, Oxford University Press, Delhi, 1987, pp. 171–92.

Prasad, S. N., *A Survey of Work Done on The Military History of India*, K. P. Bagchi and Co., Calcutta, 1976.

Raeside, Ian, *The Decade of Panipat (1751–61)*, Marathi Historical Papers and Chronicles, Popular Prakashan, Bombay, 1984.

Rai, Lala Lajpat, *Shivaji the Great Patriot*, ed. and trans. R. C. Puri, Metropolitan, Delhi, 1980.

Ralston, David B., *Importing the European Army: the Introduction of European Military Techniques and Institutions into the Extra-European World, 1600–1914*, University of Chicago Press, 1990.

Rao, K. N. P., Mukerjee, J. K., and Lahiri, A. K., 'Some Observations on the Structure of Ancient Steel from South India and its Mode of Production', *Bulletin of the Historical Metallurgy Group*, 3, no. 2 (1969).

Rao, P. S. M., *Lectures on Maratha Mughal Relations (1680–1707)*, S. V. Bagwat, Nagpur University, 1966.

Ray, Aniruddha, *The Rebel Nawab of Oudh: Revolt of Vizir Ali Khan (1799)*, K. P. Bagchi and Company, Calcutta, 1990.

Rennel, Maj. James, Surveyor-General in Bengal, *Memoir of a Map of Hindoostan; of the Mogul Empire*, London, 1792.

Richards, John F., *The New Cambridge History of India*, I.1 *The Mughal Empire*, Cambridge University Press, 1993.

'Official Revenues and Money Flows in a Mughal Province', in John F. Richards (ed.), *The Imperial Monetary System of Mughal India*, Oxford University Press, Delhi, 1987.

Roberts, Gen. Lord, V. C., *The Rise of Wellington*, London, 1895.

Roberts, P. E., *India Under Wellesley*, G. Bell and Sons, London, 1929.

Rogers, C. J. (ed.), *The Military Revolution Debate: Readings on the Military Transformation of Early Modern Europe*, Boulder, 1995.

Rogers, Col. H. C. B., *Wellington's Army*, Ian Allan, London, 1979.

Ross, Charles (ed.), *Correspondence of Charles, First Marquis Cornwallis*, 3 vols, London, 1859.

Roy, Mahendra Prakash, *Origin Growth and Suppression of the Pindaris*, Sterling Publishers, New Delhi, 1973.

Rudyerd, Charles William, *Course of Artillery at the Royal Military Academy as Established by his Grace the Duke of Richmond*, London, 1793, reprinted Museum Restoration Service, Ottawa, Canada, 1970.

Sandes, Lt-Col. E. W. C., *The Indian Sappers and Miners*, Institution of the Royal Engineers, Chatham, 1948.

Sandhu, Maj.-Gen. Gurcharn Singh, *The Indian Cavalry. History of the Indian Armoured Corps*, vol. I, Calcutta, 1981.

Sardesai, Govind Sarkharam, *English Records of Maratha History. Poona Residency Correspondence*, vol. II: *Poona Affairs 1786–1797 (Mallet's Embassy)*, Government Central Press, Bombay, 1936.

New History of the Marathas, 3 vols. 1948; reprinted Phoenix, Bombay, 1968.

(ed.), *Selections from Peshwa Daftar*, vol. XVII, Government Central Press, Bombay, 1931.

Sarkar, Sir Jadunath, *English Records of Maratha History. Poona Residency Correspondence*, vol. I, *Mahadji Sindhia and North Indian Affairs 1785–1794*, Government Central Press, Bombay, 1936.

English Records of Maratha History. Poona Residency Correspondence, vol. XIV, *Daulat Rao Sindhia and North Indian Affairs (1810–1818)*, Modern India Press, Bombay, 1951.

Fall of the Mughal Empire, 4 vols., Orient Longman, 4th edition, New Delhi, 1988–92.

Military History of India, M. C. Sarkar and Sons, Calcutta, 1960.

House of Shivaji Studies and Documents on Maratha History: Royal Period, first published 1940; reprinted Orient Longman, Delhi, 1978.

Sarkar, Jagadish Narayan, *The Art of War in Medieval India*, Munshiram Manoharla Publishers, New Delhi, 1984.

The Military Despatches of a Seventeenth Century Indian General, an English translation of the Hafat Anjuman of Munshi Udairaj alias Tale'yar Khan [Benares MS 53b-93b], Calcutta, 1969.

Some Aspects of Military Thinking and Practice in Medieval India, Ratna Prahashan, Calcutta, 1974.

Sarkesian, Sam C. (ed.), *Revolutionary Guerrilla Warfare*, Precedent Publishing Inc., Chicago, 1975.

Saxena, R. K., *The Army of the Rajputs: a Study of 18th Century Rajputana*, Saroj Prakashan, Udaipur, 1989.

Maratha Relations with the Major States of Rajputana 1761–1818, S. Chand and Co., New Delhi, 1973.

Sen, Surendra Nath, *The Military System of the Marathas*, first published 1926; reprinted Bagehi, New Delhi, 1979.

Sen, S. P., *The French in India, 1763–1816*, Munshiram Manoharlal, New Delhi, 1958.

Seth, Mesrovb Jacob, *Armenians in India*, London, 1897; reprinted Oxford University Press and IBH Publishing Co., Delhi, 1983.

Settar, S., and Sontheimer, Gunther D., *Memorial Stones: a Study of Their Origin, Significance and Variety*, published jointly by the Institute of Indian Art History, Karnatak University, Dharwad, and South Asian Institute, University of Heidelberg, Germany, Mainpal Power Press, 1982.

Shakespear, Henry, *The Wild Sports of India; with Remarks on the Breeding and Rearing of Horses and the Formation of Light Irregular Cavalry*, London, 1860.

Shakespeare, L.W., *Local History of Poona and its Battlefields*, Macmillan, London, 1916.

Shejwalkar, Tryambak Shankar (ed.), *Nagpur Affairs: Selection of Marathi Letters from the Menavli Daftar*, [vol. 1], Deccan College Monograph Series 9, Poona, published by S. M. Katre, printed by Laxmibai Narayan Chaudhari at the Nirnaya Sagar Press, Bombay, 1954.

(ed.), *Nagpur Affairs*, vol. 2, Deccan College Monograph Series 14, Poona, published by S. M. Katre, printed by V. P. Bhagwat, Mouj Printing Bureau, Bombay, 1959.

Panipat: 1761, Deccan College, Poona, 1946.

Sheppard, E. W., *Coote Bahadur: a Life of Lt. Gen. Sir Eyre Coote*, Werner Laurie, London, 1956.

Shipp, Lt J., *Memoirs of the Extraordinary Military Career of J. Shipp, Late Lieutenant in His Majesty's 87th Regt.*, London, 1843; 2nd edition T. Fisher Unwin, 1894.

Shrivastava, B. K. (ed.), *Angreys of Kolaba*, Poona, 1959.

Sinh, Raghubir (ed.), *English Records of Maratha History. Poona Residency Correspondence*, vol. x, *The Treaty of Bassein and the Anglo-Maratha War in the Deccan 1802–1804*, Sri Gouranga Press, Bombay, 1951.

Malwa in Transition, Or a Century of Anarchy, First Phase 1698–1765, D. B. Taraporevala Sons & Co., Bombay, 1936.

Studies on Maratha & Rajput History, Research Publishers, Sun-Shine Printers, Jodhpur, 1989.

Sinha, Birendra Kumar, *The Pindaris (1798–1818)*, Calcutta, 1971.

Smith, Maj. Ferdinand Lewis, *A Sketch of the Rise, Progress and Termination, of the Regular Corps, Formed & Commanded by Europeans, In the Service of the Native Princes of India, With details of the Principal Events and Actions of the Late Marhatta War*, J. Greenway, Calcutta, c. 1805.

Sontheimer, Gunther D., 'Hero and Sati-stones of Maharashtra', see main entry for Settar, and Sontheimer.

Spencer, Alfred, *Memoirs of William Hickey*, 4 vols., Hurst and Blackett, London, 1925.

Sprengel, P. N., and Hartwig, Ludwig Otto, 'Handwork and Artifice Summarized, seventh volume part three: The Gunfactory', originally printed in German (*Handwerke und Künste in Tabellen*) and published in Berlin 1771, reprinted with English translation, ed. Gary Brumfield, *Journal of Historical Armsmaking Technology*, 4 (Jan. 1991).

Spring, Col. F. W. M., *Bombay Artillery*, 1902.

Springer, William H., 'The Military Apprenticeship of Arthur Wellesley in India, 1797–1805', PhD Dissertation, Yale University, 1965. Distributed by University Microfilms International, Ann Arbor, Michigan, 1987.

Stanley, Peter, *White Mutiny: British Military Culture in India, 1825–1875*, Hurst and Co., London, 1998.

Stein, Burton, *Thomas Munro: the Origins of the Colonial State and His Vision of Empire*, Oxford University Press, Delhi, 1989.

Strachey, Henry, *A Narrative of the Mutiny of the Officers of the Bengal Army in the Year 1766*, London, 1773.

Stubbs, Maj. F. W., *History of the Organization, Equipment and War Services of the Regiment of Bengal Artillery, Compiled from Published Works, Official Records, and Various Private Sectors*, 3 vols., King and Co. and W. H. Allen and Co., London, 1877–95.

Subramanian, Lakshmi, *Indigenous Capital and Imperial Expansion: Bombay, Surat and the West Coast*, Oxford University Press, Delhi, 1996.

Syed, Anees Jahan, *Aurangzeb in Muntakhab-Al Lubab*, Somaiya, Bombay, 1977.

Thackston, Wheeler M. (trans. and ed.), *The Baburnama, Memoirs of Babur, Prince and Emperor*, Oxford University Press, 1996.

Thompson, Edward, *The Making of the Indian Princes*, Oxford University Press, 1943, reprinted Curzon Press, London, 1978.

Thorn, Maj. William, *Memoir of the War in India Conducted by General Lord Lake, Commander-in-Chief, and Major-General Sir Arthur Wellesley, Duke of Wellington; from its Commencement in 1803, to its Termination in 1806, on the Banks of the Hyphasis*, T. Egerton, London, 1818.

Tone, William Henry, 'Illustrations of Some Institutions of the Mahratta People', *The Asiatic Annual Register; or, A View of the History of Hindustan, and of the Politics, Commerce and Literature of Asia, For the Year 1799*, Miscellaneous Tracts, J. Debrett, London, 1800.

Tracy, James, *The Political Economy of Merchant Empires 1350–1750*, Cambridge University Press, 1991.

Tugwell, Lt-Col. W. B. P., *History of the Bombay Pioneers*, Sidney Press, London, 1938.

Turner, Wesley B., *British Generals in the War of 1812, High Command in the Canadas*, McGill-Queen's University Press, Montreal, 1999.

Vad, Ganesh Chimnaji, and Parasanis, D. B. (eds.), *Selections from the Satara Raja's and the Peshwa's Diaries (SSRPD), Balaji Bajirav Paishwa*, vol. IX, Deccan Vernacular Translation Society, Bombay, 1906.

Vaidya, S. G., *Peshwa Bajirao II and the Downfall of the Maratha Power*, Pragouti Prakashan, Nagpur, 1976.

Van Creveld, Martin, *Supplying War Logistics from Wallenstein to Patton*, Cambridge University Press, 1977; reprinted New York, 1987.

Vibart, Maj. H. M., *The Military History of the Madras Engineers and Pioneers, from 1743 up to the Present Time*, 2 vol. set, W. H. Allen and Co., London, 1881.

Ward, S. G. P., *Wellington's Headquarters: a Study of the Administrative Problems in the Peninsula, 1809–1814*, Amen House, London, 1957.

Wardlaw, Grant, *Political Terrorism, Theory, Tactics and Counter-measures*, 2nd edition revised and extended, Cambridge University Press, 1989.

Webster, Sir Charles (ed.), 'Some Letters of the Duke of Wellington to his Brother William Wellesley-Pole' (sect II), *Camden Miscellany*, 43, Camden Third Series vol. LXXIX, London, Royal Historical Society, 1948.

Weigley, Russell F., *The Age of Battles: the Quest for Decisive Warfare from Breitenfeld to Waterloo*, Indiana University Press, 1991.

Weller, Jac, *Wellington in India*, Longman, London, 1972.

Wellesley, Marquess Richard, *Notes Relative to the Late Transactions in the Mahratta Empire*, with Appendix of Official Documents, London, 1804.

Wellington, the Second Duke of (ed.), *Supplementary Despatches and Memoranda of Field Marshal Arthur Duke of Wellington, K. G.*, 15 vols., John Murray, London, 1858–72.

Welsh, Col. James, *Military Reminiscences Extracted from a Journal of Nearly Forty Years Active Service in the East Indies*, 2 vols., 3rd edition, Smith, Elder and Co., London, 1830.

Whiteway, R. S., *The Rise of Portuguese Power in India 1497–1550*, 1899, reprinted Asian Educational Services, New Delhi, 1989.

Wiener, Frederick Bernays, *Civilians Under Military Justice: the British Practice since 1689 Especially in North America*, University of Chicago Press, 1967.

Williams, Capt. John, *Historical Account of the Rise and Progress of the Bengal Native Infantry*, London, 1817.

Wilson, Lt-Col. W. J., *History of the Madras Army*, 5 vols., E. Keys, printed by R. Hill, Madras, 1882–88.

Wink, Andre, *Land and Sovereignty in India*, Cambridge University Press, 1986.

Wiseman, Anne and Peter (ed. and trans.), *Julius Caesar: the Battle for Gaul*, Chatto and Windus, London, 1980.

Wood, Col. M. P., *A Review of the Origin, Progress and Result of the Late Decisive War in Mysore, in a Letter from an Officer in India: with Notes and an Appendix, comprising the Whole of the Secret State Papers found in the Cabinet of Tippoo*

Sultaun, at Seringapatam; taken from the Originals, Luke Hansard, for T. Cadell, London, 1800.

Woodward, Bob, *Bush at War,* Simon and Schuster, New York, 2002.

Young, Desmond, *Fountain of the Elephants,* Collins, London, 1959.

Young, H. A., *The East India Company's Arsenals and Manufactories,* Clarendon Press, Oxford, 1937.

Yule, Col. Henry and Burnell, A. C., *Hobson-Jobson: a Glossary of Colloquial Anglo-Indian Words and Phrases, and of Kindred Terms, Etymological, Historical Geographical, and Discursive,* first published 1886; 2nd edition ed. William Crooke, 1903; new edition with a Foreword by Anthony Burgess, Routledge and Kegan Paul, London, 1985.

Index

Name order within the index: as a concession to Western readers approaching South Asian history for the first time, the names of South Asian historic figures – Hindu as well as Muslim – are listed as they appear in the text. Where applicable, descriptive political titles have been added in brackets. So '*Peshwa* Baji Rao II' appears in the index as Baji Rao II (*Peshwa*) and 'Mughal Emperor Shah Alam II' is listed below as Shah Alam II (Mughal Emperor). European names and authors' names are listed in the traditional indexing fashion by placing their last name first.

Printed by Printforce, United Kingdom